Lecture Notes in Physics

Springer-Verlag Berlin Heidelberg GmbH

The Editorial Policy for Proceedings

The series Lecture Notes in Physics reports new developments in physical research and teaching – quickly, informally, and at a high level. The proceedings to be considered for publication in this series should be limited to only a few areas of research, and these should be closely related to each other. The contributions should be of a high standard and should avoid lengthy redraftings of papers already published or about to be published elsewhere. As a whole, the proceedings should aim for a balanced presentation of the theme of the conference including a description of the techniques used and enough motivation for a broad readership. It should not be assumed that the published proceedings must reflect the conference in its entirety. (A listing or abstracts of papers presented at the meeting but not included in the proceedings could be added as an appendix.)
When applying for publication in the series Lecture Notes in Physics the volume's editor(s) should submit sufficient material to enable the series editors and their referees to make a fairly accurate evaluation (e.g. a complete list of speakers and titles of papers to be presented and abstracts). If, based on this information, the proceedings are (tentatively) accepted, the volume's editor(s), whose name(s) will appear on the title pages, should select the papers suitable for publication and have them refereed (as for a journal) when appropriate. As a rule discussions will not be accepted. The series editors and Springer-Verlag will normally not interfere with the detailed editing except in fairly obvious cases or on technical matters.
Final acceptance is expressed by the series editor in charge, in consultation with Springer-Verlag only after receiving the complete manuscript. It might help to send a copy of the authors' manuscripts in advance to the editor in charge to discuss possible revisions with him. As a general rule, the series editor will confirm his tentative acceptance if the final manuscript corresponds to the original concept discussed, if the quality of the contribution meets the requirements of the series, and if the final size of the manuscript does not greatly exceed the number of pages originally agreed upon. The manuscript should be forwarded to Springer-Verlag shortly after the meeting. In cases of extreme delay (more than six months after the conference) the series editors will check once more the timeliness of the papers. Therefore, the volume's editor(s) should establish strict deadlines, or collect the articles during the conference and have them revised on the spot. If a delay is unavoidable, one should encourage the authors to update their contributions if appropriate. The editors of proceedings are strongly advised to inform contributors about these points at an early stage.
The final manuscript should contain a table of contents and an informative introduction accessible also to readers not particularly familiar with the topic of the conference. The contributions should be in English. The volume's editor(s) should check the contributions for the correct use of language. At Springer-Verlag only the prefaces will be checked by a copy-editor for language and style. Grave linguistic or technical shortcomings may lead to the rejection of contributions by the series editors. A conference report should not exceed a total of 500 pages. Keeping the size within this bound should be achieved by a stricter selection of articles and not by imposing an upper limit to the length of the individual papers. Editors receive jointly 30 complimentary copies of their book. They are entitled to purchase further copies of their book at a reduced rate. As a rule no reprints of individual contributions can be supplied. No royalty is paid on Lecture Notes in Physics volumes. Commitment to publish is made by letter of interest rather than by signing a formal contract. Springer-Verlag secures the copyright for each volume.

The Production Process

The books are hardbound, and the publisher will select quality paper appropriate to the needs of the author(s). Publication time is about ten weeks. More than twenty years of experience guarantee authors the best possible service. To reach the goal of rapid publication at a low price the technique of photographic reproduction from a camera-ready manuscript was chosen. This process shifts the main responsibility for the technical quality considerably from the publisher to the authors. We therefore urge all authors and editors of proceedings to observe very carefully the essentials for the preparation of camera-ready manuscripts, which we will supply on request. This applies especially to the quality of figures and halftones submitted for publication. In addition, it might be useful to look at some of the volumes already published. As a special service, we offer free of charge LaTeX and TeX macro packages to format the text according to Springer-Verlag's quality requirements. We strongly recommend that you make use of this offer, since the result will be a book of considerably improved technical quality. To avoid mistakes and time-consuming correspondence during the production period the conference editors should request special instructions from the publisher well before the beginning of the conference. Manuscripts not meeting the technical standard of the series will have to be returned for improvement.

For further information please contact Springer-Verlag, Physics Editorial Department II, Tiergartenstrasse 17, D-69121 Heidelberg, Germany

Julius Wess Evgeny A. Ivanov (Eds.)

Supersymmetries and Quantum Symmetries

Proceedings of the International Seminar Dedicated to the Memory of V.I. Ogievetsky, Held in Dubna, Russia, 22-26 July 1997

Springer

Editors

Julius Wess
Sektion Physik, Universität München
Theresienstrasse 37
D-80333 Munich, Germany

Evgeny A. Ivanov
Bogoliubov Laboratory of Theoretical Physics
Joint Institute for Nuclear Research
141980 Dubna, Moscow Region, Russia

Library of Congress Cataloging-in-Publication Data.

Die Deutsche Bibliothek - CIP-Einheitsaufnahme

Supersymmetries and quantum symmetries : proceedings of International Seminar Dedicated to the Memory of V. I. Ogievetsky, held in Dubna, Russia, 22 - 26 July 1997 / Julius Wess ; Evgeny A. Ivanov (ed.).
(Lecture notes in physics ; Vol. 524)
ISBN 978-3-662-14208-0 ISBN 978-3-540-48795-1 (eBook)
DOI 10.1007/978-3-540-48795-1

ISSN 0075-8450

Originally published by Springer-Verlag Berlin Heidelberg New York in 1999
Softcover reprint of the hardcover 1st edition 1999

Typesetting: Camera-ready by the authors/editors
Cover design: *design & production*, Heidelberg

SPIN: 10644335 55/3144 - 5 4 3 2 1 0 – Printed on acid-free paper

Victor Isaakovich Ogievetsky (1928–1996)

Preface

Victor Isaakovich Ogievetsky (1928–1996)

This volume is dedicated to the memory of Professor V.I. Ogievetsky, a leading Russian theorist who made many significant contributions to theoretical elementary particle physics. The volume contains both reprints of selected papers of V.I. Ogievetsky and contributions by his friends and collaborators including the talks of the participants of the Memorial International Seminar "Supersymmetries and Quantum Symmetries" held at the Joint Institute for Nuclear Research (Dubna, Russia) 22–26 July 1997.

A short outline of V.I. Ogievetsky's scientific biography may be useful for the readers of this volume. He was born in Dnepropetrovsk on August 6, 1928 into the family of I.E. Ogievetsky, Professor of Mathematics, and started his scientific activity as a theoretician in 1950 when he became a school-teacher in Dnepropetrovsk upon graduating from Dnepropetrovsk State University. His first studies were on the problem of the penetration of gamma-rays through matter. He received his PhD degree in Physics and Mathematics at the Lebedev Physical Institute in 1954. On the recommendation of Academician I.E. Tamm, in 1955 he was admitted to the V.I. Veksler laboratory at Dubna in the group of M.A. Markov where he actively began research on elementary particle physics and quantum field theory. He moved to the Joint Institute for Nuclear Research at Dubna in 1956 at the time of its foundation and worked there for 40 years in the Laboratory of Theoretical Physics of JINR (now the Bogoliubov Laboratory of Theoretical Physics).

From the beginning of his career Victor Ogievetsky was attracted to the study of symmetries in elementary particle physics. The universality of the gauge principle and its applicability not only to quantum electrodynamics, but also to other types of interactions, were not fully understood at that time, but V.I. Ogievetsky quickly realized both the potential applicability and the beauty of gauge theories. The criterion of beauty always played an important role in all his scientific works. In a close collaboration with I.V. Polubarinov, he carried out a series of studies on the field-theoretical interpretation of gauge theories and gravity theory, works which have greatly influenced the development of this major modern direction of elementary particle theory.

The role of spin of interacting fields in gauge theories was emphasised and it was realised that gauge invariance in quantum electrodynamics and non-Abelian gauge theories is a device to ensure the consistency of interactions of spin-1 fields with each other and with the conserved vector currents of matter. In a similar way, Einstein's theory of gravity was interpreted as a gauge theory of an interacting spin-2 field coupled to a conserved tensor current. In the course of these researches, V.I. Ogievetsky and I.V. Polubarinov made discoveries whose applications became clear only years later. In 1965, they introduced the 'notoph', an antisymmetric tensor gauge field which describes helicity 0 and so is complementary to the photon describing helicities ± 1. This object was later rediscovered in the context of string theories and turned out to be a necessary ingredient in many supersymmetric field theories. Another pioneering contribution made in 1964 was the idea to regard spinors in general relativity as objects with a nonlinear (in the metric) transformation law under the diffeomorphism group. This result anticipated the theory of nonlinear realizations developed later by C. Callan, C. Coleman, D.V. Volkov, S. Weinberg, J. Wess and B. Zumino. After defending his doctor's thesis based on these results in 1966, V.I. Ogievetsky carried on with intensive work on symmetry methods in quantum field theory. In the late 1960s, he became especially interested in the above-mentioned topic of nonlinear realizations and the closely related concept of spontaneous symmetry breaking, mainly in applications to space-time symmetries including the Poincaré group as a subgroup. One of his striking results in this domain (obtained in collaboration with A. Borisov) was the interpretation of gravity as a joint nonlinear realization of two spontaneously broken finite-parameter symmetries, conformal and affine, which in their closure yield the general covariance group. The latter observation is now known as the Ogievetsky theorem (1973). According to this ideology, the graviton can be viewed not only as a gauge field but also as an analogue of the Goldstone field appearing in the nonlinear realization of an internal symmetry. Later it was proved that other gauge fields admit such a two-fold interpretation as well. This deep analogy of gravity and gauge theories with sigma models turned out to be extremely fruitful in further studies, in particular in topological field theories and in the theory of embeddings of strings and superstrings.

In the next stage of his scientific career V.I. Ogievetsky turned to supersymmetry. He was very enthusiastic about this idea from the moment the first papers on the subject appeared. This was to a great extent due to the fact that in the 1960s he and I.V. Polubarinov were actually on the verge of discovering supersymmetry when asking themselves what the theory of a conserved spin 3/2 current could describe. From the beginning V.I. Ogievetsky was preoccupied by the beauty of the idea of superspace, an extension of ordinary Minkowski space by anticommuting coordinates. One of the first reviews on supersymmetry and superfield techniques was published by V.I. Ogievetsky and L. Mezincescu in Usp.Fiz.Nauk in 1975. It is unrivalled in the lucidity and completeness of its exposition and still serves as an excellent introduction to this subject. In that period V.I. Ogievetsky organized a group of followers at Dubna who were inspired by the beauty of his geometric ideas. The geometric superfield formulation of supergravity as a supersymmetric extension of gravity theory (together with E. Sokatchev) was the main achievement of this period. This formulation is based on the superfield extension of the idea of coupling to a conserved current, this time to the supercurrent that unifies the energy–momentum tensor, the source of the graviton, with the spin 3/2 current, the source of the gravitino, a superpartner of the graviton. The consistent realization of this idea resulted in the construction of linearised superfield supergravity (in 1977) and then in the discovery of the fundamental gauge group of supergravity as the group of general coordinate transformations in a complex chiral superspace. The profound connections of supergravity with the theory of complex manifolds were thus revealed. Later on, the problem emerged of how to generalise the superfield theory of a simple ($N = 1$) supersymmetry to a more complicated extended supersymmetry that includes a non-trivial group of internal symmetries. The breakthrough came about in 1984 when the Dubna group headed by V.I. Ogievetsky (A. Galperin, E. Ivanov, S. Kalitzin, E. Sokatchev) invented the concept of harmonic superspace that involves parameters of the internal-symmetry group as coordinates. The method of harmonic superspace is at present commonly recognised as a powerful geometric approach to extended supersymmetries and supersymmetric integrable systems; it is closely related to the Penrose twistor method. During his last years, up to his untimely death, Victor Isaakovich remained faithful to subjects related to supersymmetry, and in particular to applications and generalisations of the harmonic superspace approach. In his last year he was especially interested in self-dual supersymmetric theories including self-dual supergravity and had been working on some promising ideas (these last researches were performed together with Ch. Devchand). Unfortunately, his final illness overtook him suddenly and prevented him from carrying through this programme to its conclusion.

V.I. Ogievetsky was awarded the I.E. Tamm Gold Medal of the Academy of Sciences of the USSR in 1986 and the von Humboldt Foundation award (Germany) in 1992. He was four times winner of the 1st JINR annual competition prize. V.I. Ogievetsky was a member of the editorial board of the journal Yadernaya Fizika [Sov.J.Nucl.Phys.], participated in many conferences on group-theoretical methods in physics and quantum gravity, was a permanent chairman of the organizing committee of regular international workshops on supersymmetries and quantum symmetries he had founded at Dubna. He was a member of the governing board of the International Centre for Fundamental Physics founded by the P.N. Lebedev Physical Institute and NORDITA in Moscow.

Dubna, Munich
March 1999

E.A. Ivanov, J.Wess

Contents

Part II Super p-Branes and M-Theory

Part III Supersymmetric Quantum Mechanics and Integrable Systems

List of Participants

I. Bandos	bandos@tph32.tuwien.ac.at
J. Bagger	bagger@bohr.pha.jhu.edu
I. Buchbinder	joseph@tspi.tomsk.su
A. Burinskii	qrq@ibrae.ac.ru
W-S. Chung	wschung@nongae.gsnu.ac.kr
A. Demichev	demichev@theory.npi.msu.su
A. Deriglazov	deriglaz@fma.if.usp.br
B. de Wit	B.deWit@fys.ruu.nl
S. Duplij	Steven.A.Duplij@univer.kharkov.ua
A. Filippov	filippov@thsun1.jinr.dubna.su
S.J. Gates Jr.	gates@umdhep.umd.edu
G. von Gehlen	unp02F@ibm.rhrz.uni-bonn.de
V. Gershun	gershun@kipt.kharkov.ua
N. Granda	ngranda@galois.univalle.edu.co
P. Grozman	mleites@matematik.su.se
A. Isaev	isaevap@thsun1.jinr.dubna.su
E. Ivanov	eivanov@thsun1.jinr.dubna.su
A. Kamenshchik	kamen@landau.ac.ru
A. Kapustnikov	alexandr@ff.dsu.dp.ua
S. Ketov	ketov@itp.uni-hannover.de
O. Khudaverdian	khudian@vxjinr.jinr.ru
S. Konstein	konstein@nld.yar.ru

I. Korepanov	igor@prima.tu-chel.ac.ru
S. Krivonos	krivonos@thsun1.jinr.dubna.su
S. Kuzenko	Sergei.Kuzenko@physik.uni-muenchen.de
D. Leites	mleites@matematik.su.se
J. Lukierski	lukier@proton.ift.uni.wroc.pl
S. Lyakhovich	sll@phys.tsu.tomsk.su
V. Lyakhovsky	lyakhovs@snoopy.niif.spb.su
S. Mayburov	mayburov@sci.lebedev.ru
R. Metsaev	metsaev@td.lpi.ac.ru
A. Nersessian	nerses@thsun1.jinr.dubna.su
V. Nesterenko	nestr@thsun1.jinr.dubna.su
S. Odintsov	odintsov@quantum.univalle.edu.co
O. Ogievetsky	oleg@cpt.univ-mrs.fr
B. Ovrut	ovrut@ovrut.hep.upenn.edu
E. Osipov	osipov@math.nsc.ru
O. Pashaev	pashaev@vxjinr.jinr.ru
A. Pashnev	pashnev@thsun1.jinr.dubna.su
V. Pervushin	pervush@thsun1.jinr.dubna.su
M. Plyushchay	mplyushc@lauca.usach.cl
A. Popov	popov@thsun1.jinr.dubna.su
Z. Popowicz	ziemek@proton.ift.uni.wroc.pl
S. Prokushkin	prok@lpi.ac.ru
A. Restuccia	arestu@usb.ve
M. Saveliev	saveliev@mx.ihep.su
A. Semikhatov	asemikha@td.lpi.ac.ru
A. Slavnov	slavnov@mi.ras.ru
E. Sokatchev	sokatche@lapphp0.in2p3.fr
A. Sorin	sorin@thsun1.jinr.dubna.su
D. Sorokin	sorokin@physik.hu-berlin.de

K.S. Stelle	k.stelle@ic.ac.uk
A. Sutulin	sutulin@thsun1.jinr.dubna.su
H. Terazawa	
F. Toppan	toppan@cbpfsu8.cat.cbpf.br
A. Tseytlin	a.tseytlin@ic.ac.uk
M. Tsulaia	tsulaia@thsun1.jinr.dubna.su
M. Vasiliev	vasiliev@td.lpi.ac.ru
A. Vladimirov	alvladim@thsun1.jinr.ru
J. Wess	Julius.Wess@physik.uni-muenchen.de
A. Zheltukhin	zheltukhin@kipt.kharkov.ua
P. Zhidkov	zhidkov@thsun1.jinr.dubna.su
V. Zima	aigumen@kipt.kharkov.ua
B. Zupnik	zupnik@thsun1.jinr.dubna.su

Part I

Superspace Approach to Supersymmetry

Linear and Nonlinear Supersymmetries

Jonathan Bagger and Alexander Galperin

Department of Physics and Astronomy, Johns Hopkins University
3400 N. Charles Street, Baltimore, MD 21218, USA

Abstract. In this talk we use nonlinear realizations to study the spontaneous partial breaking of rigid and local supersymmetry.

1 Introduction

The winter of 1996 was a hard one for physics, bringing the untimely deaths of Professors Dmitrij Vasilievich Volkov and Victor Isaacovitch Ogievetsky. At this symposium it seems appropriate to celebrate the memories of both men, whose scientific achievements were so closely aligned, and whose inspiring presence is already acutely missed by their many friends and colleagues across the world.

In this talk we will discuss a subject close to their hearts: supersymmetry and its nonlinear realizations. In particular, we will consider the partial breaking of extended supersymmetry. For simplicity, we will restrict our attention to the case $N = 2 \to N = 1$, but many of our results can be readily extended to the case of higher supersymmetries, spontaneously broken to $N = 1$. It is a fitting memorial to see many of the ideas pioneered by Professors Volkov and Ogievetsky come into play.

The partial breaking of supersym metry is of crucial importance to understanding the relation of theory to experiment. As theorists, we know in our bones that there is an ultimate theory, perhaps M theory, that exists at high energies. However, this theory is far removed from the physical world. To connect the two, we must integrate out the degrees of freedom associated with the high energies and construct a nonrenormalizable, effective field theory. This effective field theory should contain only those degrees of freedom that are relevant for physics in the world today.

Indeed, it is the point of view that underlies the effective field theory approach to pion dynamics. Below the scale of chiral symmetry breaking, we know that the interactions of pions and hadrons are governed by an effective field theory in which the unbroken isospin symmetry is realized linearly, but the spontaneously broken chiral symmetry is realized nonlinearly. The nonlinear symmetry is all that remains of the chiral symmetry below the scale where it is broken.

For the case at hand, we wish to construct a Lagrangian with two supersymmetries. The first supersymmetry, that of $N = 1$, is realized linearly, so it can be represented in terms of superfields. The second supersymmetry, $N = 2$, is realized nonlinearly on the superfields. In this way we can construct an effective field theory of partial supersymmetry breaking. This theory is valid up to the scale where the second supersymmetry is spontaneously broken.

At first glance, it might seem impossible to partially break $N = 2$ to $N = 1$. The argument runs as follows. Start with the $N = 2$ supersymmetry algebra,

$$\begin{aligned} \{Q_\alpha, \bar{Q}_{\dot{\alpha}}\} &= 2\sigma^m_{\alpha\dot{\alpha}} P_m \\ \{S_\alpha, \bar{S}_{\dot{\alpha}}\} &= 2\sigma^m_{\alpha\dot{\alpha}} P_m , \end{aligned} \tag{1}$$

where Q_α and its conjugate $\bar{Q}_{\dot{\alpha}}$ denote the first, unbroken supersymmetry, and S_α, $\bar{S}_{\dot{\alpha}}$ the second. Suppose that one supersymmetry is not broken, so

$$Q\,|0\rangle \;=\; \bar{Q}\,|0\rangle \;=\; 0\,. \tag{2}$$

Because of the supersymmetry algebra, this implies that the Hamiltonian also annihilates the vacuum,

$$H\,|0\rangle \;=\; 0\,. \tag{3}$$

Then, according to the supersymmetry algebra,

$$(\bar{S}S + S\bar{S})\,|0\rangle \;=\; 0\,. \tag{4}$$

The final step is to peel apart this relation and conclude that

$$S\,|0\rangle \;=\; \bar{S}\,|0\rangle \;=\; 0\,. \tag{5}$$

From this line of reasoning, one might think that partial breaking is impossible.

Fortunately, this argument has two significant loopholes. The first is that, technically-speaking, spontaneously-broken charges do not exist. Indeed, in a spontaneously broken theory, one only has the right to consider the algebra of the *currents*. For the case at hand, the current algebra can be modified as follows,

$$\begin{aligned} \{\bar{Q}_{\dot{\alpha}}, J^1_{\alpha m}\} &= 2\sigma^n_{\alpha\dot{\alpha}} T_{mn} \\ \{\bar{S}_{\dot{\alpha}}, J^2_{\alpha m}\} &= 2\sigma^n_{\alpha\dot{\alpha}} (v^4 \eta_{mn} + T_{mn}) , \end{aligned} \tag{6}$$

where the $J^i_{\alpha m}$ $(i = 1, 2)$ are the supercurrents and T_{mn} is the stress-energy tensor. Note that Lorentz invariance does not force the right-hand sides of the commutators to be the same. If there were no first supersymmetry, the v^4 term in the second commutator could be absorbed in T_{mn}; it would represent the scale of the supersymmetry breaking. Now, however, the first supersymmetry can be said to *define* the stress-energy tensor, in which case there is an extra term in the second commutator. This discrepancy prevents the current algebra from being integrated into a charge algebra, and the no-go theorem is avoided.

The second loophole involves the last step of the theorem. Even if the supercharges were to exist, it is only possible to extract (5) from (4) if the Hilbert space is positive definite. In covariantly-quantized supergravity theories, this is not the case: the gravitino $\psi_{m\alpha}$ is a gauge field with negative-norm components.

There are, by now, many examples of partial supersymmetry breaking which exploit the first loophole. The first was given by Hughes, Liu and Polchinski (1986) who showed that supersymmetry is partially broken on the world volume of an $N = 1$ supersymmetric three-brane traveling in six-dimensional superspace. Since then there has been an explosion of interest in membranes, so the number of examples has grown substantially. [For another type of example, see Antoniadis, Partouche and Taylor (1996).]

The membrane approach leaves many open questions. For example, we would like to know all possible field-theoretic realizations of partial supersymmetry breaking, even those that do not originate with branes. We would also like to know whether the $N = 2$ supersymmetry gives rise to any restrictions on matter couplings in the low-energy effective theory.

Finally, we would like to understand how partial breaking works in the presence of gravity. Gravity couples to the true stress-energy tensor, so it distinguishes between the right-hand sides of the commutators (6). Some early work on this question was done by Cecotti, Girardello and Porrati (1986) and by Zinov'ev (1987). These groups considered nonminimal cases and found that their gravitational couplings utilize the second loophole. One would like to reconcile their results with those above.

2 Coset Construction

In this talk we will take a bottom-up approach to the subject of partial supersymmetry breaking. We will use nonlinear realizations to describe the effective $N = 1$ theory which holds below the scale of the second supersymmetry breaking. We will use the formalism of Coleman, Wess and Zumino (1969), as extended by Volkov (1973), to construct theories where the $N = 1$ supersymmetry is manifest, and the second supersymmetry is nonlinearly realized.

The approach of Coleman, Wess, Zumino and Volkov is based on a coset decomposition of a symmetry group, G. We start with a group, G, of internal and spacetime symmet ries, and partition the generators of G into three classes:

- Γ_A, the generators of unbroken spacetime translations;
- Γ_a, the generators of spontaneously broken internal and spacetime symmetries; and
- Γ_i, the generators of unbroken spacetime rotations and unbroken internal symmetries.

The generators Γ_i close into the stability group, H.

Given G and H, we define the coset G/H in terms of an equivalence relation on the elements $\Omega \in G$, $\Omega \sim \Omega h$, with $h \in H$. The coset can be thought of as a section of a fiber bundle with total space, G, and fiber, H.

This equivalence relation suggests that we parametrize the coset as follows,

$$\Omega \;=\; \exp \mathrm{i} X^A \Gamma_A \; \exp \mathrm{i} \xi^a(X) \Gamma_a \; . \tag{7}$$

Physically, the X^A play the role of generalized spacetime coordinates, while the $\xi^a(X)$ are generalized Goldstone fields, defined on the generalized coordinates and valued in the set of broken generators Γ_a. There is one generalized coordinate for every unbroken spacetime translation, and one generalized Goldstone field for every spontaneously broken generator.

We define the action of the group G on the coset G/H by left multiplication, $\Omega \to g\,\Omega = \Omega'\, h$, with $g \in G$. In this expression,

$$\Omega' \;=\; \exp \mathrm{i} X'^A \Gamma_A \; \exp \mathrm{i} \xi'^a(X') \Gamma_a \tag{8}$$

and $h = \exp \mathrm{i}\alpha^i(g, X, \xi)\Gamma_i$. The group multiplication induces nonlinear transformations on the coordinates X^A and the Goldstone fields ξ^a:

$$X^A \;\to\; X'^A \;, \quad \xi^a(X) \;\to\; \xi'^a(X') \; . \tag{9}$$

These transformations realize the full symmetry group, G. Note that the field ξ^a transforms by a shift under the transformation generated by Γ_a. This confirms that ξ^a is indeed the Goldstone field corre sponding to the broken generator Γ_a.

An arbitrary G transformation induces a compensating H transformation which is required to restore the section. This transformation can be used to lift any representation, R, of H, to a nonlinear realization of the full group, G, as follows,

$$\chi(X) \;\to\; \chi'(X') \;=\; D(h)\chi(X) \; . \tag{10}$$

Here $D(h) \;=\; \exp(\mathrm{i}\alpha^i T_i)$, where α^i was defined below (8), and the T_i are generators of H in the representation R.

To proceed further, it is helpful to have a vielbein, connection and covariant derivative, built from the Goldstone fields in the following way. One first computes the Maurer-Cartan form, $\Omega^{-1}\mathrm{d}\Omega$, where d is the exterior derivative. One then expands $\Omega^{-1}\mathrm{d}\Omega$ in terms of the Lie algebra of G,

$$\Omega^{-1}\mathrm{d}\Omega \;=\; \mathrm{i}(\omega^A \Gamma_A + \omega^a \Gamma_a + \omega^i \Gamma_i) \; , \tag{11}$$

where ω^A, ω^a and ω^i are one-forms on the manifold parametrized by the coordinates X^A.

The Maurer-Cartan form transforms as follows under a rigid G transformation,

$$\Omega^{-1}d\Omega \;\to\; h(\Omega^{-1}\mathrm{d}\Omega)h^{-1} - \mathrm{d}h\, h^{-1} \; . \tag{12}$$

From this we see that the fields ω^A and ω^a transform covariantly under G, while ω^i transforms by a shift. These transformations help us identify

$$\omega^A = \mathrm{d}X^M E_M{}^A \tag{13}$$

as the covariant vielbein,

$$\omega^a = \mathrm{d}X^M E_M{}^A \mathcal{D}_A \xi^a \tag{14}$$

as the covariant derivative of the Goldstone field ξ^a, and

$$\omega^i = \mathrm{d}X^M \omega^i_M \tag{15}$$

as the connection associated with the stability group, H. With these building blocks, it is easy to construct theories invariant under the full group G.

The coset construction is very general and very powerful. For the case of internal symmetries, it allows one to prove that any H-invariant action can be lifted to be G-invariant with the help of the Goldstone bosons. For $N = 1$ supersymmetry, it can be used to show that any Lorentz-invariant action can be made supersymmetric with the help of the Goldstone fermion.

3 Nonlinear Supersymmetry

In this section we will show that any $N = 1$ supersymmetric theory can be made $N = 2$ supersymmetric with the help of an $N = 1$ Goldstone superfield. We will find that the Goldstone superfield can contain either an $N = 1$ chiral or vector multiplet (Bagger and Galperin, 1994, 1997a). [The case where the Goldstone superfield is an $N = 1$ tensor multiplet can be obtained from the chiral case by a superspace duality transformation (Bagger and Galperin, 1997b).]

It is important to emphasize that the coset construction – while very useful and very general – does not tell us anything about the underlying theory in which both supersymmetries are linearly realized. Indeed, such a theory might not even exist. Therefore we shall resolutely insist that we are working in the context of an effective field theory, and leave to others the task of finding the more fundamental theory above the supersymmetry-breaking scale.

In what follows we shall first take a minimal approach, and choose the group G to be the supergroup whose algebra is (1). We will take the subgroup H to be the supergroup generated by P_a, Q_α and $\bar{Q}_{\dot{\alpha}}$. We parametrize the coset element Ω as follows,

$$\begin{aligned}\Omega &= \exp\mathrm{i}(x^a P_a + \theta^\alpha Q_\alpha + \bar{\theta}_{\dot{\alpha}} \bar{Q}^{\dot{\alpha}}) \\ &\quad \times \exp\mathrm{i}(\Psi^\alpha S_\alpha + \bar{\Psi}_{\dot{\alpha}} \bar{S}^{\dot{\alpha}}) \ .\end{aligned} \tag{16}$$

Here x, θ and $\bar{\theta}$ are the coordinates of $N = 1$ superspace, while Ψ^α and its conjugate $\bar{\Psi}_{\dot{\alpha}}$ are Goldstone $N = 1$ superfields of (geometrical) dimension

$-1/2$. These spinor superfields contai n far too many component fields, so we need to find a set of consistent, covariant constraints to reduce the number of fields.

The correct constraints are most easily expressed in term of the $N = 2$ covariant derivatives of the Goldstone superfield. The covariant derivatives can be found following the techniques of the previous section; they can be explicitly written as follows,

$$\begin{aligned} \mathcal{D}_\alpha &= D_\alpha - \mathrm{i}(D_\alpha \Psi \sigma^a \bar{\Psi} + D_\alpha \bar{\Psi} \bar{\sigma}^a \Psi)\omega_a^{-1m}\partial_m \\ \bar{\mathcal{D}}_{\dot\alpha} &= \bar{D}_{\dot\alpha} - \mathrm{i}(\bar{D}_{\dot\alpha} \Psi \sigma^a \bar{\Psi} + \bar{D}_{\dot\alpha} \bar{\Psi} \bar{\sigma}^a \Psi)\omega_a^{-1m}\partial_m \\ \mathcal{D}_a &= \omega_a^{-1m}\partial_m \ , \end{aligned} \tag{17}$$

where $\omega_m{}^a \equiv \delta_m^a + \mathrm{i}(\partial_m \Psi \sigma^a \bar{\Psi} + \partial_m \bar{\Psi} \bar{\sigma}^a \Psi)$ and $D_\alpha,\ \bar{D}_{\dot\alpha}$ are ordinary flat $N = 1$ superspace spinor derivatives. The covariant derivatives obey the following commutation relations,

$$\begin{aligned} \{\mathcal{D}_\alpha, \mathcal{D}_\beta\} &= -2\mathrm{i}(\mathcal{D}_\alpha \Psi^\gamma \mathcal{D}_\beta \bar{\Psi}^{\dot\gamma} + (\alpha \leftrightarrow \beta))\mathcal{D}_{\gamma\dot\gamma} \\ [\mathcal{D}_\alpha, \mathcal{D}_a] &= -2\mathrm{i}(\mathcal{D}_\alpha \Psi^\gamma \mathcal{D}_a \bar{\Psi}^{\dot\gamma} + (\alpha \leftrightarrow a))\mathcal{D}_{\gamma\dot\gamma} \\ \{\mathcal{D}_\alpha, \bar{\mathcal{D}}_{\dot\beta}\} &= 2\mathrm{i}\sigma^a_{\alpha\dot\beta}\mathcal{D}_a - 2\mathrm{i}(\mathcal{D}_\alpha \Psi^\gamma \bar{\mathcal{D}}_{\dot\beta} \bar{\Psi}^{\dot\gamma} \\ &\quad + (\alpha \leftrightarrow \dot\beta))\mathcal{D}_{\gamma\dot\gamma} \ , \end{aligned} \tag{18}$$

where $\mathcal{D}_{\alpha\dot\alpha} \equiv \sigma^a_{\alpha\dot\alpha}\mathcal{D}_a$.

One set of constraints is simply (Bagger and Galperin, 1994)

$$\begin{aligned} \bar{\mathcal{D}}\bar{\mathcal{D}}\Psi_\alpha &= \mathcal{O}(\Psi^3) \\ \mathcal{D}_\alpha \Psi_\beta + \mathcal{D}_\beta \Psi_\alpha &= \mathcal{O}(\Psi^3) \ . \end{aligned} \tag{19}$$

The right-hand side of this equation must be adjusted for consistency with (18). Remarkably, this can be done using the dimensionless invariants $\bar{\mathcal{D}}_{\dot\alpha}\Psi_\alpha$ and $\mathcal{D}_\alpha \Psi_\beta$ (together with their complex conjugates). It turns out that there is a unique, consistent solution order-by-order in powers of the Goldstone field.

The solution to the constraints (19) is easy to find in perturbation theory. To lowest order, it is just the chiral multiplet Φ,

$$\begin{aligned} \Psi_\alpha &= D_\alpha \Phi + \mathcal{O}(\Psi^3) \\ \bar{D}_{\dot\alpha}\Phi &= \mathcal{O}(\Psi^3) \ . \end{aligned} \tag{20}$$

In this expression, D_α is the ordinary $N = 1$ superspace spinor derivative.

A second set of constraints is (Bagger and Galperin, 1997a)

$$\begin{aligned} \bar{\mathcal{D}}_{\dot\alpha}\Psi_\alpha &= \mathcal{O}(\Psi^3) \\ \mathcal{D}^\alpha \Psi_\alpha + \bar{\mathcal{D}}_{\dot\beta}\bar{\Psi}^{\dot\beta} &= \mathcal{O}(\Psi^3) \ . \end{aligned} \tag{21}$$

As above, the right-hand side must be adjusted for consistency with the algebra of covariant derivatives. Again, there is a unique, consistent solution. To lowest order in perturbation theory, it is

$$\begin{aligned} \Psi_\alpha &= W_\alpha + \mathcal{O}(\Psi^3) \\ W_\alpha &= -\frac{1}{4}\bar{D}\bar{D}D_\alpha V + \mathcal{O}(\Psi^3) , \end{aligned} \tag{22}$$

where V is a real $N = 1$ vector superfield. We see that the chiral and vector Goldstone multiplet can each be obtained to lowest order in perturbation theory. In fact, the consistency of the multiplets survives to all orders in perturbation theory.

The Goldstone action can be constructed order-by-order in the Goldstone fields. For the chiral case, it is simply (Bagger and Galperin, 1994)

$$S = v^4 \int d^4x d^2\theta d^2\bar{\theta}\, E\, [\Phi^+ \Phi + \mathcal{O}(\Phi^4)] . \tag{23}$$

In this expression, $E = \mathrm{Ber}(E_M{}^A)$ is the superdeterminant of the vielbein, and v is the constant of dimension one which corresponds to the scale of the supersymmetry breaking. The action (23) is invariant under the full $N = 2$ supersymmetry.

For the vector multiplet, the Goldstone action is just (Bagger and Galperin, 1997a)

$$\begin{aligned} S &= \frac{v^4}{4} \int d^4x d^2\theta\, \mathcal{E}\, W^2 \; + \text{ h.c.} \\ &\quad + \int d^4x d^4\theta\, E\, \mathcal{O}(W^4) . \end{aligned} \tag{24}$$

This action is invariant under $N = 2$ supersymmetry. It is also gauge-invariant. Curiously enough, the gauge field contribution to the Goldstone action coincides with the expansion of the Born-Infeld action.

Having constructed the $N = 2$ Goldstone action, we are now ready to add $N = 2$ covariant matter. The basic ingredients are $N = 2$ nonlinear generalizations of $N = 1$ chiral and vector superfields. The generalized chiral superfields are defined by the constraint $\bar{\mathcal{D}}_{\dot\alpha}\chi = 0$. This constraint is consistent for either type of Goldstone multiplet.

The matter action is easy to write down for either Goldstone multiplet. The kinetic term is

$$S = \int d^4x d^4\theta\, E\, K(\chi^+, \chi) \tag{25}$$

while the superpotential term is

$$S = \int d^4x d^2\theta\, \mathcal{E}\, P(\chi) . \tag{26}$$

As before, E and $\mathcal{E}$ are superdeterminants of the supervielbein $E_M{}^A$. They can be adjusted to preserve the condition

$$\int d^4x d^4\theta\, E\, F(\chi) = 0 . \tag{27}$$

This allows the matter action to be Kähler invariant, so the matter couplings are described in terms of Kähler manifolds, just as for $N = 1$.

It is not hard to generalize these results to include vector superfields. The general conclusion is that any $N = 1$ invariant theory can be lifted to be $N = 2$ supersymmetric with the help of a Goldstone superfield. Furthermore, the Goldstone superfield can be either an $N = 1$ chiral or vector multiplet.

Now that we have two explicit realizations of partial supersymmetry breaking, we can ask how they avoid the no-go argument discussed above. In each case, the nonlinear theory exploits the loophole of Hughes, Liu and Polchinski (1986). For example, in the vector case the second supercur rent goes like $J^m_\alpha \sim v^4 \sigma^m_{\alpha\dot\alpha} \bar\lambda^{\dot\alpha}$, so its commutator with the second supercharge reproduces the algebra (6).

4 Geometry

The fact that the constraints need to be adjusted order-by-order in Ψ_α hints that a deeper structure underlies partial supersymmetry breaking. The $N = 2$ supersymmetry does not provide enough symmetry to uniquely fix the covariant derivatives and the associated constraints. This intuition is borne out for the case of the chiral multiplet, where a much deeper set of symmetries acts on the Goldstone multiplet (Bagger and Galperin, 1994).

To see this, let us first extend the $N = 2$ algebra by a complex central charge, Z:

$$\begin{aligned} \{Q_\alpha, \bar Q_{\dot\alpha}\} &= 2\sigma^a_{\alpha\dot\alpha} P_a & \{S_\alpha, \bar S_{\dot\alpha}\} &= 2\sigma^a_{\alpha\dot\alpha} P_a \\ \{Q_\alpha, S_\beta\} &= 2\epsilon_{\alpha\beta} Z & \{\bar Q_{\dot\alpha}, \bar S_{\dot\beta}\} &= 2\epsilon_{\dot\alpha\dot\beta} \bar Z \ . \end{aligned} \tag{28}$$

We then consider a coset where the group G contains not only $N = 2$ supersymmetry, but also its maximal automorphism group, $SO(5,1) \times SU(2)$, where the $SU(2)$ acts on the two supersymetry generators, and $SO(5,1)$ is the $D = 6$ Lorentz group. (Under $SO(5,1)$, the generators P_a and Z form a $D = 6$ vector, while the two supercharges form a single $D = 6$ Majorana-Weyl spinor). Let us take H to be $SO(3,1) \times SO(2) \times U(1)$, where $SO(3,1) \times SO(2) \subset SO(5,1)$, $U(1) \subset SU(2)$, and $SO(3,1)$ is the $D = 4$ Lorentz group.

Our parametrization of the coset G/H involves the $N = 1$ superspace coordinates, as well as different Goldstone superfields for each of the broken symmetries,

$$\begin{aligned} \Omega &= \exp i(x^a P_a + \theta^\alpha Q_\alpha + \bar\theta_{\dot\alpha} \bar Q^{\dot\alpha}) \\ &\quad \times \exp i(\Phi Z + \bar\Phi \bar Z + \Psi^\alpha S_\alpha + \bar\Psi_{\dot\alpha} \bar S^{\dot\alpha}) \\ &\quad \times \exp i(\Lambda^a K_a + \bar\Lambda^a \bar K_a + \Xi T + \bar\Xi \bar T) \ . \end{aligned} \tag{29}$$

Here Λ^a, $\bar\Lambda^a$ are the Goldstone superfields associated with the generators K_a, $\bar K_a$ of $SO(5,1)/SO(3,1) \times SO(2)$. Similarly, Ξ, $\bar\Xi$ are the Goldstone superfields for the broken generators T, $\bar T$ of $SU(2)/U(1)$.

As before, the $N = 1$ Goldstone superfields contain far more components than the minimal Goldstone multiplet. This motivates us to impose the following consistent set of constraints:

$$\begin{aligned} \bar{\mathcal{D}}_{\dot\alpha}\Phi &= 0\,, & \mathcal{D}_\alpha\Phi &= 0\,, & \mathcal{D}_a\Phi &= 0 \\ \mathcal{D}_\alpha\Psi^\beta &= 0\,, & \bar{\mathcal{D}}_{\dot\alpha}\Psi^\beta &= 0\,. \end{aligned} \tag{30}$$

These constraints allow us to express the Goldstone superfields Ψ^α, Λ^a and $\bar\Xi$ in terms of a single superfield Φ. [This way of eliminating Goldstones was called the "inverse Higgs effect" by Ivanov and Ogievetsky (1975).] To lowest order, we find $\Psi^\alpha = -\frac{i}{2}D^\alpha\Phi$, $\Lambda_a = -\partial_a\Phi$, and $\bar\Xi = \frac{1}{4}D^2\Phi$. The constraint $\bar{\mathcal{D}}_{\dot\alpha}\Phi = 0$ reduces Φ to an $N = 1$ chiral superfield.

The remarkable fact about this construction is that it reveals a geometrical role for each component of the chiral Goldstone multiplet. The scalar field, A, is the complex Goldstone boson associated with the spontaneously broken central charge symmetry. Its derivative, $\partial_m A$, is the Goldstone boson associated with $SO(5,1)/SO(3,1) \times SO(2)$. The F-component of Φ is the complex Goldstone boson associated with the $SU(2)/U(1)$. Finally, the spinor is the Goldstone fermion that arises from the partially broken supersymmetry.

The action (23) turns out to be invariant under $SO(5,1)$, but it explicitly breaks $SU(2)$ down to $U(1)$. Furthermore, any R-invariant $N = 1$ matter action can be lifted to be $SO(5,1)$ invariant. These facts hint that the Goldstone action might be rela ted to the six-dimensional membrane of Hughes, Liu and Polchinski (1986). Indeed, it is not hard to show that the chiral Goldstone action is precisely the gauge-fixed membrane action.

The geometry that underlies the vector case is presently under study. The Born-Infeld form of the gauge action suggests that it might be related to some sort of D-brane. The fact that there are no "transverse" scalars hints that the action might be that of a space-filling D3-brane. In any case, one would like to find the Goldstone-type symmetries associated with the gauge field strength and the auxiliary field of the Goldstone multiplet.

In fact, the D-component of the Goldstone multiplet can be interpreted as the Goldstone boson associated with the following $U(1)$ subgroup of the $SU(2)$ automorphism symmetry: $\delta\theta^\alpha = i\eta\Psi^\alpha$, $\delta\Psi^\alpha = i\eta\theta^\alpha$. Under such a transformation, the D-component is shifted by the constant parameter η.

If we were to extend G in G/H by this $U(1)$, we would eliminate the dimensionless invariant $\mathcal{D}^\alpha\Psi_\alpha$ in favor of the corresponding Goldstone superfield. Even then, there would still be a dimensionless invariant associated with the gauge field strength, $\mathcal{D}_{(\alpha}\Psi_{\beta)}$. This suggests that there is an extension of $N = 2$ supersymmetry which associates a Goldstone-like symmetry with this field strength.

Moreover, gauge fields themselves can be interpreted as Goldstone fields associated with infinite-dimensional symmetry groups (Ivanov and Ogievetsky, 1976). This leads us to wonder whether the full symmetry of the new multiplet is some infinite-dimensional extension of $N = 2$ supersymmetry.

5 Supergravity

We have just seen that there are two independent Goldstone realizations of partial supersymmetry breaking in four dimensions. (A third is related by duality.) Both give rise to the current algebra (6). Because the spontaneous breaking relies on the curious shift in the "second" stress-energy tensor, one would like to see what happens when the Goldstone multiplets are coupled to supergravity.

In this section, we will work backwards, and start by constructing two Lagrangians and two sets of supersymmetry transformations for the massive $N = 1$ spin-3/2 multiplet. We will then "unHiggs" the theories by adding appropriate Goldstone fields and coupling gravity. In this way we will find the supergravities associated with each of the Goldstone multiplets. (The work in this section was done in collaboration with Richard Altendorfer and Samuel Osofsky.)

We will see that the second Lagrangian corresponds to an alternative representation for the $N = 1$ massive spin-3/2 multiplet, one which was originally found by Ogievetsky and Sokatchev (1977). When coupled to gravity, this representation gives rise to a new $N = 2$ supergravity with a modified $N = 2$ supersymmetry algebra.

5.1 The Massive $N = 1$ Spin-3/2 Multiplet

The starting point for the supergravity coupling is the massive $N = 1$ spin-3/2 multiplet. This multiplet contains six bosonic and six fermionic (on-shell) degrees of freedom, arranged in states of the following spins,

$$\begin{pmatrix} & \frac{3}{2} & \\ 1 & & 1 \\ & \frac{1}{2} & \end{pmatrix} . \tag{31}$$

The traditional representation of this multiplet contains the following fields (Ferrara and van Nieuwenhuizen, 1983): one spin-3/2 fermion, one spin-1/2 fermion, and two spin-one vectors, each of mass m. The Ogievetsky-Sokatchev representation has the same fermions, but just one vector plus one antisymmetric tensor. As we shall see, each representation has a role to play in the theory of partial supersymmetry breaking.

The traditional representation is described by the following Lagrangian (Ferrara and van Nieuwenhuizen, 1983):

$$\begin{aligned} \mathcal{L} \ = \ & \epsilon^{mnp\sigma}\overline{\psi}_m\bar{\sigma}_n\partial_p\psi_\sigma - i\bar{\zeta}\bar{\sigma}^m\partial_m\zeta - \frac{1}{4}\mathcal{A}_{mn}\bar{\mathcal{A}}^{mn} \\ & - \frac{1}{2}m^2\,\mathcal{A}_m\bar{\mathcal{A}}^m \ + \ \frac{1}{2}m\,\zeta\zeta \ + \ \frac{1}{2}m\,\bar{\zeta}\bar{\zeta} \\ & - m\,\psi_m\sigma^{mn}\psi_n - \ m\,\bar{\psi}_m\bar{\sigma}^{mn}\bar{\psi}_n \ . \end{aligned} \tag{32}$$

Here ψ_m is a spin-3/2 Rarita-Schwinger field, ζ a spin-1/2 fermion, and $\mathcal{A}_m = A_m + \mathrm{i}B_m$ a complex spin-one vector. This Lagrangian is invariant under the following $N = 1$ supersymmetry transformations,

$$\begin{aligned}
\delta_\eta \mathcal{A}_m &= 2\psi_m\eta - \mathrm{i}\frac{2}{\sqrt{3}}\bar{\zeta}\bar{\sigma}_m\eta - \frac{2}{\sqrt{3}m}\partial_m(\zeta\eta) \\
\delta_\eta\zeta &= \frac{1}{\sqrt{3}}\bar{\mathcal{A}}_{mn}\sigma^{mn}\eta - \mathrm{i}\frac{m}{\sqrt{3}}\sigma^m\bar{\eta}\mathcal{A}_m \\
\delta_\eta\psi_m &= \frac{1}{3m}\partial_m(\bar{\mathcal{A}}_{rs}\sigma^{rs}\eta + 2\mathrm{i}m\sigma^n\bar{\eta}\mathcal{A}_n) - \frac{\mathrm{i}}{2}(H_{+mn}\sigma^n + \frac{1}{3}H_{-mn}\sigma^n)\bar{\eta} \\
&\quad - \frac{2}{3}m(\sigma_m{}^n\bar{\mathcal{A}}_n\eta + \bar{\mathcal{A}}_m\eta) , \qquad (33)
\end{aligned}$$

where $H_{\pm mn} = \mathcal{A}_{mn} \pm \frac{\mathrm{i}}{2}\epsilon_{mnrs}\mathcal{A}^{rs}$.

The alternative Ogievetsky-Sokatchev representation has the following Lagrangian,

$$\begin{aligned}
\mathcal{L} &= \epsilon^{pqrs}\bar{\psi}_p\bar{\sigma}_q\partial_r\psi_s - \mathrm{i}\bar{\zeta}\bar{\sigma}^m\partial_m\zeta - \frac{1}{4}A_{mn}A^{mn} + \frac{1}{2}v^m v_m \\
&\quad - \frac{1}{2}m^2A_mA^m - \frac{1}{4}m^2B_{mn}B^{mn} + \frac{1}{2}m\zeta\zeta + \frac{1}{2}m\bar{\zeta}\bar{\zeta} \\
&\quad - m\,\psi_m\sigma^{mn}\psi_n - m\,\bar{\psi}_m\bar{\sigma}^{mn}\bar{\psi}_n , \qquad (34)
\end{aligned}$$

where A_{mn} is the field strength associated with the real vector field A_m, and $v_m = \frac{1}{2}\epsilon_{mnrs}\partial^nB^{rs}$ is the field strength for the antisymmetric tensor B_{mn}. This Lagrangian is invariant under the following $N = 1$ supersymmetry transformations,

$$\begin{aligned}
\delta_\eta A_m &= (\psi_m\eta + \bar{\psi}_m\bar{\eta}) + \frac{\mathrm{i}}{\sqrt{3}}(\bar{\eta}\bar{\sigma}_m\zeta - \bar{\zeta}\bar{\sigma}_m\eta) - \frac{1}{\sqrt{3}m}\partial_m(\zeta\eta + \bar{\zeta}\bar{\eta}) \\
\delta_\eta B_{mn} &= 2\o ver\sqrt{3}\left(\eta\sigma_{mn}\zeta + \frac{\mathrm{i}}{2m}\partial_{[m}\bar{\zeta}\bar{\sigma}_{n]}\eta\right) + \mathrm{i}\eta\sigma_{[m}\bar{\psi}_{n]} + \frac{1}{m}\eta\psi_{mn} + \text{h.c.} \\
\delta_\eta\zeta &= \frac{1}{\sqrt{3}}A_{mn}\sigma^{mn}\eta - \frac{\mathrm{i}m}{\sqrt{3}}\sigma^m\bar{\eta}A_m - \frac{1}{\sqrt{3}}m\sigma_{mn}\eta B^{mn} - \frac{1}{\sqrt{3}}v_m\sigma^m\bar{\eta} \\
\delta_\eta\psi_m &= \frac{1}{3m}\partial_m\left(A_{rs}\sigma^{rs}\eta + 2\mathrm{i}m\sigma^n\bar{\eta}A_n\right) - \frac{\mathrm{i}}{2}(H^A_{+mn}\sigma^n + \frac{1}{3}H^A_{-mn}\sigma^n)\bar{\eta} \\
&\quad - \frac{2}{3}m(\sigma_m{}^nA_n\eta + A_m\eta) + \frac{1}{3m}\partial_m\left(2v_n\sigma^n\bar{\eta} - m\sigma^{rs}\eta B_{rs}\right) \\
&\quad - \frac{2\mathrm{i}}{3}(v_m + \sigma_{mr}v^n)\eta - \frac{\mathrm{i}m}{3}(B_{mn}\sigma^n\bar{\eta} + \mathrm{i}\epsilon_{mnrs}B^{nr}\sigma^s\bar{\eta}) , \qquad (35)
\end{aligned}$$

where the square brackets denote antisymmetrization, without a factor of 1/2.

These Lagrangians describe the free dynamics of massive spin-3/2 and 1/2 fermions, together with their supersymmetric partners, massive spin-one vector and tensor fields. They can be thought of as "unitary gauge" representations of theories with additional symmetries: a second supersymmetry for the massive spin-3/2 fermion, and additional gauge symmetries associated with the massive gauge fields.

5.2 The Supergravity Coupling

To study partial breaking, we need to "unHiggs" these Lagrangians by including appropriate gauge and Goldstone fields. In each case we need to add a Goldstone multiplet and gauge the full $N = 2$ supersymmetry. The supersymmetric partners of the Goldstone fermion will turn out to be the Goldstone bosons that restore the gauge symmetries associated with the massive bosonic fields. At the end of the day, we will find two theories with $N = 2$ supersymmetry nonlinearly realized, but $N = 1$ represented linearly on the fields. The resulting effective field theories describe the physics of partial supersymmetry breaking, well be low the scale where the second supersymmetry is broken.

The trick to this construction is to add the right fields. Because $N = 1$ supersymmetry is not broken, the Goldstone fermion must belong to an $N = 1$ supersymmetry multiplet. For the two cases of interest, we shall see that the Goldstone fermion must belong to the chiral or the vector multiplet, discussed above.

Let us first consider the chiral case. Under the first supersymmetry, a complex boson ϕ transforms into a Weyl fermion χ,

$$\delta_{\eta^1}\phi \;=\; \sqrt{2}\,\eta^1\chi\,. \tag{36}$$

If χ is the Goldstone fermion, it shifts under the second supersymmetry,

$$\delta_{\eta^2}\chi \;=\; \sqrt{2}\,v^2\,\eta^2 \;+\; \dots\,, \tag{37}$$

where v is the scale of the second supersymmetry breaking. Therefore the closure of the two supersymmetries on ϕ gives

$$[\,\delta_{\eta^2},\,\delta_{\eta^1}\,]\,\phi \;=\; 2\,v^2\,\eta^1\eta^2 \;+\; \dots \tag{38}$$

The complex scalar ϕ undergoes a constant shift. This is in accord with our previous result: The field ϕ is itself a Goldstone boson, corresponding to a complex central charge. It expects to be eaten by a complex vector field, which suggests that the chiral Goldstone multiplet should be associated with the traditional representation for the massive spin-3/2 multiplet.

As shown in Figure 1(a), the degree of freedom counting works out just right. We start with the $N = 1$ chiral Goldstone multiplet and add an $N = 1$ vector multiplet. We then add the gauge fields of $N = 2$ supergravity. As we will see, the full set of fields can be used to construct a Lagrangian which is invariant under $N = 2$ supersymmetry. The final results look complicated, but they are actually very simple: In unitary gauge, the two vectors eat the two scalars, while the Rarita-Schwinger field eats one linear combination of

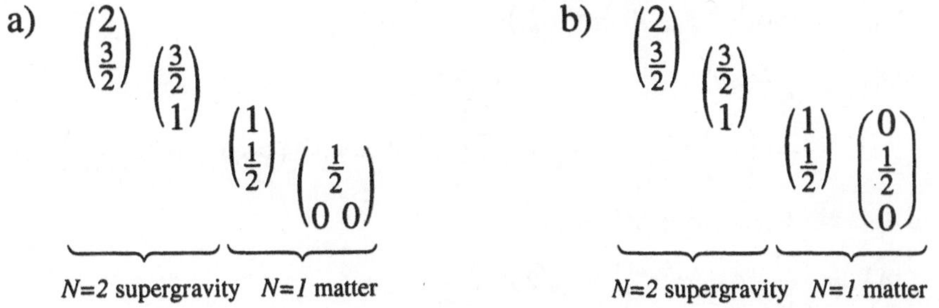

Fig. 1. The unHiggsed versions of the (a) traditional and (b) alternative representations of the $N = 1$ massive spin-3/2 multiplet. The traditional representation contains the degrees of freedom associated with an $N = 1$ chiral multiplet. The alternative representation exchanges the chiral multiplet for its dual, an $N = 1$ tensor multiplet.

the spin-1/2 fermions. This leaves the massive $N = 1$ multiplet coupled to $N = 1$ supergravity.

With that said, we now present the Lagrangian (Altendorfer, Bagger, Osofsky, 1998):

$$\begin{aligned} e^{-1}\mathcal{L} =& \\ & - \frac{1}{2\kappa^2}\mathcal{R} + \epsilon^{mnrs}\overline{\psi}_{mi}\overline{\sigma}_n D_r\psi^i_s - \mathrm{i}\overline{\chi}\,\overline{\sigma}^m D_m\chi - \mathrm{i}\overline{\lambda}\overline{\sigma}^m D_m\lambda - \mathcal{D}^m\phi\overline{\mathcal{D}_m\phi} \\ & - \frac{1}{4}\mathcal{A}_{mn}\overline{\mathcal{A}}^{mn} - \Big(\frac{1}{\sqrt{2}}m\psi^2_m\sigma^m\overline{\lambda} + \mathrm{i}m\psi^2_m\sigma^m\overline{\chi} + \sqrt{2}\mathrm{i}m\lambda\chi + \frac{1}{2}m\chi\chi \\ & + m\,\psi^2_m\sigma^{mn}\psi^2_n + \frac{\kappa}{4}\epsilon_{ij}\psi^i_m\psi^j_n\overline{H}^{mn}_{+} + \frac{\kappa}{\sqrt{2}}\chi\sigma^m\overline{\sigma}^n\psi^1_m\overline{\mathcal{D}_n\phi} \\ & + \frac{\kappa}{2\sqrt{2}}\overline{\lambda}\overline{\sigma}_m\psi^1_n\overline{H}^{mn}_{-} + \frac{\kappa}{\sqrt{2}}\epsilon^{mnrs}\overline{\psi}_{m2}\overline{\sigma}_n\psi^1_r\overline{\mathcal{D}_s\phi} + \text{h.c.}\Big)\,, \end{aligned} \tag{39}$$

where κ denotes Newton's constant, $m = \kappa v^2$, and

$$\begin{aligned} \mathcal{A}_m &= A_m + \mathrm{i}B_m \\ \mathcal{A}_{mn} &= \partial_m\mathcal{A}_m - \partial_n\mathcal{A}_m \\ H_{\pm mn} &= \mathcal{A}_{mn} \pm \frac{\mathrm{i}}{2}\epsilon_{mnrs}\mathcal{A}^{rs}\,. \end{aligned} \tag{40}$$

The supercovariant derivatives are as follows,

$$\begin{aligned} \hat{\mathcal{D}}_m\phi &= \partial_m\phi - \frac{\kappa}{\sqrt{2}}\psi^1_m\chi - \frac{1}{\sqrt{2}}\kappa v^2\mathcal{A}_m \\ \hat{\mathcal{A}}_{mn} &= \mathcal{A}_{mn} + \kappa\psi^2_{[m}\psi^1_{n]} - \frac{\kappa}{\sqrt{2}}\overline{\lambda}\overline{\sigma}_{[n}\psi^1_{m]}\,. \end{aligned} \tag{41}$$

This Lagrangian is invariant under two independent abelian gauge symmetries, as well as the following supersymmetry transformations,

$$
\begin{aligned}
\delta e_m^a &= \mathrm{i}\kappa(\eta^i\sigma^a\overline{\psi}_{mi} + \bar{\eta}_i\bar{\sigma}^a\psi_m^i) \\
\delta\psi_m^i &= \frac{2}{\kappa}D_m\eta^i \\
&\quad + \left(-\frac{\mathrm{i}}{2}\hat{H}_{+mn}\sigma^n\bar{\eta}_1 + \sqrt{2}\overline{\mathcal{D}_m\phi}\eta^1 - \kappa\psi_m^1(\bar{\chi}\bar{\eta}_1) + \mathrm{i}v^2\sigma_m\bar{\eta}_2\right)\delta_2{}^i \\
\delta\mathcal{A}_m &= 2\epsilon_{ij}\psi_m^i\eta^j + \sqrt{2}\bar{\lambda}\bar{\sigma}_m\eta^1 \\
\delta\lambda &= \frac{\mathrm{i}}{\sqrt{2}}\overline{\hat{\mathcal{A}}}_{mn}\sigma^{mn}\eta^1 - \mathrm{i}\sqrt{2}v^2\eta^2 \\
\delta\chi &= \mathrm{i}\sqrt{2}\sigma^m\hat{\mathcal{D}}_m\phi\bar{\eta}_1 + 2v^2\eta^2 \\
\delta\phi &= \sqrt{2}\chi\eta^1 \ ,
\end{aligned} \tag{42}
$$

for $i = 1, 2$. This result holds to leading order, that is, up to and including terms in the transformations that are linear in the fields. Note that this representation is irreducible in the sense that there are no subsets of fields that transform only into themselves under the supersymmetry transformations. (Because of this, the multiplet structure outlined in Fig. 1 is slightly misleading.)

Let us now consider the case of the vector Goldstone multiplet. Under the first supersymmetry, the real vector B_m of a vector multiplet transforms into a Weyl fermion λ,

$$\delta_{\eta^1} B_m \;=\; \sqrt{2}\mathrm{i}\,(\lambda\sigma_m\bar{\eta}^1 - \eta^1\sigma_m\bar{\lambda}) \ . \tag{43}$$

If λ is the Goldstone fermion, it shifts under the second supersymmetry. Therefore the closure of the two supersymmetries on B_m gives

$$[\,\delta_{\eta^2}, \delta_{\eta^1}\,]\,B_m \;=\; 2\mathrm{i}v^2\,(\eta^2\sigma_m\bar{\eta}^1 - \eta^1\sigma_m\bar{\eta}^2) \;+\; \dots \tag{44}$$

From this we see that the real vector B_m is a Goldstone boson. It expects to be eaten by an antisymmetric tensor field. This suggests that the vector Goldstone multiplet should be associated with the alternative representation for the massive spin-3/2 multiplet.

The degree of freedom counting is shown in Figure 1(b). As before, we include the $N = 2$ supergravity multiplet. This time, however, the matter fields include the $N = 1$ vector Goldstone multiplet, together with an $N = 1$ tensor multiplet. In unitary gauge, one vector eats one scalar, while the antisymmetric tensor eats the other vector. [The massless antisymmetric tensor field contains one degree of freedom. It was introduced by Ogievetsky and Polubarinov (1966), who called it the "notoph," or inverse photon.] These are the minimal set of fields that arise when coupling the Ogievetsky-Sokatchev spin-3/2 multiplet to $N = 2$ supergravity.

The Lagrangian for this system can be worked out following the same procedure described above. One finds (Altendorfer, Bagger, Osofsky, 1998):

$$
\begin{aligned}
e^{-1}\mathcal{L} \; = \;& \\
& - \frac{1}{2\kappa^2}\mathcal{R} + \epsilon^{pqrs}\bar{\psi}_{pi}\bar{\sigma}_q D_r \psi^i_s - \mathrm{i}\bar{\chi}\bar{\sigma}^m D_m\chi - \mathrm{i}\bar{\lambda}\bar{\sigma}^m D_m\lambda - \frac{1}{2}\mathcal{D}^m\phi\mathcal{D}_m\phi \\
& - \frac{1}{4}\mathcal{F}^A_{mn}\mathcal{F}^{Amn} - \frac{1}{4}\mathcal{F}^B_{mn}\mathcal{F}^{Bmn} + \frac{1}{2}v^m v_m - \Big(\frac{1}{\sqrt{2}} m\,\psi^2_m\sigma^m\bar{\lambda} + m\mathrm{i}\psi^2_m\sigma^m\bar{\chi} \\
& + \sqrt{2}m\mathrm{i}\lambda\chi + \frac{1}{2}m\chi\chi + m\,\psi^2_m\sigma^{mn}\psi^2_n + \frac{\kappa}{2\sqrt{2}}\epsilon_{ij}\psi^i_m\psi^j_n\mathcal{F}^{Amn}_- \\
& + \frac{\kappa}{2}\chi\sigma^m\bar{\sigma}^n\psi^1_m\mathcal{D}_n\phi + \frac{\kappa}{2}\bar{\lambda}\bar{\sigma}_m\psi^1_n\mathcal{F}^{Bmn}_+ + \frac{\kappa}{2}\epsilon^{pqrs}\bar{\psi}^2_p\bar{\sigma}_q\psi^1_r\mathcal{D}_s\phi \\
& - \mathrm{i}\frac{\kappa}{2}\chi\sigma^m\bar{\sigma}^n\psi^1_m v_n - \mathrm{i}\frac{\kappa}{2}\epsilon^{pqrs}\bar{\psi}^2_p\bar{\sigma}_q\psi^1_r v_s s \;+\; \text{h.c.}\Big) \qquad (45)
\end{aligned}
$$

where, as before, $m = \kappa v^2$, and

$$
\begin{aligned}
\mathcal{D}_m\phi &= \partial_m\phi - \frac{m}{\sqrt{2}}(A_m + B_m) \\
\mathcal{F}^A_{mn} &= \partial_{[m}A_{n]} + \frac{m}{\sqrt{2}}B_{mn} \\
\mathcal{F}^B_{mn} &= \partial_{[m}B_{n]} - \frac{m}{\sqrt{2}}B_{mn} \,. \qquad (46)
\end{aligned}
$$

This Lagrangian is invariant under an ordinary abelian gauge symmetry, an antisymmetric tensor gauge symmetry, as well as the following two supersymmetries,

$$
\begin{aligned}
\delta_\eta e^a_m &= \mathrm{i}\,\kappa(\eta^i\sigma^a\bar{\psi}_{mi} + \bar{\eta}_i\bar{\sigma}^a\psi^i_m) \\
\delta_\eta\psi^1_m &= \frac{2}{\kappa}D_m\eta^1 \\
\delta_\eta A_m &= \sqrt{2}\epsilon_{ij}(\psi^i_m\eta^j + \bar{\psi}^i_m\bar{\eta}^j) \\
\delta_\eta B_m &= \bar{\eta}^1\bar{\sigma}_m\lambda + \bar{\lambda}\bar{\sigma}_m\eta^1 \\
\delta_\eta B_{mn} &= 2\eta^1\sigma_{mn}\chi + \mathrm{i}\,\eta^1\sigma_{[m}\bar{\psi}^2_{n]} + \mathrm{i}\,\eta^2\sigma_{[m}\bar{\psi}^1_{n]} \;+\; \text{h.c.} \\
\delta_\eta\lambda &= \mathrm{i}\,\hat{\mathcal{F}}^B_{mn}\sigma^{mn}\eta^1 - \mathrm{i}\sqrt{2}v^2\eta^2 \\
\delta_\eta\chi &= \mathrm{i}\,\sigma^m\bar{\eta}^1\hat{\mathcal{D}}_m\phi - \hat{v}_m\sigma^m\bar{\eta}^1 + 2v^2\eta^2 \\
\delta_\eta\psi^2_m &= \frac{2}{\kappa}D_m\eta^2 + \mathrm{i}v^2\sigma_m\bar{\eta}^2 - \frac{\mathrm{i}}{\sqrt{2}}\hat{\mathcal{F}}^A_{+mn}\sigma^n\bar{\eta}^1 \\
&\quad + \hat{\mathcal{D}}_m\phi\eta^1 + \kappa\left((\bar{\psi}^1_m\bar{\chi})\eta - (\bar{\chi}\bar{\eta})\psi^1_m\right) - \mathrm{i}\,\hat{v}_m\eta^1 \\
\delta_\eta\phi &= \chi\eta^1 + \bar{\chi}\bar{\eta}^1 \qquad (47)
\end{aligned}
$$

up to linear order in the fields. The supercovariant derivatives are given by

$$
\begin{aligned}
\hat{\mathcal{D}}_m\phi &= \mathcal{D}_m\phi - \frac{\kappa}{2}(\psi^1_m\chi + \bar{\psi}^1_m\bar{\chi}) \\
\hat{\mathcal{F}}^A_{mn} &= \mathcal{F}^A_{mn} + \frac{\kappa}{\sqrt{2}}(\psi^2_{[m}\psi^1_{n]} + \bar{\psi}^2_{[m}\bar{\psi}^1_{n]})
\end{aligned}
$$

$$\begin{aligned}\hat{\mathcal{F}}^{B}_{mn} &= \mathcal{F}^{B}_{mn} - \frac{\kappa}{2}(\bar\lambda\bar\sigma_{[n}\psi^1_{m]} + \bar\psi^1_{[m}\bar\sigma_{n]}\lambda)\\ \hat{v}_m &= v_m + \left(\mathrm{i}\kappa\psi^1_n\sigma_m{}^n\chi - \frac{\mathrm{i}\kappa}{2}\epsilon_m{}^{nrs}\psi^1_n\sigma_r\bar\psi^2_s + \text{h.c.}\right) . \end{aligned} \tag{48}$$

These fields form an irreducible representation of the $N = 2$ algebra.

5.3 The SuperHiggs Effect

Each of the two Lagrangians presented above has a full $N = 2$ supersymmetry (up to the appropriate order). The first supersymmetry is realized linearly, so it is not broken. The second is realized nonlinearly, so it is spontaneously broken. In each case, the transformations imply that

$$\zeta = \frac{1}{\sqrt{3}}(\chi - \mathrm{i}\sqrt{2}\lambda) \tag{49}$$

does not shift, while

$$\nu = \frac{1}{\sqrt{3}}(\sqrt{2}\chi + \mathrm{i}\lambda) \tag{50}$$

does. Therefore ν is the Goldstone fermion for $N = 2$ supersymmetry, spontaneously broken to $N = 1$.

In the chiral case, we find

$$\begin{aligned}[\delta_{\eta_1}, \delta_{\eta_2}]\,\phi &= 2\sqrt{2}\,v^2\,\eta_1\eta_2\\ [\delta_{\eta_1}, \delta_{\eta_2}]\,A_m &= \frac{4}{\kappa}\partial_m\,\eta_1\eta_2 .\end{aligned} \tag{51}$$

The complex scalar ϕ is indeed the Goldstone boson for a gauged central charge. Moreover, in unitary gauge, where

$$\phi = \nu = 0 , \tag{52}$$

this Lagrangian reduces to the usual representation for a massive $N = 1$ spin-3/2 multiplet.

In the vector case, we have

$$\begin{aligned}[\delta_{\eta^2}, \delta_{\eta^1}]\,A_m &= \frac{2\sqrt{2}}{\kappa}\partial_m(\eta^1\eta^2 + \bar\eta^1\bar\eta^2) - \sqrt{2}\mathrm{i}\,v^2\,(\eta^2\sigma_m\bar\eta^1 - \eta^1\sigma_m\bar\eta^2)\\ [\delta_{\eta^2}, \delta_{\eta^1}]\,B_m &= \sqrt{2}\mathrm{i}\,v^2\,(\eta^2\sigma_m\bar\eta^1 - \eta^1\sigma_m\bar\eta^2)\\ [\delta_{\eta^2}, \delta_{\eta^1}]\,B_{mn} &= \frac{2\mathrm{i}}{\kappa}D_{[m}(\eta^2\sigma_{n]}\bar\eta^1 - \eta^1\sigma_{n]}\bar\eta^2) .\end{aligned} \tag{53}$$

We see that the real vector $-(A_m - B_m)/\sqrt{2}$ is the Goldstone boson for a gauged it vectorial central extension of the $N = 2$ algebra. In addition, the real scalar ϕ is the Goldstone boson associated with a single *real* gauged central charge. In unitary gauge, with

$$-\frac{1}{\sqrt{2}}(A_m - B_m) \;=\; \phi \;=\; \nu \;=\; 0\,, \tag{54}$$

this Lagrangian reduces to the Ogievetsky-Sokatchev representation for the massive $N = 1$ spin-3/2 multiplet.

Now that we have two explicit realizations of partial supersymmetry breaking, we can go back and see how they avoid the no-go argument presented in the introduction. We first compute the second supercurrent. In each case it turns out to be

$$J^2_{m\alpha} \;=\; v^2\,(\sqrt{6}\,\mathrm{i}\,\sigma_{\alpha\dot\alpha m}\bar\nu^{\dot\alpha} + 4\,\sigma_{\alpha\beta mn}\psi^{2n\beta}) \tag{55}$$

plus higher-order terms. Computing, we find

$$\begin{aligned} \{\,\bar Q_{\dot\alpha},\, J^1_{m\alpha}\,\} \;&=\; 2\,\sigma^n_{\alpha\dot\alpha}\,T_{mn} \\ \{\,\bar S_{\dot\alpha},\, J^2_{m\alpha}\,\} \;&=\; 2\,\sigma^n_{\alpha\dot\alpha}\,T_{mn}\,. \end{aligned} \tag{56}$$

In the presence of supergravity, there is no confusion about the stress-energy tensor. There is just one such tensor, and it shows up on the right-hand side of the current algebra.

For the case at hand, however, $J^i_{\alpha m}$ and T_{mn} contain contributions from *all* of the fields, including the second gravitino. When covariantly-quantized, the second gravitino gives rise to states of negative norm. Indeed, it is not hard to check that

$$(\bar S S + S\bar S)\,|0\rangle \;=\; 0\,, \tag{57}$$

even though

$$S\,|0\rangle \;\neq\; 0 \qquad\quad \bar S\,|0\rangle \;\neq\; 0\,. \tag{58}$$

The supergravity couplings exploit the second loophole to the no-go theorem!

The Lagrangian in the chiral case is a truncation of the supergravity coupling found by Cecotti, Girardello and Porrati (1986) and by Zinov'ev (1987). Their results were based on *linear* $N = 2$ supersymmetry; they involved $N = 2$ vector- and hyper-multiplets. The Lagrangian for the vector case is new. It contains a new realization of $N = 2$ supergra vity. In each case, the couplings presented here are minimal and model-independent. They describe the superHiggs effect in the low-energy effective theories that arise from partial supersymmetry breaking.

Thus we have seen that there is no obstacle to partial supersymmetry breaking in the presence of gravity. Indeed, each of the two Goldstone multiplets give rise to its own massive spin-3/2 multiplet. Of course, the connection between these results and the theory of membranes and D-branes is an urgent and open question.

It is a pleasure to thank Sam Osofsky and Richard Altendorfer for collaboration on the supergravity couplings presented here. We would also like to acknowledge our debt to Victor Isaacovitch Ogievetsky for teaching us the physics that underlies this work, and so much more. This work was supported by the National Science Foundation, grant NSF-PHY-9404057.

References

Altendorfer, R., Bagger, J., Osofsky, S. (1998): In preparation.
See also Altendorfer, R., Bagger, J. (1998): New Supersymmetry Algebras from Partial Supersymmetry Breaking. hep-th/9809171.

Antoniadis, I., Partouche, H., Taylor, T. (1996): Spontaneous Breaking of $N = 2$ Global Supersymmetry. Phys. Lett. **B372**, 83.

Bagger, J., Galperin, A. (1994): Matter Couplings in Partially Broken Extended Supersymmetry. Phys. Lett. **B336**, 25.

Bagger, J., Galperin, A. (1997a): New Goldstone Multiplet for Partially Broken Supersymmetry. Phys. Rev. **D55**, 1091.

Bagger, J., Galperin, A. (1997b): The Tensor Goldstone Multiplet for Partially Broken Supersymmetry. Phys. Lett. **B412**, 296.

Cecotti, S., Girardello, L., Porrati, M. (1986): An Exceptional $N = 2$ Supergravity with Flat Potential and Partial SuperHiggs. Phys. Lett. **168B**, 83.

Coleman, S., Wess, J., Zumino, B. (1969): Structure of Phenomenological Lagrangians. 1. Phys. Rev. **177**, 2239.

Ferrara, S., van Nieuwenhuizen, P. (1983): Noether Coupling of Massive Gravitinos to $N = 1$ Supergravity. Phys. Lett. **B127**, 70.

Hughes, J., Liu, J., Polchinski, J. (1986): Supermembranes. Phys. Lett. **180B**, 370. See also Hughes, J., Polchinski, J. (1986): Partially Broken Global Supersymmetry and the Superstring. Nucl. Phys. **B278**, 147.

Ivanov, E., Ogievetsky, V. (1975): The Inverse Higgs Phenomenon in Nonlinear Realizations. Teor. Mat. Fiz. **25**, 164.

Ivanov, E., Ogievetsky, V. (1976): Gauge Theories as Theories of Spontaneous Breakdown. JETP Lett. **23**, 606.

Ogievetsky, V., Sokatchev, E. (1976): On Gauge Spinor Superfield. JETP Lett. **23**, 58.
See also Gates, S.J. (1977): Spinor Yang-Mills Superfields. Phys. Rev. **D16**, 1727.

Ogievetsky, V., Polubarinov, I. (1966): The Notoph and Its Possible Interactions. Sov. J. Nucl. Phys. **4**, 156.

Volkov, D.V. (1973): Phenomenological Lagrangians. Sov. J. Particles and Nuclei **4**, 3. See also Ogievetsky, V.I. (1974): Nonlinear Realizations of Internal and Spacetime Symmetries, in *Proceedings of the X-th Winter School of Theoretical Physics in Karpacz,* (Wroclaw), 227.

Zinov'ev, Yu M. (1987): Spontaneous Symmetry Breaking in $N = 2$ Supergravity. Sov. J. Nucl. Phys. **46**, 540.

Covariant Harmonic Supergraphity for $N = 2$ Super Yang–Mills Theories

Ioseph Buchbinder[1], Sergei Kuzenko[2], and Burt Ovrut[3 4]

[1] Department of Theoretical Physics, Tomsk State Pedagogical University, Tomsk 634041, Russia
[2] Department of Physics, Tomsk State University, Lenin Ave. 36, Tomsk 634050, Russia
[3] Department of Physics, University of Pennsylvania, Philadelphia, PA 19104-6396, USA
[4] School of Natural Sciences, Institute for Advanced Study, Olden Lane, Princeton, NJ 08540, USA

Abstract. We review the background field method for general $N = 2$ super Yang-Mills theories formulated in the $N = 2$ harmonic superspace. The covariant harmonic supergraph technique is then applied to rigorously prove the $N = 2$ non-renormalization theorem as well as to compute the holomorphic low-energy action for the $N = 2$ $SU(2)$ pure super Yang-Mills theory and the leading non-holomorphic low-energy correction for $N = 4$ $SU(2)$ super Yang-Mills theory.

1 Introduction

Manifest covariance is one of the imperative principles in modern theoretical physics. It means that any physical theory possessing some symmetries must be formulated and studied in such a form where all the symmetries are manifest both at the classical and quantum levels.

The present paper is a brief review of recent progress in constructing the manifestly covariant quantum formulation for the $N = 2$ supersymmetric Yang-Mills (SYM) theories (Buchbinder et al. (1998b,c,d)) on the base of the $N = 2$ harmonic superspace developed by V.I. Ogievetsky and collaborators (Galperin et al. (1984)). As we understand now, the harmonic superspace approach is an elegant and universal setting to formulate general $N = 2$ SYM theories (Galperin et al. (1984)) and $N = 2$ supergravity (Galperin et al. (1987a,b)) in a manifestly supersymmetric way. Its universality follows simply from the fact that all known $D = 4$, $N = 2$ supersymmetric theories can be naturally realized in harmonic superspace. In particular, the formulations for $N = 2$ SYM theories in the conventional $N = 2$ superspace (Howe et al. (1984)) and in the $N = 2$ projective superspace (Lindström et al. (1990)) turn out to be gauge fixed and truncated versions, respectively, of that in harmonic superspace. It is the $N = 2$ harmonic superspace which allows us to realize the general $N = 2$ SYM theories in terms of unconstrained superfields. Therefore, just the harmonic superspace approach is an adequate and convenient base for developing $N = 2$ supersymmetric quantum field theory.

The manifestly $N = 2$ supersymmetric Feynman rules in harmonic superspace have been developed by Galperin et al. (1985a,b). One of the basic purposes of the present paper is to extend these rules in order to have manifest gauge invariance along with $N = 2$ supersymmetry. As is well known, the most efficient way to realize such a goal is the background field method.

The paper is organized as follows. In section 2 we review the (harmonic) superspace formulation for the $N = 2$ SYM theories. Section 3 is devoted to the presentation of the background field method for such theories. In section 4 we use the background field formulation developed to prove the $D = 4$, $N = 2$ non-renormalization theorem. The structure of the one-loop effective action is discussed in section 5. Finally, in section 6 we compute the low-energy holomorphic action for the pure $N = 2$ $SU(2)$ SYM theory as well as the non-holomorphic action for the $N = 4$ $SU(2)$ SYM theory.

2 $N = 2$ super Yang-Mills Theories in Superspace

We start with a brief review of $N = 2$ SYM theories in superspace.

2.1 $N = 2$ SYM in Standard Superspace

The constrained geometry of $N = 2$ super Yang-Mills field is formulated in standard $N = 2$ superspace $\mathbf{R}^{4|8}$ with coordinates $z^M \equiv (x^m, \theta^\alpha_i, \bar\theta^i_{\dot\alpha})$ in terms of the gauge covariant derivatives

$$\mathcal{D}_M \equiv (\mathcal{D}_m, \mathcal{D}^i_\alpha, \overline{\mathcal{D}}^{\dot\alpha}_i) = D_M + \mathrm{i}\mathcal{A}_M \; , \qquad \mathcal{A}_M = \mathcal{A}^a_M(z) T^a \tag{1}$$

satisfying the algebra (Grimm et al. (1978))

$$\begin{gathered} \{\mathcal{D}^i_\alpha, \overline{\mathcal{D}}_{\dot\alpha j}\} = -2\mathrm{i}\delta^i_j \mathcal{D}_{\alpha\dot\alpha} \; , \\ \{\mathcal{D}^i_\alpha, \mathcal{D}^j_\beta\} = 2\mathrm{i}\varepsilon_{\alpha\beta}\varepsilon^{ij}\overline{\mathcal{W}} \; , \qquad \{\overline{\mathcal{D}}_{\dot\alpha i}, \overline{\mathcal{D}}_{\dot\beta j}\} = 2\mathrm{i}\varepsilon_{\dot\alpha\dot\beta}\varepsilon_{ij}\mathcal{W} \; . \end{gathered} \tag{2}$$

Here $D_M \equiv (\partial_m, D^i_\alpha, \overline{D}^{\dot\alpha}_i)$ are the flat covariant derivatives, T^a are the generators of the gauge group. The covariantly chiral strength $\mathcal{W}$ satisfies the Bianchi identities

$$\overline{\mathcal{D}}_{\dot\alpha i}\mathcal{W} = 0 \; , \qquad \mathcal{D}^{\alpha(i}\mathcal{D}^{j)}_\alpha \mathcal{W} = \overline{\mathcal{D}}^{(i}_{\dot\alpha}\overline{\mathcal{D}}^{j)\dot\alpha}\overline{\mathcal{W}} \; . \tag{3}$$

The covariant derivatives and a matter superfield multiplet $\varphi(z)$ transform as follows

$$\mathcal{D}'_M = \mathrm{e}^{\mathrm{i}\tau}\mathcal{D}_M \mathrm{e}^{-\mathrm{i}\tau} \; , \qquad \varphi' = \mathrm{e}^{\mathrm{i}\tau}\varphi \tag{4}$$

under the gauge group. Here $\tau = \tau^a(z)T^a$, and $\tau^a = \bar\tau^a$ are unconstrained real parameters. The set of all transformations (4) is said to form the τ-group.

The gauge invariant action of the $N = 2$ pure SYM theory reads (Grimm et al. (1978))

$$S_{\rm SYM} = \frac{1}{2g^2}\mathrm{tr}\int \mathrm{d}^4x\mathrm{d}^4\theta\, \mathcal{W}^2 = \frac{1}{2g^2}\mathrm{tr}\int \mathrm{d}^4x\mathrm{d}^4\bar{\theta}\, \overline{\mathcal{W}}^2 \,. \tag{5}$$

2.2 N = 2 SYM in Harmonic Superspace

To realize the $N = 2$ pure SYM theory as a theory of unconstrained dynamical superfields, Galperin et al. (1984) extended the original superspace to $N = 2$ harmonic superspace $\mathbf{R}^{4|8} \times S^2$. A natural global parametrization of $S^2 = SU(2)/U(1)$ is that in terms of the harmonic variables $(u_i{}^-, u_i{}^+) \in SU(2)$ which parametrize the automorphism group of $N = 2$ supersymmetry,

$$u_i^+ = \varepsilon_{ij}u^{+j}\,, \qquad \overline{u^{+i}} = u_i^-\,, \qquad u^{+i}u_i^- = 1\,. \tag{6}$$

Tensor fields over S^2 are in a one-to-one correspondence with functions on $SU(2)$ possessing definite harmonic $U(1)$-charges. A function $\Psi^{(p)}(u)$ is said to have the harmonic $U(1)$-charge p if

$$\Psi^{(p)}(\mathrm{e}^{\mathrm{i}\varphi}u^+, \mathrm{e}^{-\mathrm{i}\varphi}u^-) = \mathrm{e}^{\mathrm{i}p\varphi}\Psi^{(p)}(u^+, u^-)\,, \qquad |\mathrm{e}^{\mathrm{i}\varphi}| = 1\,.$$

A function $\Psi^{(p)}(z,u)$ on $\mathbf{R}^{4|8} \times S^2$ with $U(1)$-charge p is called a harmonic $N = 2$ superfield.

Introducing a new basis of covariant derivatives

$$\mathcal{D}^\pm_\alpha = \mathcal{D}^i_\alpha u^\pm_i\,, \qquad \overline{\mathcal{D}}^\pm_{\dot\alpha} = \overline{\mathcal{D}}^i_{\dot\alpha} u^\pm_i \tag{7}$$

the covariant derivative algebra (2) implies

$$\{\mathcal{D}^+_\alpha, \mathcal{D}^+_\beta\} = \{\overline{\mathcal{D}}^+_{\dot\alpha}, \overline{\mathcal{D}}^+_{\dot\beta}\} = \{\mathcal{D}^+_\alpha, \overline{\mathcal{D}}^+_{\dot\beta}\} = 0 \tag{8}$$

and, hence,

$$\mathcal{D}^+_\alpha = \mathrm{e}^{-\mathrm{i}\Omega} D^+_\alpha \mathrm{e}^{\mathrm{i}\Omega}\,, \qquad \overline{\mathcal{D}}^+_{\dot\alpha} = \mathrm{e}^{-\mathrm{i}\Omega} \overline{D}^+_{\dot\alpha} \mathrm{e}^{\mathrm{i}\Omega} \tag{9}$$

for some Lie-algebra valued harmonic superfield $\Omega = \Omega^a(z,u)T^a$ with vanishing $U(1)$-charge, which is called the 'bridge'.

As a consequence of (8), one can define covariantly analytic superfields constrained by

$$\mathcal{D}^+_\alpha \Phi^{(p)} = \overline{\mathcal{D}}^+_{\dot\alpha} \Phi^{(p)} = 0\,, \tag{10}$$

where $\Phi^{(p)}(z,u)$ carries $U(1)$-charge p and can be represented as follows

$$\Phi^{(p)} = \mathrm{e}^{-\mathrm{i}\Omega}\phi^{(p)}\,, \qquad D^+_\alpha \phi^{(p)} = \overline{D}^+_{\dot\alpha}\phi^{(p)} = 0\,. \tag{11}$$

The superfield $\phi^{(p)}$ is, in general, an unconstrained function over an analytic subspace of the harmonic superspace (Galperin et al. (1984)) parametrized by

$$\zeta \equiv \{x^m_A, \theta^{+\alpha}, \bar\theta^+_{\dot\alpha}, u^\pm_i\}\,, \qquad \phi^{(p)}(z,u) \equiv \phi^{(p)}(\zeta)\,, \tag{12}$$

where (Galperin et al. (1984))

$$x^m_A = x^m - 2\mathrm{i}\theta^{(i}\sigma^m\bar{\theta}^{j)}u^+_i u^-_j \,, \qquad \theta^\pm_\alpha = \theta^i_\alpha u^\pm_i \,, \qquad \bar{\theta}^\pm_{\dot\alpha} = \bar{\theta}^i_{\dot\alpha} u^\pm_i \,. \tag{13}$$

That is why such superfields are called analytic.

The analytic subspace (12) is closed under $N = 2$ supersymmetry transformations and real with respect to the generalized conjugation $\breve{} \equiv \bar{*}$ (Galperin et al. (1984)), where the operation $*$ is defined by

$$(u^+_i)^* = u^-_i \,, \qquad (u^-_i)^* = -u^+_i \quad \Rightarrow \quad (u^\pm_i)^{**} = -u^\pm_i \,. \tag{14}$$

A remarkable property of this generalized conjugation (called below the 'smile-conjugation') is that it allows us to consistently define real analytic superfields with even $U(1)$-charge.

Without loss of generality, the bridge Ω (9) can be chosen to be real with respect to the smile-conjugation, $\breve{\Omega} = \Omega$. The bridge possesses a richer gauge freedom than the original τ-group. Its transformation law reads

$$\mathrm{e}^{\mathrm{i}\Omega'} = \mathrm{e}^{\mathrm{i}\lambda}\mathrm{e}^{\mathrm{i}\Omega}\mathrm{e}^{-\mathrm{i}\tau} \tag{15}$$

with an unconstrained analytic gauge parameter $\lambda = \lambda^a(\zeta)T^a$ being real with respect to the smile-conjugation, $\breve{\lambda}^a = \lambda^a$. The set of all λ-transformations form the so-called λ-group (Galperin et al. (1984)). The τ-group acts on $\Phi^{(p)}$ and leaves $\phi^{(p)}$ unchanged while the λ-group acts only on $\phi^{(p)}$ as follows

$$\phi'^{(p)} = \mathrm{e}^{\mathrm{i}\lambda}\phi^{(p)} \,. \tag{16}$$

The superfields $\Phi^{(p)}$ and $\phi^{(p)}$ are said to correspond to the τ- and λ-frames respectively.

The λ-frame is most useful to work with the covariantly analytic superfields. At the same time, it is the λ-frame in which a single unconstrained prepotential of the $N = 2$ SYM theory naturally emerges. Let us, first of all, introduce the harmonic derivatives (Galperin et al. (1984))

$$D^{\pm\pm} = u^{\pm i}\frac{\partial}{\partial u^{\mp i}} \,, \qquad D^0 = u^{+i}\frac{\partial}{\partial u^{+i}} - u^{-i}\frac{\partial}{\partial u^{-i}} \,, \tag{17}$$

where $D^{\pm\pm}$ are two independent derivatives on S^2, and D^0 is the operator of $U(1)$ charge, $D^0\Phi^{(p)} = p\Phi^{(p)}$. Operators $\mathcal{D}_{\underline{M}} \equiv (\mathcal{D}_M, D^{++}, D^{--}, D^0)$ form a full set of gauge covariant derivatives in the τ-frame. The λ-frame is defined by the following transform

$$\mathcal{D}_{\underline{M}} \;\to\; \nabla_{\underline{M}} = \mathrm{e}^{\mathrm{i}\Omega}\mathcal{D}_{\underline{M}}\mathrm{e}^{-\mathrm{i}\Omega} \,, \qquad \Phi^{(p)} \;\to\; \phi^{(p)} = \mathrm{e}^{\mathrm{i}\Omega}\Phi^{(p)} \tag{18}$$

$$\nabla^+_\alpha = D^+_\alpha \,, \qquad \bar{\nabla}^+_{\dot\alpha} = \bar{D}^+_{\dot\alpha} \,, \qquad \nabla^{\pm\pm} = D^{\pm\pm} + \mathrm{i}V^{\pm\pm} \,. \tag{19}$$

Since $[\nabla^{++}, \nabla^+_\alpha] = [\nabla^{++}, \bar{\nabla}^+_{\dot\alpha}] = 0$, the connection $V^{++} = V^{++a}T^a$ is a real analytic superfield, $\breve{V}^{++} = V^{++}$, $D^+_\alpha V^{++} = \bar{D}^+_{\dot\alpha}V^{++} = 0$, and its transformation law is

$$V'^{++} = e^{i\lambda} V^{++} e^{-i\lambda} - i e^{i\lambda} D^{++} e^{-i\lambda} . \tag{20}$$

The analytic superfield V^{++} turns out to be the single unconstrained prepotential of the pure $N = 2$ SYM theory and all other objects are expressed in terms of it. In particular, action (5) can be rewritten via V^{++} as follows (Zupnik (1987))

$$S_{\rm SYM} = \frac{1}{g^2} {\rm tr} \sum_{n=2}^{\infty} \frac{(-i)^n}{n} \int d^{12}z \, d^n u \frac{V^{++}(z, u_1) \dots V^{++}(z, u_n)}{(u_1^+ u_2^+) \dots (u_n^+ u_1^+)} . \tag{21}$$

The rules of integration over $SU(2)$ as well as the properties of harmonic distributions were given by Galperin et al. (1984) and Galperin et al. (1985a).

2.3 Supersymmetric Matter

Harmonic superspace provides us with two possibilities to describe $N = 2$ supersymmetric matter in terms of unconstrained analytic superfields (Galperin et al. (1984)). A charge hypermultiplet, transforming in a complex representation R_q of the gauge group, is described by an unconstrained analytic superfield $q^+(\zeta)$ and its conjugate $\breve{q}^+(\zeta)$ (q-hypermultiplet). A neutral hypermultiplet, transforming in a real representation R_ω of the gauge group, is described by an unconstrained analytic real superfield $\omega(\zeta)$, $\breve{\omega} = \omega$, (ω-hypermultiplet). The matter action reads

$$S_{\rm MAT} = - \int d\zeta^{(-4)} \, \breve{q}^+ \nabla^{++} q^+ - \frac{1}{2} \int d\zeta^{(-4)} \, \nabla^{++} \omega^{\rm T} \, \nabla^{++} \omega , \tag{22}$$

where the integration is carried out over the analytic subspace (12).

3 Background Field Quantization

To quantize the pure $N = 2$ SYM theory, we split V^{++} into *background* V^{++} and *quantum* v^{++} parts

$$V^{++} \quad \rightarrow \quad V^{++} + g \, v^{++} . \tag{23}$$

Then, the original infinitesimal gauge transformations (20) can be realized in two different ways:

(i) *background transformations*

$$\delta_{\rm B} V^{++} = -D^{++}\lambda - i[V^{++}, \lambda] = -\nabla^{++}\lambda , \qquad \delta_{\rm B} v^{++} = i[\lambda, v^{++}] \tag{24}$$

(ii) *quantum transformations*

$$\delta_{\rm Q} V^{++} = 0 , \qquad \delta_{\rm Q} v^{++} = -\frac{1}{g} \nabla^{++}\lambda - i[v^{++}, \lambda] . \tag{25}$$

It is worth pointing out that the form of the background-quantum splitting (23) and the corresponding background and quantum transformations (24), (25) are much more analogous to the conventional Yang-Mills theory than to the $N = 1$ non-abelian SYM model. Our aim now is to construct an effective action as a gauge-invariant functional of the background superfield V^{++}.

Upon the splitting (23), the classical action (21) takes the form

$$S_{\rm SYM} = S_{\rm SYM}[V^{++}] + \frac{1}{4g}{\rm tr}\int {\rm d}\zeta^{(-4)}\, v^{++}(\overline{D}^{+})^2\overline{W}_\lambda + \Delta S\,, \qquad (26)$$

where $\Delta S[v^{++}, V^{++}]$ reads

$$\Delta S = -{\rm tr}\sum_{n=2}^{\infty}\frac{(-{\rm i}g)^{n-2}}{n}\int {\rm d}^{12}z\,{\rm d}^n u \frac{v_\tau^{++}(z,u_1)\dots v_\tau^{++}(z,u_n)}{(u_1^+u_2^+)\dots(u_n^+u_1^+)}\,. \qquad (27)$$

Here W_λ and v_τ^{++} denote the λ- and τ-transforms of W and v^{++}, respectively, with the bridge Ω corresponding to the background covariant derivatives constructed on the base of the background connection V^{++}. The quantum action ΔS given in (27) depends on V^{++} via the dependence of v_τ^{++} on Ω, the latter being a complicated function of V^{++}. Each term in the action (26) is manifestly invariant with respect to the background gauge transformations. The linear in v_τ^{++} term in (26) determines the equations of motion. This term should be dropped when considering the effective action.

To construct the effective action, we can use the Faddeev-Popov Ansatz. Within the framework of the background field method, we should fix only the quantum transformations (25). Let us introduce the gauge fixing function in the form

$$\mathcal{F}^{(4)} = \nabla^{++}v^{++}\,, \qquad \delta_{\rm Q}\mathcal{F}^{(4)} = \frac{1}{g}\left\{\nabla^{++}(\nabla^{++}\lambda + {\rm i}g[v^{++},\lambda])\right\}\,. \qquad (28)$$

Eq. (28) leads to the Faddeev-Popov determinant

$$\Delta_{\rm FP}[v^{++}, V^{++}] = {\rm Det}\left\{\nabla^{++}(\nabla^{++} + {\rm i}gv^{++})\right\}\,. \qquad (29)$$

To get a path-integral representation for $\Delta_{\rm FP}[v^{++}, V^{++}]$, we introduce two analytic fermionic ghosts $\mathbf{b}$ and $\mathbf{c}$, in the adjoint representation of the gauge group, and the corresponding ghost action

$$S_{\rm FP}[\mathbf{b},\mathbf{c},v^{++},V^{++}] = {\rm tr}\int {\rm d}\zeta^{(-4)}\,\mathbf{b}\nabla^{++}(\nabla^{++}\mathbf{c} + {\rm i}g\,[v^{++},\mathbf{c}])\,. \qquad (30)$$

As a result, we arrive at the effective action $\Gamma_{\rm SYM}[V^{++}]$ in the form

$$\begin{aligned} {\rm e}^{{\rm i}\Gamma_{\rm SYM}[V^{++}]} &= {\rm e}^{{\rm i}S_{\rm SYM}[V^{++}]}\int \mathcal{D}v^{++}\mathcal{D}\mathbf{b}\mathcal{D}\mathbf{c} \\ &\times {\rm e}^{{\rm i}(\Delta S[v^{++},V^{++}] + S_{\rm FP}[\mathbf{b},\mathbf{c},v^{++},V^{++}])}\delta[\mathcal{F}^{(4)} - f^{(4)}]\,, \end{aligned} \qquad (31)$$

where $f^{(4)}(\zeta)$ is an external Lie-algebra valued analytic real superfield, and $\delta[\mathcal{F}^{(4)}]$ is the proper functional analytic delta-function.

To bring the effective action to a form more adapted for calculations, we average (31) with the weight

$$\Xi[V^{++}]\,\exp\left\{\frac{\mathrm{i}}{2\alpha}\mathrm{tr}\int d^{12}z du_1 du_2\, f_\tau^{(4)}(z,u_1)\frac{(u_1^- u_2^-)}{(u_1^+ u_2^+)^3} f_\tau^{(4)}(z,u_2)\right\} . \quad (32)$$

Here α is an arbitrary (gauge) parameter, and $f_\tau^{(4)}$ is the τ-transform of $f^{(4)}$. The functional $\Xi[V^{++}]$ is represented as follows (Buchbinder et al. (1998b))

$$\Xi[V^{++}] = \left(\mathrm{Det}_{(4,0)}\,\widehat{\Box}\right)^{\frac{1}{2}} \int \mathcal{D}\phi\, e^{\mathrm{i}S_{\mathrm{NK}}[\phi,V^{++}]}$$

$$S_{\mathrm{NK}}[\phi,V^{++}] = -\frac{1}{2}\mathrm{tr}\int d\zeta^{(-4)}\,\nabla^{++}\phi\nabla^{++}\phi \quad (33)$$

with the integration variable ϕ being a bosonic real analytic superfield, with its values in the Lie algebra of the gauge group, and presenting itself a Nielsen-Kallosh ghost for the theory. The gauge-covariant operator $\widehat{\Box}$ defined by [1]

$$\widehat{\Box} = -\frac{1}{2}(\nabla^+)^4(\nabla^{--})^2 = -\frac{1}{2}(D^+)^4(\nabla^{--})^2 \quad (34)$$

moves every harmonic superfield into an analytic one, and it is equivalent to the second-order differential operator

$$\widehat{\Box}_\tau = e^{-\mathrm{i}\Omega}\,\widehat{\Box}\,e^{\mathrm{i}\Omega} = \mathcal{D}^m\mathcal{D}_m + \frac{\mathrm{i}}{2}(\mathcal{D}^{+\alpha}\mathcal{W})\mathcal{D}^-_\alpha + \frac{\mathrm{i}}{2}(\overline{\mathcal{D}}^+_{\dot\alpha}\overline{\mathcal{W}})\overline{\mathcal{D}}^{-\dot\alpha}$$
$$- \frac{\mathrm{i}}{4}(\mathcal{D}^{+\alpha}\mathcal{D}^+_\alpha\mathcal{W})\mathcal{D}^{--} + \frac{\mathrm{i}}{8}[\mathcal{D}^{+\alpha},\mathcal{D}^-_\alpha]\mathcal{W} + \frac{1}{2}\{\overline{\mathcal{W}},\mathcal{W}\} \quad (35)$$

when acting on the covariantly analytic superfields. This operator is said to be the analytic d'Alambertian. The functional $\mathrm{Det}_{(4,0)}\,\widehat{\Box}$, which enters the first line of eq. (33), is defined by the following path integral

$$\left(\mathrm{Det}_{(4,0)}\,\widehat{\Box}\right)^{-1} = \int \mathcal{D}\rho^{(+4)}\mathcal{D}\sigma\,\exp\left\{-\mathrm{i}\,\mathrm{tr}\int d\zeta^{(-4)}\,\rho^{(+4)}\widehat{\Box}\,\sigma\right\} \quad (36)$$

over unconstrained bosonic analytic real superfields $\rho^{(+4)}$ and σ.

Upon averaging the effective action with the weight (32), for $\alpha = -1$ one gets the following path integral representation (Buchbinder et al. (1998b))

$$e^{\mathrm{i}\Gamma_{\mathrm{SYM}}[V^{++}]} = e^{\mathrm{i}S_{\mathrm{SYM}}[V^{++}]}\left(\mathrm{Det}_{(4,0)}\,\widehat{\Box}\right)^{\frac{1}{2}}$$
$$\times \int \mathcal{D}v^{++}\mathcal{D}\mathbf{b}\mathcal{D}\mathbf{c}\mathcal{D}\phi\, e^{\mathrm{i}S_{\mathrm{Q}}[v^{++},\mathbf{b},\mathbf{c},\phi,V^{++}]} , \quad (37)$$

[1] We use the notation $(D^+)^4 = \frac{1}{16}(D^+)^2(\overline{D}^+)^2$, $(D^\pm)^2 = D^{\pm\alpha}D^\pm_\alpha$, $(\overline{D}^\pm)^2 = \overline{D}^\pm_{\dot\alpha}\overline{D}^{\pm\dot\alpha}$ and similar notation for the gauge-covariant derivatives.

where action S_Q reads

$$S_Q[v^{++}, \mathbf{b}, \mathbf{c}, \phi, V^{++}] = S_2 + S_{\text{int}} \tag{38}$$

$$S_2 = \text{tr} \int \mathrm{d}\zeta^{(-4)} \left\{ -\frac{1}{2} v^{++} \,\widehat{\Box}\, v^{++} + \mathbf{b}(\nabla^{++})^2 \mathbf{c} + \frac{1}{2}\phi(\nabla^{++})^2 \phi \right\} \tag{39}$$

$$S_{\text{int}} = -\text{tr} \int \mathrm{d}^{12}z \mathrm{d}^n u \sum_{n=3}^{\infty} \frac{(-ig)^{n-2}}{n} \frac{v_\tau^{++}(z,u_1)\dots v_\tau^{++}(z,u_n)}{(u_1^+ u_2^+)\dots(u_n^+ u_1^+)}$$
$$-ig\, \text{tr} \int \mathrm{d}\zeta^{(-4)} \nabla^{++}\mathbf{b}\, [v^{++}, \mathbf{c}]\,. \tag{40}$$

Eqs. (37–40) completely determine the structure of the perturbation expansion for calculating the effective action $\Gamma_{\text{SYM}}[V^{++}]$ of the pure $N = 2$ SYM theory in a manifestly supersymmetric and gauge invariant form.

So far we have considered the pure $N = 2$ SYM theory only. In the general case, the classical action contains not only the pure SYM part given by (5) (or, what is equivalent, by (21)), but also the matter action (22). Our previous consideration can be easily extended to the case of the general $N = 2$ SYM theory. The only non-trivial new information, however, is the explicit structure of the matter superpropagators associated with the action (22). They read as follows

$$\mathrm{i} < q^+(1)\,\breve{q}^+(2) > \tag{41}$$
$$= -\frac{1}{\widehat{\Box}} (\overrightarrow{D_1^+})^4 \left\{ \delta^{12}(z_1 - z_2) \frac{1}{(u_1^+ u_2^+)^3} \mathrm{e}^{\mathrm{i}\Omega(1)} \mathrm{e}^{-\mathrm{i}\Omega(2)} \right\} (\overleftarrow{D_2^+})^4$$

$$\mathrm{i} < \omega(1)\,\omega^{\mathrm{T}}(2) > \tag{42}$$
$$= \frac{1}{\widehat{\Box}} (\overrightarrow{D_1^+})^4 \left\{ \delta^{12}(z_1 - z_2) \frac{(u_1^- u_2^-)}{(u_1^+ u_2^+)^3} \mathrm{e}^{\mathrm{i}\Omega(1)} \mathrm{e}^{-\mathrm{i}\Omega(2)} \right\} (\overleftarrow{D_2^+})^4\,.$$

The Green's functions (41) and (42) are to be used for loop calculations in the background field approach.

The propagators of the gauge and ghost superfields follow from (39). For the gauge superfield one get

$$\mathrm{i} < v^{++}(1)\, v^{++}(2) > = \frac{1}{\widehat{\Box}} (D_1^+)^4 \left\{ \delta^{12}(z_1 - z_2) \delta^{(-2,2)}(u_1, u_2) \right\} \tag{43}$$

with $\delta^{(-2,2)}(u_1, u_2)$ being the proper harmonic delta-function (Galperin et al. (1985a)). The propagator of the Faddeev-Popov ghosts $\mathbf{b}$ and $\mathbf{c}$ is completely analogous to the ω-hypermultiplet propagator (42). The third ghost ϕ contributes at the one-loop level only.

4 The $D = 4$, $N = 2$ Non-renormalization Theorem

Let us apply the covariant harmonic supergraph technique to analyse the divergence structure of the theory. The result is formulated as the $D = 4$, $N = 2$ non-renormalization theorem: there are no ultraviolet divergences beyond the one-loop level ((Howe, Stelle and Townsend (1984)), Buchbinder et al. (1998c)).

Consider the loop expansion of the effective action within the background field formulation. Then, the effective action is given by vacuum diagrams (that is, diagrams without external lines) with background field dependent propagators and vertices. In our case, the corresponding propagators are defined by eqs. (41–43), and the vertices can be read off from eqs. (22) and (40). It is obvious that any such diagram can be expanded in terms of background fields, and leads to a set of conventional diagrams with an arbitrary number of external legs.

As follows from eqs. (22) and (40), the gauge superfield vertices are given by integrals over the full superspace, while the matter vertices and the Faddeev-Popov ghosts vertices are given by integrals over the analytic subspace. Note, however, that propagators (41–43) contain factors of $(D^+)^4$, which can be used to transform integrals over the analytic subspace into integrals over the full superspace if we make use of the identity

$$\int \mathrm{d}\zeta^{(-4)}\,(D^+)^4\mathcal{L} = \int \mathrm{d}^{12}z\,\mathrm{d}u\,\mathcal{L}\,. \tag{44}$$

The cost of doing this is, as a rule, the removal of one of the two $(D^+)^4$-factors entering each matter and ghost propagator (41,42). There is, however, one special case. Let us consider a vertex with two external ω-legs, and start to transform the corresponding integral over the analytic subspace into an integral over the full superspace. To do this, we should remove the factor $(D^+)^4$ from one of the two gauge superfield propagators (43) associated with this vertex. As a result of transforming all integrals over the analytic subspace into integrals over the full superspace, each of the remaining propagators will contain, at most, one factor of $(D^+)^4$. Thus, any supergraph contributing to the effective action is given in terms of the integrals over the full $N = 2$ harmonic superspace. Since this conclusion is true for each conventional supergraph in the expansion of a given background field supergraph, we see that an arbitrary background field supergraph is also given by integrals over the full $N = 2$ harmonic superspace. This is in complete analogy with $N = 1$ supersymmetric field theories.

Once we have constructed the supergraphs with all vertices integrated over the full $N = 2$ harmonic superspace, we can perform all but one of the integrals over the θ's, step by step and loop by loop, due to the spinor delta-functions $\delta^8(\theta_i - \theta_j)$ contained in the propagators (41–43). To do this, we remove the $(D^+)^4$-factors acting on the spinor delta-functions in the propagators by making an integration by parts. This allows one to obtain spinor delta-functions without $(D^+)^4$-factors. One can then perform the integrals over the θ's. We note that in the process of integration by parts, some of the $(D^+)^4$-factors can act on the external legs of the supergraph. To obtain a non-zero result in the case of an L-loop supergraph, we should remove $2L$ factors of $(D^+)^4$ attached to some of the propagators using the identity

$$\delta^8(\theta_1 - \theta_2)(D_1^+)^4(D_2^+)^4\,\delta^8(\theta_1 - \theta_2) = (u_1^+ u_2^+)^4 \delta^8(\theta_1 - \theta_2)\,. \tag{45}$$

Thus, any supergraph contributing to the effective action is given by a single integral over $\mathrm{d}^8\theta$.

Now, it is not difficult to calculate the superficial degree of divergence for the theory under consideration. Let us consider an L-loop supergraph G with P propagators, N_{MAT} external matter legs and an arbitrary number of gauge superfield external legs. We denote by N_D the number of spinor covariant derivatives acting on the external legs as a result of integration by parts in the process of transformating the contributions to a single integral over $\mathrm{d}^8\theta$. Then, the superficial degree of divergence $\omega(G)$ of the supergraph G turns out to be (Buchbinder et al. (1998c))

$$\omega(G) = -N_{\mathrm{MAT}} - \frac{1}{2} N_D\,. \tag{46}$$

We see immediately that all supergraphs with external matter legs are automatically finite. As to supergraphs with pure gauge superfield legs, they are clearly finite only if some non-zero number of spinor covariant derivatives acts on the external legs. Let us now show that this is always the case beyond one loop.

The Feynman rules for $N = 2$ supersymmetric field theories in the harmonic superspace approach have been formulated in the λ-frame, where the propagators are given by eqs. (41–43). As we have noticed, all vertices in the background field supergraphs, including the vertices of matter and Faddeev-Popov ghosts superfields, can be given in a form containing integrals over the full $N = 2$ harmonic superspace only. To be more precise, this property is stipulated by the identity in the λ-frame

$$(D^+)^4\, \widehat{\Box} = \widehat{\Box}\, (D^+)^4\,. \tag{47}$$

This identity allows one to operate with factors $(D^+)^4$ as in case without background field, and use them to transform the integrals over the analytic subspace into integrals over the full superspace directly in background field supergraphs. Let us consider the structure of the propagators in the λ-frame (41–43). The background field V^{++} enters these propagators via both $\widehat{\Box}$ and the background bridge Ω. The form of the propagators (41–43) has one drawback: if we use this form, we can not say how many spinor derivatives act on the external legs since the explicit dependence of Ω on the background field is rather complicated. To clarify the situation when a number of spinor derivatives act on external legs, we use a completely new step (in comparison with the conventional harmonic supergraph approach developed by Galperin et al. (1985a,b)) and transform the supergraph to the τ-frame (after restoring the full superspace measure at the matter and ghost vertices). The propagators in the τ-frame are given by (Buchbinder et al. (1998c))

$$
\begin{aligned}
\mathrm{i} < q^+_\tau(1)\,\breve{q}^+_\tau(2) > &= -\frac{1}{\widehat{\Box}_\tau}(\overrightarrow{\mathcal{D}^+_1})^4 \left\{\delta^{12}(z_1 - z_2)\frac{1}{(u^+_1 u^+_2)^3}\right\}(\overleftarrow{\mathcal{D}^+_2})^4 \\
\mathrm{i} < \omega_\tau(1)\,\omega^{\mathrm{T}}_\tau(2) > &= \frac{1}{\widehat{\Box}_\tau}(\overrightarrow{\mathcal{D}^+_1})^4 \left\{\delta^{12}(z_1 - z_2)\frac{(u^-_1 u^-_2)}{(u^+_1 u^+_2)^3}\right\}(\overleftarrow{\mathcal{D}^+_2})^4 \\
\mathrm{i} < v^{++}_\tau(1)\,v^{++}_\tau(2) > &= \frac{1}{\widehat{\Box}_\tau}(\mathcal{D}^+_1)^4 \left\{\delta^{12}(z_1 - z_2)\delta^{(-2,2)}(u_1, u_2)\right\}. \qquad (48)
\end{aligned}
$$

They contain, at most, one factor of $(\mathcal{D}^+)^4$ after restoring the full superspace measure at the matter and ghost vertices. The essential feature of these propagators is that they contain the background field V^{++} only via the $\widehat{\Box}_\tau$ and $\mathcal{D}^+$-factors; that is, only via the u-independent connections $\mathcal{A}_M$ (1). But all connections $\mathcal{A}_M$ contain at least one spinor covariant derivative acting on the background superfield V^{++} (Galperin et al. (1984)). Therefore, if we expand any background field supergraph in the background superfield V^{++}, we see that each external leg must contain at least one spinor covariant derivative. Thus, the number N_D in eq. (46) must be greater than or equal to one. As a consequence, $\omega(G) < 0$ and, hence, all supergraphs are ultravioletly finite beyond the one-loop level. This completes the proof of the non-renormalization theorem.

5 The One-Loop Effective Action

As is clear from the above analysis, the one-loop effective action requires a separate investigation. In what follows, we restrict our attention to the part $\Gamma[V^{++}]$ of effective action, which depends on the gauge superfield only. It is $\Gamma[V^{++}]$ which (i) determines the one-loop ultraviolet divergences; (ii) constitutes the effective dynamics in the Coulomb branch of $N = 2$ SYM theories.

It follows from eqs. (22,37,39) that the one-loop effective action $\Gamma^{(1)}[V^{++}]$ of the general $N = 2$ SYM theory reads

$$
\begin{aligned}
\Gamma^{(1)}[V^{++}] = {} & \frac{\mathrm{i}}{2}\,\mathrm{Tr}_{(2,2)}\ \ln\widehat{\Box} - \frac{\mathrm{i}}{2}\,\mathrm{Tr}_{(4,0)}\ \ln\widehat{\Box} \\
& - \frac{\mathrm{i}}{2}\,\mathrm{Tr}_{ad}\ \ln(\nabla^{++})^2 \\
& + \mathrm{i}\,\mathrm{Tr}_{R_q}\ \ln(\nabla^{++}) + \frac{\mathrm{i}}{2}\,\mathrm{Tr}_{R_\omega}\ \ln(\nabla^{++})^2 . \qquad (49)
\end{aligned}
$$

Here the contribution in the first line comes not only from the overal factor in (37), but also from the gauge superfield,

$$\left(\mathrm{Det}_{(2,2)}\,\widehat{\Box}\right)^{-\frac{1}{2}} = \int \mathcal{D}v^{++} \exp\left\{-\frac{\mathrm{i}}{2}\,\mathrm{tr}\int \mathrm{d}\zeta^{(-4)}\, v^{++}\widehat{\Box}\, v^{++}\right\}\,. \tag{50}$$

The second line in (49) represents the joint contribution from the Faddeev-Popov ghosts **b**, **c** and the third ghost ϕ. Finally, the third line includes the contributions from the matter q- and ω-hypermultiplets.

The joint contribution of the Faddeev-Popov ghosts and the third ghost differs only in sign from that of an ω-hypermultiplet in the adjoint representation of the gauge group. In case of the $N = 4$ SYM theory realized in the $N = 2$ harmonic superspace, the matter sector is formed by a single ω–hypermultiplet in the adjoint representation (Galperin et al. (1985b)), and the classical action reads

$$S^{N=4}_{\mathrm{SYM}} = \frac{1}{2g^2}\mathrm{tr}\int \mathrm{d}^4x\mathrm{d}^4\theta\, \mathcal{W}^2 - \frac{1}{2g^2}\,\mathrm{tr}\int \mathrm{d}\zeta^{(-4)}\, \nabla^{++}\omega\nabla^{++}\omega\,. \tag{51}$$

Therefore, the corresponding one-loop effective action is given by the first line of eq. (49),

$$\Gamma^{(1)}_{N=4}[V^{++}] = \frac{\mathrm{i}}{2}\,\mathrm{Tr}_{\,(2,2)}\,\ln\widehat{\Box} - \frac{\mathrm{i}}{2}\,\mathrm{Tr}_{\,(4,0)}\,\ln\widehat{\Box}\,. \tag{52}$$

It is the contributions in the second and third lines of (49) which (i) are responsible for all the ultraviolet divergences of the theory and (ii) generate the low-energy holomorphic action (see Buchbinder et al. (1998c) for more detail). By now, we have a well elaborated perturbative scheme to compute such quantum hypermultiplet corrections (Buchbinder et al. (1997), Buchbinder et al. (1998a)). As concerns the $N = 4$ SYM effective action (52), it is free of ultraviolet divergences, but its calculation turns out to be a nontrivial technical problem. The point is that the one-loop supergraphs contributing to $\Gamma^{(1)}_{N=4}[V^{++}]$ in the harmonic superspace approach contain coinciding harmonic singularities, that is harmonic distributions at coinciding points. The problem of coinciding harmonic singularities in the framework of harmonic supergraph Feynman rules was first discussed by Galperin et al. (1987c). Such singularities have no physical origin, in contrast to ultraviolet divergences. They can appear only at intermediate stages of calculation and should cancel each other in the final expressions for physical quantities. The origin of this problem is an infinite number of internal degrees of freedom associated with the bosonic internal coordinates.

To get rid of the one-loop coinciding harmonic singularities, Buchbinder et al. (1998b) introduced, as is generally accepted in quantum field theory, some regularization of harmonic distributions. Unfortunately, this regularization proved to be unsuccessful; its use led us to the wrong conclusion $\Gamma^{(1)}_{N=4}[V^{++}] = 0$. In a sense, the situation in hand is similar to that with the well-known supersymmetric regularization via dimensional reduction which leads to obstacles at higher loops. The harmonic regularization we used turned out to be improper already at the one-loop level.

We would like to emphasize that the problem of coinciding harmonic singularities is associated only with perturbative calculations of the effective action and has no direct relation to the $N = 2$ background field method itself. The problem of coinciding harmonic singularities has been solved by Buchbinder et al. (1998d) for a special $N = 2$ SYM background

$$\mathcal{D}^{\alpha(i}\mathcal{D}^{j)}_{\alpha}\mathcal{W} = 0 \,. \tag{53}$$

In this case the effective action can be equivalently represented in the form

$$\exp\left\{\mathrm{i}\,\Gamma^{(1)}_{N=4}\right\} = \frac{\int \mathcal{D}\mathcal{G}^{++} \exp\left\{-\frac{\mathrm{i}}{2}\,\mathrm{tr}\int \mathrm{d}\zeta^{(-4)}\,\mathcal{G}^{++}\widehat{\Box}\,\mathcal{G}^{++}\right\}}{\int \mathcal{D}\mathcal{G}^{++} \exp\left\{\frac{\mathrm{i}}{2}\,\mathrm{tr}\int \mathrm{d}\zeta^{(-4)}\,\mathcal{G}^{++}\,\mathcal{G}^{++}\right\}} \tag{54}$$

where the analytic integration variable $\mathcal{G}^{++}$ is constrained by

$$\nabla^{++}\mathcal{G}^{++} = 0 \,. \tag{55}$$

Representation (54) allows us to perturbatively compute $\Gamma^{(1)}_{N=4}$. Moreover, it can to used to prove equivalence of the $N = 2$ covariant supergraph technique to the famous $N = 1$ background field formulation for the $N = 4$ SYM (Grisaru et al. (1979)), when the lowest $N = 1$ superspace component of the $N = 2$ vector multiplet is switched off (Buchbinder et al. (1998d)).

6 Low-Energy Effective Action

In the Coulomb branch of the $N = 2$ SYM theory, the matter hypermultiplets are integrated out and the gauge superfield lies along a flat direction of the $N = 2$ SYM potential. If the gauge group is $SU(2)$, only the $U(1)$ gauge symmetry survives, upon the spontaneous breakdown of $SU(2)$, and the gauge superfield $V^{++} = V^{++a}T^a$ ($T^a = \frac{1}{\sqrt{2}}\sigma^a$, $a = 1,2,3$) takes the form

$$V^{++} = V^{++3}T^3 \equiv \mathbf{V}^{++}T^3 \,. \tag{56}$$

Here $\mathbf{V}^{++}$ consists of two parts, $\mathbf{V}^{++} = \mathbf{V}_0^{++} + \mathbf{V}_1^{++}$, where $\mathbf{V}_0^{++}$ corresponds to a constant strength $\mathbf{W}_0 = \text{const}$, and $\mathbf{V}_1^{++}$ is an abelian gauge superfield. The presence of $\mathbf{V}_0^{++}$ leads to the appearance of mass $|\mathbf{W}_0|^2$ for matter multiplets (see Buchbinder et al. (1997)).

Since the effective action $\Gamma[\mathbf{V}^{++}]$ is gauge invariant, it presents itself a functional of the chiral strength $\mathbf{W}$ and its conjugate $\overline{\mathbf{W}}$. Assuming the validity of momentum expansion, one can present the effective action $\Gamma[\mathbf{W},\overline{\mathbf{W}}]$ in the form

$$\Gamma[\mathbf{W},\overline{\mathbf{W}}] = \left(\int \mathrm{d}^4x\mathrm{d}^4\theta\,\mathcal{L}^{(c)}_{\mathrm{eff}} + \text{c.c.}\right) + \int \mathrm{d}^4x\mathrm{d}^8\theta\,\mathcal{L}_{\mathrm{eff}} \,. \tag{57}$$

Here the chiral effective Lagrangian $\mathcal{L}^{(c)}_{\mathrm{eff}}$ is a local function of $\mathbf{W}$ and its space-time derivatives, $\mathcal{L}^{(c)}_{\mathrm{eff}} = F(\mathbf{W}) + \ldots$, and the higher-derivative effective

Lagrangian $\mathcal{L}_{\text{eff}}$ is a real function of $\mathbf{W}$, $\overline{\mathbf{W}}$ and their covariant derivatives, $\mathcal{L}_{\text{eff}} = H(\mathbf{W}, \overline{\mathbf{W}}) + \dots$

At the one-loop level, it is the Faddeev-Popov ghosts, the third ghost and the matter hypermultiplets which contribute to $F(\mathbf{W})$. As concerns the quantum correction in the first line of (49), it contributes to the higher-derivative action $H(\mathbf{W}, \overline{\mathbf{W}})$. A general analysis of covariant harmonic supergraphs given by Buchbinder et al. (1998c) shows that the holomorphic action $F(\mathbf{W})$ is completely generated by the one-loop contribution. Another consequence of such an analysis is that there is no two-loop contribution to $H(\mathbf{W}, \overline{\mathbf{W}})$.

The covariant harmonic supergraph technique allows us to easily compute the holomorphic effective action. Let us restrict, for simplicity, our consideration to the case of the pure $N = 2$, $SU(2)$ SYM theory. If we are interested in the low-energy holomorphic action, it is proper to use the following approximation

$$\Gamma^{(1)}_{SU(2)}[V^{++}] \approx -\Gamma_{\phi}[V^{++}] \tag{58}$$

with $\Gamma_{\phi}[V^{++}]$ the effective action of a real ω-hypermultiplet in the adjoint representation of $SU(2)$ coupled to the external gauge superfield V^{++} :

$$e^{\mathrm{i}\,\Gamma_{\phi}[V^{++}]} = \int \mathcal{D}\phi \exp\left\{-\frac{\mathrm{i}}{2}\,\mathrm{tr} \int \mathrm{d}\zeta^{(-4)}\, \nabla^{++}\phi\, \nabla^{++}\phi\right\}$$
$$\phi = \phi^a T^a\,, \qquad \nabla^{++}\phi = D^{++}\phi + \mathrm{i}[V^{++}, \phi]\,. \tag{59}$$

Since the gauge superfield has the form (56), ϕ^3 completely decouples. Unifying ϕ^1 and ϕ^2 in to the complex ω-hypermultiplet $\omega = \phi^1 - \mathrm{i}\phi^2$, we observe

$$\nabla^{++}\omega = D^{++}\omega + \mathrm{i}\sqrt{2}\mathcal{V}^{++}\omega\,, \tag{60}$$

hence the $U(1)$-charge of ω is $e = \sqrt{2}$. As was shown by Buchbinder et al. (1997), the effective actions of the charged complex ω-hypermultiplet and the charged q-hypermultiplet, interacting with background $U(1)$ gauge superfield $\mathbf{V}^{++}$, are related by $\Gamma_{\omega}[\mathbf{V}^{++}] = 2\Gamma_q[\mathbf{V}^{++}]$ and the leading contribution to $\Gamma_q[\mathbf{V}^{++}]$ in the massive theory is given by

$$\Gamma_q[\mathbf{V}^{++}] = \int \mathrm{d}^4x\mathrm{d}^4\theta\; F(\mathbf{W}) + \text{c.c.}\,, \qquad F(\mathbf{W}) = -\frac{e^2}{64\pi^2}\mathbf{W}^2 \ln\frac{\mathbf{W}^2}{\Lambda^2}\,. \tag{61}$$

Here e is the charge of q^+ (it coincides with the charge of ω in the above correspondence), Λ is the renormalization scale. Since in our case $e = \sqrt{2}$, from eqs. (58,61) we finally obtain the perturbative holomorphic of the $N = 2$ $SU(2)$ SYM theory

$$F^{(1)}_{SU(2)}(\mathbf{W}) = \frac{1}{16\pi^2}\,\mathbf{W}^2 \ln\frac{\mathbf{W}^2}{\Lambda^2}\,. \tag{62}$$

This is exactly Seiberg's low-energy effective action (Seiberg (1988)) found by integrating the $U(1)$ global anomaly and using the component analysis.

Let us finally turn to the $N = 4$ $SU(2)$ SYM theory (51). Here the non-holomorphic action $H(\mathbf{W}, \overline{\mathbf{W}})$ constitutes the leading low-energy quantum correction. Its calculation is based on the representation (54). Using the technique developed in our paper (Buchbinder et al. (1998d)) and under additional restrictions on the background superfields, one can represent the effective action $\Gamma^{(1)}_{N=4}$ by a path integral over an unconstrained $N = 1$ complex superfield V and its conjugate

$$\exp\{\mathrm{i}\,\Gamma^{(1)}_{N=4}\} = \int \mathcal{D}\bar{V}\mathcal{D}V \exp\left\{\frac{\mathrm{i}}{2}\,\mathrm{tr}\int \mathrm{d}^8 z\,\bar{V}\Delta V\right\}$$
$$\Delta = \mathcal{D}^a\mathcal{D}_a - eW^\alpha\mathcal{D}_\alpha + e\overline{W}_{\dot\alpha}\overline{\mathcal{D}}^{\dot\alpha} + e^2|\phi|^2\,. \tag{63}$$

Here ϕ and W_α are the $N = 1$ projections of $\mathbf{W}$: $\phi = \mathbf{W}|$, $2\mathrm{i}W_\alpha = \mathcal{D}^2_\alpha\mathbf{W}|$. Being rewritten in terms of the $N = 1$ projections, the leading non-holomorphic correction to $\Gamma^{(1)}_{N=4}$ takes the form

$$\int \mathrm{d}^{12}z\; H(W,\overline{W}) = \int \mathrm{d}^8 z\, W^\alpha W_\alpha \overline{W}_{\dot\alpha}\overline{W}^{\dot\alpha}\; \frac{\partial^4 H(\phi,\bar\phi)}{\partial\phi^2\partial\bar\phi^2} + \cdots \tag{64}$$

To calculate $\partial^4 H(\phi,\bar\phi)/\partial\phi^2\partial\bar\phi^2$, we use a superfield proper-time technique introducing the Schwinger kernel for the operator Δ (63). Then one gets $\partial^4 H(\phi,\bar\phi)/\partial\phi^2\partial\bar\phi^2 = (4\pi\phi\bar\phi)^{-2}$. One can easily find a general solution to this equation. Since the effective action of the $N = 4$ SYM theory should be scale and chiral invariant, we finally get

$$H^{(1)}_{N=4}(\mathbf{W},\overline{\mathbf{W}}) = \frac{1}{4(4\pi)^2}\ln\frac{\mathbf{W}^2}{\Lambda^2}\ln\frac{\overline{\mathbf{W}}^2}{\Lambda^2}\,. \tag{65}$$

The details can be found in our work (Buchbinder et al. (1998d)). This action was independently computed by Periwal et al. (1998) and Gonzalez et al. (1998). The possibility of quantum corrections of the form (65) in the effective action for the $N = 4$ $SU(2)$ SYM theory was first argued by Dine et al. (1997).

Acknowledgements. We are grateful to E. Buchbinder and E. Ivanov for fruitful collaboration and numerous discussions. We are thankful to B. de Wit, N. Dragon, J. Gates, M. Grisaru, M. Roček, E. Sokatchev, S. Theisen, and B. Zupnik for critical remarks and comments. I. B. and S. K. acknowledge a partial support from RFBR grant, project No 96-02-1607; RFBR-DFG grant, project No 96-02-00180; INTAS grant, INTAS-96-0308. B. O. acknowledges the DOE Contract No. DE-AC02-76-ER-03072 and is grateful to the Alexander von Humboldt Foundation for partial support.

References

Buchbinder, I., Buchbinder, E., Ivanov, E., Kuzenko, S., Ovrut, B. (1997): Effective action of the $N = 2$ Maxwell multiplet in harmonic superspace. Phys. Lett. **B 412**, 309–319

Buchbinder, E., Buchbinder, I., Ivanov, E., Kuzenko, S. (1998a): Central charge as the origin of holomorphic effective action in $N =$ gauge theory. Mod. Phys. Lett. **A 13**, 1071–1082

Buchbinder, I., Buchbinder, E., Kuzenko, S., Ovrut, B. (1998b): The background field method for $N = 2$ super Yang-Mills theories in harmonic superspace. Phys. Lett. **B 417**, 61–71

Buchbinder, I., Kuzenko, S., Ovrut, B. (1998c): On the $D = 4$, $N = 2$ non-renormalization theorem. Phys. Lett. **B 433**, 335–345

Buchbinder, I., Kuzenko, S. (1998d): Comments on the background field method in harmonic superspace: non-holomorphic corrections in $N = 4$ SYM. Mod. Phys. Lett. **A 13**, 1623–1635

Dine, M., Seiberg, N. (1997): Comments on higher derivative operators in some SUSY field theories. Phys. Lett. **B 409**, 239–247

Galperin, A., Ivanov, E., Kalitzin, S., Ogievetsky, V., Sokatchev, E. (1984): Unconstrained $N = 2$ matter, Yang-Mills and supergravity theories in harmonic superspace. Class. Quant. Grav. **1**, 469–498

Galperin, A., Ivanov, E., Ogievetsky, V., Sokatchev, E. (1985a): Harmonic supergraghs: Green functions. Class. Quant. Grav. **2**, 601–616

Galperin, A., Ivanov, E., Ogievetsky, V., Sokatchev, E. (1985b): Harmonic supergraghs: Feynman rules and examples. Class. Quant. Grav. **2**, 617–630

Galperin, A., Ky, N. A., Sokatchev, E. (1987a): $N = 2$ supergravity in superspace: solution to the constraints and the invariant action. Class. Quant. Grav. **4**, 1235–1253

Galperin, A., Ivanov, E., Ogievetsky, V., Sokatchev, E. (1987b): $N = 2$ supergravity in superspace: different versions and matter couplings. Class. Quant. Grav. **4**, 1255–1265

Galperin, A., Ky, N. A., Sokatchev, E. (1987c): Coinciding harmonic singularities in harmonic supergraphs. Mod. Phys. Lett. **A 2**, 33–39

Gonzalez, F., Roček, M. (1998): Nonholomorphic $N = 2$ terms in $N = 4$ SYM: one-loop calculation in $N = 2$ superspace. hep-th/9804010

Grimm, R., Sohnius, M., Wess, J. (1978): Extended supersymmetry and gauge theories. Nucl. Phys. **B 133**, 275–284

Grisaru, M. T., Roček, M., Siegel, W. (1979): Improved methods for supergraphs. Nucl. Phys. **B 159**, 429–450

Howe, P. S., Stelle, K. S., Townsend, P. K. (1984): Miraculous ultraviolet cancellations in supersymmetry made manifest. Nucl. Phys. **B 236**, 125–166

Lindström, U., Roček, M. (1990): $N = 2$ super Yang-Mills theory in projective superspace. Commun. Math. Phys. **128**, 191–196

Periwal, V., von Unge, R. (1998): Accelerating D-branes. Phys. Lett. **B 430**, 71–76

Seiberg, N. (1988): Supersymmetry and nonperturbative beta functions. Phys. Lett. **B 206**, 75–80

Zupnik, B. M. (1987): The action of the supersymmetric $N = 2$ gauge theory in harmonic superspace. Phys. Lett. **B 183**, 175–176

Rigid $N=2$ Superconformal Hypermultiplets

Bernard de Wit[1], Bas Kleijn[1], and Stefan Vandoren[2]

[1] Institute for Theoretical Physics, Utrecht University
3508 TA Utrecht, The Netherlands
[2] Department of Physics, University of Wales, Swansea
SA2 8PP Swansea, U.K.

Abstract. We discuss superconformally invariant systems of hypermultiplets coupled to gauge fields associated with target-space isometries.

1 Introduction

Hypermultiplets played an important role in the work of Victor I. Ogievetsky, to whose memory this meeting is dedicated. We remember him as a devoted scientist, but above all as a dear friend and colleague who is sorely missed.

In this contribution[1] we discuss hypermultiplets coupled to gauge fields whose action is invariant under rigid $N = 2$ superconformal symmetries. This study is both motivated by recent interest in superconformal theories (Maldacena (1997)) and by our attempts to understand the coupling of hypermultiplets to supergravity in a way that is more in parallel with the special-geometry formulation of vector multiplets (De Jaegher et al. (1998)). In this respect it is important that we employ an on-shell treatment of hypermultiplets (to avoid an infinite number of fields), while the vector multiplets and the superconformal theory are considered fully off-shell. This implies that the algebra of the superconformal and gauge symmetries is known up to the hypermultiplet field equations.

2 Hypermultiplet Lagrangians

Hyper-Kähler spaces serve as target spaces for nonlinear sigma models based on hypermultiplets (Bagger and Witten (1983)). We start here by summarizing some results on the formulation of these theories following De Jaegher et al. (1998). With respect to the results of Bagger and Witten (1983) this formulation differs in that it incorporates both a metric g_{AB} for the hyper-Kähler target space and a metric $G_{\bar\alpha\beta}$ for the fermions. Here we assume that the n hypermultiplets are described by $4n$ real scalars ϕ^A, $2n$ positive-chirality spinors $\zeta^{\bar\alpha}$ and $2n$ negative-chirality spinors ζ^α. The latter two are related by complex conjugation (so that we have $2n$ Majorana spinors) under which indices are converted according to $\alpha \leftrightarrow \bar\alpha$, while SU(2) indices $i, j, \ldots$ are raised

[1] Talk given by B. de Wit. The content of this contribution is related to the actual presentation at the meeting, but takes into account more recent developments.

and lowered. The presence of the fermionic metric is important in obtaining the correct transformation rules under symplectic transformations induced by the so-called **c**-map from the electric-magnetic duality transformations on a corresponding theory of vector multiplets. In formulations based on $N=1$ superfields (such as in Hull et al. (1986)) one naturally has a fermionic metric but of a special form.

The supersymmetry transformations are parametrized in terms of certain ϕ-dependent quantities γ^A and V_A as

$$\delta_Q\phi^A = 2(\gamma^A_{i\bar\alpha}\,\bar\epsilon^i\zeta^{\bar\alpha} + \bar\gamma^{Ai}_{\alpha}\,\bar\epsilon_i\zeta^{\alpha})\,, \qquad \begin{aligned} \delta_Q\zeta^\alpha &= V^\alpha_{A\,i}\,\partial\!\!\!/\phi^A\epsilon^i - \delta_Q\phi^A\,\Gamma_A{}^\alpha{}_\beta\,\zeta^\beta\,, \\ \delta_Q\zeta^{\bar\alpha} &= \bar V^{i\bar\alpha}_{A}\,\partial\!\!\!/\phi^A\epsilon_i - \delta_Q\phi^A\,\bar\Gamma_A{}^{\bar\alpha}{}_{\bar\beta}\,\zeta^{\bar\beta}\,. \end{aligned} \tag{1}$$

Observe that these variations are consistent with a U(1) chiral invariance under which the scalars remain invariant, which we will denote by $\mathrm{U(1)_R}$ to indicate that it is a subgroup of the automorphism group of the supersymmetry algebra. In Sect.4 this U(1) will be included as one of the conformal gauge groups. However, for generic γ^A and V_A, the $\mathrm{SU(2)_R}$ part of the automorphism group cannot be realized consistently on the fields. In the above, we only used that ζ^α and $\zeta^{\bar\alpha}$ are related by complex conjugation.

The Lagrangian takes the following form

$$\mathcal{L} = -\tfrac{1}{2}g_{AB}\,\partial_\mu\phi^A\partial^\mu\phi^B - G_{\bar\alpha\beta}(\bar\zeta^{\bar\alpha}\,D\!\!\!\!/\,\zeta^\beta + \bar\zeta^\beta\,D\!\!\!\!/\,\zeta^{\bar\alpha}) - \tfrac{1}{4}W_{\bar\alpha\beta\bar\gamma\delta}\,\bar\zeta^{\bar\alpha}\gamma_\mu\zeta^\beta\,\bar\zeta^{\bar\gamma}\gamma^\mu\zeta^\delta\,, \tag{2}$$

where we use the covariant derivatives

$$D_\mu\zeta^\alpha = \partial_\mu\zeta^\alpha + \partial_\mu\phi^A\,\Gamma_A{}^\alpha{}_\beta\,\zeta^\beta\,, \quad D_\mu\zeta^{\bar\alpha} = \partial_\mu\zeta^{\bar\alpha} + \partial_\mu\phi^A\,\bar\Gamma_A{}^{\bar\alpha}{}_{\bar\beta}\,\zeta^{\bar\beta}\,. \tag{3}$$

Besides the Riemann curvature R_{ABCD} we will be dealing with another curvature $R_{AB}{}^\alpha{}_\beta$ associated with the connections $\Gamma_A{}^\alpha{}_\beta$, which takes its values in $sp(n) \cong usp(2n;\mathbf{C})$. The tensor W is defined by

$$W_{\bar\alpha\beta\bar\gamma\delta} = R_{AB}{}^{\bar\epsilon}{}_{\bar\gamma}\,\gamma^A_{i\bar\alpha}\,\bar\gamma^{iB}_{\beta}\,G_{\bar\epsilon\delta} = \tfrac{1}{2}R_{ABCD}\,\gamma^A_{i\bar\alpha}\,\bar\gamma^{iB}_{\beta}\,\gamma^C_{j\bar\gamma}\,\bar\gamma^{jD}_{\delta}\,. \tag{4}$$

Most of these quantities are not independent, as we shall specify below, and the models are entirely characterized by the target-space geometry (for instance, encoded in the metric g_{AB}) and the $\mathrm{Sp}(n)\times\mathrm{Sp}(1)$ one-forms $V_i^\alpha = V^\alpha_{A\,i}\,\mathrm{d}\phi^A$. The Sp(1) factor is associated with the indices $i,j,\ldots$, and coincides with the $\mathrm{SU(2)_R}$ group mentioned above.

The metric g_{AB}, the tensors γ^A, V_A and the fermionic metric $G_{\bar\alpha\beta}$ are all covariantly constant with respect to the Christoffel connection and the connections $\Gamma_A{}^\alpha{}_\beta$. Furthermore we note the following relations,

$$\begin{aligned} &\gamma^A_{i\bar\alpha}\,\bar V^{j\bar\alpha}_B + \bar\gamma^{A\,j}_{\alpha}\,V^\alpha_{B\,i} = \delta^j_i\,\delta^A_B\,, \\ &g_{AB}\,\gamma^B_{i\bar\alpha} = G_{\bar\alpha\beta}\,V^\beta_{A\,i}\,, \qquad \bar V^{i\bar\alpha}_A\,\gamma^A_{j\bar\beta} = \delta^i_j\,\delta^{\bar\alpha}_{\bar\beta}\,. \end{aligned} \tag{5}$$

These conditions define a number of useful relations between bilinears[2] which include three antisymmetric covariantly constant target-space tensors,

$$J^{ij}_{AB} = \gamma_{Ak\bar\alpha}\, \varepsilon^{k(i}\bar V^{j)\bar\alpha}_{B}\,, \tag{6}$$

that span the complex structures of the hyper-Kähler target space. They satisfy

$$(J^{ij}_{AB})^* = \varepsilon_{ik}\varepsilon_{jl}\, J^{kl}_{AB}\,, \qquad J^{ijC}_{A}\, J^{kl}_{CB} = \tfrac12 \varepsilon^{i(k}\varepsilon^{l)j}\, g_{AB} + \varepsilon^{(i(k}\, J^{l)j)}_{AB}\,. \tag{7}$$

In addition we note the following useful identities,

$$\gamma_{Ai\bar\alpha}\, \bar V^{j\bar\alpha}_{B} = \varepsilon_{ik} J^{kj}_{AB} + \tfrac12 g_{AB}\,\delta^j_i\,, \qquad J^{ij}_{AB}\,\gamma^{B}_{\bar\alpha k} = -\delta^{(i}_{k}\varepsilon^{j)l}\,\gamma_{Al\bar\alpha}\,. \tag{8}$$

We also note the existence of covariantly constant antisymmetric tensors,

$$\Omega_{\bar\alpha\bar\beta} = \tfrac12 \varepsilon^{ij}\, g_{AB}\,\gamma^{A}_{i\bar\alpha}\,\gamma^{B}_{j\bar\beta}\,, \qquad \bar\Omega^{\bar\alpha\bar\beta} = \tfrac12 \varepsilon_{ij}\, g^{AB}\,\bar V^{i\bar\alpha}_{A}\,\bar V^{j\bar\beta}_{B}\,, \tag{9}$$

satisfying $\Omega_{\bar\alpha\bar\gamma}\,\bar\Omega^{\bar\gamma\bar\beta} = -\delta^{\bar\beta}_{\bar\alpha}$.

The existence of the covariantly constant tensors implies a variety of integrability conditions for the curvature tensors. For instance, one proves that the Riemann curvature and the Sp(n) curvature are related, as indicated in (4). The tensor W defined in (4) can also be written as $W_{\alpha\beta\gamma\delta}$ by contracting with the metric G and the antisymmetric tensor Ω. It then follows that $W_{\alpha\beta\gamma\delta}$ is symmetric in symmetric index pairs $(\alpha\beta)$ and $(\gamma\delta)$. Using the Bianchi identity for Riemann curvature, which implies $g_{D[A}R_{BC]}{}^{\bar\beta}{}_{\bar\alpha}\,\gamma^{D}_{i\bar\beta} = 0$, one shows that it is in fact symmetric in all four indices. For further results and discussion we refer to De Jaegher et al. (1998). In the next section we consider the gauging of invariances of the hypermultiplet action. Such invariances are related to isometries of the hyper-Kähler manifold. These isometries have been studied earlier in the literature (Sierra and Townsend (1984), Hull et al. (1986), Bagger et al. (1988), D'Auria et al. (1991), Andrianopoli et al. (1997)) but our purpose is to incorporate them into the set-up discussed in this section.

3 Hypermultiplets With Gauged Target-Space Isometries

The above Lagrangian and transformation rules are subject to two classes of equivalence transformations associated with the target space. One class consists of the target-space diffeomorphisms associated with $\phi \to \phi'(\phi)$. The other refers to reparametrizations of the fermion 'frame' of the form $\zeta^\alpha \to S^\alpha{}_\beta(\phi)\,\zeta^\beta$, and similar redefinitions of other quantities carrying indices α or $\bar\alpha$. For example, the fermionic metric transforms as

[2] Such as $\bar\gamma^{j}_{A\alpha}\, V^{\alpha}_{Bi} = \gamma_{Bi\bar\alpha}\,\bar V^{j\bar\alpha}_{A} = -\bar\gamma^{j}_{B\alpha}\, V^{\alpha}_{iA} + \delta^j_i\, g_{AB}$.

$G_{\bar{\alpha}\beta} \to [\bar{S}^{-1}]^{\bar{\gamma}}{}_{\bar{\alpha}}\,[S^{-1}]^{\delta}{}_{\beta}\,G_{\bar{\gamma}\delta}$. Under these rotations the quantities $\Gamma_A{}^{\alpha}{}_{\beta}$ play the role of connections.

The above transformations do not constitute invariances of the theory. This is only the case when the metric g_{AB} and and the Sp(n)×Sp(1) one-form V_i^{α} (and thus the related geometric quantities) are left invariant under (a subset of) them. To see how this works, let us consider the scalar fields transforming under a certain isometry (sub)group G characterized by a number of Killing vectors $k_I^A(\phi)$, with parameters θ^I. Hence under infinitesimal transformations,

$$\delta_{\rm G}\phi^A = g\,\theta^I k_I^A(\phi)\ , \tag{10}$$

where g is the coupling constant and the $k_I^A(\phi)$ satisfy the Killing equation

$$D_A k_{IB} + D_B k_{IA} = 0\,. \tag{11}$$

The quantities such as V_{Ai}^{α} that carry Sp(n) indices are only required to be invariant under isometries up to fermionic equivalence transformations. Thus $-g(k_I^B\,\partial_B V_{Ai}^{\alpha} + \partial_A k_I^B\, V_{Bi}^{\alpha})$ must be cancelled by a suitable infinitesimal rotation on the index α. Here we make the important assumption that the effect of the diffeomorphism is entirely compensated by a rotation that affects the indices α. In principle, one can also allow a compensating Sp(1) transformation acting on the indices $i, j, \ldots$. However, we will not do this here, as this would imply that the isometry group would neither commute with Sp(1) nor with supersymmetry. Instead we return to this option in the next section.

Let us parametrize the compensating transformation acting on the Sp(n) indices by $\delta_{\rm G}\zeta^{\alpha} = g[t_I - k_I^A\,\Gamma_A]^{\alpha}{}_{\beta}\,\zeta^{\beta}$, where the ($\phi$-dependent) matrices $t_I(\phi)$ remain to be determined,

$$-k_I^B\,\partial_B V_{A\,i}^{\alpha} - \partial_A k_I^B\, V_{B\,i}^{\alpha} + (t_I - k_I^B\,\Gamma_B)^{\alpha}{}_{\beta}\, V_{A\,i}^{\beta} = 0\,. \tag{12}$$

Obviously similar equations apply to the other geometric quantities, but as those are not independent we do not need to consider them.

Subsequently we derive the main consequences of the two equations (11) and (12). First of all the isometries must constitute an algebra with certain structure constants. This is expressed by

$$k_I^B\partial_B k_J^A - k_J^B\partial_B k_I^A = -f_{IJ}{}^K\, k_K^A\ , \tag{13}$$

where our definitions are such that the gauge fields that are needed once the θ^I become spacetime dependent, transform according to $\delta_{\rm G} W_\mu^I = \partial_\mu\theta^I - g f_{JK}{}^I\, W_\mu^J\,\theta^K$. The Killing equation implies the following property

$$D_A D_B k_{IC} = -R_{BCAD}\, k_I^D\,. \tag{14}$$

Then, using the covariant constancy of V_A, we find from (12),

$$(t_I)^{\alpha}{}_{\beta} = \tfrac{1}{2} V_{Ai}^{\alpha}\,\bar{\gamma}_{\beta}^{Bi}\, D_B k_I^A\,. \tag{15}$$

Target-space scalars satisfy algebraic identities, e.g.,

$$(\bar{t}_I)^{\bar{\gamma}}{}_{\bar{\alpha}}\, G_{\bar{\gamma}\beta} + (t_I)^{\gamma}{}_{\beta}\, G_{\bar{\alpha}\gamma} = (t_I)^{\bar{\gamma}}{}_{[\bar{\alpha}}\, \Omega_{\bar{\beta}]\bar{\gamma}} = 0\,, \tag{16}$$

which shows that the field-dependent matrices t_I take values in $sp(n)$. An explicit calculation, making use of the equations (11) and (14), shows that

$$D_A t_I{}^{\alpha}{}_{\beta} = k_I^B\, R_{AB}{}^{\alpha}{}_{\beta}\,, \tag{17}$$

for any infinitesimal isometry. From the group property of the isometries it follows that the matrices t_I satisfy the commutation relation

$$[\,t_I, t_J\,]^{\alpha}{}_{\beta} = f_{IJ}{}^{K}\,(t_K)^{\alpha}{}_{\beta} + k_I^A\, k_J^B\, R_{AB}{}^{\alpha}{}_{\beta}\,. \tag{18}$$

The apparent lack of closure represented by the presence of the infinitesimal Sp(n) holonomy transformation is related to the fact that the coordinates ϕ^A on which the matrices depend, transform under the action of the group. One can show that this result is consistent with the Jacobi identity.

Furthermore we derive from (12) that the complex structures J^{ij}_{AB} are invariant under the isometries,

$$k_I^C\, \partial_C J^{ij}_{AB} - 2\partial_{[A} k_I^C\, J^{ij}_{B]C} = 0\,. \tag{19}$$

This means that the isometries are *tri-holomorphic*. From (19) one shows that $\partial_A(J^{ij}_{BC}\, k_I^C) - \partial_B(J^{ij}_{AC}\, k_I^C) = 0$, so that, locally, one can associate three Killing potentials (or moment maps) P_I^{ij} to every Killing vector, according to

$$\partial_A P_I^{ij} = J^{ij}_{AB}\, k_I^B\,. \tag{20}$$

Observe that this condition determines the moment maps up to a constant. Up to constants one can also derive the equivariance condition,

$$J^{ij}_{AB}\, k_I^A k_J^B = -f_{IJ}{}^{K}\, P_K^{ij}\,, \tag{21}$$

which implies that the moment maps transform covariantly under the isometries,

$$\delta_{\rm G} P_I^{ij} = \theta^J\, k_J^A\, \partial_A P_I^{ij} = -f_{JI}{}^{K}\, P_K^{ij}\, \theta^J\,. \tag{22}$$

Summarizing, the invariance group of the isometries acts as follows,

$$\delta_{\rm G}\phi = g\,\theta^I\, k_I^A\,, \qquad \delta_{\rm G}\zeta^{\alpha} = g\,(\theta^I t_I)^{\alpha}{}_{\beta}\,\zeta^{\beta} - \delta_{\rm G}\phi^A\,\Gamma_A{}^{\alpha}{}_{\beta}\,\zeta^{\beta}\,. \tag{23}$$

When the parameters of these isometries become spacetime dependent we introduce corresponding gauge fields and fully covariant derivatives,

$$D_\mu\phi^A = \partial_\mu\phi^A - gW_\mu^I\, k_I^A\,, \quad D_\mu\zeta^{\alpha} = \partial_\mu\zeta^{\alpha} + \partial_\mu\phi^A\,\Gamma_A{}^{\alpha}{}_{\beta}\zeta^{\beta} - gW_\mu{}^{\alpha}{}_{\beta}\zeta^{\beta}\,, \tag{24}$$

where $W_\mu{}^{\alpha}{}_{\beta} = W_\mu^I\,(t_I)^{\alpha}{}_{\beta}$. The covariance of $D_\mu\zeta^{\alpha}$ depends crucially on (17) and (18). The gauge fields W_μ^I are accompanied by complex scalars X^I, spinors Ω_i^I and auxiliary fields Y_{ij}^I, constituting off-shell $N=2$ vector multiplets. For our notation of vector multiplets, the reader may consult De Jaegher et al. (1998).

The minimal coupling to the gauge fields requires extra terms in the supersymmetry transformation rules for the hypermultiplet spinors as well as in the Lagrangian, in order to regain $N=2$ supersymmetry. The extra terms in the transformation rules are

$$\delta'_Q\zeta^\alpha = 2gX^I k_I^A V_{Ai}^\alpha \varepsilon^{ij}\epsilon_j\,, \qquad \delta'_Q\zeta^{\bar\alpha} = 2g\bar X^I k_I^A \bar V_A^{\bar\alpha i}\,\varepsilon_{ij}\epsilon^j\,. \tag{25}$$

These terms can be conveniently derived by imposing the commutator of two supersymmetry transformations on the scalars, as this commutator should yield the correct field-dependent gauge transformation.

We distinguish three additional couplings to the Lagrangian. The first one is quadratic in the hypermultiplet spinors and reads

$$\mathcal{L}_g^{(1)} = g\bar X^I \bar\gamma_\alpha^{Ai}\epsilon_{ij}\bar\gamma_\beta^{Bj}\, D_B k_{AI}\,\bar\zeta^\alpha\zeta^\beta + \text{h.c.} = 2g\bar X^I t_I{}^\gamma{}_\alpha\,\Omega_{\beta\gamma}\,\bar\zeta^\alpha\zeta^\beta + \text{h.c.}\,. \tag{26}$$

The second one is proportional to the vector multiplet spinor Ω^I and takes the form

$$\mathcal{L}_g^{(2)} = -2gk_I^A V_{Ai}^\alpha \Omega_{\alpha\beta}\,\bar\zeta^\beta\Omega^{Ii} + \text{h.c.} = 2gk_I^A\bar\gamma_{A\alpha}^i\epsilon_{ij}\,\zeta^\alpha\Omega^{Ij} + \text{h.c.}\,. \tag{27}$$

Finally there is a potential given by

$$\mathcal{L}_g^{\text{scalar}} = -2g^2k_I^Ak_J^B\,g_{AB}\,X^I\bar X^J + g\,P_I^{ij}\,Y_{ij}^I\,, \tag{28}$$

where P_I^{ij} is the triplet of moment maps on the hyper-Kähler space. These terms were determined both from imposing the supersymmetry algebra and from the invariance of the action. To prove (28), one has to make use of the equivariance condition (21). Actually, gauge invariance, which is prerequisite to supersymmetry, already depends on (22).

4 Superconformally Invariant Hypermultiplets

In this last section we determine the restrictions from superconformal invariance on the hypermultiplets by evaluating some of the couplings to the fields of $N=2$ conformal supergravity. At this stage we have only a modest goal, namely to determine the restrictions on the hyper-Kähler geometry that arise from requiring invariance under rigid superconformal transformations. This is the situation that arises when freezing all the fields of conformal supergravity to zero in a flat spacetime metric. In that case the superconformal transformations acquire an explicit dependence on the spacetime coordinates.

We start by implementing the $N=2$ superconformal algebra (De Wit et al. (1980)) on the hypermultiplet fields. We assume that the scalars are invariant under special conformal and special supersymmetry transformations, but they transform under Q-supersymmetry and under the additional bosonic symmetries of the superconformal algebra, namely chiral $[\mathrm{SU}(2)\times\mathrm{U}(1)]_{\mathrm{R}}$ and dilatations denoted by D. At this point we do not assume that these transformations are symmetries of the action and we simply parametrize them as follows,

$$\delta\phi^A = \theta_{\mathrm{D}}\, k_{\mathrm{D}}^A(\phi) + \theta_{\mathrm{U(1)}}\, k_{\mathrm{U(1)}}^A(\phi) + (\theta_{\mathrm{SU(2)}})^i{}_k\, \varepsilon^{jk} k_{ij}^A(\phi)\,, \tag{29}$$

where the k^A are left arbitrary. Note that $k_{ij}^A(\phi)$ is assigned to the same symmetric pseudoreal representation of SU(2) as the complex structures, while $\theta_{\mathrm{SU(2)}}$ is antihermitean and traceless.

An important difference with the situation described in the previous section, is that in the conformal superalgebra the dilatations and chiral transformations do not appear in the commutator of two Q-supersymmetries, but in the commutator of a Q- and an S-supersymmetry. To evaluate the S-supersymmetry variation of the fermions, we use that $\delta_{\mathrm{S}}\phi^A = \delta_{\mathrm{K}}\zeta^\alpha = 0$ and covariantize the derivative in the fermionic transformations with respect to dilatations. Subsequently we impose the commutator, $[\,\delta_{\mathrm{K}}(\Lambda_{\mathrm{K}}), \delta_{\mathrm{Q}}(\epsilon)\,] = -\delta_{\mathrm{S}}(\not{\Lambda}_{\mathrm{K}}\epsilon)$ on the spinors. This expresses the S-supersymmetry variations in terms of k_{D}^A,

$$\delta_{\mathrm{S}}(\eta)\,\zeta^\alpha = V_{iA}^\alpha\, k_{\mathrm{D}}^A\, \eta^i\,, \qquad \delta_{\mathrm{S}}(\eta)\,\zeta^{\bar\alpha} = \bar V_A^{i\bar\alpha}\, k_{\mathrm{D}}^A\, \eta_i\,. \tag{30}$$

With this result we first evaluate the commutator of an S- and a Q-supersym- metry transformation on the scalars. This yields

$$[\,\delta_{\mathrm{S}}(\eta), \delta_{\mathrm{Q}}(\epsilon)\,]\,\phi^A = (\bar\epsilon^i\eta_i + \bar\epsilon_i\eta^i)\, k_{\mathrm{D}}^A + 2J_{ik}{}^A{}_B\,\varepsilon^{kj}\,(\bar\epsilon^i\eta_j - \bar\epsilon_j\eta^i)\, k_{\mathrm{D}}^B\,. \tag{31}$$

This result can be confronted with the universal result from $N=2$ conformal supergravity, which reads

$$\begin{aligned}[\,\delta_{\mathrm{S}}(\eta), \delta_{\mathrm{Q}}(\epsilon)\,] = {}& \delta_{\mathrm{M}}(2\bar\eta^i\sigma^{ab}\epsilon_i + \mathrm{h.c.}) + \delta_{\mathrm{D}}(\bar\eta_i\epsilon^i + \mathrm{h.c.}) \\ & +\delta_{\mathrm{U(1)}}(i\bar\eta_i\epsilon^i + \mathrm{h.c.}) + \delta_{\mathrm{SU(2)}}(-2\bar\eta^i\epsilon_j - \mathrm{h.c.}\,;\ \text{traceless})\,.\end{aligned} \tag{32}$$

Comparison thus shows that $k_{\mathrm{U(1)}}^A$ vanishes and that the SU(2) vectors satisfy

$$k_{ij}^A = J_{ij}{}^A{}_B\, k_{\mathrm{D}}^B\,. \tag{33}$$

Now we proceed to impose the same commutator on the fermions, where on the right-hand side we find a Lorentz transformation, a U(1) transformation and a dilatation, iff we assume the following condition on k_{D}^A,

$$D_A k_{\mathrm{D}}^B = \delta_A^B\,. \tag{34}$$

This condition suffices to show that the kinetic term of the scalars is scale invariant, provided one includes a spacetime metric or, in flat spacetime, include corresponding scale transformations of the spacetime coordinates. Nevertheless, observe that k_{D}^A is *not* a Killing vector of the hyper-Kähler space, but a special example of a conformal homothetic Killing vector (we thank G. Gibbons for an illuminating discussion regarding such vectors). An immediate consequence of (34) is that $k_{\rm D}^A$ can (locally) be expressed in terms of a potential $\chi_{\rm D}$, according to $k_{{\rm D}\,A} = \partial_A \chi_{\rm D}$. Another consequence is that the SU(2) vectors k_{ij}^A, as expressed by (33), are themselves Killing vectors, because their derivative is proportional to the corresponding antisymmetric complex structure,

$$D_A k_B^{ij} = -J_{AB}^{ij} \,. \qquad (35)$$

Therefore, the bosonic action is also invariant under these SU(2) transformations.

From the $[\delta_{\rm S}, \delta_{\rm Q}]$ commutator we also find the fermionic transformation rules under the chiral transformations and the dilatations,

$$\begin{aligned} &\delta_{\rm SU(2)}\, \zeta^\alpha + \delta_{\rm SU(2)}\phi^A\, \Gamma_A{}^\alpha{}_\beta \zeta^\beta = 0\,, \\ &\delta_{\rm U(1)}\zeta^\alpha + \delta_{\rm U(1)}\phi^A\, \Gamma_A{}^\alpha{}_\beta \zeta^\beta = -\tfrac{1}{2} i\, \theta_{\rm U(1)}\zeta^\alpha\,, \qquad \delta_{\rm D}\zeta^\alpha + \delta_{\rm D}\phi^A\, \Gamma_A{}^\alpha{}_\beta \zeta^\beta = \tfrac{3}{2}\, \theta_{\rm D}\zeta^\alpha\,. \end{aligned} \qquad (36)$$

Note that the U(1) transformation further simplifies because $\delta_{\rm U(1)}\phi^A = 0$.

To establish that the model as a whole is now invariant under the superconformal transformations it remains to be shown that the tensor V_{Ai}^α is invariant under the diffeomorphisms generated by k_{ij}^A, $k_{\rm U(1)}^A$ and $k_{\rm D}^A$ up to compensating transformations that act on the Sp(n)×Sp(1) indices in accordance with the transformations of the ζ^α given above and the symmetry assignments of the supersymmetry parameters ϵ^i. To emphasize the systematics we ignore the fact that $k_{\rm U(1)}^A$ actually vanishes and we write

$$\begin{aligned} -k_{kl}^B\, \partial_B V_{Ai}^\alpha - \partial_A k_{kl}^B\, V_{Bi}^\alpha - k_{kl}^B\, \Gamma_B{}^\alpha{}_\beta V_{Ai}^\beta + [-\delta_{(k}^j \varepsilon_{l)i}]\, V_{Aj}^\alpha &= 0\,, \\ -k_{\rm U(1)}^B \partial_B V_{Ai}^\alpha - \partial_A k_{\rm U(1)}^B\, V_{Bi}^\alpha + [-\tfrac{1}{2} i\, \delta_\beta^\alpha - k_{\rm U(1)}^B\, \Gamma_B{}^\alpha{}_\beta] V_{Ai}^\beta + [\tfrac{1}{2} i\, \delta_i^j]\, V_{Aj}^\alpha &= 0\,, \\ -k_{\rm D}^B\, \partial_B V_{Ai}^\alpha - \partial_A k_{\rm D}^B\, V_{Bi}^\alpha + [\tfrac{3}{2}\delta_\beta^\alpha - k_{\rm D}^B\, \Gamma_B{}^\alpha{}_\beta]\, V_{Ai}^\beta + [-\tfrac{1}{2}\, \delta_i^j]\, V_{Aj}^\alpha &= 0 \end{aligned} \qquad (37)$$

In these equations the first two terms on the left-hand side represent the effect of the isometry, the third terms represent a uniform scale and chiral U(1) transformation on the indices associated with the Sp(n) tangent space, and the last terms represent an SU(2), a U(1) and a scale transformation, respectively, on the indices associated with the Sp(1) target space. Eq. (37) should be regarded as a direct extension of (12).

We close with a few comments. First of all, the SU(2) isometries induce a rotation on the complex structures,

$$k_{kl}^C\, \partial_C J_{AB}^{ij} - 2\partial_{[A} k_{kl}^C\, J_{B]C}^{ij} = -2 J_{klC[A}\, J_{B]}^{ijC} = 2\delta_{(k}^{(i}\, \varepsilon_{l)m}\, J_{AB}^{j)m}\,, \qquad (38)$$

as should be expected. Secondly, one can verify that the isometries discussed in section 3 commute with the scale and chiral transformations, provided that $k_I^A = k_{\rm D}^B\, D_B k_I^A$. This condition is also required for the scale invariance of the full action. Observe also that $k_{\rm D}^A$ satisfies (14), in spite of the fact that it is not a Killing vector. Finally, it is straightforward to write down actions for the vector multiplets that are invariant under rigid $N = 2$ superconformal transformations. Those are based on a holomorphic function that is homogeneous of degree two.

References

L. Andrianopoli, M. Bertolini, A. Ceresole, R. D'Auria, S. Ferrara, P. Fré and T. Magri, J. Geom. Phys. **23** (1997) 111 (hep-th/9605032).

J. Bagger and E. Witten, Nucl. Phys. **B222** (1983) 1.

J.A. Bagger, A.S. Galperin, E.A. Ivanov and V.I. Ogievetsky, Nucl. Phys. **B303** (1988) 522.

R. D'Auria, S. Ferrara and P. Fré, Nucl. Phys. **B359** (1991) 705.

J. De Jaegher, B. de Wit, B. Kleijn and S. Vandoren, Nucl. Phys. **B514** (1998) 553 (hep-th/9707262).

B. de Wit, J.W. van Holten and A. Van Proeyen, Nucl. Phys. **B167** (1980) 186.

C.M. Hull, A. Karlhede, U. Lindström and M. Roček, Nucl. Phys. **B266** (1986) 1.

J. Maldacena, *The large-N limit of superconformal field theories and supergravity* (hep-th/9711200).

G. Sierra and P.K. Townsend, Nucl. Phys. **B233** (1984) 289.

Ectoplasm Has No Topology: The Prelude

S. James Gates, Jr.[1]

Department of Physics
University of Maryland at College Park
College Park, MD 20742-4111, USA
E-mail: gates@umdhep.umd.edu

Abstract. Preliminary evidence is presented that a long overlooked and critical element in the fundamental definition of a general theory of integration over curved Wess-Zumino superspace lies with the imposition of "the Ethereal Conjecture" which states the necessity of the superspace to be topologically "close" to its purely bosonic sub-manifold. As a step in proving this, a new theory of integration based on closed super p-forms is proposed. [1, 2]

Presently 'Salam-Strathdee superspace (Salam and Strathdee (1978))' is almost universally accepted as the requisite mathematical setting for describing supersymmetrical field theories. Even so, there remain a fairly large number of open questions about superspace, particularly with regard to those with large values of N ($\equiv N_F$) or D ($\equiv N_B$). There are also particularly pointed questions that remain largely unanswer about a general theory of integration on the curved versions of these spaces also known as 'Wess-Zumino superspace (Wess and Zumino (1977)).' To answer some of these questions, extensions such as 'harmonic superspace' have been developed especially by the late Dr. Ogievetsky and collaborators. Although my discussion today will only tangentially touch on such constructions, I wish to dedicate this talk to Victor Isaakovich's memory.

Near the beginning of research using superspace, more mathematically motivated investigators such as Rogers (Rogers (1980,1981,1985)) asked a question we may paraphrase as,

> "Is it possible to construct a superspace whose topological properties are significantly different from those of its purely bosonic subspace?"

In all cases of interest to physicists to date the answer appears to be, "No!" The emphasis on this negation is mine own because I believe that there is a hidden message in this answer.

In establishing a nomenclature appropriate to researching these issues, one often finds the 'spiritualist' denotations (see for example (De Witt (1992)))

monomials in $x \equiv$ body of the superspace,
monomials in x and θ or purely $\theta \equiv$ soul of the superspace.

In deference to this convention, I may call the 'basic substance' of which the soul is composed, the "ectoplasm" of superspace.

---

[1] Research supported by NSF grant # PHY-96-43219
[2] Supported in part by NATO Grant CRG-93-0789

There is a peculiar sense in which the question of how to construct integration measures over curved superspaces is unanswered. Arnowitt, Nath and Zumino (Arnowitt, Zumino and Nath (1975)) first suggested such integration measures should be written as

$$\int d\mu \equiv \int d^{N_B+N_F} z \, \mathrm{E}^{-1} = \int d^{N_B+N_F} z \, [sdet \, (\mathrm{E}_{\underline{A}}{}^{\underline{M}}(\theta, x) \,)]^{-1} \quad . \tag{1}$$

for a superspace of N_B bosonic coordinate and N_F fermionic coordinates.

In principle this is perfectly consistent. In practice, however, for any theory with large N_B or N_F ($N_F = 4$ is large), this becomes an impractical way to obtain component results in a supergravity theory of 'physical' interest. The impracticality arises because the complete θ-expansion of the superdeterminant of the inverse vielbein $[sdet \, (\mathrm{E}_{\underline{A}}{}^{\underline{M}}(\theta, x) \,)]^{-1}$ is complicated to calculate[3]. For practical calculations an alternative to the method of Arnowitt, Nath and Zumino is required. To my knowledge, only two such alternatives exist in the literature. They have been discussed in three books listed by authors below.

a. "Covariant Theta Expansion" - Wess & Bagger, (Wess and Bagger (1983))
b. "Density Projectors" - Gates, Grisaru, Roček & Siegel, (Gates, Grisaru, Roček and Siegel (1983a))
- Buchbinder & Kuzenko. (Buchbinder and Kuzenko (1995a))

I will obviously speak on the second of these because I have recently found increasing and unexpected indications that it is directly connected to more general issues of the calculus and topology of curved supermanifolds with torsion.

I begin by writing the "Ectoplasmic Integration Theorem" (or E.I.T.). There should exist an operator $\mathcal{D}^{N_F}$ such that

$$\int d^{N_B+N_F} z \, \mathrm{E}^{-1} \mathcal{L} = \int d^{N_B} z \, \mathrm{e}^{-1} [\, \mathcal{D}^{N_F} \mathcal{L} | \,] \quad , \tag{2}$$

independent of the superfield $\mathcal{L}$ that appears in this equation and where

$$\mathrm{e}^{-1} \equiv [det \, (\mathrm{e}_{\underline{a}}{}^{\underline{m}}(x) \,)]^{-1} \quad , \quad \mathcal{D}^{N_F} \mathcal{L} | \equiv \lim_{\theta \to 0} \, (\mathcal{D}^{N_F} \mathcal{L}) \quad . \tag{3}$$

This theorem is of a similar form to that of the standard Gauss', Green's or Stoke's Theorems of multi-variable calculus. It is different, however, because the operator $\mathcal{D}^{N_F}$ appears on the "wrong" side of the equation from the standard multi-variable calculus analogs. The E.I.T. is also the natural extension

[3] To my knowledge, this calculation has only been done explicitly by no more than six physicists to this date for relatively simple case of 4D, N = 1 supergravity.

of the Berezinian definition of integrating over Grassmann numbers (Berezin (1966)).

To see why this is a practical improvement in calculational matters, let me consider the case of flat 4D, N = 1 superspace where the E.I.T. becomes

$$\int d^4x\, d^2\theta\, d^2\bar{\theta}\, \mathcal{L} \equiv \tfrac{1}{2}\Big\{ \int d^4x\, [\, D^2\, \overline{D}^2\, \mathcal{L}\,|\,] + \text{h.c.} \Big\} \quad , \tag{4}$$

where

$$D_\alpha \equiv \partial_\alpha + i\frac{1}{2}\bar{\theta}^{\dot{\alpha}}\partial_{\underline{a}} \quad , \quad \overline{D}_{\dot{\alpha}} \equiv \bar{\partial}_{\dot{\alpha}} + i\frac{1}{2}\theta^{\alpha}\partial_{\underline{a}} \quad . \tag{5}$$

Anyone familiar with rigid supersymmetry can attest to the practical utility of the above equation. For example, if I define $\mathcal{L} \equiv \overline{\Phi}\Phi$ where $\overline{D}_{\dot{\alpha}}\Phi = 0$ use the component field definitions $A(x) \equiv \Phi|$, $\psi_\alpha(x) \equiv D_\alpha\Phi|$ and $F(x) \equiv D^2\Phi|$, apply the E.I.T. and use of the Leibnitz rule for differentiation, it is simple to show

$$\int d^4x\, d^2\theta\, d^2\bar{\theta}\, \overline{\Phi}\Phi = \int d^4\, x\, [\, -\frac{1}{2}(\partial^{\underline{a}}\overline{A})\,(\partial_{\underline{a}}A) - i\bar{\psi}^{\dot{\alpha}}\partial_{\underline{a}}\psi^{\alpha} + F\overline{F}\,] \ . \tag{6}$$

No explicit θ-expansion was required at any point to derive this component result. Thus, it should be obvious why it is calculationally superior to use the E.I.T. By using techniques that are essentially the same as above, we simple by-pass the need to know the explicit structure of the θ-expansion of $[sdet\, (\mathrm{E}_{\underline{A}}{}^{\underline{M}}(\theta, x)\,)]^{-1}$!

From this viewpoint, the whole problem becomes how to develop a theory for the calculation of the operator $\mathcal{D}^{N_F}$ that appears in equation (2). The expression $e^{-1}[\mathcal{D}^{N_F}\mathcal{L}|\,]$ is called "the density projection operator" or "density projector" (see '*Superspace*' (Gates, Grisaru, Roček and Siegel (1983b)) or '*Ideas*' (Buchbinder and Kuzenko (1995b))). It should be clear that this operator, in the general case, can be written as

$$\int d^{N_B}z\, \mathrm{e}^{-1}\left[\, \mathcal{D}^{N_F}\mathcal{L}\,|\,\right] = \int d^{N_B}z\, \mathrm{e}^{-1}\left[\, \sum_{i=0}^{N_F} c_{(N_F-i)}\, (\nabla \cdots \nabla)^{N_F-i}\, \mathcal{L}|\,\right] \quad , \tag{7}$$

in terms of some field-dependent coefficients $c_{(N_F-i)}$ and powers of the spinorial superspace supergravity covariant derivative ∇_α. How are these coefficients to be found?

In '*Superspace*' (Gates, Grisaru, Roček and Siegel (1983b)) it was shown that given the local supersymmetry variations of some matter superfield, it is possible to re-construct these coefficients. In '*Ideas*' (Buchbinder and Kuzenko (1995b)), it was shown that the density projector follows after solving the constraints to find the basic supergravity pre-potentials. Neither of these approaches is a theory[4] for $\mathcal{D}^{N_F}$. In the early to middle eighties, Zumino was

[4] We may think of the approach in (Gates, Grisaru, Roček and Siegel (1983b)) as a 'handicraft' method for summarizing component results. The fact that it was required to go component at all, was equivalent to an admission that we did not have an *a priori* theoretical basis for this result.

the first to raise the question of a purely *theoretical* basis for this operator. This bring us to the point of my presentation.

In the rest of my presentation, I will attempt to convince the reader that the answer can be found in the study of super topology similar to the investigations by Rogers. I will argue that local supergravity theories (as a principle) obey what I call "The Ethereal Conjecture" which largely determines the form of $\mathcal{D}^{N_F}$.

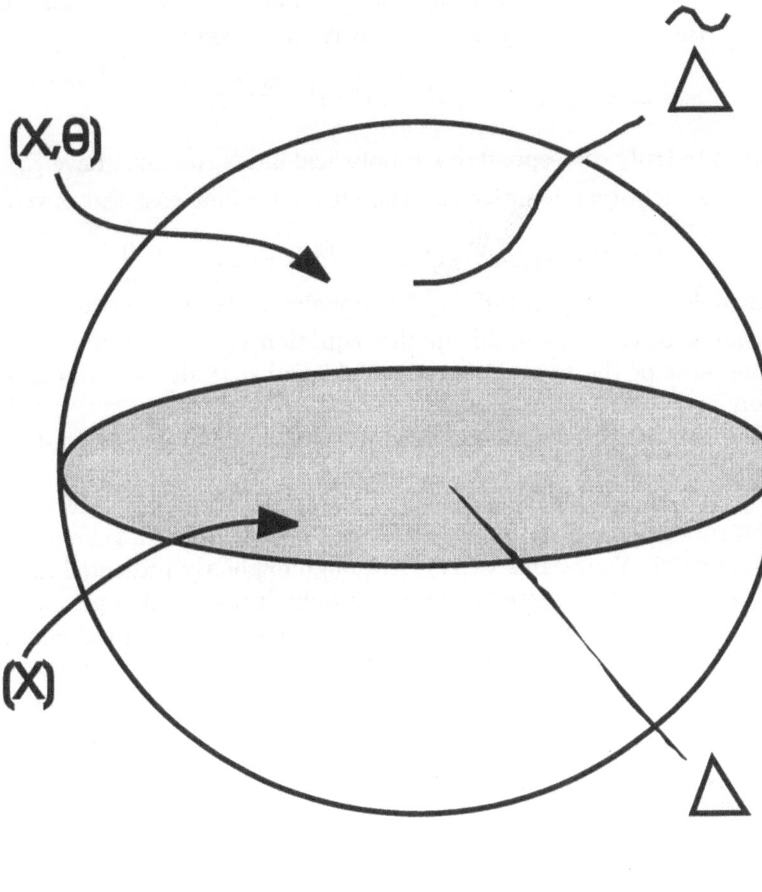

Fig. 1. A Representation of Superspace
The volume of the sphere represents the entirety of superspace and the equatorial plane represents the bosonic sub-space.

In terms of the topological indices represented in the diagram I may formulate the Ethereal Conjecture (E. C.) as:

For all Wess-Zumino superspaces, the operator $\mathcal{D}^{N_F}$ that appears in the E. I.T. has the property that it insures that $\widehat{\Delta} \simeq \Delta$.

I will later give some meaning to $\simeq$ that appears in this relation. This is the result to which I was alluding in the other-worldly sounding title. Stated another way, this relation says that the local integration measure of a Wess-Zumino superspace is such that the superspace is topologically "close" to its underlying bosonic manifold. Or alternately, the local Grassmann integration measure of a Wess-Zumino superspace is actually to be detemined from topological considerations.

For an ordinary N_B-form $f_{\underline{a}_1 \cdots \underline{a}_{N_B}}$ defined over an ordinary bosonic manifold, the calculation of the index Δ is just given by

$$\Delta \equiv (N_B!)^{-1} \int d^{N_B} z \, e^{-1} \, \epsilon^{\underline{a}_1 \cdots \underline{a}_{N_B}} f_{\underline{a}_1 \cdots \underline{a}_{N_B}} \quad , \tag{8}$$

and for Δ to truly correspond to a topological index, we must have $f_{\underline{a}_1 \cdots \underline{a}_{N_B}}$ satisfy a Bianchi-type identity (for the purely bosonic case this is trivial)

$$e_{[\underline{a}_1|} f_{|\underline{a}_2 \cdots \underline{a}_{1+N_B}]} - c_{[\underline{a}_1 \, \underline{a}_2|}{}^{\underline{d}} f_{\underline{d}|\underline{a}_3 \cdots \underline{a}_{1+N_B}]} = 0 \quad , \tag{9}$$

and as well $f_{\underline{a}_1 \cdots \underline{a}_{N_B}}$ must <u>not</u> be globally defined over the entire N_B-dimensional bosonic manifold. In this equation $e_{\underline{a}}$ denotes the local frame field operator of the bosonic sub-manifold and $c_{\underline{a}\underline{b}}{}^{\underline{c}}$ denotes the associated anholonomy.

What are the corresponding structures available over a Salam-Strathdee superspace?

In 1981 (Gates (1981)), I proposed the initial formulation of <u>irreducible</u> off-shell super p-forms for Salam-Strathdee superspace as a generalization of supersymmetric Yang-Mills theory. Although explicitly presented in a 4D, N = 1 superspace, the general structure is ubiquitous to all superspaces. The distinctive feature of the 1981 proposal was that it showed the constraints, Bianchi identities, and pre-potential solutions all exist for a simplex of super p-forms in exactly the same way as in supersymmetric Yang-Mills theory.

In 1983 (Gates (1983)), I was able to derive a further interesting result. If a super N_B-form like $F_{\underline{A}_1 \cdots \underline{A}_{N_B}}$ satisfies a set of Bianchi identities (i.e. it is super-closed), it follows that <u>independent</u> of its constraints and in the presence of supergravity, there exist a W-Z gauge where

$$\begin{aligned}\left(F_{\underline{a}_1 \cdots \underline{a}_{N_B}}|\right) = \Big[\, & \widetilde{f}_{\underline{a}_1 \cdots \underline{a}_{N_B}} + \lambda^{(N_B,1)} \psi_{[\underline{a}_1|}{}^{\alpha_1} \left(F_{\alpha_1 |\underline{a}_2 \cdots \underline{a}_{N_B}]}| \right) \\ & + \lambda^{(N_B,2)} \psi_{[\underline{a}_1|}{}^{\alpha_1} \psi_{|\underline{a}_2|}{}^{\alpha_2} \left(F_{\alpha_1 \alpha_2 |\underline{a}_3 \cdots \underline{a}_{N_B}]}| \right) \cdots \\ & + \lambda^{(N_B,N_B)} [\psi_{\underline{a}_1}{}^{\alpha_1} \psi_{\underline{a}_2}{}^{\alpha_2} \cdots \psi_{\underline{a}_{N_B}}{}^{\alpha_{N_B}}] \left(F_{\alpha_1 \alpha_2 \cdots \alpha_{N_B}}| \right) \Big] \quad ,\end{aligned} \tag{10}$$

here $\widetilde{f}_{\underline{a}_1 \cdots \underline{a}_{N_B}}$ is an ordinary bosonic closed N_B-form, $\psi_{\underline{a}}{}^{\alpha}$ denotes the component gravitino field and $\lambda^{(N_B,i)}$ are a set of constants that are easily derivable.

My original derivation of this was in the context of 4D, N = 4 supergravity but that derivation can easily be extended to all values of N_B and N_F.

Now the interesting thing about this equation is that I can isolate $\widetilde{f}_{\underline{a}_1\cdots\underline{a}_{N_B}}$ (which can differ from $f_{\underline{a}_1\cdots\underline{a}_{N_B}}$ by exact terms) to find

$$\begin{aligned}\widetilde{f}_{\underline{a}_1\cdots\underline{a}_{N_B}} = \Big[\Big(F_{\underline{a}_1\cdots\underline{a}_{N_B}} | \Big) - \lambda^{(N_B,1)} \psi_{[\underline{a}_1|}{}^{\alpha_1} \Big(F_{\alpha_1|\underline{a}_2\cdots\underline{a}_{N_B}]} | \Big) \\ - \lambda^{(N_B,2)} \psi_{[\underline{a}_1|}{}^{\alpha_1} \psi_{|\underline{a}_2|}{}^{\alpha_2} \Big(F_{\alpha_1\alpha_2|\underline{a}_3\cdots\underline{a}_{N_B}]} | \Big) \cdots \\ - \lambda^{(N_B,N_B)} [\psi_{\underline{a}_1}{}^{\alpha_1} \psi_{\underline{a}_2}{}^{\alpha_2} \cdots \psi_{\underline{a}_{N_B}}{}^{\alpha_{N_B}}] \Big(F_{\alpha_1\alpha_2\cdots\alpha_{N_B}} | \Big) \Big] \ .\end{aligned} \tag{11}$$

Upon multiplying by an ϵ-tensor and integrating $(1/N_B!) \int d^{N_B} z \, \mathrm{e}^{-1}$, I find

$$\widetilde{\Delta} = \widehat{\Delta} \ , \tag{12}$$

$$\widetilde{\Delta} \equiv (N_B!)^{-1} \int d^{N_B} z \, \mathrm{e}^{-1} \, \epsilon^{\underline{a}_1\cdots\underline{a}_{N_B}} \, \widetilde{f}_{\underline{a}_1\cdots\underline{a}_{N_B}} \ , \tag{13}$$

$$\begin{aligned}\widehat{\Delta} \equiv \int d^{N_B} z \, \mathrm{e}^{-1} \, \epsilon^{\underline{a}_1\cdots\underline{a}_{N_B}} \\ \Big[(N_B!)^{-1} \Big(F_{\underline{a}_1\cdots\underline{a}_{N_B}} | \Big) - \lambda^{(N_B,1)} \psi_{\underline{a}_1}{}^{\alpha_1} \Big(F_{\alpha_1\underline{a}_2\cdots\underline{a}_{N_B}} | \Big) \\ - \lambda^{(N_B,2)} \psi_{\underline{a}_1}{}^{\alpha_1} \psi_{\underline{a}_2}{}^{\alpha_2} \Big(F_{\alpha_1\alpha_2\underline{a}_3\cdots\underline{a}_{N_B}} | \Big) \cdots \\ - \lambda^{(N_B,N_B)} (N_B!)^{-1} [\psi_{\underline{a}_1}{}^{\alpha_1} \psi_{\underline{a}_2}{}^{\alpha_2} \cdots \psi_{\underline{a}_{N_B}}{}^{\alpha_{N_B}}] \Big(F_{\alpha_1\alpha_2\cdots\alpha_{N_B}} | \Big) \Big].\end{aligned} \tag{14}$$

I now define the supertopological index $\widehat{\Delta}$ that was introduced into the diagram by asserting that equation (14) *is* the correct definition of how to integrate the closed super N_B-form $F_{\underline{A}_1\cdots\underline{A}_{N_B}}$ over the entirety of the superspace[5]! Since $\widetilde{f}_{\underline{a}_1 \cdots \underline{a}_{N_B}}$ typically differs from $f_{\underline{a}_1 \cdots \underline{a}_{N_B}}$ by exact terms we have

$$\widehat{\Delta} = \Delta + \cdots \ . \tag{15}$$

So the definition above certainly enforces the Ethereal Conjecture but how does this solve the problem of finding $\mathcal{D}^{N_F}$?

The answer lies in the fact that the field strengths superfields in $\widehat{\Delta}$ (i.e. the F's) must be chosen to be subject to the constaints implied by irreducibility of the super N_B-form. In this case a number of the F's vanish and the remaining ones, via the solution of their Bianchi identities, are related by ∇_α, the spinorial derivative. When this solution for the various components of F

[5] For a previous proposal to define the integration theory of closed super p-forms, see the work of ref. (Zupnik (1991)).

is inserted into $\widehat{\Delta}$, as if by magic the operator $\mathcal{D}^{N_F}$ appears in all the cases I have studied. Let me show by some explicit examples how this topological tool works.

The simplest of all 2D supergravity theories is (1,0) or heterotic supergravity (Brooks, Gates, and Muhammad (1986), Evans and Ovrut (1986), Nelson and Moore (1986)) which is described by a set of covariant derivatives (∇_+, $\nabla_=$, $\nabla_{\ddagger}$) satisfying the commutator algebra and single differential equation below

$$[\,\nabla_+\,,\nabla_+\,\} = i2\nabla_{\ddagger}\ ,\ \ [\,\nabla_+\,,\nabla_{\ddagger}\} = 0\ ,\ \ \nabla_+\Sigma^+ = \tfrac{1}{2}\mathcal{R}\ ,$$
$$[\,\nabla_+\,,\nabla_=\,\} = -i2\Sigma^+\mathcal{M}\ ,\ \ [\,\nabla_{\ddagger},\nabla_=\,\} = -(\,\Sigma^+\nabla_+ + \mathcal{R}\mathcal{M})\ . \qquad (16)$$

The quantities Σ^+ and $\mathcal{R}$ are field strength superfields and $\mathcal{M}$ denotes the generator of the 2D Lorentz group defined to act according to the rules; $[\mathcal{M},\psi_+] = \frac{1}{2}\psi_+$, $[\mathcal{M},\psi_-] = -\frac{1}{2}\psi_-$, $[\mathcal{M},e_{\ddagger}] = e_{\ddagger}$ and $[\mathcal{M},e_=] = -e_=$. On defining $\Sigma^+\,|$ as the limit of Σ^+ as the Grassmann coordinate is taken to zero and similarly for $\mathcal{R}\,|$, we find

$$\Sigma^+\,| = -\psi_{\ddagger,\,=}{}^+ = -[\,e_{\ddagger}\psi_={}^+ - e_=\psi_{\ddagger}{}^+ - c_{\ddagger,\,=}{}^{\ddagger}\psi_{\ddagger}{}^+ - c_{\ddagger,\,=}{}^{=}\psi_={}^+\,]\ , \qquad (17)$$

$$r_{\ddagger,\,=}(\omega) = -[\,e_{\ddagger}\omega_= - e_=\omega_{\ddagger} - c_{\ddagger,\,=}{}^{\ddagger}\omega_{\ddagger} - c_{\ddagger,\,=}{}^{=}\omega_=\,]\ ,$$
$$\nabla_+\Sigma^+\,| = -\tfrac{1}{2}[\,r_{\ddagger,\,=}(\omega) + i2\psi_{\ddagger}{}^+\psi_{\ddagger,\,=}{}^+\,]\ , \qquad (18)$$

$$\widehat{\nabla}_{\ddagger} \equiv e_{\ddagger} + \omega_{\ddagger}\mathcal{M}\ ,\qquad \widehat{\nabla}_= \equiv e_= + \omega_=\mathcal{M}\ ,\qquad \omega_{\ddagger} = c_{\ddagger,\,=}{}^{=}\ ,$$
$$\omega_= = c_{\ddagger,\,=}{}^{\ddagger} + i2\psi_{\ddagger}{}^+\psi_={}^+\ ,\ \ e_a \equiv e_a{}^m\partial_m\ ,\ \ [e_a,e_b] = c_{a,b}{}^c e_c\ . \qquad (19)$$

But $\mathcal{R}$ is the vector-vector component of the super 2-form R_{AB}. Thus, we take the first equality in (17) and use it to replace the $\psi_{\ddagger,\,=}{}^+$ term on the last line of (18) to find,

$$-\tfrac{1}{2}\,r_{\ddagger,\,=}(\omega) = \Big[\Big(\nabla_+ - i\psi_{\ddagger}{}^+\Big)\Sigma^+\,|\Big]\ . \qquad (20)$$

This is a special case of (11) and following the general discussion we enforce the E.C. by defining

$$\widetilde{\Delta} \equiv -\tfrac{1}{2}\int d^2\sigma\, e^{-1}\, r_{\ddagger,\,=}(\omega(e,\psi))\ , \qquad (21)$$

$$\widehat{\Delta} \equiv \int d^2\sigma\, e^{-1}\Big[\Big(\nabla_+ - i\psi_{\ddagger}{}^+\Big)\Sigma^+\,|\Big]\ . \qquad (22)$$

and according to the E.I.T. and E.C. it must also be the case that

$$\begin{aligned}\int d^2\sigma\, d\zeta^-\, E^{-1}\mathcal{L}_- &\equiv \int d^2\sigma\, e^{-1}\Big[\mathcal{D}_+\mathcal{L}_-\,|\Big] \\ &= \int d^2\sigma\, e^{-1}\Big[\Big(\nabla_+ - i\psi_{\ddagger}{}^+\Big)\mathcal{L}_-\,|\Big]\ ,\end{aligned} \qquad (23)$$

exactly as stated in the first work of reference (Brooks, Gates, and Muhammad (1986), Evans and Ovrut (1986), Nelson and Moore (1986)).

Now the expression for $\widetilde{\Delta}$ in (21) allows us to calculate the form of the terms in $\widetilde{\Delta} = \Delta + ...$ This is done by observing that

$$r_{\ddagger, =}(\omega(e,\psi)) = r_{\ddagger, =}(\omega(e,0)) + i2\{\nabla_{\ddagger}(e)[\psi_{\ddagger}{}^{+}\psi_{=}{}^{+}]\} \quad , \tag{24}$$

so that

$$\widetilde{\Delta} \equiv -\tfrac{1}{2}\int d^2\sigma\, e^{-1}\, r_{\ddagger, =}(\omega(e,0)) - i\int d^2\sigma\, e^{-1}\{\partial_m[e_{\ddagger}{}^{m}(\psi_{\ddagger}{}^{+}\psi_{=}{}^{+})]\} \quad . \tag{25}$$

We see that the first term above is $\Delta = 2\pi(g-1)$ (where g is the genus of the manifold), the usual topological index on a 2-manifold,

$$\Delta \equiv -\tfrac{1}{2}\int d^2\sigma\, e^{-1}\, r_{\ddagger, =}(\omega(e,0)) \quad , \tag{26}$$

and the second term in (25) is what was indicated by ... in the E.C.

Perhaps the reader was not impressed by the (1,0) example. So let's repeat all of this in the more complicated case of 3D, N = 1 superspace. In 1979 (Brown and Gates (1979)) the superspace description of 3D, N = 1 irreducible off-shell supergravity was first given

$$\begin{aligned}
[\nabla_\alpha\,,\,\nabla_\beta\} &= i2(\gamma^c)_{\alpha\beta}[\nabla_c - R\mathcal{M}_c] \quad ,\\
[\nabla_\alpha\,,\,\nabla_b\} &= i(\gamma_b)_\alpha{}^\delta[\,\tfrac{1}{2}R\nabla_\delta + (\Sigma_\delta{}^d + i\tfrac{2}{3}(\gamma^d)_\delta{}^\epsilon(\nabla_\epsilon R))\mathcal{M}_d\,]\\
&\quad + (\nabla_\alpha R)\mathcal{M}_b \quad ,\\
[\nabla_a\,,\,\nabla_b\} &= -\tfrac{1}{2}\epsilon_{abc}[\,\Sigma^{\alpha c} + i\tfrac{2}{3}(\gamma^c)^{\alpha\beta}(\nabla_\beta R)\,]\nabla_\alpha\\
&\quad - \epsilon_{abc}[\,\mathcal{R}^{cd} + \tfrac{2}{3}\eta^{cd}(\nabla^2 R - \tfrac{3}{2}R^2)\,]\mathcal{M}_d \quad ,
\end{aligned} \tag{26}$$

where $\mathcal{R}^{ab} - \mathcal{R}^{ba} = \eta_{ab}\mathcal{R}^{ab} = (\gamma_d)^{\alpha\beta}\Sigma_\beta{}^d = 0$ and

$$\nabla_\alpha\Sigma_\beta{}^c = i(\gamma_b)_{\alpha\beta}\mathcal{R}^{bc} - \tfrac{2}{3}[\,C_{\alpha\beta}\eta^{cd} + i\tfrac{1}{2}(\gamma_b)_{\alpha\beta}\epsilon^{bcd}\,](\nabla_d R) \quad . \tag{27}$$

In writing these results, their form was simplified by replacing the usual Lorentz generator according to: $\mathcal{M}_{bc} \to \epsilon_{bc}{}^a\mathcal{M}_a$, so that when acting on a spinor ψ_α or a vector v_a we have

$$[\,\mathcal{M}_a\,,\,\psi_\alpha\,] = i\tfrac{1}{2}(\gamma_a)_\alpha{}^\beta\psi_\beta \quad , \quad [\,\mathcal{M}_a\,,\,v_b\,] = \epsilon_{ab}{}^c v_c \quad . \tag{28}$$

Now R is a component of a super 2-form, but in 3D we need a super 3-form. Using the formalism of the 1981 work (Gates (1981)), it is easy to show that an irreducible 3D, N = 1 closed super 3-form is described by $\mathcal{G}_{ABC}$ where

$$\begin{aligned}
\mathcal{G}_{\alpha\beta\gamma} &= 0 \quad , \quad \mathcal{G}_{\alpha\beta c} = i2(\gamma_c)_{\alpha\beta}\mathcal{G} \quad ,\\
\mathcal{G}_{\alpha bc} &= i\epsilon_{abc}(\gamma^a)_\alpha{}^\beta(\nabla_\beta\mathcal{G}) \quad , \quad \mathcal{G}_{abc} = \epsilon_{abc}[\nabla^2\mathcal{G} - R\mathcal{G}] \quad .
\end{aligned} \tag{29}$$

For this theory, the general result in (11) takes the form

$$\mathcal{G}_{abc}| = \widetilde{g}_{abc} + \epsilon_{abc}\Big[i\psi_d{}^\alpha(\gamma^d)_\alpha{}^\beta(\nabla_\beta \mathcal{G}|) - i\epsilon^{def}\psi_d{}^\alpha(\gamma_e)_{\alpha\beta}\psi_f{}^\beta(\mathcal{G}|)\Big] \quad , \tag{30}$$

so that after substitution of the last result from (29) into the lhs of (30) it follows that

$$\tfrac{1}{6}\epsilon^{abc}\widetilde{g}_{abc} = (\mathcal{D}^2\mathcal{G}|) \quad , \tag{31}$$

where the explicit form of the operator $\mathcal{D}^2$ is given by

$$\mathcal{D}^2 \equiv \nabla^2 - i\psi_a{}^\alpha(\gamma^a)_\alpha{}^\beta\nabla_\beta - R + i\epsilon^{abc}\psi_a{}^\alpha(\gamma_b)_{\alpha\beta}\psi_c{}^\beta \quad . \tag{32}$$

Therefore I am to define

$$\widetilde{\Delta} \equiv \tfrac{1}{6}\int d^3x\, \mathrm{e}^{-1}\, \epsilon^{abc}\widetilde{g}_{abc} \quad , \tag{33}$$

$$\widehat{\Delta} \equiv \int d^3x\, \mathrm{e}^{-1}\Big(\mathcal{D}^2\mathcal{G}|\Big) \quad , \tag{34}$$

and again using the E.I.T. and E.C. to define

$$\int d^3x\, d^2\theta\, \mathrm{E}^{-1}\mathcal{L} \equiv \int d^3x\, \mathrm{e}^{-1}\Big[\mathcal{D}^2\mathcal{L}|\Big] \quad . \tag{35}$$

Since $\mathcal{D}^2$ was derived via the E.C., we might want to check it on another choice of $\mathcal{L}$ such as $\mathcal{L} = R$. It is known for this choice that $S_{SG} \propto \int d^3x\, d^2\theta \mathrm{E}^{-1}R$ is the correct answer. After the usual projection techniques, I find

$$\int d^3x d^2\theta\, \mathrm{E}^{-1}\, R = \int d^3x \mathrm{e}^{-1}\Big[-\frac{1}{2}\epsilon^{abc}\left(\mathcal{R}_{abc}(\omega) + \psi_{a\alpha}\Psi_{bc}{}^\alpha\right) - B^2\Big], \tag{36}$$

where the $\Psi_{ab}{}^\beta$ is the usual component level gravitino field strength and the spin-connection is given by,

$$\omega_a{}^b = \tfrac{1}{4}\epsilon^{bcd}\left[C_{cda} - 2C_{acd} + i4\left(\psi_c{}^\alpha(\gamma_a)_{\alpha\beta}\psi_d{}^\beta + \psi_a{}^\alpha(\gamma_c)_{\alpha\beta}\psi_d{}^\beta\right)\right] - \tfrac{1}{2}B\delta_a{}^b. \tag{37}$$

Finally I have checked this same procedure using old minimal off-shell supergravity in 4D, N = 1 Wess-Zumino superspace to calculate the topological index $\widehat{\Delta}$ associated with the 4D, N = 1 super 4-form multiplet described in the 1981 paper (Gates (1981)). I find the result

$$\widehat{\Delta} = \int d^4x\, \mathrm{e}^{-1}\Big[-i(\mathcal{D}^2\mathcal{F}|) + \mathrm{h.c.}\Big] \quad , \tag{38}$$

here the operator $\mathcal{D}^2$ is defined by

$$\mathcal{D}^2 \equiv \nabla^2 + i\overline{\psi}^{\underline{a}}{}_{\dot\alpha}\nabla_\alpha + 3\overline{R} + \tfrac{1}{2}C^{\alpha\beta}\overline{\psi}_{\underline{a}}{}^{(\dot\alpha}\overline{\psi}_{\underline{b}}{}^{\dot\beta)} \quad , \tag{39}$$

and $\mathcal{F}$ is the lowest non-trivial super 4-form field strength component

$$F_{\alpha\beta\underline{c}\underline{d}} = C_{\dot{\gamma}\dot{\delta}} C_{\alpha(\gamma} C_{\delta)\beta} \overline{\mathcal{F}} \quad . \tag{40}$$

I note that the form of (38) suggests the formula

$$\int d\mu \, \mathcal{L}_{Gen} = \tfrac{1}{2} \int d\mu_c \, \mathcal{L}_c + \text{h.c.} \quad . \tag{41}$$

So that the E.C. and E.I.T. imply

$$\int d\mu_c \, \mathcal{L}_c = \int d^4x \, \mathrm{e}^{-1} \left[\mathcal{D}^2 \mathcal{L}_c \, | \right] \quad , \tag{42}$$

acting on a chiral superfield (such as $\mathcal{F}$). More generally

$$\mathcal{L}_c = (\overline{\nabla}^2 + R) \, \mathcal{L}_{Gen} \quad , \tag{43}$$

so that we may define

$$\begin{aligned} \int d\mu \, \mathcal{L}_{Gen} &= \tfrac{1}{2} \Big\{ \int d^4x \, \mathrm{e}^{-1} \left[\mathcal{D}^2 \, (\overline{\nabla}^2 + R) \, \mathcal{L}_{Gen} \, | \right] + \text{h.c.} \Big\} \\ &\equiv \int d^4x \, \mathrm{e}^{-1} \left[\mathcal{D}^4 \, \mathcal{L}_{Gen} \, | \right] \quad , \end{aligned} \tag{44}$$

where the operator $\mathcal{D}^4$ is defined by

$$\mathcal{D}^4 = \tfrac{1}{2} \left[\mathcal{D}^2 \, (\overline{\nabla}^2 + R) + \overline{\mathcal{D}}^2 \, (\nabla^2 + \overline{R}) \right] \quad . \tag{45}$$

This final result can be seen to coincide exactly with the result in our book '*Superspace*' where it was 'derived by a handicraft' argument.

Thus, I see that there is excellent support for the E.I.T. and E.C. from a number of explicit cases. I have also found numerous other examples. I am still checking even more examples in an attempt to understand if there are any limitations on this method.

I think the E.C. is a universal feature of all supergravity theories that has escaped our notice since the beginning of the era of using Wess-Zumino superspace! I also have some evidence that the E. C. plays an even more important role than I presented here. In some examples I know, there occur topological obstructions to the imposition of the E. C. The most interesting point about these obstructions is that they take the form of the supergravity constraints themselves! It is perhaps not too optimistic to hope that at last we have begun to grasp the 'deep' reason why constraints must be imposed in supersymmetrical theories. The answer seems to be to enforce the E. C.

I further conjecture that the E.C. will ultimately be found to apply to even covariant string field theory! The reasoning goes as follows. In a fully covariant and geometrical approach to string and superstring field theory, one must be confronted with calculating the integral $\int d^D\mathbf{X}(\sigma) \, d\mathbf{B}(\sigma) \, d\mathbf{C}(\sigma)$ (here we consider the bosonic string for the sake of simplicity). It ought to be possible to write an equation like

$$\int d^D\mathbf{X}(\sigma)\; d\mathbf{B}(\sigma)\; d\mathbf{C}(\sigma) \;=\; \int d^D\mathbf{X}(0)\; \mathcal{D}^{(\infty)} \quad , \tag{46}$$

so that the string coordinate zero-modes define the manifold of an ordinary appearing field theory. The oscillator modes (of all types) define the ectoplasm of the string space. Thus in a fully geometrical approach to covariant string field theory I expect that there should exist an operator $\mathcal{D}^{(\infty)}$ that appears in a 'stringy' E.I.T.

If this conjecture proves to be true, it provides an elegantly simple basis for understanding why even if we live in a universe described by fiber bundles, Kaluza-Klein spaces, Wess-Zumino superspace, strings, superstrings, heterotic strings, branes, M-theory, F-theory etc., the topological triviality of all the extra "coordinates" may forbid their having direct physical consequences (at least in the absence of strong coupling given the current views of strong/weak duality). The Ethereal Conjecture, properly interpreted, may be a physical principle.

Acknowledgment I wish to thank the organizers of the International Seminar on "Supersymmetries and Quantum Symmetries" for their kind invitation to deliver this presentation at this memorial meeting to honor Dr. Ogievetsky. Additional thanks go to T. Húbsch and M. Luty for their assistance in the preparation of this manuscript.

References

A. Salam and J. Strathdee (1978): *Introduction to Supersymmetry.* Fortschr. Phys. **26** 57.

J. Wess and B. Zumino (1977): Phys. Lett., **66B** 361.

A. Rogers (1980): J. Math. Phys. **21** 1352; ibid. **22** (1981) 443, 939; ibid., **26** (1985) 2749.

B. De Witt (1992): *Supermanifolds*, Cambridge University Press, Cambridge, 1992.

R. Arnowitt, B. Zumino and P. Nath (1975): Phys. Lett., **56B** 81.

J. Wess and J. Bagger (1983): *Supersymmetry and Supergravity*, Princeton University Press, Princeton, NJ (1983).

S. J. Gates, Jr., M. T.Grisaru, M. Roček and W. Siegel (1983): *Superspace*, Benjamin-Cummings (Addison-Wesley) Reading, MA, (1983).

I. L. Buchbinder and S. M. Kuzenko (1995): *Ideas and Methods of Supersymmetry and Supergravity*, Institute of Physics Publishing, Bristol and Philadelphia, PA, (1995).

F. A. Berezin (1966): *The Method of Second Quantization*, Academic Press, New York, (1966).

S. J. Gates, Jr., M. T.Grisaru, M. Roček and W. Siegel (1983): *Superspace*, pp. 332-333.

I. L. Buchbinder and S. M. Kuzenko (1995) *Ideas and Methods of Supersymmetry and Supergravity*, pp. 476-478.

S. J. Gates, Jr. (1981): Nucl. Phys. **B184** 381.

S. J. Gates, Jr. (1983): Nucl. Phys. **B213** 409.

B. Zupnik (1991): Theor. Math. Phys. **89**, Vol. 2 (1991) 1191; idem. Phys. Lett. **254B** (1991) 127.

R. Brooks, S. J. Gates, Jr. and F. Muhammad (1986): Nucl. Phys. **B268** (1986) 599; M. Evans and B. Ovrut (1986): Phys. Lett. **175B** 145; P. Nelson and G. Moore (1986): Nucl. Phys. **B274** 509.

M. Brown and S. J. Gates, Jr. (1979): Annals of Physics, Vol. **122**, No.2 443.

From Supergravity to Ballbearings

Pavel Grozman, Dimitry Leites

Dept. of Math., Univ. of Stockholm, Roslagsv. 101, Kräftriket hus 6, S-106 91, Stockholm, Sweden; e-mail: mleites@matematik.su.se

Abstract. The analog of the Riemann tensor of the phase spaces of the nonholonomic (with constrained velocities) dynamical systems on manifolds and supermanifolds is proposed. These tensors are needed to perform H. Hertz's formulation of mechanics (without the notion of force) and to write supergravity equations on any N-extended Minkowski superspace. Our approach provides one also with a method to select coset superspaces of $\mathcal{SL}(N|4)$ for the role of *the* N-extended Minkowski superspaces. For $N = 1, 2, 4, 8$ certain most symmetric examples are considered. The method is applicable as well to M. Vasiliev's models with $N > 8$.

Introduction

The details of this paper will be given elsewhere. Here we briefly explain how to derive supergravity equations, SUGRA(N) — the analogues of Einstein's equations (EE) on an N-extended Minkowski superspace $\mathcal{M}(N)$ and what are our the criteria for distinguishing suitable Minkowski superspaces among other supermanifolds. As in twistor theory, we consider complex case, the physical reality to be recovered on a suitable real form of $\mathcal{M}(N)$. Our requirements: SUGRA(N) should be (A) a differential equation of order ≤ 2 on the components; (B) the component expantion of SUGRA(N) should contain the ordinary Einstein's equations. For simplicity we assume that the supergroup of motions of $\mathcal{M}(N)$ is $\mathcal{G} = \mathcal{SL}(N|4)$ (though other possibilities can not be eliminated, cf. (Leites e.a. (1998), Manin (1997))), so we wonder: what is the stationary subgroup $\mathcal{P}$ for which $\mathcal{M}(N) = \mathcal{SL}(N|4)/\mathcal{P}$?

As N grows, it becomes clear that to justify the above requirements we have to diminish $\mathcal{P}$, as GIKOS did, cf. (Galperin e.a. (1984)). Then for $N > 3$ we see that the underlying manifold of $\mathcal{M}(N)$ is the direct product of several copies of the Minkowski space M (times, perhaps, an auxiliary space of a yet unclear merit) and SUGRA splits into the usual Einstein equations on each copy of M glued together by odd superfields. In particular, for $N = 4$ there are two copies of M.

For $N = 8$ there are three copies of M, one of them distinguished (say, "our world"), the other two – perfectly interchangeable (in the model considered here; there are other possibililties) – mirrowing, say, "heaven" and "hell"). These extra copies of the Universe (they MUST appear in our approach) embody an idea first, perhaps, voiced in (Sakharov (1986)). Another feature of nonholonomic nature of Minkowski *super*space is the prefered direction of time, an observation we derive by directly looking at the rattleback;

the universality of this observation for nonholonomic systems follows from recent studies by A. Nordmark (1997).

We start with a presentation of Einstein's equations in a form convenient to us — as equations on conformally noninvariant components of the Riemann tensor represented as a section of the bundle whose fiber is certain *Lie algebra cohomology*. This is equivalent to the standard modern treatment of G-structures in differential geometry that uses *Spencer homology* (Sternberg (1985)) but allows a generalization embracing nonholonomic structures, such as SUGRA. *Nonholonomic* manifolds, i.e., manifolds with nonintegrable distributions, see (Hertz (1956)), are encountered quite often. The applications range from the Cat's Problem to electro-mechanical devices. For a moving account of nonholonomic problems and their history see (Vershik and Gershkovich (1994)).

One can apply (Sternberg (1985)) to any supermanifold with a G-structure (such attempts are numerous in the literature) but the tensors obtained do not match the one physicists consider, cf. (Wess and Bagger (1983)). We also offer a general method to derive constraints — analogues of Wess–Zumino constaraints — for any N. We discover that one of the conventional WZ-constraints for $N = 1$ is redundant: its cohomology class is zero. This demonstrates that a computer-aided study (Grozman and Leites (1997)) is a must here: the amount of computations is too vast for a human not to make a slip.

Though the notion of supermanifolds will soon celebrate their 25-th birthday (see (Leites (1974)) for the first definition of supervariety), certain basics are, regrettably, insufficiently known yet. So we will recall them.

1 Structure Functions for Nonholonomic Structures

1.1 Nonholonomic (Super)Manifolds

Let M be a manifold with a distribution D. Let

$$D = D_1 \subset D_2 \subset D_3 \subset \ldots \subset D_d \tag{1}$$

be the sequence of strict inclusions, where $D_i(x) = D_{i-1}(x)+[D_1(x), D_{i-1}(x)]$ for every $x \in M$ and d is the least number for which the sequence (1.0) stabilizes, i.e., such that $D_d(x) \cup [D_1(x), D_d(x)] = D_d(x)$. In case $D_d = TM$ the manifold M is called *completely nonholonomic*. Let $n_i(x) = \dim\ D_i(x)$. The distribution D is called *regular* if all the dimensions n_i are constant functions on M. Each pair: (M, D) with a nonintegrable D will be referred to as a *nonholonomic manifold* if $d \neq 1$. We will only consider completely nonholonomic (super)manifolds with regular distributions.

With the tangent bundle over a nonholonomic manifold (M, D) we can naturally associate a sheaf of nilpotent Lie algebras as follows. At point $x \in M$ set

$$\mathfrak{n}(x) = \bigoplus_{-d\leq i\leq -1} \mathfrak{n}_i(x), \quad \text{where } \mathfrak{n}_{-i}(x) = D_i(x)/D_{i-1}(x), \quad D_0 = 0. \tag{2}$$

Clearly, $\mathfrak{n}(x)$ is a nilpotent Lie algebra.

1.1.1. A flat $(G, \mathfrak{n}(x))$-structure. Let $M = \mathbb{R}^n$ with a nilpotent $\mathbb{Z}$-graded Lie algebra structure, call it $\mathfrak{n}$. Let G be a subgroup of homogeneous (preserving the grading (2)) automorphisms of $\mathfrak{n}$. Let us identify the tangent space at a point $m \in M$ with M by means of a translation from G. The preimages $\mathfrak{n}_{-1}(m)$ of $\mathfrak{n}_{-1}$ under this identification determine a distribution on M. This distribution together with the G-action on the accompanying flag $\mathfrak{n}(m)$ at each $m \in M$ will be called a *flat* $(G, \mathfrak{n})$*-structure.*

1.2 Generalized Cartan's Prolongs

Given a $\mathbb{Z}$-graded nilpotent Lie algebra $\mathfrak{g}_- = \bigoplus_{0>i\geq -d} \mathfrak{g}_i$ and a Lie subalgebra $\mathfrak{g}_0 \subset \mathfrak{der}\, \mathfrak{g}_-$ which preserves the $\mathbb{Z}$-grading of $\mathfrak{g}_-$, define the i-th prolong of $(\mathfrak{g}_-, \mathfrak{g}_0)$ for $i > 0$ to be (here $S^\bullet = \oplus S^k$ and V^* is the dual of V):

$$\mathfrak{g}_i = [(S^\bullet(\mathfrak{g}_-)^* \otimes \mathfrak{g}_0) \cap (S^\bullet(\mathfrak{g}_-)^* \otimes \mathfrak{g}_-)]_i, \tag{3}$$

where the subscript in the rhs singles out the component of degree i and the intersection is well-defined thanks to the fact that $\mathfrak{g}_0 \subset \mathfrak{der}\, \mathfrak{g}_- \subset \mathfrak{g}_-^* \otimes \mathfrak{g}_-$.

Define the *generalized Cartan's prolong*: $(\mathfrak{g}_-, \mathfrak{g}_0)_* = \bigoplus_{i\geq -d} \mathfrak{g}_i$. By the routine arguments, $(\mathfrak{g}_-, \mathfrak{g}_0)_*$ is a Lie algebra. By the same arguments as for the G-structures, cf. (Sternberg (1985), Goncharov (1987)), the space $H^2(\mathfrak{g}_-; (\mathfrak{g}_-, \mathfrak{g}_0)_*)$ is the space of obstructions to flatness of the nonholonomic supermanifold (M, D) and the elements of $H^2(\mathfrak{g}_-; (\mathfrak{g}_-, \mathfrak{g}_0)_*)$ will be called (as for the case $d = 1$) *structure functions.*

The space of structure functions naturally splits into homogeneous components whose degree is induced by the $\mathbb{Z}$-grading of $(\mathfrak{g}_-, \mathfrak{g}_0)_*$. Let $C^s(\mathfrak{g}_-; (\mathfrak{g}_-, \mathfrak{g}_0)_*) = \bigoplus_k C^{k,s}(\mathfrak{g}_-; (\mathfrak{g}_-, \mathfrak{g}_0)_*)$ be this splitting on the cochain level; the corresponding cohomology $H^{k,s}(\mathfrak{g}_-; (\mathfrak{g}_-, \mathfrak{g}_0)_*)$ are precisely the analogues of the Spencer cohomology and coinside with them for $\mathfrak{g}_- = \mathfrak{g}_{-1}$. Sign Rule carries superization.

2 Structure Functions of the N-Extended Minkowski Supermanifold

Recall that the ground field is $\mathbb{C}$. The G-structure of the Minkowski space can be viewed as either (a) (pseudo) Riemannian or, equivalently, (b) twistor structure. "Straightforward" superizations of these structures are distinct. They are considered in (a) (Leites e.a. (1998)) and (b) (Manin (1997)) or (Grozman and Leites (1997)), respectively. Generally, neither of these superizations gives rise to what is *accepted* as supergravity. The reason is that a Minkowski superspace is still another superization of the Minkowski space and is naturally endowed with a nonholonomic structure.

2.1 What is an N-Extended Complexified Minkowski Supermanifold

Recall that the "physical reasons" for the restrictions $N \leq 4$ for the Yang-Mills and $N \leq 8$ for the supergravity theories were put to doubt in (Vasiliev (1995)).

Recapitulations (Leites (1983)). A *supermatrix* is a rectangular table with elements from a supercommutative superalgebra C with given sets of parities P_{row} and P_{col} of its rows and columns. The *size* of a matrix is $P_{row} \times P_{col}$. Usually, the parities are chosen so that the even rows and columns come first followed by the odd ones; such matrices are said to be of the *standard format.* For the square matrices we will only consider the cases when $P_{row} = P_{col}$ and will denote this set of parities by Par. The parity of the matrix unit — the matrix with an element $c \in C$ in the (i,j)-th slot and 0's elsewhere — is defined to be $p(c) + P_{row}(i) + P_{col}(j)$. Hereafter in this paper $C = \mathbb{C}$.

Let $\mathtt{gl}\ (Par)$ be the set of square matrices of size $Par \times Par$; let p be the number of 0's and q the number of 1's in Par. It is immediately clear, that for distinct Par's with the same p and q the Lie superalgebras $\mathtt{gl}\ (Par)$ have *nonisomorphic* maximal nilpotent (say, upper triangular) subalgebras, though the algebras $\mathtt{gl}\ (Par)$ themselves are isomorphic. It often suffices to consider the standard format only, and $\mathtt{gl}\ (Par)$ is abbreviated to $\mathtt{gl}(p|q)$. In supergravity we MUST consider nonstandard formats as well. Generally, we separate collections of even and odd positions in Par, say, $\mathtt{gl}(a|b|c)$ or $\mathtt{gl}(a|b|c|d)$, etc.

Since $P_{row} = P_{col}$, the Lie subsuperalgebra of upper triangular matrices in $\mathtt{gl}\ (Par)$ is isomorphic to the Lie subsuperalgebra of lower triangular matrices and we will confine ourselves to one of them, denoted by $\mathfrak{n}$. The generators of $\mathfrak{n}$ are the elements just above (below) the main diagonal; we will denote the even generators of $\mathfrak{n}$ by white nodes and the odd generators by "grey" nodes, the nodes corresponding to commuting generators are disconnected, otherwise they are joined by a segment. For instance, for the standard format, i.e., $\mathtt{gl}(p|q)$, we have:

$$\underbrace{0 - \ldots - 0}_{p-1 \text{ nodes}} - \otimes - \underbrace{0 - \ldots - 0}_{q-1 \text{ nodes}} \tag{4}$$

Consider the Lie supergroup $\mathcal{SL}(N|4)$ and its parabolic subsupergroup $\mathcal{P}$ corresponding to the two marked odd simple roots in the following system of simple roots (this means that $\mathcal{P}$ is generated by all the simple roots except the marked negative ones):

$$0 - - - \overset{+}{\otimes} - - - \underbrace{0 - - \ldots - - 0}_{N-1 \text{ nodes}} - - - \overset{+}{\otimes} - - - 0 \tag{5}$$

The Lie group corresponding to the 0-th term of the $\mathbb{Z}$-grading described by diagram (5) is $G = SL(N) \times SL_L(2) \times SL_R(2) \times \mathbb{C}^*$, i.e., the degree of a marked root is equal to 1 the other simple roots being of degree 0; here indices L and R distinguish the "left" copy of $SL(2)$ from its "right" twin. In this case $\mathfrak{g}_0 = \mathfrak{g} = \mathfrak{o}(4) \oplus \mathfrak{gl}(N) = \mathfrak{sl}_L(2) \oplus \mathfrak{sl}_R(2) \oplus \mathfrak{gl}(N)$, $\mathfrak{g}_- = \bigoplus\limits_{-1 \geq i \geq -2} \mathfrak{g}_i$ with $\mathfrak{g}_{-1} = (id_L \otimes Id) \oplus (id_R \otimes Id^*)$, $\mathfrak{g}_{-2} = id_L \otimes id_R^*$, where id_j is the space of the standard (identity) representation of $\mathfrak{sl}_j(2)$, $j = L, R$; and Id is the space of the identity representation of $\mathfrak{sl}(N)$. The corresponding matrix representaion of $\mathfrak{p} = Lie(\mathcal{P})$ is of format $2|N|2$.

The *N-extended Minkowski superspace* $\mathcal{M}(N)$ is $\mathcal{SL}(N|4)/\mathcal{P}$ endowed with the natural $(G, \mathfrak{g}_-)$-structure. The conventional versions of the Minkowski superspace correspond to a certain real form of the (complex) superspace $\hat{\mathcal{M}}(N)$ with the reduced $(G, \mathfrak{g}_-)$-structure, $(\hat{G}, \mathfrak{g}_-)$-structure for which $\hat{G}$ is semisimple. Clearly,

$$\hat{\mathcal{M}}(N) = \mathcal{P}/\hat{\mathcal{G}}, \text{ where } \hat{G} = SL(N) \times SL_L(2) \times SL_R(2). \tag{6}$$

GIKOS guessed (and we can prove) that these $\mathcal{M}(N)$ never satisfy our requirements (A) and (B) on SUGRA for $N > 1$. GIKOS considered an enlargement $\hat{\mathcal{R}}(N)$ of $\hat{\mathcal{M}}(N)$ defined $\hat{\mathcal{R}}(N) = \mathcal{P}/\hat{\mathcal{G}}'$, where $\hat{G}' = Q \times SL_L(2) \times SL_R(2)$ and Q is a parabolic subgroup of $SL(N)$. In other words, from $\mathcal{P}$ we pass to a smaller parabolic subsupergroup, $\mathcal{P}'$, whose diagram has several middle roots marked as well.

To satisfy (A) and (B), we have to test various $\mathcal{P}'$s. This is impossible without a computer.

We can prove that diagram (4) can not satisfy requirements (A) and (B) regardless of the number of middle roots marked. In particular, it is well known to physicists that even for $N = 1$ the standard format does not satisfy (B). So in order to satisfy requirements (A) and (B) we consider nonstandard formats.

For $N = 8$ the following two possibilities seem to be distinguished:

$$0-\overset{+}{\otimes}-0-\overset{+}{0}-0-\overset{+}{0}-0-\overset{+}{0}-0-\overset{+}{\otimes}-0 \quad Par = (001111111100)$$

$$0-\overset{+}{0}-0-\overset{+}{\otimes}-0-\overset{+}{0}-0-\overset{+}{\otimes}-0-\overset{+}{0}-0 \quad Par = (000011110000)$$

Elsewhere we will discuss the assumptions of Haag-Lopuszanski-Sohnius' theorem which lead to the Poincare supergroup and its "twistor enlargment", $\mathcal{SL}(N|4)$. The stationary subgroups we consider here are the simplest ones: the Lie groups, subgroups of $O(4) \times SL(N)$. Notice that even for the same stationary subgroup $O(4) \times SL(N)$ we can consider several realizations.

The experience with the analogues of Einstein's equaitions on symmetric spaces (Leites e.a. (1998)) teaches us to consider the models of Minkowski superspace whose stationary subgroup is smaller: a product of several copies of $SL(2)$: otherwise the equations will be of order > 2.

With all these conventions, we consider the following examples of Minkowski superspaces which we will denote more puristically by $\hat{\mathcal{M}}(Par)$ rather than $M(N)$.

Theorem. *In Table (below) there are listed all the orders and weights of all the structure functions for the indicated $\hat{\mathcal{M}}(Par)$.*

The corresponding cocycles are listed in a detailed version.

Clearly, there are more candidates for the role of $\hat{\mathcal{M}}(Par)$ for the N we have considered; Grozman's package (described in (Grozman and Leites (1997))) allows one to perform corresponding calculations for any model.

To interpret the supergravity in the same way as we have treated the Einstein Equations (Leites e.a. (1998)), define the supergravity equations as follows. On $\hat{\mathcal{M}}(N)$, the stationary subgroup (i.e., $\hat{G}$) of the point preserves $\varepsilon_L \otimes \varepsilon_R \otimes \varepsilon_1 \otimes \ldots \otimes \varepsilon_k$, where ε_i is the volume preserved by $SL_i(2)$, the i-th copy of $SL(2)$, in the 2-dimensional identity representation and $2k = N$.

If there are several, say s, tensors of weight 0 — "scalar curvatures" — we can take for $\mathbf{R}$ their linear combination and the coefficients of this combination determine a parameter which runs over the projective space $\mathbb{C}P^{s-1}$.

Example: $N = 1$. The tensor $\mathbf{R}$ depends on a parameter, the ratio $a : b$ which runs over the projective line $\mathbb{C}P^1$. Physicists call this parameter the *Gates–Sigel* parameter. On $\mathfrak{g}_{-1}$, there is the inner product given by the bracket. Notice that this product is even for $Par = (00100)$ and odd for $Par = (00001)$. (The tacit choice was $Par = (00100)$.) For $Par = (00100)$ the metric g on the Minkowski space M is the product of spinorial metrics ε_L and ε_R on the the maximal isotropic (with respect to the pairing) subspaces of $\mathfrak{g}_{-1}$.

The equation on scalar curvatures takes the form

$$a\mathbf{R}_1(00) + b\mathbf{R}_2(00) = \lambda g. \tag{7}$$

(The numbers $(w_1, \ldots)$ in Table are the components of the highest weight of the irreducible $SL_L(2) \times SL_R(2) \times SL_1 \times \ldots$-modules $R(w_1, \ldots)$. Each SL_i corresponds to a neighboring pair 11 in Par.)

From the explicit form of the cocycles it is clear that the component expansion of the above equation does not contain the usual equation on the scalar curvature for $Par = (00100)$, and only $\mathbf{R}_1(00)$ for $Par = (00001)$ has the right expansion.

Notice immediately, that for (7) to be well-defined, we must demand that all structure functions of orders > 2 vanish. These conditions are called the *Wess-Zumino constraints.* We have fewer of them than, say, in (Wess and Bagger (1983)): one of the constraints is "harmless", its cohomology class is zero (like torsion of the Levi–Civita conneciton).

What shall we take for analogs of Ricci flatness? As for $N = 0$, these should be the vanishing conditions on the part of the Riemann tensor which does not belong to the conformal, i.e., the analog of Weyl, tensor. For $N > 0$ there are several such components and we can equate to zero either or all of them. Different choices correspond to different supergravities (minimal, flexible, etc.). The equations are well-defined provided the constraints vanish.

If all structure functions of order 2 (and of lesser orders) vanish, the higher obstructions are well-defined and we can write an equation on them. For example, for $Par = (00100)$ we can equate to zero one or both of the tensors:

$$\mathbf{R}(3,0) = 0, \quad \mathbf{R}(0,3) = 0. \tag{8}$$

Observe that the "flatness" and the obstructions to flatness we introduced differ drastically from their conventional counterparts. E.g., each contact manifold or supermanifold is flat in our sence, but it is endowed with a connection (whose form is the contact form) with nonzero, moreover, nondegenerate curvature form.

2.2 Table

$\deg(SF)\backslash Par$	(0000)	(00100)	(00001)
$\boxed{0}$	not defined	$(3,1), (1,3)$	—
1	—	$(1,0), (0,1)$	$(1,1)^{\boxed{1}+1}, (0,1)_1, \boxed{(0,3)_1}$
2	$(2,2), (0,0)$ $\boxed{(4,0), (0,4)}$	$(1,1), (0,0)^2$	$(0,0)^2, (2,2),$ $(1,0)_1, \boxed{(3,2)_1}$
3	not defined	$(3,0), (0,3)$	—

$\deg(SF)\backslash Par$	(001100)	(100001)	(000011)
$\boxed{-1}$	not defined	$(0,2), (2,0)$	not defined
$\boxed{0}$	$(222), (123), (321)$	$(0,1)_1, (1,0)_1$	$(222), (123)_1, (301)_1$
1	$(110)_1, (011)_1$	$\boxed{(0,0)}, \boxed{(2,2)^2}$ $(1,1)^2, (1,0)_1^2, (0,1)_1^2$	$\boxed{(110)}, \boxed{(031)_1}$ $(011)_1$
2	$(020)^2, \boxed{\boxed{(200)}}$ $(101), \boxed{\boxed{(002)}}$	$(2,2), (0,0)^3$ $(1,0)_1, (0,1)_1$	$(000), (220), (002)$ $\boxed{(200)}, (020), (101)_1$
≥ 3	—	not defined	—

$\deg(SF)\backslash Par$	(00111100)	(00001111)
$\boxed{-1}$	$(1021), (1023), (1201), (2112),$ $(3201), (1122)_1, (2211)_1$	$(1122), (2211), (1001)_1$ $(1221)_1, (2112)_1, (3001)_1, (1003)_1$
0	—	—
1	$(0110), (0011)_1^{\boxed{2}+1}, (1100)_1^{\boxed{2}+1}$	$(0011)^{\boxed{2}+1}, (1100)^2, (0110)_1^{\boxed{2}+1}$
2	$(0000), (0220), (2000)$ $(1111), (0101)_1, (10010)_1,$ $(0002), (0020), (0200)$	$(0000)^2, (0022), (2200),$ $(0200), (0020)$ $(0101)_1, (1010)_1$
≥ 3	—	—

$\deg(SF)\backslash Par$	(111100001111)
$\boxed{-3}$	$(100001)^2, (100003), (300001), (100201), (102001)$ $(100221), (122001), (101112), (211101), (210012), (111111)$
$\boxed{-2}$	$(100122), (221001), (101211), (112101), (111012)$ $(210111), (030001), (100030)$
-1	—
0	—
1	$(000011)^4, (001100)^4, (110000)^4, (000110)_1^4, (011000)_1^4$
2	$(000000)^3, (000020), (000022)$ $(000200), (002200), (002000), (020000)$ $(220000), (000101)_1, (001010)_1, (010100)_1, (101000)_1$
≥ 3	—

Notations and remarks. Clearly, the cocycles from Table are invariant under the change of parities $1 \longleftrightarrow 0$ in *Par*. The cocycles which also correspond to the "conformal" case — on shell — are $\boxed{\text{boxed}}$; the cocycles of small orders are all conformally invariant, such $\boxed{\text{orders}}$ are boxed; the cocycles *which only exist in the conformal case*, off shell — unheard of in the absence of super — are $\boxed{\boxed{\text{doubleboxed}}}$; the suscript 1 singles out odd cocycles; the exponent denotes the multiplicity of the cocycle; the multiplicity of conformally invariant vectors is boxed.

The entry "not defined" in the Table refers to the "Riemannian" case, i.e., to the semisimple stationary subgroup. It so happened that in these cases there are no conformal cohomologies. The dash — indicates that either there are no structure functions in this order or (if the degree is > 2 and the case is the "Riemannian" one) they are not defined. In the cases considered (but not generally!) "Riemannian" and "on shell" are synonyms.

References

Fuks (Fuchs), D., (1986) *Cohomology of infinite dimensional Lie algebras*, Consultants Bureau, NY

Galperin, A., Ivanov, E., Kalitzin, S., Ogievetsky, V., Sokatchev, E. Unconstrained off-shell $N = 3$ supersymmetric Yang–Mills theory. Classical Quantum Gravity **2** (1985), no. 2, 155–166. id., Corrigendum: "Unconstrained $N = 2$ matter, Yang–Mills and supergravity theories in harmonic superspace". Classical Quantum Gravity **2** (1985), no. 1, 127. id., Unconstrained $N = 2$ matter, Yang–Mills and supergravity theories in harmonic superspace. Classical Quantum Gravity **1** (1984), no. 5, 469–498.

Goncharov, A., Infinitesimal structures related to hermitian symmetric spaces, Funct. Anal. Appl., **15**, 3, 1981, 23–24 (Russian); for details see: id., Generalized conformal structures on manifolds. Selecta Math. Soviet. **6**, 1987, no. 4, 307–340

Grozman, P., Leites, D., *Mathematica*-aided study of Lie algebras and their cohomology. From supergravity to ballbearings and magnetic hydrodynamics In: Keränen V. (ed.) (1997) *The second International Mathematica symposium*, Rovaniemi, 185–192

Hertz, H., (1956) *The principles of mechanics in new relation*, NY, Dover

Leites, D. Spectra of graded commutative rings. Russian Math. Surveys, **30**, 3, 1974, 209–210 (in Russian)

Leites, D. A. (1983) *Supermanifold theory* Karelia Branch of the USSR Acad. Sci., Petrozavodsk, (in Russian); an expanded version in: id., (ed.) *Seminar on Supermanifolds*, ##1–34, Reports of Dept. of Math. of Stockholm Univ., 1987–90, 2100 pp.; *Introduction to supermanifold theory*, Russian Math. Surveys, **35**, 1, 1980, 3–53; Quantization. Supplement 3. In: Berezin, F., Shubin, M. *Schrödinger equation*, Kluwer, Dordrecht, 1991

Leites, D., Poletaeva, E., Serganova, V. On Einstein equations on manifolds and supermanifolds; Grozman, P., Leites, D., A new twist of Penrose' twistor theory (to appear)

Manin, Yu., (1997) *Gauge fields and complex geometry*, 2nd ed., Springer

Nordmark, A., Essén, H., Systems with preferred spin direction, Proc. Royal Soc., London (submitted)

Onishchik, A. L., Vinberg, E. B., (1990) *Seminar on algebraic groups and Lie groups*, Springer, Berlin e.a.

Sakharov, A. D., Evaporation of black mini-holes and high energy physics, ZhETPh Lett., **44**, (6), 1986, 295–298 (in Russian)

Sternberg, S., (1985) *Lectures on differential geometry*, Chelsey, 2nd edition

Vasiliev, M. A., Higher-spin gauge theories in four, three and two dimensions. The Sixth Moscow Quantum Gravity Seminar (1995), Internat. J. Modern Phys. D **5** (1996), no. 6, 763–797; id., Higher-spin-matter gauge interactions in $2+1$ dimensions. Theory of elementary particles (Buckow, 1996), Nuclear Phys. B Proc. Suppl. **56B** (1997), 241–252

Vershik, A., Gershkovich, V., (1994) *Encyclop. of Math. Sci., Dynamical systems–7*, Springer

Wess, J., Bagger, J., (1983)*Supersymmetry and supergravity*, Princeton Univ. Press

On Harmonic Superspace

Paul Howe

Mathematics, King's College, London, UK.

Abstract. Some aspects of harmonic superspace are discussed.

1 Introduction

Victor Ogievetsky made many significant contributions to theoretical physics during a long and distinguished career often conducted in difficult circumstances. Perhaps his most interesting work, at least from this author's point of view, was the development of a deeper understanding of the geometry of supersymmetry. This began in the seventies with his and Sokatchev's interpretation of the geometry of $N = 1$ supergravity as the generalisation of chirality to the curved superspace context. Some years later he and his group, now expanded to include Galperin, Ivanov and Kalitzin as well as Sokatchev (GIKOS), succeeded in finding the most useful generalisation of chirality to the case of flat space $N = 2$ supersymmetry, namely $N = 2$ harmonic superspace. This new approach to the geometry of supersymmetry has turned out to have wide applications to both geometry and physics. In this note I shall briefly review flat harmonic superspace, its relation to twistor theory, its generalisation to curved superspace and its potentially interesting application to some on-shell supermultiplets which arise naturally in string theory in the context of the Maldacena conjecture.

2 Harmonic Superspace

In extended supersymmetry flat harmonic superspaces [Galperin et al (1984)] are superspaces of the form $M_H = M \times \mathbb{F}$ where M is the corresponding Minkowski superspace and $\mathbb{F}$ is a coset space of the internal symmetry group which is chosen to be a compact, complex manifold, in fact a flag manifold. For example, in $D = 4$, the internal symmetry group is $U(N)$ (or sometimes just $SU(N)$), the relevant flags sequences $V_{k_1} \subset V_{k_2} \subset .... \subset \mathbb{C}^N$ of subspaces with dimension k_i, and the flag manifold is the space whose points correspond to such flags. In particular, the flags $V_p \subset V_{N-q} \subset \mathbb{C}^N$, $p+q \leq N$, define (N,p,q) harmonic superspace [Hartwell and Howe (1995)] for which $\mathbb{F} = H\backslash U(N)$, the isotropy group being $U(p) \times U(N-(p+q)) \times U(q)$. This family generalises in a natural way the harmonic superspaces for $N = 2$ and $N = 3$ supersymmetry introduced by GIKOS, which are respectively

(2,1,1) [Galperin et al (1984)] and (3,1,1) [Galperin et al (1985)] harmonic superspace in the above notation. [1]

The harmonic superspace approach of GIKOS emphasises the group theoretic aspects of fields defined on such spaces rather than the holomorphic aspects. For this reason it has become standard practice to work on the space $\hat{M}_H = M \times G$ (where $G = U(N)$ or whatever internal symmetry group is appropriate). This is equivalent to working on M_H provided that the fields are restricted to being equivariant with respect to the subgroup H. Such a field is a map $F : \hat{M}_H \to V$, where V is a representation space for the group H, such that $F(z, hu) = M(h)F(z, u)$, for all $u \in G$, $h \in H$ and $z \in M$, where $M(h)$ denotes the action of H on V. An equivariant field of this type defines in a natural way a section of a vector bundle E over M_H with typical fibre V and the two types of object are in one-to one correspondence.

In $D = 4$ (N, p, q) harmonic superspace u is written in index notation as $u_I{}^i$ where the i index is acted on by $U(N)$ and the I index by H, and the inverse u^{-1} is denoted $u_i{}^I$. The index I splits under H as (r, R, r') where $r = 1, \ldots p$, $R = p+1, \ldots N-q$, $r' = N-q+1, \ldots N$. The following derivatives can be defined on equivariant fields F,

$$D_{\alpha r} = u_r{}^i D_{\alpha i}; \qquad D^{r'}_{\dot{\alpha}} = u_i{}^{r'} \bar{D}^i_{\dot{\alpha}} \tag{1}$$

These derivatives anticommute and so allow the introduction of generalised chiral fields, or Grassmann-analytic (G-analytic) fields, F, which satisfy

$$D_{\alpha r} F = D^{r'}_{\dot{\alpha}} F = 0 \tag{2}$$

Moreover, for any fixed values of the $u's$ the derivatives $\{D_{\alpha r}, D^{r'}_{\dot{\alpha}}\}$ define a CR structure on M; that is, they are basis vectors for an involutive subundle $\bar{K}$ of the complexified tangent bundle T_c and their complex conjugates are linearly independent of t hem at any point in M.[2] The space $\mathbb{F}$ can thus be viewed as the space of all CR structures of this type on M. In addition, the derivatives can be combined with a subset of the right-invariant vector fields $D_I{}^J$ on $U(N)$ to define a CR structure on M_H. The right-invariant vector fields decompose under H into the following subsets:

$$D_I{}^J = \{D_r{}^s, D_R{}^S, D_{r'}{}^{s'}\}; \quad \{D_r{}^S, D_r{}^{s'}, D_R{}^{s'}\}; \quad \{D_R{}^s, D_{r'}{}^s, D_{r'}{}^S\} \tag{3}$$

The first subset correponds to the isotropy group, the second to the components of the usual $\bar{\partial}$ operator on $\mathbb{F}$ and the third to its conjugate ∂. The CR structure on M_H is then specified by the derivatives $\{D_{\alpha r}, D^{r'}_{\dot{\alpha}}, D_r{}^S, D_r{}^{s'}, D_R{}^{s'}\}$. Since the $D_I{}^J$ are characterised by

[1] (3,2,1) superspace was first discussed by Galperin et al (1987a)

[2] The rôle of CR structures in harmonic superspace was first emphasised by Rosly and Schwarz (1986).

$$D_I{}^J u_K{}^k = \delta_K{}^J u_I{}^k; \; D_I{}^J u_k{}^K = -\delta_I{}^K u_k{}^J \tag{4}$$

it is straightforward to verify that this set of derivatives does indeed specify a CR structure. Finally, when $p = q$, there is a real structure on harmonic superspace defined by $u \mapsto \epsilon u$ where

$$\epsilon = \begin{pmatrix} 0 & 0 & 1_p \\ 0 & 1_{N-2p} & 0 \\ -1_p & 0 & 0 \end{pmatrix} \tag{5}$$

This can be combined with complex conjugation to act on equivariant fields by

$$F(z,u) \mapsto \widetilde{F(z,u)} = \overline{(F(z,\epsilon u))} \tag{6}$$

For fields transforming under certain representations of the isotropy group it is possible to demand that they be real with respect to this conjugation.

3 Relation to Twistor Theory

The above discussion is reminiscent of Euclidean twistor theory for $\mathbb{R}^4$ where $\mathbb{CP}^1$ parametrises the space of (anti)-self-dual complex structures on $\mathbb{R}^4$ and where the antipodal map of $\mathbb{CP}^1$ defines the real structure (as it does in (2,1,1) harmonic superspace) (see, for example [Penrose and Rindler (1986), Ward and Wells (1990)]). Indeed, particular ly in cases where the appropriate symmetry group is the superconformal group, flat-space harmonic superspace theory can be viewed as an example of the group-theoretic approach to twistor theory [Baston and Eastwood (1989)]. The full power of this method only becomes apparent in complexified spacetime (or superspace), in which case the generalised twistor method starts with the selection of two parabolic subgroups P_1, P_2 of a complex simple group G. Since $P_{12} \equiv P_1 \cap P_2$ is also parabolic there are three (complex) homogeneous spaces $\mathbb{M}_k = P_k \backslash G$, $k = 1, 2$ *or* 12 which fit together into a double fibration $\mathbb{M}_2 \stackrel{\pi_2}{\leftarrow} \mathbb{M}_{12} \stackrel{\pi_1}{\rightarrow} \mathbb{M}_1$ such that each projection injects the fibres of the other one and thereby sets up a correspondence between points of $m_2 \in \mathbb{M}_2$ and subsets $\pi_1 \circ \pi^{-1}(m_2) \subset \mathbb{M}_1$ and vice versa. The basic idea of twistor theory applied to physics is that one of the spaces, say $\mathbb{M}_1$, is spacetime or Minkowski superspace and that information about field theory on (super) Minkowski space can be encoded as holomorphic data on $\mathbb{M}_2$. $N = 1$ supersymmetry in four dimensions provides an exceptional case where $\mathbb{M}_1$ and $\mathbb{M}_2$ are taken to be left and right chiral superspaces respectively and $\mathbb{M}_{12}$ is super Minkowski space [Manin (1988)].

In four dimensions the complex superconformal group is $SL(4|N)$. The parabolic isotropy groups define flag supermanifolds where the flags are now made of sequences of super subspaces of N-extended twistor space $\mathbb{T}_N =$

$\mathbb{C}^{4|N}$, and the superspaces of interest are constructed as certain open subsets of these supermanifolds [Manin (1988), see also Howe and Hartwell (1995)]. (The flag supermanifolds themselves have compact bodies). As an example consider $N = 2$. The group $SL(4|2)$ consists of supermatrices of the form

$$g = \left(\begin{array}{c|c} A & \Gamma \\ \hline \Delta & B \end{array}\right) \tag{7}$$

where A and B are 4×4 and 2×2 even matrices, Γ and Δ are 4×2 and 2×4 respectively odd matrices, and where $\mathrm{sdet} g = 1$. $N = 2$ super Minkowski space is the space of flags of type (i.e. dimension) $((2|0), (2|2))$ specified by the isotropy group consisting of matrices of the form

$$\left(\begin{array}{cccc|cc} \times & \times & & & & \\ \times & \times & & & & \\ \times & \times & \times & \times & \times & \times \\ \times & \times & \times & \times & \times & \times \\ \hline \times & \times & & & \times & \times \\ \times & \times & & & \times & \times \end{array}\right) \tag{8}$$

where the crosses denote entries which are not necessarily zero. The body of this superspace can be identified from the top left part of the diagram which represents ordinary complex Minkowski space. Complex $N = 2$ harmonic superspace $\mathbb{M}_H$ is the space of flags of type $((2|2), (2|1), (2|2)) \subset \mathbb{T}_2$ specified by the subgroup of matrices of the form

$$\left(\begin{array}{cccc|cc} \times & \times & & & & \\ \times & \times & & & & \\ \times & \times & \times & \times & \times & \times \\ \times & \times & \times & \times & \times & \times \\ \hline \times & \times & & & \times & \\ \times & \times & & & \times & \times \end{array}\right) \tag{9}$$

Note that this differs from super Minkowski space only in the bottom right-hand corner. Thus, $N = 2$ harmonic superspace has the same odd dimensionality as $N = 2$ super Minkowski space and its body is locally $\mathbb{M} \times \mathbb{CP}^1$. These two superspaces fit into a double fibration with analytic superspace $\mathbb{M}_A$ which the space of flags of type $(2|1) \subset \mathbb{T}_2$. The isotropy group consists of matrices of the form

$$\left(\begin{array}{cccc|cc} \times & \times & & & \times & \\ \times & \times & & & \times & \\ \times & \times & \times & \times & \times & \times \\ \times & \times & \times & \times & \times & \times \\ \hline \times & \times & & & \times & \\ \times & \times & \times & \times & \times & \times \end{array}\right) \tag{10}$$

The odd dimensionality of analytic superspace is half that of harmonic superspace but its body is the same. The double fibration is $\mathbb{M}_A \overset{\pi_2}{\leftarrow} \mathbb{M}_H \overset{\pi_1}{\rightarrow} \mathbb{M}$. Points of $\mathbb{M}$ correspond to twistor lines, copies of the fibres of π_1, in $\mathbb{M}_A$, while points of $\mathbb{M}_A$ parametrise certain planes of dimension $(0|4) \subset \mathbb{M}$, and these are copies of the fibres of π_2. Note that analytic superspace is essentially a complex space; it does not exist as a coset space of the real $N = 2$ superconformal group $SU(2,2|2)$. Furthermore, it is not a subspace of complex harmonic superspace, but rather a quotient space defined by the projection π_2. The antipodal map on $\mathbb{CP}^1$ defines an anti-holomorphic involution of $\mathbb{M}_A$ and the real twistor lines in $\mathbb{M}_A$ are parametrised by points in real Minkowski superspace. When super Minkowski space is taken to be real, i.e. when $\mathbb{M}$ is replaced by M, real harmonic superspace $M_H = {\pi_1}^{-1}(M)$ and π_2 embeds M_H as a submanifold of $\mathbb{M}_A$. In the real case, therefore, the double fibration can be replaced by the single fibration $M_H \to M$ while holomorphic fields on $\mathbb{M}_A$ are replaced by CR-analytic fields on M_H [Howe and Hartwell (1995)].

The above discussion generalises in a natural way to (N,p,q) harmonic superspace [Hartwell and Howe (1995), Howe and Hartwell (1995)]. An entirely different family of supermanifolds can be constructed by using isotropy groups which are completely filled in the bottom right corner, but which have different structures to Minkowski space in the top left corner. Such superspaces, which are usually referred to as supertwistor spaces and which have conventional twistor spaces as bodies, have also been used in the context of supersymmetric field theory (see, for example, [5], Manin (1988)).

4 Curved Harmonic Superspace

The notion of harmonic superspace generalises to curved superspace relatively straightforwardly [Galperin et al (1987b), Galperin et al (1987c), Hartwell and Howe (1995)]. Perhaps the simplest geometries that can be discussed in this context are $D = 4$ superconformal geometries ($N \leq 4$) [Howe (1981)]. A superconformal structure on a real $(4|4N)$ supermanifold is a choice of odd tangent bundle F (rank $(0|4N)$) such that the Frobenius tensor, defined by evaluating the commutator of two odd vector fields (sections of F) modulo F has the following components with respect to bases $\{E_{\alpha i}, \bar{E}^i_{\dot{\alpha}}\}$ of F and $\{E^a\}$ for B^* where $B = T/F$:

$$T_{\alpha i \beta j}{}^c = T^{ij}_{\dot{\alpha}\dot{\beta}}{}^c = 0; \qquad T_{\alpha i \dot{\beta}}{}^{jc} = -i\delta_i{}^j(\sigma^c)_{\alpha\dot{\beta}} \tag{11}$$

The Frobenius tensor is just the dimension zero component of the usual torsion tensor, although it is not necessary to introduce a connection to define it. With the components satisfying (11) above it is invariant under the group $\mathbb{R}^+ \times SL(2,\mathbb{C}) \cdot U(N)$ acting on F. The above constraints can be shown to lead to the usual constraints of off-shell conformal supergravity if standard conventional choices are made for the even tangent bundle B and for the

$SL(2,\mathbb{C})\cdot U(N)$ connection [Howe (1981)]. For $N = 4$ an additional constraint must be imposed arising from the fact that the true internal gauge symmetry group in this case is $SU(4)$.

In the curved case the space $\hat{M}_H$ is the (perhaps locally defined) principal $U(N)$ bundle associated with the $U(N)$ factor of the structure group of F, and harmonic superspace M_H itself is then given as $H\backslash\hat{M}_H$ where H is the relevant isotropy group, so that $\hat{M}_H$ is also a principal H-bundle over M_H.

For $N = 1$ harmonic superspace is not necessary and the constraints (11) above imply that a real $(4|4)$ supermanifold has a superconformal structure if and only if it has a CR structure with involutive CR bundle of rank $(0|2)$ such that the Frobenius tensor completely specifies B. This is Ogievetsky-Sokatchev supergeometry [Ogievetsky and Sokatchev (1980)].

For $N = 2$, the appropriate harmonic superspace is $(2,1,1)$ superspace Galperin et al (1987c). There is a natural bundle $\bar{K}$ on M_H defined by the set of basis vector fields ${D_1}^2$ and the horizontal lifts of the odd basis vector fields on M projected in the $(\alpha 1)$ and $\begin{pmatrix} 2 \\ \dot{\alpha} \end{pmatrix}$ directions. To define the latter it is necessary to introduce a $U(2)$ connection. This bundle defines a CR structure if and only if the contraints of $N = 2$ superconformal geometry are satisfied [Hartwell and Howe (1995)]. Note, however, that there is no gauge-invariant notion of G-analyticity. This problem was circumvented in [Galperin et al (1987c)] by allowing ${u_1}^i$ and ${u_2}^i$ to be independent, but the geometrical status of this strategy could perhaps be made more precise.

For $N = 3$ in $(3,1,1)$ harmonic superspace a similar CR structure is compatible with the constraints of conformal supergravity but does not imply them [3], although demanding that the lift of the Frobenius tensor to M_H should vanish along the $(\alpha 1)$ and $\begin{pmatrix} 3 \\ \dot{\alpha} \end{pmatrix}$ directions does. In addition, at the linearised level, the conformal supergravity potential in harmonic superspace is a dimension -2 G-analytic field ${V_1}^3$. This field can be interpreted as a 3-form along the antiholomorphic fibre directions, but a clear geometrical understanding is lacking at present.

Finally, for $N = 4$, it is possible to use $(4,2,2)$ harmonic superspace in which case the natural CR structure is equivalent to the constraints given above in (11). However, as noted previously, an additional constraint has to be imposed and this should eventually lead to the construction of a prepotential which is known to be a G-analytic 4-form along the antiholomorphic fibre directions in the linearised theory [Hartwell and Howe (1995)]. The case of $N = 4$ conformal supergravity is particularly interesting in view of its relevance to the Maldacena conjecture.

---

[3] This point has been observed by others (E. Sokatchev, private communication).

5 Superconformal Fields

In many supersymmetric theories for which there is no known off-shell formulation the fundamental field may be considered to be a field strength tensor, and in many cases this field turns out to have a simple description as a single-component CR-analytic field W on an appropriate harmonic superspace. The hope is that this point of view may be useful in constructing, or attempting to construct, non-perturbative correlation functions of gauge-invariant composite operators made from powers of W by exploiting CR-analyticity, particularly when the theory under consideration is superconformal. This approach to quantum supersymmetry has been studied extensively in the case of $N = 4$ Yang-Mills theory (see, for example, Howe and West (1995), Howe and West (1997) and Howe et al (1998)) and is currently highly topical in that it is directly relevant to the Maldacena conjecture which relates supergravity theories on anti-de Sitter backgrounds to conformal field theories on the boundary, i.e. (super) Minkowski space [Maldacena (1997)].

The multiplets to be discussed here are matter (i.e. non-gravitational) multiplets with the maximal number of supersymmetries, sixteen. The basic multiplets are the $D = 4, N = 4$ Yang-Mills field strength multiplet, the $D = 6, (2,0)$ tensor multiplet and the $D = 3, N = 8$ scalar multiplet. They are relevant to IIB string theory on $AdS_5 \times S^5$, M-theory on $AdS_7 \times S^4$ and M-theory on $AdS_4 \times S^7$ respectively, although the latter two cases are less well understood as these multiplets are only known at the non-interacting level. Nevertheless they are extremely interesting multiplets, corresponding as they do to the multiplets of the M-theory 5-brane and 2-brane respectively.

5.1 $D = 4, N = 4$ Yang-Mills

The field strength superfield $W_{ij} = -W_{ij}$ in super Minkowski space tra nsforms according to the six-dimensional representation of the internal symmetry group $SU(4)$ and satisfies the following conditions:

$$\nabla_{\alpha i} W_{jk} = \nabla_{\alpha[i} W_{jk]} \tag{12}$$

$$\bar{\nabla}^i_{\dot{\alpha}} W_{jk} = -\frac{2}{3}\delta_{[j}{}^i \bar{\nabla}^l_{\dot{\alpha}} W_{k]l} \tag{13}$$

$$W^{*ij} = \frac{1}{2}\epsilon^{ijkl} W_{kl} \tag{14}$$

The appropriate superspace in this case is $(4,2,2)$ harmonic superpace specified by the internal flag manifold $\mathbb{F} = S(U(2)\times U(2))\backslash SU(4)$. An element $u \in SU(4)$ is written $u_I{}^i = (u_r{}^i, u_{r'}{}^i)$, $r = 1,2$; $r' = 3,4$. The right-invariant derivatives on $SU(4)$ are $\{D_r{}^s, D_{r'}{}^{s'}, D_o; D_r{}^{s'}; D_{r'}{}^s\}$, where $\{D_r{}^s, D_{r'}{}^{s'}, D_o\}$ correspond to the isotropy algebra $\mathfrak{su}(2) \oplus \mathfrak{su}(2) \oplus \mathfrak{u}(1)$, $D_r{}^{s'}$ corresponds to the $\bar{\partial}$ operator on $\mathbb{F}$ and $D_{r'}{}^s$ to its conjugate ∂. The normalisation is

$$D_o u_r{}^i = \frac{1}{2} u_r{}^i; \; D_o u_{r'}{}^i = -\frac{1}{2} u_{r'}{}^i \tag{15}$$

The CR-structure on harmonic superspace is specified by the set of derivatives $\{D_{\alpha r}, D_{\dot\alpha}^{r'}, D_r{}^{s'}\}$.

The claim is that W_{ij} is equivalent to a charge 1 field W on M_H which is covariantly G-analytic and ordinarily $\mathbb{F}$-analytic; it is also real with respect to the real structure discussed in section 2 above, where covariantly G-analytic means that

$$\nabla_{\alpha r} W = \nabla_{\dot\alpha}^{r'} W = 0 \tag{16}$$

with $\nabla_{\alpha r} = u_r{}^i \nabla_{\alpha i}$, etc. This claim is easily verified Hartwell and Howe (1995).

The gauge-invariant operators $A_q = \mathrm{tr}(W^q)$ are G-analytic in the usual sense and hence CR-analytic. These are the conformal fields of interest in the Maldacena conjecture. Since they are in short representations of $SU(2,2|4)$, the integer q cannot be affected by quantum corrections and so, since this integer determines the dimension, they do not have anomalous dimensions. This family of operators was introduced in [Howe and West (1995)] and it has been shown that it is in one-to-one correspondence with the Kaluza-Klein spectrum of IIB supergravity on $AdS_5 \times S^5$ [Andrianopoli and Ferrara (1998)]. In particular, the family of operators includes the energy-momentum tensor $T = A_2$ [Howe et al (1981)]. In complex spacetime A_q becomes a holomorphic section of the qth power of a certain homogeneous line bundle $\mathcal{L}$ over analytic superspace. This superspace is in fact the super-Grassmannian whose points are (2|2) planes in $\mathbb{T}_4 = \mathbb{C}^{4|4}$.

5.2 The $D = 6, (2,0)$ Tensor Multipet

In six-dimensional $(n, 0)$ supersymmetry the internal symmetry group is $Sp(n)$ and the spinors are taken to be symplectic Majorana-Weyl. The (2,0) tensor multiplet is a scalar superfield W_{ij} which transforms according to the real 5-dimensional representation of $Sp(2)$ (vector representation of $SO(5)$). It satisfies the constraints [Howe et al (1983)]

$$W_{ij} = -W_{ji}; \qquad \eta^{ij} W_{ij} = 0 \tag{17}$$

$$\bar{W}^{ij} = \eta^{ik} \eta^{jl} W_{kl} \tag{18}$$

$$D_{\alpha i} W_{jk} = 2\eta_{ij} \lambda_{\alpha k} - 2\eta_{ik} \lambda_{\alpha j} + \eta_{jk} \lambda_{\alpha i} \tag{19}$$

where $\lambda_{\alpha i} = \frac{1}{5} \eta^{jk} D_{\alpha j} W_{ki}$, and where η_{ij} is the antisymmetric $Sp(2)$ invariant tensor. The components of W_{ij} consist of five scalars, four spinors and a self-dual three-form field strength tensor.

The appropriate harmonic superspace in this case is $M_H = M \times \mathbb{F}$ where $\mathbb{F} = U(2) \backslash Sp(2)$ [Howe (1998)]. Elements u of the group $Sp(2)$ are written

$u_I{}^i = (u_r{}^i, u_{r'}{}^i)$, $r = 1, 2$; $r' = 3, 4$. In addition to being unitary the matrices u also satisfy $u_I{}^i u_J{}^j \eta_{ij} = \eta_{IJ}$. The matrix η has components

$$\eta_{rs} = \eta_{rs'} = 0; \ \eta_{rs'} = -\eta_{s'r} = \delta_{rs'} \tag{20}$$

The right-invariant vector fields $\hat{D}_{IJ}$ on the group are real, symmetric and satisfy

$$\hat{D}_{IJ} u_K{}^i = -\eta_{IK} u_J{}^i - \eta_{JK} u_I{}^i \tag{21}$$

The derivatives split into the isotropy group derivatives $\hat{D}_{rs'}$ and the coset derivatives $\hat{D}_{rs} \equiv D_{rs}$ and $\hat{D}_{r's'} \equiv D_{r's'}$. The isotropy group derivatives may be separated into a $\mathfrak{u}(1)$ derivative $D_o = \frac{1}{2}\eta^{rs'} D_{rs'}$ and $\mathfrak{su}(2)$ derivatives $D_{rs'} = \hat{D}_{rs'} - \eta_{rs'} D_o$. On u, $D_o u_r{}^i = u_r{}^i; D_o u_{r'}{}^i = -u_{r'}{}^i$. The element $u_r{}^i$ transforms as a doublet under $\mathfrak{su}(2)$ while $u_{r'}{}^i$ transforms as the conjugate doublet.

The CR structure on M_H is specified by the derivatives $D_{\alpha r} = u_r{}^i D_{\alpha i}$ and D_{rs}, the latter corresponding to the components of $\bar{\partial}$ on $\mathbb{F}$ as usual.

In a very similar manner to the $D = 4, N = 4$ Yang-Mills case discussed above it can easily be shown that the field W_{ij} is equivalent to a single-component CR-analytic field W on M_H with $W = \frac{1}{2}\epsilon^{rs} u_r{}^i u_s{}^j W_{ij}$. It is also real with respect to a suitably defined real structure. Although there is no known non-Abelian version of this multiplet it is still possible to use it to discuss conformal field theory in the abstract [Aharony et al (1998)]. The powers of W give conformal fields in short representations, in particular $W^2 = T$ is again the energy-momentum tensor multiplet (conformal supercurrent) [Howe et al (1983)]; a few of them were listed in [Leigh and Rozali (1998)]. It is likely that the complete set of such fields corresponds to the Kaluza-Klein spectrum of eleven-dimensional supergravity on $AdS_7 \times S^4$.

5.3 D=3, N=8 Scalar Multiplet

The scalar multiplet in three dimensions is slightly trickier to describe than the preceeding two cases because the internal symmetry group is $SO(8)$ and it is necessary to use both spinor and vector representations of the group to describe the multiplet. Nevertheless it can be done and the result is simply stated: the $D = 3, N = 8$ scalar multiplet can be represented by a single-component CR-analytic superfield defined on the harmonic superspace $M_H = M \times U(4)\backslash Spin(8)$ [Howe (1998)].

The multiplet is defined in $D = 3, N = 8$ Minkowski space by the equation

$$D_A W_I = (\Sigma_I)_{AA'} \lambda^{A'} \tag{22}$$

where I, A and A' transform according to the vector, spinor and primed spinor representations of $Spin(8)$, Σ denotes the spin matrices and where the

$D=3$ spinor indices on both D and λ hav e been suppressed. The spinor representations decompose into 4 and $\bar{4}$ representations under $SU(4)$ with particular charges under $U(1)$ while the vector representation decomposes into the six-dimensional representation of $SU(4)$, neutral under $U(1)$ and a charged singlet and its conjugate under $U(1)$. That is $v_I \to (v_{++}, v_{--}, v_i)$, $i=1\ldots 6$, where $v_{++}=\frac{1}{2}(v_7 - iv_8), v_{--}=\bar{v}_{++}$, while for a $D=8$ Majorana spinor Ψ, with

$$\Psi = \begin{pmatrix} \psi \\ \chi \end{pmatrix} \tag{23}$$

the indices are split as follows,

$$\psi_A \to \begin{pmatrix} \psi_{\alpha+} \\ \bar{\psi}^{\alpha}_{-} \end{pmatrix} \qquad \chi^{A'} \to \begin{pmatrix} \chi_{\alpha-} \\ \bar{\chi}^{\alpha}_{+} \end{pmatrix} \tag{24}$$

The lower (upper) α indices correspond to the $4(\bar{4})$ representations of $SU(4)$.

The right-invariant derivatives on the group split into the isotropy derivatives $\{D_{ij}, D_o\}$ and the coset derivatives $\{D_{++i}, D_{--i}\}$. The CR structure is specified by the derivatives $D_{\alpha+} = u_{\alpha+}{}^A D_A$ and D_{++i}. The field $W_{++} = u_{++}{}^I W_I$ is easily seen to be CR-analytic, as well as being real with respect to an appropriate real structure.

In both of the latter two cases the superconformal group in complex spacetime is $OSp(8|2)$ and the analytic superspaces are the same, although the line bundles for the field strengths are different. However, the underlying twistor spaces are $\mathbb{C}^{4|8}$ for $D=3$ and $\mathbb{C}^{8|4}$ for $D=6$, the two being related by the Grassmann parity flip operation Π. This may be a reflection of "electromagnetic" duality in $D=11$; the 2-brane couples to the 4-form field strength of $D=11$ supergravity while the 5-brane couples to its 7-form dual. In case (a) on the other hand, the D3-brane is self-dual, as is $D=4, N=4$ twistor space under Π.

References

Andrianopoli L, Ferrara S (1998): K-K excitations on $AdS_5 \times S^5$ and $N=4$ primary superfields. Phys. L ett. **B430** 248-253.

Aharony O, Berkooz M, Seiberg N (1998): Light-cone descritption of (2,0) conformal field theories in six dimensions. Adv. Theor. Math. Phys. **2** 119-153.

Baston R, Eastwood M (1989): *The Penrose Transform* (Oxford University Press).

Galperin A, Ivanov E, Kalitzin S, Ogievetsky V, Sokatchev E (1984): Unconstrained $N=2$ matter, Yang-Mills and supergravity theories in harmonic superspace. Class. Quantum Grav. **1** 469.

Galperin A, Ivanov E, Kalitzin S, Ogievetsky V, Sokatchev E (1985): Unconstrained off-shell $N=3$ supersymmetric Yang-Mills theory. Class. Quantum Grav. **2** 155.

Galperin A, Ivanov E, Ogievetsky V (1987): Superspaces for $N=3$ supersymmetry. Soviet Journal for Nuclear Physics **46** 543.

Galperin A, Nguyen Ahn Ky, Sokatchev E (1987): $N = 2$ supergravity: solution to the constraints. Class. Quantum Grav. **4** 1235-1254.

Galperin A, Ivanov E, Ogievetsky O, Sokatchev E (1987): $N = 2$ supergravity in superspace: different versions and matter couplings. Class. Quantum Grav. **4** 1255.

Hartwell G G, Howe P S (1995): (N, p, q) harmonic superspace. Int J. Mod. Phys **10**, 3901-3919.

Howe P S (1981): A superspace approach to conformal supergravity. Phys. Lett. **100B** 389-392.

Howe P S (1998): Superconformal fields in diverse dimensions, in preparation.

Howe P S, Hartwell G G (1995): A superspace survey. Class. Quant. Grav. **12** 1823-1880.

Howe P S, Sierra G, Townsend P K (1983): Supersymmetry in six dimensions. Nucl. Phys. **B221** (1983) 331-348.

Howe P S, Stelle K S, Townsend P K (1982): Supercurrents. Nucl. Phys. **B192** 332.

Howe P S, Sokatchev E, West P C (1998): Three-point functions in $N = 4$ Yang-Mills. Hep-th/9808162.

Howe P S, West P C (1995): Nonperturbative Green's functions in theories with extended superconformal symmetry. Hep-th/9509140.

Howe P S, West P C (1995): Superconformal invariants and extended supersymmetry. Phys. Lett. **B400** 307-313.

Leigh R G, Rozali M (1998): The large N limit of (2,0) superconformal field theory. Phys. Lett. **B431** 311-316.

Maldacena J (1997): The large N limit of superconformal field theories and supergravity. Hep-th/9711200.

Manin Y I (1988): *Gauge Field Theory and Complex Geometry* (Springer-Verlag).

Ogievetsky O, Sokatchev E (1980): Soviet J. of Nucl. Phys. **32** 447.

Penrose R, Rindler W (1986): *Spinors and Spacetime II* (Cambridge University Press).

Rosly A A, Schwarz A S (1986): Supersymmetry in spaces with auxiliary variables. Comm. Math. Phys. **105**, (1986) 645.

Ward R S, Wells R O (1990): *Twistor Geometry and Field Theory* (Cambridge University Press).

Witten E (1978): Phys. Lett. **77B**:394–398.

Harmonic Approach and Quaternionic Taub-NUT Metric

Evgeny Ivanov[1] and Galliano Valent[2]

[1] Bogoliubov Laboratory of Theoretical Physics, JINR, Dubna, 141 980 Moscow region, Russia
[2] Laboratoire de Physique Théorique et des Hautes Energies, Unité associée au CNRS URA 280, Université Paris 7, 2 Place Jussieu, 75251 Paris Cedex 05, France

Abstract. We use the harmonic space technique to construct explicitly a quaternionic extension of the Taub-NUT metric. It depends on two parameters, the first being the Taub-NUT 'mass' and the second one the cosmological constant. We compare the metric constructed with those available in the literature.

1. An efficient way to explicitly construct hyper-Kähler and quaternionic-Kähler metrics is provided by the harmonic (super)space method (Galperin et al (1984) - Galperin, Ivanov and Ogievetsky (1994)). It was V.I. Ogievetsky who has played the decisive role in inventing and further developing of this approach.

It was firstly introduced in the context of $N = 2$ supersymmetry (Galperin et al (1984)). The basic idea was to extend the standard $N = 2$ superspace by a set of internal ('harmonic') variables $u^{\pm i}, u^{+i}u_i^- = 1$, parametrizing the automorphism group $SU(2)$ of $N = 2$ superalgebra. It was shown in (Galperin et al (1984)) that all $N = 2$ theories admit a manifestly supersymmetric off-shell description in terms of unconstrained superfields given on an analytic subspace of the $N = 2$ harmonic superspace, harmonic *analytic* superfields.

It was soon realized that the harmonics are also relevant to some purely bosonic geometric problems. As is shown in (Galperin et al (1988)), the constraints defining the hyper-Kähler (HK) geometry can be given an interpretation of the integrability conditions for the existence of analytic fields in a $SU(2)$ harmonic extension of the original $4n$-dimensional HK manifold $\{x^{i\mu}\}, (i = 1, 2; \mu = 1, ...2n)$. This time, the $SU(2)$ to be 'harmonized' is an extra $SU(2)$ rotating three complex structures of the HK manifold. The analytic subspace is spanned by the harmonic variables $u^{\pm i}$ and half of the initial x-coordinates, $x^{+\mu}$. The constraints of HK geometry can be solved via an unconstrained analytic HK potential $\mathcal{L}^{+4}(x^{+\mu}, u^{\pm i})$. It encodes (at least, locally) all the information about the associated metric. Remarkably, it allows one to *explicitly* construct the HK metrics by simple rules (Galperin et al (1988)).

In (Galperin, Ivanov and Ogievetsky (1994)), a generalization of this approach to the quaternionic-Kähler (QK) manifolds was given. It was shown in (Galperin, Ivanov and Ogievetsky (1994)), that the QK geometry constraints can be also solved in terms of some unconstrained potential $\mathcal{L}^{+4}$ living on the analytic subspace parametrized by $SU(2)$ harmonics and half of the original coordinates. It is interesting to consider some examples in order to see in detail how the machinery proposed in (Galperin, Ivanov and Ogievetsky (1994)) works. Only the simplest case of the homogeneous QK manifold $Sp(n+1)/Sp(1)\times Sp(n)$ (corresponding to $\mathcal{L}^{+4}=0$) was considered in ref. (Galperin, Ivanov and Ogievetsky (1994)).

The aim of this paper is to demonstrate the power of the harmonic geometric approach on the example of less trivial QK metric, a quaternionic generalization of the well-known four-dimensional Taub-NUT (TN) metric (Eguchi, Gilkey and Hanson (1980)). Like in the HK case (Galperin et al (1986)), the computations are greatly simplified due to the $U(1)$ isometry of the quaternionic TN metric. The metric depends on two parameters, the TN 'mass' parameter and the constant $SU(2)$ curvature parameter which can be interpreted as the inverse 'radius' of the corresponding 'flat' QK background $\sim Sp(2)/Sp(1)\times Sp(1)$. We perform the identification of the metric with those known in literature.

2. We first recall basic features of the construction of (Galperin, Ivanov and Ogievetsky (1994)). One starts with a $4n$-dimensional Riemann manifold parametrized by local coordinates $\{x^{\mu m}\}, \mu=1,2,...,2n;\ m=1,2$, and uses a vielbein formalism. The QK geometry can be defined as a restriction of the general Riemannian geometry in $4n$-dimensions, such that the holonomy group of the corresponding manifold is a subgroup of $Sp(1)\times Sp(n)$ [1]. Thus one can choose the tangent space group from the very beginning to be $Sp(1)\times Sp(n)$ and define the QK geometry via appropriate restrictions on the curvature tensor lifted to the tangent space (taking into account that the holonomy group is generated by this tensor). As explained in (Galperin, Ivanov and Ogievetsky (1994)), for the QK manifold of generic dimension the defining constraints can be written as a restriction on the commutator of two covariant derivatives

$$\left[\mathcal{D}_{\alpha(i},\mathcal{D}_{\beta k)}\right]=-2\Omega_{\alpha\beta}R\Gamma_{(ik)}\ . \tag{1}$$

Here

$$\mathcal{D}_{\alpha i}=e^{\mu m}_{\alpha i}(x)\mathcal{D}_{\mu m}=e^{\mu m}_{\alpha i}(x)\ \frac{\partial}{\partial x^{\mu m}}+[Sp(1)\times Sp(n)-\text{connections}]\ , \tag{2}$$

$e^{\mu m}_{\alpha i}(x)$ being the $4n\times 4n$ vielbein with the indices $\alpha=1,2,...2n$ and $i=1,2$ rotated, respectively, by the tangent local $Sp(n)$ and $Sp(1)$ groups, $\Omega_{\alpha\beta}$ is the $Sp(n)$-invariant skew-symmetric tensor serving to raise and lower the $Sp(n)$ indices ($\Omega_{\alpha\beta}\Omega^{\beta\gamma}=\delta^{\gamma}_{\alpha}$), $\Gamma_{(ik)}$ are the $Sp(1)$ generators, and R is a constant,

[1] For the 4-dimensional case this definition has to be replaced by the requirement that the totally symmetric part of the $Sp(1)$ component of the curvature tensor lifted to the tangent space is vanishing.

remnant of the $Sp(1)$ component of the Riemann tensor (its constancy is a consequence of the QK geometry constraint and Bianchi identities). The scalar curvature coincides with R up to a positive numerical coefficient, so the cases $R > 0$ and $R < 0$ correspond to compact and non-compact manifolds, respectively. In the limit $R = 0$ eq. (1) is reduced to the constraint defining the HK geometry (Galperin et al (1988)), in accord with the interpretation of HK manifolds as a degenerate subclass of the QK ones.

Like in the HK case (Galperin et al (1988)), in order to explicitly figure out which kind of restrictions is imposed by (1) on the vielbein $e^{\mu m}_{\alpha i}(x)$ and, hence, on the metric

$$g^{\mu m\, \nu s} = e^{\mu m}_{\alpha i}\, e^{\nu s\, \alpha i} \;, \quad g_{\mu m\, \nu s} = e_{\mu m\, \alpha i}\, e^{\alpha i}_{\nu s} \;, \tag{3}$$

one should solve the constraints (1) by regarding them as integrability conditions along some complex directions in a harmonic extension of the original manifold.

Due to the non-vanishing r.h.s. in (1), the road to such an interpretation in the QK case is more involved. Modulo these peculiarities, the basic step still consists in extending $\{x^{\mu m}\}$ by a set of some harmonic variables, $\{x^{\mu m}\} \to \{x^{\mu m}, w^{\pm}_i\}$, $w^{+\,i} w^{-}_i = 1$. Then, following the general strategy (Galperin et al (1988), Galperin, Ivanov and Ogievetsky (1994)), one passes to a new ('analytic') basis in $\{x^{\mu m}, w^{\pm}_i\}$

$$\{x^{\mu m}, w^{\pm}_i\} \Rightarrow \{x_A^{+\mu}, x_A^{-\mu}, w_A^{\pm\, i}\} \tag{4}$$

$$x_A^{\pm\mu} = x^{\mu\, i} w^{\pm}_i + v^{\pm\mu}(x,w) \;, w_A^{+\, i} = w^{+\, i} - R v^{++}(x,w) w^{-\, i} \;,$$
$$w_A^{-\, i} = w^{-\, i} \;, \tag{5}$$

where the 'bridges' $v^{\pm\mu}(x,w), v^{++}(x,w)$ are chosen so as to make the w^+-projection of $\mathcal{D}_{\alpha i}$ in this basis to be proportional to the partial derivative with respect to $x^{-\mu}$

$$\mathcal{D}^+_\alpha \sim w^{+i}\mathcal{D}_{\alpha i} = w^{+i} e^{\mu m}_{\alpha i}(x)\partial_{\mu m} + \ldots = E^\mu_\alpha(x,w)\frac{\partial}{\partial x_A^{-\mu}} = E^\mu_\alpha(x,w)\partial^+_\mu \tag{6}$$

(simultaneously, one performs an appropriate $Sp(n)$ rotation of the tangent space index α by a matrix $Sp(n)$-'bridge'). The possibility to reduce $\mathcal{D}^+_\alpha$ to this 'short' form amounts to the possibility to define *analytic* fields living on the analytic subspace $\{x_A^{+\mu}, w_A^{\pm\, i}\}$. The original QK geometry constraints prove to be equivalent to the existence of such analytic fields and subspace (Galperin, Ivanov and Ogievetsky (1994)). An essential difference of the QK case from the HK case (Galperin et al (1988)) is the necessity to shift the harmonic variables with the new bridge v^{++}.

Besides the opportunity to make $\mathcal{D}^+_\alpha$ short, the passing to the harmonic extension of $\{x^{\mu n}\}$ and further to the analytic basis and frame ('the λ-world') allows one to exhibit the fundamental unconstrained objects of the QK geometry, the QK potential. While in the original formulation ('the τ-world') the basic geometric objects are the vielbeins $e^{\mu m}_{\alpha k}(x)$ properly constrained by eq. (1), in the analytic basis such objects are the harmonic vielbeins covariantizing the derivatives with respect to the harmonic variables. In the original basis the harmonic derivatives are $D^{\pm\pm} = \partial_w^{\pm\pm} = w^{\pm i}\partial/\partial w^{\mp i}$, $D^0 = \partial^0_w = w^{+i}\partial/\partial w^{+i} - w^{-i}\partial/\partial w^{-i}$, $[\partial_w^{++}, \partial_w^{--}] = \partial^0$, i.e. they contain no partial derivatives with respect to the variables $x^{\mu n}$, because the harmonic space $\{x^{\mu m}, w_i^{\pm}\}$ has the structure of the direct product $\{x^{\mu m}\} \otimes \{w_i^{\pm}\}$. After passing to the analytic basis by eqs. (5), the derivatives $D^{\pm\pm}$ acquire terms proportional to $\partial_\mu^{\pm} \equiv \partial/\partial x^{\mp\mu}$. Besides, in D^{++} there emerges a term proportional to $\partial_{w_A}^{--}$. These new terms appear with the appropriate vielbein components $H^{+3\mu}$, $H^{--\pm\mu}$, H^{+4} which are related to the bridges as follows

$$(\partial_w^{++} + Rv^{++})x_A^{+\mu} = H^{+3\mu} \,, \tag{7}$$

$$(\partial_w^{++} + Rv^{++})v^{++} = -H^{+4} \,, \tag{8}$$

$$\frac{1}{1 - R\partial_w^{--}v^{++}}\, \partial_w^{--} x_A^{\pm\mu} = H^{--\pm\mu} \,. \tag{9}$$

Note that $x^{-\mu}$ is determined in terms of $x^{+\mu}$ by the equation

$$(\partial_w^{++} - Rv^{++})x_A^{-\mu} = x_A^{+\mu} \,. \tag{10}$$

The original QK geometry constraints require $H^{+3\mu}$, H^{+4} to be analytic

$$\partial_\mu^{+} H^{+3\mu} = \partial_\mu^{+} H^{+4} = 0 \quad \Rightarrow H^{+3\mu} = H^{+3\mu}(x_A^{+}, w_A) \,,\; H^{+4} = H^{+4}(x_A^{+}, w_A) \,, \tag{11}$$

and express $H^{+3\mu}$ in terms of H^{+4}. The unconstrained QK potential $\mathcal{L}^{+4}$ is related to H^{+4} as (after properly fixing the λ-world gauge freedom)(Galperin, Ivanov and Ogievetsky (1994))

$$H^{+4}(x_A^{+}, w_A) = \mathcal{L}^{+4}(x_A^{+}, w_A) + x_\mu^{+} H^{+3\mu}(x_A^{+}, w_A) \,, \qquad x_\mu^{+} \equiv \Omega_{\mu\nu}x^{+\nu} \,, \tag{12}$$

and

$$H^{+3\nu} = \frac{1}{2}\Omega^{\nu\mu}\hat{\partial}_\mu^{-}\mathcal{L}^{+4} \,, \qquad \hat{\partial}_\mu^{-} \equiv \partial_\mu^{-} + Rx_\mu^{+}\partial_A^{--} \,. \tag{13}$$

It can be shown that the only constraint to be satisfied by $\mathcal{L}^{+4}$ is its analyticity, so this object encodes all the information about the relevant QK geometry and metrics, whence its name 'QK potential'. Choosing one or another explicit $\mathcal{L}^{+4}$, and substituting (12), (13) into eqs. (7) - (10), one can solve the latter for $x^{\pm\mu}$ and v^{++} as functions of harmonics and the τ-world coordinates $x^{\mu m}$. Having at hand the explicit form of the variable change (5), it remains to find the appropriate expression of the λ-world vielbeins in terms of $\mathcal{L}^{+4}$ in order to be able to restore the τ-world vielbein and hence the QK metric itself.

Skipping intermediate steps (they can be found in (Galperin, Ivanov and Ogievetsky (1994))), the non-vanishing components of the λ-world inverse QK metric are given by the following expressions

$$g_{(\lambda)}^{\mu+\ \nu-} = g_{(\lambda)}^{\nu-\ \mu+} = \Omega^{\mu\rho}(\partial\hat{H})^{-1}{}_{\rho}^{\ \nu}\ ,$$
$$g_{(\lambda)}^{\mu-\ \nu-} = -2\,\Omega^{\rho\sigma}(\partial\hat{H})^{-1}{}_{\sigma}^{\ \omega}(\partial\hat{H})^{-1}{}_{\rho}^{\ (\mu}\partial_{\omega}^{+}\hat{H}^{-3\nu)}\ , \tag{14}$$

where

$$(\partial\hat{H})_{\nu}^{\ \mu} \equiv \partial_{\nu}^{+}\hat{H}^{--+\mu}\ , \quad \hat{H}^{--\pm\mu} \equiv \frac{1}{1-R(x\cdot H)}\,H^{--\pm\mu}\ ,$$
$$x\cdot H \equiv x_{\mu}^{+}H^{--+\mu}\ . \tag{15}$$

Then the τ-world metric can be obtained via the change of variables inverse to (4)

$$g^{\mu m\ \nu s} = g_{(\lambda)}^{\omega-\ \sigma-}\partial_{\omega}^{+}x^{\mu m}\partial_{\sigma}^{+}x^{\nu s} + g_{(\lambda)}^{\omega+\ \sigma-}\left(\hat{\partial}_{\omega}^{-}x^{\mu m}\partial_{\sigma}^{+}x^{\nu s} + \hat{\partial}_{\omega}^{-}x^{\nu m}\partial_{\sigma}^{+}x^{\mu s}\right)\ . \tag{16}$$

In the case of 4-dimensional QK manifolds we will deal with in the sequel ($\mu,\nu = 1,2$) the τ-basis metric (16), after some algebra, can be put in the form

$$g^{\mu m\ \nu s} = \frac{1}{\det(\partial\hat{H})}\,\frac{1}{[1-R(x\cdot H)](1-R\partial_{w}^{--}v^{++})}\,G^{\mu m\ \nu s}\ , \tag{17}$$
$$G^{\mu m\ \nu s} = \epsilon^{\lambda\rho}\left[\,\partial_{w}^{--}X_{\lambda}^{+\mu m}\,X_{\rho}^{+\nu s} + (\mu m \leftrightarrow \nu s)\,\right]\ . \tag{18}$$

Here

$$X_{\rho}^{+\mu m} \equiv \partial_{\rho}^{+}x^{\mu m} \tag{19}$$

are solutions of the system of algebraic equations

$$X_{\rho}^{+\mu m}\nabla_{\mu m}x^{-\nu} = \partial_{\rho}^{+}x^{-\nu} = \delta_{\rho}^{\nu}, \qquad X_{\rho}^{+\mu m}\nabla_{\mu m}x^{+\nu} = \partial_{\rho}^{+}x^{+\nu} = 0\ , \tag{20}$$
$$\nabla_{\mu m} \equiv \partial_{\mu m} + \frac{R}{1-R\partial_{w}^{--}v^{++}}(\partial_{\mu m}v^{++})\partial_{w}^{--}\ . \tag{21}$$

We see that the problem of calculating the QK metric (17), (18) is reduced to solving the differential equations (7), (10), (8) which define, by the known $\mathcal{L}^{+4}(x^{+\mu}, w_{A}^{\pm i})$, $x^{\pm\mu}$ and v^{++} as functions of the τ-basis coordinates $x^{\mu m}$ and $w^{\pm i}$. This difficult task is simplified for the QK metrics with isometries, like in the HK case (Galperin et al (1986)). We will demonstrate this on the example of the QK analog of the Taub-NUT metric.

3. The QK counterpart of the TN manifold is characterized by the same $\mathcal{L}^{+4}$ (Bagger et al (1988))

$$\mathcal{L}^{+4} = \left(2i\lambda x^{+}\bar{x}^{+}\right)^{2} \equiv \left(\phi^{++}\right)^{2}\ . \tag{22}$$

Here we introduced the notation [2]

[2] We adopt the convention $\epsilon_{12} = -\epsilon^{12} = 1$. The complex conjugation is always understood as a generalized one, i.e. the product of the ordinary conjugation and Weyl reflection of harmonics (Galperin et al (1984)).

$$(x^{+1}, x^{+2}) = (x^+, -\bar{x}^+)\,, \quad \bar{x}^+ = \overline{(x^+)}\,, \quad \overline{(\bar{x}^+)} = -x^+\,. \tag{23}$$

We also assume

$$\bar{\lambda} = \lambda \quad \Rightarrow \quad \overline{\phi^{++}} = \phi^{++}\,. \tag{24}$$

The basic equations (8), (7), (10) for the given case take the form

$$\partial^{++} v^{++} + R(v^{++})^2 = (\phi^{++})^2\,, \tag{25}$$
$$(\partial^{++} + Rv^{++})x^+ = 2i\lambda x^+ \phi^{++}\,, \tag{26}$$
$$(\partial^{++} - Rv^{++})x^- = x^+ \tag{27}$$

(together with their conjugates). These equations are covariant under two rigid symmetries preserving the analytic subspace $\{x^+, \bar{x}^+, w_A^{\pm i}\}$: $U(1)$ Pauli-Gürsey (PG) symmetry

$$x^{+\,\prime} = e^{i\alpha}\, x^+\,, \qquad \bar{x}^{+\,\prime} = e^{-i\alpha}\, \bar{x}^+\,, \tag{28}$$

and $SU(2)$ symmetry which uniformly rotates the doublet indices of the harmonic variables ($x^{\pm}$ and v^{++} are scalars with respect to this $SU(2)$). They constitute the $U(2)$ isometry group of the QK TN metric.

We will firstly solve eq. (25). Defining

$$v^{++} = \partial^{++} v\,, \quad \omega \equiv e^{Rv}\,, \quad \hat{x}^+ \equiv \omega\, x^+\,, \quad \hat{\phi}^{++} = 2i\lambda \hat{x}^+ \bar{\hat{x}}^+ = \omega^2\, \phi^{++}\,, \tag{29}$$

we rewrite (25), (26) as

$$(\partial^{++})^2 \omega = R\, \frac{(\hat{\phi}^{++})^2}{\omega^3}\,, \tag{30}$$
$$\partial^{++} \hat{x}^+ = 2i\lambda x^+ \frac{\hat{\phi}^{++}}{\omega^2} \equiv 2i\lambda x^+ \kappa^{++}\,. \tag{31}$$

From eq. (31) and the definition of $\hat{\phi}^{++}$ one immediately finds

$$\partial^{++} \hat{\phi}^{++} = 0 \quad \Rightarrow \quad \hat{\phi}^{++} = \hat{\phi}^{ik}(x) w_i^+ w_k^+\,. \tag{32}$$

We observe that eq. (30) coincides with the pure harmonic part of the equation defining the Eguchi-Hanson metric in the harmonic superspace approach (Galperin et al (1986a)). Its general solution was given in (Galperin et al (1986a)), it depends on four arbitrary integration constants, that is, in our case, on four arbitrary functions of $x^{\mu i}$. However, these harmonic constants turn out to be unessential due to four hidden gauge symmetries of the set of equations (25) - (27). One of them is the scale invariance $v' = v + \beta(x)$, while three remaining ones form an extra local $SU(2)$ symmetry (Ivanov and Valent). Using this gauge freedom one can gauge away four integration constants in ω and write a solution to eq. (30) in the following simple form

$$\omega = \sqrt{1 + R\hat{\phi}^2} \quad \Rightarrow \quad v = \frac{1}{2R} \ln(1 + R\hat{\phi}^2) \,, \tag{33}$$

$$v^{++} = \partial^{++} v = \frac{\hat{\phi}\, \hat{\phi}^{++}}{1 + R\hat{\phi}^2} \,, \qquad \hat{\phi} \equiv \hat{\phi}^{(ik)}(x) w_i^+ w_k^- \,. \tag{34}$$

One can restore the general form of the solution as it was given in (Galperin et al (1986a)), acting on (33) by a finite form of the aforementioned hidden symmetry transformations. In (Ivanov and Valent) we demonstrate that the whole effect of the full gauge $SU(2)$ transformation is reduced to the rotation of the τ-world metric corresponding to the fixed-gauge solution (34) by some harmonic-independent non-singular matrix which becomes identity upon restriction to x- independent $SU(2)$ transformations. Thus in what follows we can stick to this solution.

Now we turn to solving eq. (26) (or (31)). This can be done in a full analogy with the hyper-Kähler TN case (Galperin et al (1986)), based essentially upon the PG invariance (28). Using (34), we obtain

$$\begin{aligned} &\kappa^{++} = \partial^{++} \kappa \,, \\ &(1)\; R > 0 \,,\; \kappa = \frac{1}{\sqrt{R}} \arctan \sqrt{R}\, \hat{\phi} \,; \\ &(2)\; R < 0 \,,\; \kappa = \frac{1}{\sqrt{|R|}} \operatorname{arctanh} \sqrt{|R|}\, \hat{\phi}. \end{aligned} \tag{35}$$

Without loss of generality, in what follows we will choose the solution (1) in (35). Then, redefining

$$\hat{x}^+ = \exp\{2i\kappa\} \tilde{x}^+ \,, \qquad \overline{\hat{x}^+} = \exp\{-2i\kappa\} \bar{\tilde{x}}^+ \,, \tag{36}$$

we reduce (31) to

$$\begin{aligned} &\partial^{++} \tilde{x}^+ = 0 \quad \Rightarrow \\ &\tilde{x}^+ = x^i w_i^+ \,,\; \bar{\tilde{x}}^+ = \bar{x}_i w^{+i} = -\bar{x}^i w_i^+ \,,\; \hat{\phi} = -2i\lambda x^{(i} \bar{x}^{k)} w_i^+ w_k^- \,, \end{aligned} \tag{37}$$

where, in expressing $\hat{\phi}$, we essentially made use of the PG symmetry (28).

Combining eqs. (29), (33), (36) and (37) we can now write the expressions for x^+, $\bar{x}^+$ in the following form

$$x^+ = \frac{1}{\sqrt{1 + R\hat{\phi}^2}} \exp\{2i\kappa\}\, x^i w_i^+ \,, \quad \bar{x}^+ = -\frac{1}{\sqrt{1 + R\hat{\phi}^2}} \exp\{-2i\kappa\}\, \bar{x}^i w_i^+ \,, \tag{38}$$

where κ and $\hat{\phi}$ are expressed through $x^i, \bar{x}^i$ according to eqs. (35), (37). Comparing (38) with the general definition of the x -bridges (5), we can identify $x^i, \bar{x}^i$ with the τ- world coordinates, i.e. with the coordinates of the initial 4-dimensional QK manifold.

We still need to find $x^-, \bar{x}^-$ as functions of $x^i, \bar{x}^i$ and harmonics $w_i^\pm$ by solving eq. (27) and its conjugate. Dropping intermediate technical steps (they involve a number of redefinitions), its general solution can be presented in the following form

$$x^- = \frac{1}{2\lambda}\,\frac{\sqrt{1+R\hat{\phi}^2}}{(\lambda s)-i\hat{\phi}}\left[\mathrm{e}^{-2i\kappa(i\lambda s)}-\mathrm{e}^{2i\kappa(\hat{\phi})}\right]\tilde{x}^- \ ,$$

$$\bar{x}^- = \frac{1}{2\lambda}\,\frac{\sqrt{1+R\hat{\phi}^2}}{(\lambda s)+i\hat{\phi}}\left[\mathrm{e}^{-2i\kappa(i\lambda s)}-\mathrm{e}^{-2i\kappa(\hat{\phi})}\right]\tilde{\bar{x}}^- \ . \tag{39}$$

Here

$$\tilde{x}^- = x^i w_i^- \ , \tilde{\bar{x}}^- = -\bar{x}^i w_i^- \ , s = x^i\bar{x}_i \ , \kappa(i\lambda s) \equiv \kappa_0 = \frac{i\lambda}{\sqrt{R}}\ \mathrm{arctanh}\ \sqrt{R}(\lambda s) \ . \tag{40}$$

For what follows it will be convenient to define

$$A(s) \equiv 1 - R\lambda^2 s^2 \ , \ B(s) \equiv 1 + 4\lambda^2 s + R\lambda^2 s^2 \ , \ C(s) \equiv 1 + Rs + R\lambda^2 s^2 \tag{41}$$

At this step we can find explicit expressions for the two important quantities entering the τ-metric (17), (18):

$$1 - R\partial^{--}v^{++} = A\,\frac{1-R\hat{\phi}^2}{(1+R\hat{\phi}^2)^2} \ , \quad 1 - R(x\cdot H) = \frac{C}{A}\,\frac{1+R\hat{\phi}^2}{1-R\hat{\phi}^2} \ . \tag{42}$$

As a next step towards the QK Taub-NUT metric, one needs to find the entries of the matrix $X_\nu^{+\mu i} \equiv \partial_\nu^+ x^{\mu i}$ by solving the set of algebraic equations (20). In the complex notation, this set is divided into the two mutually conjugated ones, each consisting of four equations. It is clearly enough to consider one such set, e.g.

$$X^{+\rho k}\nabla_{\rho k}x^- = 1, \quad X^{+\rho k}\nabla_{\rho k}x^+ = 0, \quad X^{+\rho k}\nabla_{\rho k}\bar{x}^\pm = 0, \tag{43}$$

where $X^{+\rho k} \equiv X_1^{+\rho k}, (\bar{X}^{+\rho k} \equiv -X_2^{+\rho k})$. It is convenient to work with

$$\hat{X}^{+\rho k} = e^{Rv}X^{+\rho k} = \sqrt{1+R\hat{\phi}^2}\ X^{+\rho k} \ . \tag{44}$$

It remains to calculate the transition matrix elements $\nabla_{\rho k}x^\pm, \nabla_{\rho k}\bar{x}^\pm$ entering eqs. (20). This can be done straightforwardly, the corresponding expressions look rather involved and by this reason we do not quote them here explicitly (more details are given in (Ivanov and Valent)). Surprisingly, the expressions for $\hat{X}^{+\rho k}$ prove to be much simpler:

$$\hat{X}^{+1k} = \frac{1}{4}\left[(3A+B)\epsilon^{kl} - 4\lambda^2(A+C)x^{(k}\bar{x}^{l)}\ \right] w_l^+ e^{2i\kappa_0} \ ,$$

$$\hat{X}^{+2k} = (\partial^+\hat{\bar{x}}^k) = \lambda^2\ (A+C)\ (\bar{x}^k\bar{x}^l)w_l^+ e^{2i\kappa_0} \tag{45}$$

(the remaining components can be obtained by conjugation).

It will be convenient to rewrite the metric (17), (18) through $\hat{X}^{+\rho i}_{\mu}$

$$g^{\rho i,\lambda k} = \frac{1}{C\det(\partial\hat{H})}\,\hat{G}^{\rho i,\lambda k}\,, \tag{46}$$

$$\hat{G}^{\rho i,\lambda k} = (1+R\hat{\phi}^2)\,G^{\rho i,\lambda k} = \epsilon^{\omega\beta}\,[\,\partial^{--}\hat{X}^{+\rho i}_{\omega}\,\hat{X}^{+\lambda k}_{\beta} + (\rho i \leftrightarrow \lambda k)\,]\,. \tag{47}$$

As the last step, one should compute $\det(\partial\hat{H})$. Afters a rather cumbersome, though straightforward computation one eventually gets the simple expression for this quantity

$$\det(\partial\hat{H}) = A^2\,\frac{B}{C^3}\,e^{4i\kappa_0} = (1-R\lambda^2s^2)^2\,\frac{1+2\lambda^2 s+\lambda^2 s(2+sR)}{(1+Rs+R\lambda^2s^2)^3}\,e^{4i\kappa_0}\,. \tag{48}$$

The harmonic dependence disappeared in $\det(\partial\hat{H})$, as it should be.

Once this has been done, the computation of the τ basis inverse metric amounts to the computation of entries of the matrix $\hat{G}^{\rho i,\lambda l}$. The final answer for the metric tensor is as follows

$$\begin{cases} g_{1k,1t} = \dfrac{D}{C^2B}\,(\bar{x}_k\bar{x}_t)\,, & A = 1-R\lambda^2s^2, \\ g_{2k,2t} = \dfrac{D}{C^2B}\,(x_k x_t)\,, & B = 1+4\lambda^2 s+R\lambda^2s^2, \\ g_{1k,2t} = \dfrac{1}{C^2B}\,[B^2\epsilon_{kt}+D(\bar{x}_k x_t)\,]\,, & C = 1+Rs+R\lambda^2s^2. \end{cases} \tag{49}$$

Here $D \equiv \lambda^2(A+C)(A+B) = 2\lambda^2(2+Rs)(1+2\lambda^2 s)$.

4. To compare to the results in the literature (Eguchi, Gilkey and Hanson (1980)) one has to use (Delduc and Valent (1993))

$$dx^i = x^i\left(\frac{ds}{2s}+i\frac{\sigma_3}{2}\right) - \bar{x}^i\left(\frac{\sigma_2-i\sigma_1}{2}\right), \qquad d\sigma_i = \frac{1}{2}\epsilon_{ijk}\,\sigma_j\wedge\sigma_k. \tag{50}$$

Using the notation $A\cdot B \equiv A^i\,B_i$ relation (50) implies

$$-\bar{x}\cdot dx = \frac{ds}{2}+is\frac{\sigma_3}{2}, \qquad dx\cdot d\bar{x} = \frac{ds^2}{4s}+\frac{s}{4}(\sigma_1^2+\sigma_2^2+\sigma_3^2), \qquad s = x\cdot\bar{x}. \tag{51}$$

The metric given by (49) becomes

$$\frac{1}{2}\left[\frac{B}{sC^2}\,ds^2+\frac{sB}{C^2}\,(\sigma_1^2+\sigma_2^2)+\frac{sA^2}{C^2B}\,\sigma_3^2\right]. \tag{52}$$

The most general Bianchi IX euclidean Einstein metrics can be deduced from Carter's results (Carter (1968)). A convenient standardization (Chave and Valent (1996)) is the following

$$d\tau^2 = l^2\left\{\frac{r^2-1}{\Delta(r)}(dr)^2+4\,\frac{\Delta(r)}{r^2-1}\,\sigma_3^2+(r^2-1)(\sigma_1^2+\sigma_2^2)\right\}, \tag{53}$$

with

$$\Delta(r) = \frac{-\Lambda l^2}{3} r^4 + (1 + 2\Lambda l^2) r^2 - 2\,M\,r + 1 + \Lambda l^2. \tag{54}$$

These metrics are Einstein, with Einstein constant Λ and isometry group $U(2)$. If we take $M = 4/3\Lambda l^2 + 1$ the metric simplifies to

$$d\tau^2(Q) = l^2 \left\{ \frac{r+1}{r-1} \frac{(dr)^2}{\Sigma(r)} + 4\,\frac{r-1}{r+1}\,\Sigma(r)\,\sigma_3^2 + (r^2-1)(\sigma_1^2 + \sigma_2^2) \right\}, \tag{55}$$

where now

$$\Sigma(r) = 1 - \frac{\Lambda\, l^2}{3}(r-1)(r+3). \tag{56}$$

The identifications

$$\frac{r-1}{2} = (4\lambda^2 - R)\frac{s}{1 + Rs + R\lambda^2 s^2}, \qquad \frac{4}{3}\Lambda l^2 = \frac{R}{4\lambda^2 - R}, \tag{57}$$

give the relation

$$4(4\lambda^2 - R)\left[\frac{B}{sC^2}\,ds^2 + \frac{sB}{C^2}\,(\sigma_1^2 + \sigma_2^2) + \frac{sA^2}{C^2 B}\,\sigma_3^2\right] = \frac{d\tau^2(Q)}{l^2}. \tag{58}$$

The quaternionic metric (55) is complete for $\Lambda < 0$ and is asymptotically Anti de Sitter. It has been considered recently in (Hawking, Hunter and Page (1998)) under the name Taub-NUT-AdS metric and reveals itself a useful background for computing black-holes entropy.

5. In this note we made the first practical use of the harmonic space formulation of the QK geometry (Galperin, Ivanov and Ogievetsky (1994)) to compute a non-trivial QK metric, the four-dimensional quaternionic Taub-NUT metric. As we were convinced, the harmonic space techniques, like in the HK case (Galperin et al (1986), Galperin et al (1988), Galperin et al (1986a)), allows one to get the *explicit* form of the QK metric starting from a given QK potential and following a generic set of rules. It would be interesting to apply this approach to find the QK analogs of some other interesting 4- and higher-dimensional HK metrics, in particular, the quaternionic Eguchi-Hanson metric and the quaternionic generalization of the multicenter metrics of Gibbons and Hawking (Gibbons and Hawking (1978)).

A first simple generalization of the HK Taub-NUT potential would be to add the dipolar breaking (Gibbons et al (1988))

$$\mathcal{L}^{+4} = \eta^{--}(\phi^{++})\,, \quad \eta^{--} = \eta^{(ik)} u_i^- u_k^-\,,\ \eta^{(ik)} = const\,.$$

When trying to compute the *quaternionic* metric with the same $\mathcal{L}^{+4}$, one encounters a difficulty already at the step of solving the equation for the bridge v^{++}. Performing the same changes of variables as those leading to eq. (30), we have in this case

$$(\partial^{++})^2\,\omega = 2\,R\,\eta^{--}\,\frac{(\hat{\phi}^{++})^3}{\omega^5}\,, \quad \partial^{++}\hat{\phi}^{++} + 0\,.$$

This equation is not so easy to solve as compared to (30). Thus, in order to treat the multicenter case, some ways around should be perhaps developed.

Finally, we note that the HK Taub-NUT metric and its multicenter generalizations play an important role in the modern p-branes realm, yielding an essential part of one of the fundamental brane-like classical solutions of $D = 11$ supergravity, the so-called 'Kaluza-Klein monopole' (see, e.g. (Stelle (1998))). It would be of interest to reveal possible brane implications of the QK Taub-NUT metric constructed here.

Acknowledgement. E.I. thanks the Directorate of LPTHE, Université Paris 7, for the hospitality extended to him during the course of this work. His work was partly supported by the grant of Russian Foundation of Basic Research RFBR 96-02-17634 and by INTAS grants INTAS-93-127-ext, INTAS-96-0538.

References

A. Galperin, E. Ivanov, S. Kalitzin, V. Ogievetsky, and E. Sokatchev, Class. Quantum Grav. **1** (1984) 469.

A. Galperin, E. Ivanov, V. Ogievetsky and E. Sokatchev, Commun. Math. Phys. **103** (1986) 515.

A. Galperin, E. Ivanov, V. Ogievetsky and E. Sokatchev, Ann. Phys. (N.Y.) **185** (1988) 22.

A. Galperin, E. Ivanov and O. Ogievetsky, Ann. Phys. (N.Y.) **230** (1994) 201.

T. Eguchi, B. Gilkey and J. Hanson, Physics Reports, **66**, No. 6 (1980) 213.

J.A. Bagger, A.S. Galperin, E.A. Ivanov and V.I. Ogievetsky, Nucl. Phys. **B 303** (1988) 522.

E. Ivanov, G. Valent, 'Harmonic Space Construction of the Quaternionic Taub-NUT metric', in preparation.

A. Galperin, E. Ivanov, V. Ogievetsky and P.K. Townsend, Class. Quantum Grav. , **7** (1986) 625.

F. Delduc and G. Valent, Class. Quantum Grav. , **10** (1993) 1201.

B. Carter, Commun. Math. Phys. **10** (1968) 280.

T. Chave and G. Valent, Class. Quantum Grav., **13** (1996) 2097.

H. Pedersen, Math. Ann., **274** (1986) 35.

G. Gibbons and S.W. Hawking, Phys. Lett., **B 78** (1978) 430.

S.W. Hawking, G.C. Hunter and D.N. Page, 'Nut Charge, Anti-de Sitter Space and Entropy', `hep-th/9809035`.

G.W. Gibbons, D. Olivier, P.J. Ruback and G. Valent, Nucl. Phys., **B 296** (1988) 679.

K.S. Stelle, 'BPS Branes in Supergravity', CERN-TH/98-80, Imperial/TP/97-98/30; `hep-th/9803116`.

Supergeometry in Equivariant Cohomology

Armen Nersessian[12]

[1] Bogolyubov Laboratory of Theoretical Physics, JINR,
Dubna, 141980, Russia
[2] Department of Theoretical Physics, Yerevan State University,
A. Manoukian st., 5, Yerevan, 375012 Armenia
E-mail: nerses@thsun1.jinr.ru

Abstract. We analyze S^1 equivariant cohomology from the supergeometrical point of view. For this purpose we equip the external algebra of given manifold with equivariant even super(pre)symplectic structure, and show, that its Poincare-Cartan invariant defines equivariant Euler classes of surfaces. This allows to derive localization formulae by use of superanalog of Stockes theorem.

1 Introduction

Since the late eighties localization formulae attract permanent interest in the physical community due their application to evaluation of path integrals in quantum mechanics, topological and supersymmetric field theories (see [1] and refs therein).

The original localization formula [2](Duistermaat and Heckman) states, that if (M, ω, S^1) is $2n$-dimensional compact symplectic manifold, and $H(x)$ is the Hamiltonian, generating S^1-group symplectic action, the classical partition function is localized over the critical points of Hamiltonaian

$$\left(\frac{i\phi}{2\pi}\right)^{2n} \int_M \mathrm{e}^{i\phi H} \frac{\omega^n}{n!} = \sum_{dH=0} \frac{\mathrm{e}^{i\phi H}\sqrt{\det \omega}}{\sqrt{\det \mathrm{Pf} H}}. \tag{1}$$

Application of this formula to path integrals gives an elegant way for their evaluation, as well as the conditions of exactness of stationary phase approximation.

Localization formula of Duistermaat and Heckman can be naturally interpreted in terms of equivariant cohomology [1] [3](Atiah and Bott). This allowed one to obtain other localization formulae, related with both Abelian (see, e.g. [4]), and non-Abelian [5] equivariant cohomologies. Also other features of equivariant cohomologies relevant to various aspects of quantum field theory were found (see [1], [5], [6] and refs therein).

The language of supersymmetry simplifies greatly formulation of equivariant cohomologies, openings new horizons in their study by use of advanced supermatematical technique [7](Schwarz and Zaboronsky). It gives the shortest

[1] G-equivariant cohomology of the G-manifold (M, G) called the G-invariant cohomology of the factormanifold M/G

way for incorporation of equivariant cohomology in supersymmetric theories and BRST quantization methods.

In the framework of supermathematical description of equivariant cohomology the exterior algebra ΛM of the given manifold M is considered a supermanifold, parametrized by local coordinates $z^A = (x^i, \theta^i)$, where x^i are local coordinates of M and θ^i are basic 1-forms dx^i, $p(\theta^i) = 1$. Thus, differential forms on M can be considered as functions defined on a supermanifold. The operators of exterior derivative, Lie derivative and inner product are represented by vector fields. Divergency operator is represented by odd second-order differential operator, known as "Batalin-Vilkovisky Δ-operator".

Most convenient model of S^1–equivariant cohomology, the so-called Cartan model, is formulated in terms of simplest superalgebra

$$[\hat{E}, \hat{E}]_+ = 2\hat{X}, \tag{2}$$

realized by the vector fields:

$$\hat{X} = \xi^i(x)\frac{\partial}{\partial x^i} + \xi^i_{,k}(x)\theta^k\frac{\partial}{\partial \theta^i}, \quad \hat{E} = \xi^i(x)\frac{\partial}{\partial \theta^i} + \theta^i\frac{\partial}{\partial x^i} \tag{3}$$

where vector field $\xi = \xi^i \frac{\partial}{\partial x^i}$ defines infinitesimal S^1-action on M.

It is obvious that vector $\hat{X}$ corresponds to Lie derivative of differential forms. The vector $\hat{E}$ corresponds to sum of operators of exterior derivative and inner product. Expression (2) is nothing else but the homotopy formula $L_\xi = d\imath_\xi + \imath_\xi d$. Thus, restriction of $\hat{E}$ field (equivariant differential) to the (sub)space of $\hat{X}$-invariant functions is nilpotent, and its cohomologies on this subspace defines the equivariant cohomologies of (M, S^1).

To relate equivariant cohomologies with localization formulae, let consider the following functional

$$Z^\lambda(A) = \int_{\Lambda(M)} A(x,\theta)\mathrm{e}^{-\lambda\hat{E}\Psi}\mathcal{D}(x,\theta), \tag{4}$$

where $\mathcal{D}$ is volume element, A and Ψ are respectively even and odd functions on ΛM, λ is arbitrary parameter.

This functional is $\hat{E}$-invariant, if requires:

$$\hat{E}A = 0, \quad \hat{X}\Psi = 0, \quad div_{\mathcal{D}}\hat{E} = 0. \tag{5}$$

Applying standard BRST analyzes, one can find, that the functional (4) is λ-independent, i.e. only S^1-equivariant forms have contribution to its value.

In order to derive localization formulae, we have to choose the "gauge fermion"

$$\Psi = \xi^i g_{ij}\theta^i, \tag{6}$$

where g is an S^1-invariant Riemann metric: $\mathcal{L}_\xi g = 0$.

Taking into account the following representation of δ-function

$$\delta(\xi) = \frac{\lambda^{\dim M}}{\pi^{\dim M}\sqrt{\det g_{ij}}} \lim_{\lambda\to\infty} e^{-\lambda\xi^i g_{ij}\xi^j} \tag{7}$$

we find, that in $\lambda \to \infty$ limit the functional (4) localizes in the critical points of ξ.

For example, for a compact symplectic manifold (M, ω, S^1), where $\xi = \omega^{-1}(dH,\ )$ one can choose $\mathcal{D} = 1$ and A is equivariant Chern class $Ch(H, \omega)$ defined as follows

$$Ch(H,\omega) \equiv \exp i\phi(H - \frac{1}{2}\omega_{ij}\theta^i\theta^j). \tag{8}$$

Substituting these expressions in (4), we will get Duistermaat-Heckman formula (1).

There is also another $\hat{E}$-invariant function, $Eu(\xi, g)$

$$Eu(\xi, g) \equiv \sqrt{\det\left(\xi^i_{;j} + R^i_{jkl}\theta^k\theta^l\right)}, \tag{9}$$

called equivariant Euler class.

Substituting this function in (4), one gets that in $\lambda \to 0$ limit we recognize the statement of Poincare-Hopf theorem , while in $\lambda \to \infty$ limit: Gauss-Bonnet Theorem.

Derivation of another existing localization formulae is analogous.

In present paper we study S^1 equivariant cohomology from the point of view of supersymplectic geometry. The key observation is that on ΛM there exists $\hat{E}$-invariant presymplectic structure, whose Poincare-Cartans ivariants define equivariant Euler classes of the surfaces in given manifold. This allows one to formulate an analog of the functional (4) on surfaces in ΛM, and to derive localization formulae by use of generalization of Stokes theorem for even symplectic supermanifolds [8] (Khudaverdian, Schwarz and Tyupkin), and relate them with topological numbers of surfaces. In a contrast to equivariant Euler classes, equivariant Chern classes are superfunctions, related with the odd symplectic structure, constructed on the exterior algebra of symplectic manifold [12].

Remarkable point of actual approach is that it is based on the odd symplectic geometry, which gives transparent parallels with Batalin-Vilkovisky formalism [9].

This make possible the application to equivariant cohomologies the results, concerning the geometry of Batalin-Vilkovisky formalism ([10]). From the other hand, using of recently developed technique of quantization of antibrackets [11](Batalin and Marnelius,1998) seems fruitful in application of equivariant cohomology to path integrals.

2 Equivariant Euler Class

Let us consider (M, S^1) manifold with the S^1-invariant metric g. Using this metrics, let us construct on ΛM an odd $\hat{X}$-invariant one-form and corresponding odd symplectic structure:

$$\mathcal{A} = \theta^i g_{ij} dx^j, \qquad \mathcal{L}_{\hat{X}}\mathcal{A} = 0; \tag{10}$$

$$\Omega_1 = d\mathcal{A} = g_{ij} dx^i \wedge D\theta^j, \mathcal{L}_{\hat{X}}\Omega_1 = 0, \tag{11}$$

where

$$D\theta^j \equiv g_{ij} dx^i \wedge (d\theta^i + \Gamma^i_{kl}\theta^k dx^l), \tag{12}$$

and Γ^i_{kl} denote Cristoffel symbols of the metrics g.

From this follows, that $\hat{X}$ is hamiltonian vector field one with respect to (11), and the gauge fermion (6), plays the role of Hamiltonian:

$$d\Psi = \Omega_1(\hat{X}, \quad), \quad \Psi = \xi^i g_{ij}\theta^j. \tag{13}$$

Forms $\mathcal{A}$ and Ω_1 possess remarkable property: being $\hat{X}-$ invariant, they are not $\hat{E}$-invariant:

$$\mathcal{E} \equiv \mathcal{L}_{\hat{E}}\mathcal{A} \neq 0; \quad \Omega_0 \equiv \mathcal{L}_{\hat{E}}\Omega_1 \neq 0. \tag{14}$$

Explicitly,

$$\mathcal{E} = \xi_i dx^i + \theta^i g_{ij} D\theta^j, \tag{15}$$

$$\Omega_0 = \tfrac{1}{2}(\xi_{[i,j]} + g_{in} R^n_{jkl}\theta^k\theta^l) dx^i \wedge dx^j + g_{ij} D\theta^i \wedge D\theta^j. \tag{16}$$

From Eqs. (15) and (16) it follows immediately, that the forms $\mathcal{E}$ and Ω are $\hat{E}$-invariant. Indeed

$$\mathcal{L}_{\hat{E}}\mathcal{E} = \mathcal{L}_{\hat{E}}\mathcal{L}_{\hat{E}}\mathcal{A} = 2\mathcal{L}_{\hat{X}}\mathcal{A} = 0.$$

Similarly one can check $\hat{E}$-invariance of the second structure.

Since the two-form Ω is closed, we conclude, that *the expression (16) defines the S^1-equivariant even pre-symplectic structure on ΛM.*

The Hamiltonians of the vector fields, given by Eq.(3) read:

$$\Omega_0(\hat{X}, \quad) = d\mathcal{H}, \quad \mathcal{H} = \hat{E}\Psi = \xi^i g_{ij}\xi^j - \xi_{i;j}\theta^i\theta^j. \tag{17}$$

$$\Omega_0(\hat{E}, \quad) = d\Psi, \quad \Psi = \xi_i\theta^i. \tag{18}$$

Consider now a closed surface $\Gamma \subset \Lambda M$, $\Omega|_\Gamma \neq 0$, parametrized by the equations $z^A = z^A(w)$, where w^μ are local coordinates of Γ.

One can construct a closed density on Γ, invariant under canonical transformations of Ω_0 (Poincare-Cartan invariant), which defines characteristic class of Γ [8]:

$$\mathcal{D}_\Gamma(z(w), \partial z(w)) = \sqrt{Ber \frac{\partial z^A}{\partial w^\mu}\Omega_{(0)AB}\frac{\partial z^B}{\partial w^\nu}}. \tag{19}$$

Taking into account, that $\hat{E}$-field generates canonical transformation, we conclude, that the density, given by Eq.(19) is $\hat{E}-$invariant. Thus, it defines an equivariant characteristic class of Γ surface. When Γ coincides with ΛM, this density is just *equivariant Euler class* (9) .

Let us give some explicit realization of density, given by Eq. (19). Consider surface $\Gamma \subset \Lambda M$, given by the equations

$$z^A = z^A(w): \quad x^i = x^i(y^a), \quad \theta^i = P^i_\alpha(y)\eta^\alpha, \qquad p(\eta) = p(y) + 1. \qquad (20)$$

Here $w^\mu = (y^a, \eta^\alpha)$ are local coordinates on Γ, where y^a are local coordinates on $N \subset M$ (notice, that after change of the grading $p(\eta) \to p(\eta) + 1$ lead to vector bundle $V(N) \subset T(M)$, associated with Γ).

The restriction of presymplectic structure (16), on Γ_0 looks as follows

$$\Omega_0|_{\Gamma_0} = \frac{1}{2}(\xi_{[a,b]} + g_{\alpha\delta}R^\delta_{\beta ab}\eta^\alpha\eta^\beta)dy^a \wedge dy^b + g_{\alpha\beta}D\eta^\alpha \wedge D\eta^\beta, \qquad (21)$$

where

$$D\eta^\alpha \equiv d\eta^\alpha + A^\alpha_\beta\eta^\beta, \quad A^\alpha_\beta = g^{\alpha\delta}P^i_\delta g_{ij}\left(P^j_{\beta,a} + \Gamma^j_{ik}P^k_\beta\frac{\partial x^l}{\partial y^a}\right)dy^a,$$

defines connection one-form, compatible with the induced metric on fiber, given by the following expression

$$g_{\alpha\beta} = P^i_\alpha g_{ij}P^j_\beta,$$

$R^\delta_{\beta ab}dy^a \wedge dy^b$ is the curvature of the connection A^α_β, while

$$\xi_{[a,b]}dy^a \wedge dy^b \equiv \xi_{i;j}\frac{\partial x^i}{\partial y^a}\frac{\partial x^j}{\partial y^b}dy^a \wedge dy^b$$

is induced (pre)symplectic structure on N.

Thus, on the surface Γ_0 the density (19) takes the following form

$$\mathcal{D}(\Omega|\Gamma_0, w) = \left(\frac{\det\,(\xi_{[a,b]} + g_{\alpha\delta}R^\delta_{\beta ab}\eta^\alpha\eta^\beta)}{\det g_{\alpha\beta}}\right)^{\frac{1}{2}}. \qquad (22)$$

3 Equivariant Chern Class

The supergeometrical origin of equivariant Chern class, given by Eq.(8) considerably simpler, than Euler one. Chern class is a function on the exterior algebra of symplectic manifold (M, ω, S^1), with S^1-group action, generated by the Hamiltonian H: $\xi = \omega^{-1}(dH, \,)$.

This algebra can be equipped with odd $\hat{E}$-invariant symplectic structure Ω, in contrast to Ω_1, given by Eq.(11) [12]. Indeed,

$$\mathcal{L}_{\hat{X}}\omega = 0, \quad \mathcal{L}_{\hat{E}}\omega \equiv \Omega \neq 0, \quad \Rightarrow \mathcal{L}_{\hat{E}}\Omega = \frac{1}{2}\mathcal{L}_{\hat{X}}\omega = 0, \quad d\Omega = 0.$$

Explicitly, Ω looks as follows

$$\Omega = \omega_{ij}dx^i \wedge d\theta^j + \frac{1}{2}\omega_{ij,k}\theta^k dx^i \wedge dx^j. \tag{23}$$

The Hamiltonians of the vector fields $\hat{E}$ and $\hat{X}$ are defined by the following expressions:

$$d(H - F) = \Omega(\hat{E}, \ ), \quad Q = \Omega(\hat{X}, \ ), \tag{24}$$

where

$$F = \frac{1}{2}\omega_{ij}\theta^i\theta^j, \quad Q = \hat{E}H = \theta^i \partial H/\partial x^i.$$

Functions F, Q, H, form superalgebra

$$\begin{gathered} \{H \pm F, H \pm F\} = \pm 2Q, \quad \{H + F, H - F\} = 0 \\ \{H \pm F, Q\} = \{Q, Q\} = 0, \end{gathered} \tag{25}$$

where $\{\ ,\ \}$ denotes the antibracket, corresponding to the odd symplectic structure Ω given by Eq. (23).

The functions H, F, Q forms algebra (25) with respect to odd symplectic structure given by Eq.(11), only if (M, g, ω) is a Kahler manifold. This is the only case, when antibrackets, corresponding to these symplectic structures are compatible in the sense of bi-Hamiltonian mechanics. Other relations of equivariant Chern classes with supersymmetric mechanics can be found in Ref.[12].

4 Localization Formulae

Above we constructed $\hat{E}$-invariant closed density (19), corresponding to equivariant Euler classes. Using this density, we can define the following functional

$$Z^{\lambda}(\Gamma, A) = \int_{\Gamma} A(z) e^{-\lambda \hat{E}\Psi} \mathcal{D}_{\Gamma} \tag{26}$$

on the surface Γ.

This functional become $\hat{E}$-invariant, if we assume, as in Introduction, $\hat{E}A = \hat{X}\Psi = 0$. From $\hat{E}$-invariance follows its λ-independence.

Since the density (19) is closed, it doesn't change under smooth deformation of Γ if $A = 1$ [8], i.e. it is defines a topological invariant. In $\lambda \to 0$ and $\lambda \to \infty$ limits one reproduces the statements of Gauss-Bonnet and Poincare-Hopf theorems for Euler characters of the surfaces. Choosing $A = Ch(H, \omega)$ one obtains a generalization of these statements, while $A = Ch(H, \omega)Eu^{-1}(M, g)$ corresponds to Duistermaat-Heckman formula for a surface.

Using the functional (26), one can easily derive localization formulae for the case, when the critical points of ξ-field form some submanifold M_0 [14](Niemi and Palo, 1994). Consider the decomposition $M = M_0 \cup N$, where M_0 is the set of critical points of ξ and N is normal bundle to M_0. Under appropriate choice of $A(z)$ discussed above, one comes to formulae, derived in [14].

Finally, let us give the dual construction for the functional (26), when critical points of ξ are isolated. Let Γ be parametrized by the set equations $f^a(z) = 0$, $a = 1, \ldots \mathrm{codim}\, \Gamma$. In this case the functional (26) can be presented in the following form

$$Z = \int_{\Lambda M} A(z)\delta(f^a)\sqrt{\mathrm{Ber}\{f^a, f^b\}_0}\mathcal{D}_{\Lambda M}(z), \tag{27}$$

where $\{f(z), g(z)\}_0$ is Poisson bracket, corresponding to the (pre)symplectic 2-form (16).

Acknowledgments The author would like to thank I.A.Batalin, O. M. Khudaverdian, A. Niemi for the interest to work and useful comments.

Special thanks to C.Sochichiu for careful reading of manuscript and numerous remarks.

This work has been supported in part by grants INTAS-RFBR No.95-0829, INTAS-96-538 and INTAS-93-127-ext .

References

1 A.J.Niemi, O.Tirkkonen, *Ann. Phys.* **235**(1994), 318;

2 J.J. Duistermaat, G.H.Heckman, *Inv. Math.* **69**(1982) 259; *ibid***72** (1983), 153

3 M.F. Atiah, R.Bott, *Topology*, **23** (1984), 1

4 N.Berline, E.Getzler, M.Vergne, *Heat Kernel and Dirac Operators*(Springer Verlag, Berlin, 1991)

5 E.Witten, *J.Geom.Phys.*, **9**(1992), 303
L.C.Jefferey, F.C. Kirwan, alg-geom/9307001

6 M.F.Atiah, L.Jefferey, *J.Geom.Phys.*, **7**(1990), 119
M.Blau, F.Hussain, G.Tompson, *Nucl.Phys.* **B488**(1997), 541

7 A.Schwarz, O.Zaboronsky, *Comm. Math. Phys.* **183**(1996), 463

8 O.M.Khudaverdian, A.S.Schwarz, Yu.S.Tyupkin, *Lett.Math.Phys.* **5**(1981), 517

9 I.A.Batalin, G.A.Vilkovisky, *Phys.Lett.* **B102**(1981), 27; Phys.Rev. **D28**(1983) 2563; Nucl.Phys.**B234**(1984), 106

10 A.Schwarz, *Comm. Math. Phys.*, **155**(1993), 249; *ibid.*, **158**(1993), 373
O.M.Khudaverdian, dg-ga/9706004, *Comm.Math.Phys.*(1998) (in press)

11 I.A.Batalin, R. Marnelius,*Phys.Lett.* **B434**(1998), 312; hep-th/9809208

12 A.Nersessian, *JETP Lett.* **58**, No. 1 (1993), 66 (hep-th/9305181)

13 A.Nersessian, *NATO ASI Series B:Physics*, **331**, 353 (hep-th/9306111); hep-th/9310013

14 A.J.Niemi, K.Palo, hep-th/940668

Harmonics, Notophs and Chiral Bosons

Paolo Pasti[1], Dmitri Sorokin[123], and Mario Tonin[1]

[1] Università Degli Studi Di Padova, Dipartimento Di Fisica "Galileo Galilei" ed INFN, Sezione Di Padova
Via F. Marzolo, 8, 35131 Padova, Italia
[2] Humboldt-Universität zu Berlin, Institut für Physik
Invalidenstrasse 110, D-10115 Berlin, Germany
[3] Alexander von Humboldt fellow.
On leave from Kharkov Institute of Physics and Technology, Kharkov, 310108, Ukraine.

Abstract. A way of covariantizing duality symmetric actions is described.

The presence of self–dual fields or, in more general case, duality–symmetric fields in field–theoretical and string models reflects their duality properties whose extreme importance for understanding a full quantum theory has been appreciated during an impetuous development of the duality field happened during last few years. The knowledge of duality–symmetric effective actions is useful for carrying out more systematic study of the classical and quantum properties of the theory, and in this memorial contribution we would like to demonstrate how fruitful physical ideas and mathematical techniques which Victor Isakovich Ogievetsky and his colleagues have developed helped us to construct a covariant Lagrangian formulation applicable to all known models with duality–symmetric fields in space–time of Lorentz signature.

The problem of constructing and studying models described by duality–invariant actions has a rather long history. It goes back to time when Poincare and later on Dirac noticed electric–magnetic duality symmetry of the free Maxwell equations, and, Dirac (1931),(1948) assumed the existence of magnetically charged particles (monopoles and dyons) admitting the duality symmetry to be also held for the Maxwell equations in the presence of charged sources. To describe monopoles and dyons on an equal footing with electrically charged particles one should have a duality–symmetric form of the Maxwell action. This problem was studied (among others) by Schwinger and Zwanziger, and Zwanziger (1971) proposed a duality–symmetric action for Maxwell fields interacting with dyonic sources. An alternative duality–symmetric Maxwell action was proposed by Deser and Teitelboim (1976). The two actions, which proved to be dual to each other Maznytsia et. al. (1998), are not manifestly Lorentz–invariant. This feature turned out to be a general one. Duality and space–time symmetries hardly coexist in one and the same action.

Later on this problem arose in multidimensional supergravity theories in space–time of a dimension $D = 4p + 2$ where one would like to know how to construct an action for self–dual tensor fields (chiral bosons) which are present in some versions of supergravity and in the heterotic string. One of the ways of solving this problem is to sacrifice Lorentz covariance in favour of duality symmetry. A non–covariant action for $D = 2$ chiral bosons was constructed by Floreanini and Jackiw (1987), and Henneaux and Teitelboim (1987),(1988) proposed non–covariant actions for self–dual fields in higher dimensional $D = 4p+2$ space–time. In a context of modern aspects of duality Tseytlin considered a duality–symmetric action for a string. Finally, Schwarz and Sen (1994) constructed non–covariant duality–symmetric actions for dual tensor fields in any space–time dimension.

There have also been developed covariant approachs to the construction of duality–symmetric actions. These use auxiliary fields. The first covariant Lagrangian formulation of chiral bosons was proposed by Siegel (1984) and its modification was considered by Kavalov, Mkrtchyan (1990) in application to D=6 and D=10 chiral supergravities. Another covariant approach is based on the use of an infinite number of auxiliary fields (McClain, Wu and Yu (1990), Wotzasek (1991), Martin and Restuccia (1994), Devecchi and Henneaux (1996), Bengtsson and Kleppe (1997), Berkovits (1996)). It might be interesting that an effective self–dual action of this kind was extracted from a type IIB string field theory (Berkovits (1996)).

The third formulation was proposed in (Pasti, Sorokin and Tonin (1995), (1997)). In its minimal version only one scalar auxiliary field is used to ensure space–time covariance of duality–symmetric actions. This approach turned out to be the most appropriate for the construction of the worldvolume action for the M-theory five–brane (Pasti, Sorokin and Tonin (1997), Bandos et. al. (1997), Aganagic et. al. (1997)), duality–symmetric D=11 supergravity (Bandos, Berkovits and Sorokin (1998)) and D=10 IIB supergravity (Dall'Agata et. al. (1997), (1998)).

Below we will use Maxwell theory to demonstrate how this third approach was developed with promptings provided by works of V. I. Ogievetsky.

It is well known that the standard action for a free Maxwell field is not invariant under duality transformations of its electric and magnetic strength vectors. To have a duality symmetry at the level of action one should double the number of gauge fields (A^α_m, $\alpha = 1,2$, m=0,1,2,3) (Zwanziger (1971), Deser and Teitelboim (1976), Schwarz and Sen (1994)) and construct an action in such a way that one of the gauge fields becomes an auxiliary field upon solving equations of motion. The duality symmetric action of refs. (Deser and Teitelboim (1976), Schwarz and Sen (1994)) can be written in the following form:

$$S = \int d^4x \left(-\frac{1}{8}F^\alpha_{mn}F^{mn\alpha} + \frac{1}{4}\mathcal{F}^\alpha_{0i}\mathcal{F}^\alpha_{i0}\right), \qquad (i = 1,2,3) \tag{1}$$

where

$$\mathcal{F}^\alpha_{mn} = \mathcal{L}^{\alpha\beta}F^\beta_{mn} - \frac{1}{2}\epsilon_{mnlp}F^{lp\alpha} = \frac{1}{2}\epsilon_{mnlp}\mathcal{F}^{lp\beta}\mathcal{L}^{\alpha\beta}, \tag{2}$$

($\mathcal{L}^{12} = -\mathcal{L}^{21} = 1$) is the self–dual combination of the field strengths.

The Zwanziger (1971) action differs from (1) by the sign in front of the second term and in that, instead of the time–coordinate index, one of the spatial indices is separated in the analogous term of the Zwanziger action.

Duality symmetry is a discrete subgroup of $SO(2)$ rotations of A^α_m ($A^\alpha_m \to \mathcal{L}^{\alpha\beta}A^\beta_m$).

Note that because of the self–duality property (2) $\mathcal{F}^\alpha_{mn}\mathcal{F}^{\alpha mn} \equiv 0$, and the best thing which one can do is to take the square of only a part of the components of $\mathcal{F}^\alpha_{mn}$ for the construction of the second term of the action (1), and this breaks manifest Lorentz invariance.

Here is a place to explain why the signature of space–time is important for the possibility of applying the Lagrangian approach considered to the description of chiral bosons. It is crucial for this approach that the "square" of a self–dual tensor is zero, which holds, for instance, in $D = 2p + 2$ spaces of a Lorentz signature. Then taking the square of an appropriate part of the components of the self–dual tensor (as in (1)) one gets the desirable result. On the contrary, for instance, in $D = 4$ space of Euclidian signature the square of the self–dual combination of a gauge field–strength is no–zero and reproduces (up to a total derivative) the standard Maxwell Lagrangian, and no reasonable choice of its components is known in these cases to construct actions analogous to (1).

We have seen that the method we used to get the action breaks manifest Lorentz invariance, however, beside the manifest spatial rotations the action (1) is invariant under the following modified space–time transformations of A^α_i (in the gauge $A^\alpha_0 = 0$)

$$\delta A^\alpha_i = x^0 v^k \partial_k A^\alpha_i + v^k x^k \partial_0 A^\alpha_i + v^k x^k \mathcal{L}^{\alpha\beta}\mathcal{F}^\beta_{0i}, \tag{3}$$

where the first two terms describe the ordinary Lorentz boosts along a constant velocity v^i and the third term vanishes on the mass shell since an additional local symmetry of the action (1)

$$\delta A^\alpha_0 = \varphi^\alpha(x) \tag{4}$$

allows one to reduce the equations of motion

$$\frac{\delta S}{\delta A^\alpha_i} = \epsilon^{ijk}\partial_i \mathcal{F}^\alpha_{k0} = 0 \tag{5}$$

to the duality condition

$$\mathcal{F}^\alpha_{mn} = \mathcal{L}^{\alpha\beta}F^\beta_{mn} - \frac{1}{2}\epsilon_{mnlp}F^{lp\alpha} = 0 \tag{6}$$

which, on the one hand, leads to the Maxwell equations

$$\partial_m \mathcal{F}^{mn\alpha} = \partial_m F^{mn\alpha} = 0 \tag{7}$$

and, on the other hand, completely determines one of the gauge fields through another one. For instance, using the relation $\frac{1}{2}\epsilon_{mnlp}F^{2mn} = F^1_{mn}$ we can exclude $A^2_m(x)$ from (1) and get the conventional Maxwell action.

One can admit that the action (1) arose as a result of some gauge fixing which specifies time direction in a Lorentz covariant action (Pasti, Sorokin and Tonin (1995), (1997)).

The first step is to covariantize the self–dual part of the action (1). For people who are acquainted with harmonic techniques served for similar covariantization purposes in supersymmetric theories (GIKOS) the first thing which comes into mind is to introduce an auxiliary harmonic–like vector field

$$l_m(x) \equiv \frac{u_m(x)}{\sqrt{-u_n u^n}}, \qquad l_m l^m = -1, \tag{8}$$

and to write the action as follows:

$$S_A = \int d^4x(-\frac{1}{8}F^{\alpha}_{mn}F^{\alpha mn} + \frac{1}{4(-u_l u^l)}u^m \mathcal{F}^{\alpha}_{mn}\mathcal{F}^{\alpha np}u_p). \tag{9}$$

The field (8) is harmonic in the sense that if supplemented with space–like fields $l^i_m(x)$, $l^i_m l^{jm} = \delta^{ij}$, $l^m l^i_m = 0$ (which do not enter the action) the set of the four vector fields form a matrix of the Lorentz group $SO(1,3)$ and can be used to contract Lorentz indices of other fields in a covariant way. For the analysis of properties of the action it was proved more convenient to work with the field $u(x)$ rather then with its normalized form $l(x)$, and at the same time to use harmonic properties of the latter.

The main problem is to find a local symmetry which would permit to choose a gauge where u_m is a constant vector, in particular,

$$u_m(x) = \delta^0_m. \tag{10}$$

Then the action (9) can reduce to (1). Note that a spatial gauge, for instance $u_m(x) = \delta^3_m$ is equally admissible and leads to a non-covariant action which also produces the duality condition (6).

The search for this symmetry turns out to be connected with another problem, namely, the problem of preserving a local symmetry under (4). In the covariant version this transformation should be replaced by

$$\delta A^{\alpha}_m = u_m \varphi^{\alpha}. \tag{11}$$

To keep this symmetry is important (as we have already seen) for getting the duality condition (6).

To have the invariance under transformations (11 one should add to the Lorentz invariant action (9) another term

$$S_B = -\int d^4x \epsilon^{mnpq} u_m \partial_n B_{pq}, \tag{12}$$

where $B_{mn}(x)$ is an antisymmetric tensor field. Then the variation of (9) under (11) is canceled by the variation of (12) under

$$\delta B_{mn} = -\frac{\varphi^\alpha}{u^2}(\mathcal{F}^\alpha_{mp}u^p u_n - \mathcal{F}^\alpha_{np}u^p u_m). \tag{13}$$

The equation of motion of B_{mn}

$$\partial_{[m}u_{n]} = 0 \quad \rightarrow \quad u_m(x) = \partial_m a(x) \tag{14}$$

reads that $u_m(x)$ is the derivative of a scalar field. Note also that (12) is invariant under

$$\delta B_{mn} = \partial_{[m}b_{n]}(x). \tag{15}$$

As in the case of the action (1), the local symmetry (11) allows one to fix a gauge on the mass shell in such a way that the duality condition (6) takes place. To arrive at eq. (6) harmonic techniques found to be rather useful. Let us sketch the derivation of (6).

The equation of motion of A^α_m produced by (9) is

$$\epsilon^{mnpq}\partial_l(l_p\mathcal{F}^\alpha_q) = 0, \tag{16}$$

where $\mathcal{F}^\alpha_q \equiv \mathcal{F}^\alpha_{qp}l^p$ and l_p is defined by (8) and (14).

From (16) it follows that

$$l_{[p}\mathcal{F}^\alpha_{q]} = \partial_{[p}\Phi^\alpha_{q]}, \quad \mathcal{F}^\alpha_q = l^p\partial_p\Phi^\alpha_q - l^p\partial_q\Phi^\alpha_p, \tag{17}$$

where Φ^α_q are two vector functions. Projecting (17) onto harmonics l^p_i ($i = 1,2,3$) orthogonal to l_p we get

$$l^p_{[i}l^q_{j]}\partial_p\Phi^\alpha_q = 0,$$

which implies that

$$l^q_i\Phi^\alpha_q = l^q_i\partial_q\varphi^\alpha(x) \quad \Rightarrow \quad \Phi^\alpha_q = l_q\Phi^\alpha + \partial_q\varphi^\alpha(x). \tag{18}$$

Substituting (18) into (17) and taking into account that the last term in (18) can be neglected since (17) is invariant under gauge transformations $\Phi^\alpha_q \rightarrow \Phi^\alpha_q + \partial_q\varphi^\alpha(x)$ we obtain

$$\mathcal{F}^\alpha_q = \partial_q\Phi^\alpha(x) + l^p\partial_p(l_q\Phi^\alpha). \tag{19}$$

Now the transformation (11) can be used to put in (19) $\Phi^\alpha = 0$ as a gauge fixing condition. Then we have $\mathcal{F}^\alpha_q = 0$ which, because of the self-duality of $\mathcal{F}^\alpha_{pq}$, implies (6).

Thus, we again remain with only one independent Maxwell field and get the duality between its electric and magnetic strength vector. In view of the vanishing condition for the self-dual strength tensor the equations of motion of u_m reduce to:

$$\frac{\delta(S_A + S_B)}{\delta u_m} = \epsilon^{mnlp}\partial_n B_{lp} = 0 \quad \rightarrow \quad B_{mn} = \partial_{[m}b_{n]}, \tag{20}$$

which means that B_{mn} is completely auxiliary and can be eliminated by use of the corresponding local transformations (15).

The only thing which has remained to show is that u_m itself does not carry physical degrees of freedom and can be gauge fixed to $u_m = \delta^0_m$. For this we have to find a corresponding local symmetry. And here an analogy of the antisymmetric field B_{mn} with the 'notoph' of Ogievetsky and Polubarinov (1967) helps us to get the corresponding symmetry transformations.

The form of the action (12) containing B_{mn} reminds a term which one encounters in a dual formulation of a pseudoscalar ('axion') field as an antisymmetric notoph field (see, for instance,(Ogievetsky and Polubarinov (1967), Kalb and Ramond (1974), Coleman, Preskill and Wilczek (1992)))

$$S = \int d^4x \left(-\frac{1}{2}(\partial_m a(x) - u_m(x))(\partial^m a(x) - u^m(x)) - \epsilon^{pqmn} u_p \partial_q B_{mn} \right). \tag{21}$$

The action (21) is invariant under local Peccei–Quinn transformations

$$\delta a(x) = \varphi(x), \qquad \delta u_m = \partial_m \varphi(x), \tag{22}$$

(u_m being the corresponding gauge field) and produces dual versions of the axion action[1]:

$$L = -\frac{1}{2}\partial_m a(x)\partial^m a(x), \qquad L = \frac{1}{3!}\partial_{[m}B_{np]}\partial^{[m}B^{np]}. \tag{23}$$

The duality relation between the pseudoscalar field $a(x)$ and the antisymmetric tensor field B_{mn}

$$\partial_l a(x) = \epsilon_{lmnp}\partial^m B^{np} \tag{24}$$

is a consequence of the equations of motion of u_m obtained from (21).

Now one can assume that the action (9)+(12) is also invariant under the transformations (22. This is indeed the case provided A^α_m and B_{mn} transform as follows

$$\delta A^\alpha_m = \frac{\varphi(x)}{u^2}\mathcal{L}^{\alpha\beta}\mathcal{F}^\beta_{mn}u^n, \qquad \delta B_{mn} = \frac{\varphi(x)}{(u^2)^2}\mathcal{F}^{\alpha r}_m u_r \mathcal{F}^{\beta s}_n u_s \mathcal{L}^{\alpha\beta}. \tag{25}$$

Then, taking into account (14) and requiring that $u^2 \neq 0$ (to escape singularities), we can use the local transformation (22) to put $u_m = \delta^0_m$. In this gauge the manifestly Lorentz invariant duality–symmetric action

$$S = \int d^4x \left(-\frac{1}{8}F^\alpha_{mn}F^{\alpha mn} - \frac{1}{4(u_l u^l)}u^m \mathcal{F}^\alpha_{mn}\mathcal{F}^{\alpha np}u_p - \epsilon^{mnpq}u_m\partial_n B_{pq}\right) \tag{26}$$

[1] We denoted the scalar field in (14 with the same letter $a(x)$ as the axion field to point to their formal "generic roots".

reduces to (1), and the local transformations of A^α_m (25) (with $\varphi(x) = x^i v^i$) are combined with the corresponding Lorentz transformations and produce the modified space–time symmetry (3) of the action (1).

One can reduce the number of the auxiliary fields in the action (26) to one scalar field by substituting into (26 the solution of the equation of motion (14). The resulting minimal form of the covariant duality symmetric action

$$S = \int d^4x(-\frac{1}{8}F^\alpha_{mn}F^{\alpha mn} - \frac{1}{4(\partial_l a \partial^l a)}\partial^m a(x)\mathcal{F}^\alpha_{mn}\mathcal{F}^{\alpha np}\partial_p a(x)), \qquad (27)$$

which remains the same in all space–time dimensions, has been used for the description of chiral bosons in various theoretical models (see (Pasti, Sorokin and Tonin (1997), Bandos et. al. (1997), Aganagic et. al. (1997), Bandos, Berkovits and Sorokin (1998), Dall'Agata et. al. (1997), (1998)) and references therein).

We have thus obtained a covariant Lagrangian formulation of Maxwell theory which is also invariant under electric–magnetic duality. The action was shown to produce in the temporal gauge (10) the duality–symmetric action (1) of (Deser and Teitelboim (1976),Schwarz and Sen (1994)).

As has been mentioned in the introduction the first duality–symmetric action for the Maxwell fields was constructed by (Zwanziger (1971)) and that it differs from (1) in the sign of the second term. This difference leads to essentially different symmetry properties of the Zwanziger action (Maznytsia et. al. (1998)). However, since both actions describe one and the same physical model there should be a relationship between them. This relation is established through a duality transform of the auxiliary scalar field $a(x)$ into the auxiliary 'notoph' field B_{mn} (Maznytsia et. al. (1998)). For this consider (26) as a master action which produces different dual actions depending on which auxiliary fields are integrated out. The action (27) is one of these dual actions. Another one is obtained by varying (26) with respect to u_m, which gives an expression for the dual field strength $v^m = \epsilon^{mnpq}\partial_n B_{pq}$ in terms of u_m and the Maxwell field strengths, solving this expression for the vector field u_m in terms of v_m and substituting the result back into the action (26). The action now contains only the dual field strength v_m of the auxiliary field B_{mn} and has the form

$$S = \int d^4x(-\frac{1}{8}F^\alpha_{mn}F^{\alpha mn} + \frac{1}{4(v_l v^l)}v^m\mathcal{F}^\alpha_{mn}\mathcal{F}^{\alpha np}v_p), \qquad (28)$$

the sign of the second term being changed with respect to the analogous term in (26) and (27). It can be shown that the action (28) is invariant under local transformations which allow one to fix $v_m(x)$ to be a constant time–like or space–like vector upon which (28) reduces to the Zwanziger action (see (Maznytsia et. al. (1998)) for details).

In conclusion we have shown how the covariant Lagrangian approach to the description of duality–symmetric fields unifies different non–covariant formulations. This approach is also related to the infinite field approach (McClain, Wu and Yu (1990), Wotzasek (1991), Martin and Restuccia (1994), Devecchi and Henneaux (1996), Bengtsson and Kleppe (1997)) being a consistent truncation of the latter (see the last ref. in (Pasti, Sorokin and Tonin (1995), (1997))).

From the action (27) one can *formally* obtain the Siegel (1984) action by replacing $\frac{\partial_m a(x)\partial_p a(x)}{(\partial_l a \partial^l a)}$ with a Lagrange multiplier field $\Lambda_{mp}(x)$. But this relation is only a formal one since "hiding" derivatives of fields in other fields is not an innocent trick. The properties of the two actions are very different, a main difference being that the duality condition (6) is obtained from the actions (1, (26)–(28) as a consequence of equations of motion of the gauge fields $A^\alpha_m(x)$, while in the Siegel formulation it arises as a "square root" of the constraint produced by the Lagrange multiplier Λ_{mn} equation of motion. This results in a different structure of Hamiltonian constraints and, as a consequence, leads to different ways of quantizing chiral bosons. We also note that for more complicated cases of self–interacting chiral bosons, such as the M–theory five–brane (Pasti, Sorokin and Tonin (1997), Bandos et. al. (1997), Aganagic et. al. (1997)) an effective action in the Siegel form is not know yet.

Acknowledgements. This work was partially supported by the European Commission TMR Programme ERBFMPX-CT96-0045 to which the authors are associated, and by the INTAS Grant 96-308.

References

Dirac, P. A. M.: *Proc. R. Soc.* **A133** (1931) 60 ; *Phys. Rev.* **74** (1948) 817 .
Zwanziger, D.: Phys. Rev. **D3** (1971) 880.
Maznytsia, A., Preitschopf C. R. and Sorokin, D.: Duality of self–dual actions, hep–th/9805110.
Deser, S. and Teitelboim, C.: Phys. Rev. **D13** (1976) 1592 .
Floreanini, R. and Jackiw R.: Phys. Rev. Lett. **59** (1987) 1873.
Henneaux, M. and Teitelboim, C.: in Proc. Quantum mechanics of fundamental systems 2, Santiago, 1987, p. 79; Phys. Lett. **B206** (1988) 650 .
Tseytlin, A. Phys. Lett. **B242** (1990) 163 ; Nucl. Phys. **B350** (1991) 395.
Schwarz, J. H. and Sen, A.: Nucl. Phys. **B411** (1994) 35 .
Siegel, W.: Nucl. Phys. **B238** (1984) 307.
Kavalov, A. R. and Mkrtchyan, R. L.: Nucl. Phys. **B331** (1990) 391.
McClain, B. Wu, Y. S. and Yu, F.: Nucl. Phys. **B343** (1990) 689 ;
Wotzasek, C.: Phys. Rev. Lett. **66** (1991) 129;
Martin, I. and Restuccia, A.: Phys. Lett. **B323** (1994) 311 ;
Devecchi, F. P. and Henneaux, M.: Phys. Rev **D45** (1996) 1606;
Bengtsson, I. and Kleppe, A. *Int. J. Mod. Phys.* **A12** (1997) 3397.
Berkovits, N. Phys. Lett. **B388** (1996) 743.
Pasti, P. Sorokin, D. and Tonin, M.: Phys. Lett. **B352** (1995) 59; Phys. Rev. **D52** (1995) R4277; Phys. Rev. **D55** (1997) 6292.

Pasti, P. Sorokin, D. and Tonin, M. Phys. Lett. **398B** (1997) 41;
Bandos, I. Lechner, K. Nurmagambetov, A. Pasti, P. Sorokin, D. and Tonin, M.: *Phys. Rev. Lett.* **78** (1997) 4332;
Aganagic, M. Park, J. Popescu, C. and Schwarz, J. H.: Nucl. Phys. **B496** (1997) 191.
Bandos, I. Berkovits, N. and Sorokin, D.: *Nucl. Phys.* **B522** (1998) 214.
Dall'Agata, G. Lechner, K. and Sorokin, D.: Class. Quant. Grav. **14** (1997) L195 ;
Dall'Agata, G. Lechner, K. and Tonin, M.: D=10, N=IIB supergravity: Lorentz–invariant actions and duality, hep–th/9806140.
Galperin, A. Ivanov, E. Kalitzin, S. Ogievetsky, V. and Sokatchev, E.: *Class. Quantum Grav.* **1** (1984) 498; *Class. Quantum Grav.* **2** (1985) 155.
Ogievetsky, V. I. and Polubarinov, I.: V. Soviet. J. Nucl. Phys. **4** (1967) 156.
Kalb, M. and Ramond, P.: Phys. Rev. **D9** (1974) 2273.
Coleman, S. Preskill, J. and Wilczek, F.: Nucl. Phys. **B378** (1992) 175.

3-Point Functions in $\mathcal{N} = 4$ Yang-Mills in Harmonic Superspace

E. Sokatchev[1]

Laboratoire d'Annecy-le-Vieux de Physique Théorique** LAPTH, Chemin de Bellevue, B.P. 110, F-74941 Annecy-le-Vieux, France

Abstract. Three-point two-loop functions of analytic operators in $\mathcal{N} = 4$ Yang-Mills theory in four dimensions are calculated using the harmonic superspace formulation of this theory.

Although there was no known example of a four dimensional conformally invariant quantum field theory in the 1960's and 1970's, the properties of such theories were investigated. It was realised that conformal invariance could be used to determine the two- and three-point Green's functions up to constants in any dimension and the space-time dependence of many such correlators were found (For a review see Todorov, Mintchev and Petkova (1978)). With the discovery of supersymmetry, examples of conformally invariant quantum field theory were found. The first such theory to be found was the $\mathcal{N} = 4$ Yang-Mills theory (Sohnius and West (1981), Mandelstam (1983), Howe, Stelle and Townsend (1983,1984), Brink, Lindgren and Nilsson (1983)). Conformal invariance was very successfully exploited (Belavin, Polyakov and Zamolodchikov (1984)) in two dimensions to determine Green's functions for higher-point functions in certain theories. However, these developments in two dimensions relied on the infinite nature of the two dimensional conformal group and the existence of null vectors in certain representations of this algebra. In four dimensions, the conformal group is only a finite dimensional group and it appeared that, unlike in two dimensions, one would not be able to exploit conformal invariance to solve for higher-point Green's functions. One indication to the contrary concerned the Green's functions that involved $\mathcal{N} = 2$ chiral superfields of the same chirality in the two dimensional $\mathcal{N} = 2$ minimal model series. Since these correlators belong to a minimal conformal field theory, it was to be expected that one could solve for these Green's functions explicitly, but in (Howe and West (1989) it was shown that one could do this using only the globally defined superconformal group and chirality. As it is the globally defined part of the two dimensional conformal group that generalises to higher dimensions, this work lead to the hope (Howe and West (1995)),(Conlong and West) that the constrained nature of the superfields that describe supersymmetric theories when combined with conformal invariance might be sufficient, even in higher dimensions, to solve for more than just the two- and three-point Green's functions. A related

** URA 1436 associée à l'Université de Savoie

example of this phenomenon is the simple relation between the anomalous weight of a chiral superfield and its R weight in any superconformal theory, which in many cases allows one to deduce the anomalous weight of the chiral superfield (Conlong and West (1993)).

In fact all four dimensional supersymmetric theories of interest are described by constrained superfields. The Wess-Zumino model and the $\mathcal{N} = 1$ and $\mathcal{N} = 2$ Yang-Mills field strengths are described by chiral superfields. The remaining theories of extended rigid supersymmetry are the $\mathcal{N} = 2$ matter and the $\mathcal{N} = 4$ Yang-Mills theory. The $\mathcal{N} = 2$ matter is best described by a harmonic superspace formulation (A. Galperin, E. Ivanov, S. Kalitzin, V. Ogievetsky and E. Sokatchev (1984,1985)) in which it is represented by a single component superfield q^+ which satisfies an analyticity condition. Here, analytic means that q^+ is both Grassmann analytic, i.e. it depends on only half of the fermionic coordinates in a similar manner to a chiral superfield, and analytic on the internal (bosonic) space, a compact, complex manifold which is used to extend standard ($\mathcal{N} = 2$) Minkowski superspace to the harmonic superspace of interest. The $\mathcal{N} = 4$ Yang-Mills theory also has a succinct description when formulated on an appropriate harmonic superspace (Hartwell and Howe (1995)),(Howe and West (1995)). Explicitly, the Yang-Mills field strength multiplet is described by a single-component analytic superfield W (taking its values in the Lie algebra of $SU(N)$). Although, in the non-Abelian case, W is covariantly analytic (with respect to the gauge group), the gauge invariant operators, A_q, defined by

$$A_q = \mathrm{tr}(W^q) \tag{1}$$

are analytic fields in the strict sense.

In references (Howe and West (1996a)), (P. Howe and P. West (1997)), (Howe and West (1996b)) the constraints due to superconformal invariance on four dimensional Green's functions involving chiral or harmonic superfields were found. It was clear that these constraints were very strong and it was suggested that they were sufficiently powerful to determine, up to constants, a class of these Green's functions. These included all the Green's functions of operators composed of $\mathcal{N} = 2$ matter of sufficiently low dimension in the $\mathcal{N} = 2$ supersymmetric theories and operators composed of gauge invariant polynomials of the harmonic superfield W, also of sufficiently low dimension, in the $\mathcal{N} = 4$ Yang-Mills theory. In the latter case these are just Green's functions of the above operators $A_q = \mathrm{tr}(W^q)$ for sufficiently small q. The calculations required to establish this result are complicated, but have been successfully completed for the four-point Green's functions involving $\mathcal{N} = 2$ matter (Eden, Howe, Pickering, Sokatchev and West). This result encourages us to believe that some four-point functions in the $\mathcal{N} = 4$ theory may be amenable to a similar analysis. Some other four-point calculations from a different viewpoint have appeared more recently (Liu and Tseytlin (1998)), (Freedman, Mathur, Matusis and Rastelli (1998a)).

Recently there has been considerable interest in the Maldacena conjecture which relates string theory on AdS backgrounds to conformal field theory on the boundary (Maldacena (1997), Gubser, Klebanov and Polyakov (1998), Witten (1998)). In the most studied example it is conjectured that classical IIB supergravity on $AdS_5 \times S^5$ is equivalent to the large N limit of $\mathcal{N} = 4$ $SU(N)$ Yang-Mills theory on the boundary which in this case is four-dimensional Minkowski spacetime. A key ingredient in this conjecture is the fact that the symmetry groups of the supergravity background and the conformal field theory are the same, namely $SU(2,2|4)$. Although not all gauge-invariant operators in the $\mathcal{N} = 4$ theory are of the type given in equation (1) it turns out that it is precisely this set of operators that is relevant to the Maldacena conjecture in its simplest form. The spectrum of IIB supergravity on $AdS_5 \times S^5$ consists of the gauged $D = 5, \mathcal{N} = 8$ supergravity multiplet together with the massive Kaluza-Klein multiplets. These all fall into short representations of $SU(2,2|4)$ with maximum spin 2 and are in one-to-one correspondence with the superfields A_q introduced above (Andrianopoli and Ferrara (1998)).

An important example of this type of operator is the supercurrent $T = A_2 = \mathrm{tr}(W^2)$. This multiplet has $128 + 128$ components and contains amongst them the traceless, conserved energy-momentum tensor, four gamma-traceless, conserved supersymmetry currents and fifteen conserved currents corresponding to the internal $SU(4)$ symmetry of the theory.

In the present talk we shall focus on the three-point functions of the $\mathcal{N} = 4$ theory. On general grounds it is to be expected that one should be able to solve for the two- and three-point functions of an arbitrary conformal field theory in any dimension (For a review see Todorov, Mintchev and Petkova (1978)). However, the advantage of our formalism is that it allows us to solve for the complete superfield correlation functions in a very simple way, not least because the operators we are interested in all have only one component. This means that the tensor structures which arise in a component approach are dealt with automatically. In addition, for those three-point functions which have non-zero leading terms in a θ expansion, it is easy to show that the solutions we obtain are unique. In fact, the form of the two- and three-point function can be found as a special case of the general formula given in ref. (Howe and West (1995)) for any Green's function for which there are no corresponding superconformal invariants. The procedure to determine certain of the higher-point Green's functions works in essentially the same way, but the details are very much more complicated.

Here we shall present an explicit calculation of the two-loop contribution to the three-point function $\langle +2+3+3 \rangle$ in an $\mathcal{N} = 2$ theory consisting of complex (Fayet-Sohnius) hypermultiplets coupled to Yang-Mills (as is well-known, if the matter is in the adjoint representation, such a model describes $\mathcal{N} = 4$ Yang-Mills in terms of $\mathcal{N} = 2$ superfields). The main purpose of this example is to show that the assumption of harmonic analyticity made earlier is indeed justified. We also show that the two-loop contribution to this class of correlators actually vanishes, in line with the results of ref. (D'Hoker, Freedman and Skiba (1998)).

The $\mathcal{N}=2$ matter and Yang-Mills multiplets will be described in a way which maintains the $SU(2)$ symmetry manifest (A. Galperin, E. Ivanov, S. Kalitzin, V. Ogievetsky and E. Sokatchev (1984,1985)). For this purpose we introduce the Grassmann-analytic (G-analytic) harmonic superspace with coordinates

$$x_A^\mu, \theta^{+\alpha}, \bar\theta^{+\dot\alpha}, u_i^\pm \ . \tag{2}$$

Here $u_i^\pm$ are the harmonic variables parametrising the coset space $SU(2)/U(1) \sim S^2$, i.e. one is to regard $u_i^\pm$ as the two columns of an $SU(2)$ matrix; the index i transforms under the (right) $SU(2)$ and $\pm$ are its harmonic (left) $U(1)$ projections. As a consequence, they have the defining properties:

$$u_i^- = (u^{+i})^* \ , \quad u^{+i}u_i^- = 1 \ . \tag{3}$$

The Grassmann variables $\theta^{+\alpha}, \bar\theta^{+\dot\alpha}$ are $U(1)$ harmonic projections of the odd coordinates of $N=2$ superspace,

$$\theta^{+\alpha,\dot\alpha} = u^{+i}\theta_i^{\alpha,\dot\alpha} \ . \tag{4}$$

The G-analytic space-time coordinate $x_A^{\alpha\dot\alpha}$ is obtained by shifting x^μ:

$$x_A^\mu = x^\mu - 2i\theta^{\alpha(i}\sigma^\mu_{\alpha\dot\alpha}\bar\theta^{\dot\alpha j)}u_i^+u_j^- \ . \tag{5}$$

Under Q-supersymmetry it transforms into the $+$ projections θ^+ and not their complex conjugates θ^-. This is the reason why the superspace (2) is called G-analytic. In what follows we shall always work in the G-analytic superspace, therefore we shall drop the index A of x_A^μ. Given two points in x space, $x_{1,2}$, one can define the Q-supersymmetry invariant difference

$$\hat x_{12} = x_{12} + \frac{2i}{(12)}[(1^-2)\theta_1^+\bar\theta_1^+ - (12^-)\theta_2^+\bar\theta_2^+ + \theta_1^+\bar\theta_2^+ + \theta_2^+\bar\theta_1^+] \ . \tag{6}$$

Here (12), (1^-2), (12^-) are short-hand notations for contractions of harmonic variables:

$$(12) \equiv u_1^{+i}u_{2i}^+ \ , \quad (1^-2) \equiv u_1^{-i}u_{2i}^+ \ , \quad (12^-) \equiv u_1^{+i}u_{2i}^- \ .$$

In fact, (12) and $\hat x_{12}$ are the $SU(2)$ covariant counterparts of the variables y_{12} and $\hat x_{12}$.

The matter and Yang-Mills multiplets are described by the analytic superfields $q_r^+(x,\theta^+,\bar\theta^+,u)$ and $V_a^{++}(x,\theta^+,\bar\theta^+,u)$, with r and a being indices of the matter and ajoint representations of the gauge group, respectively. The details can be found in (A. Galperin, E. Ivanov, S. Kalitzin, V. Ogievetsky and E. Sokatchev (1984,1985)), here we only give a brief summary of the Feynman rules (Galperin, Ivanov, Ogievetsky and Sokatchev (1985)). The matter propagator Π, the gluon propagator P in the Fermi-Feynman gauge and the only vertex relevant to our calculation are indicated below:

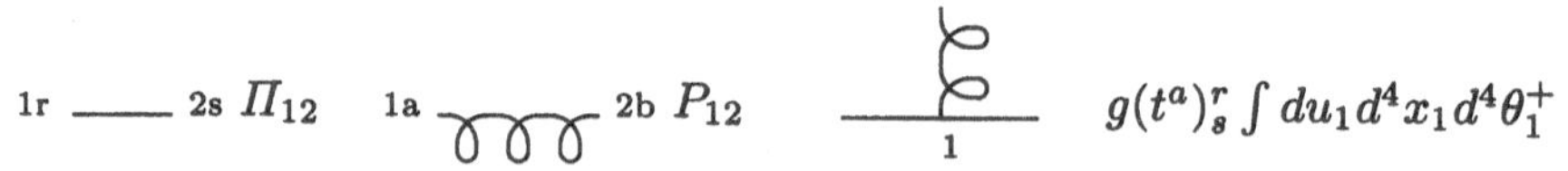

Figure 1

The expression of the matter propagator is

$$(\Pi_{12})^r_s = \langle \tilde{q}^{+r}(1)|q^+_s(2)\rangle = \frac{(12)}{\hat{x}^2_{12}}\delta^r_s \,. \qquad (7)$$

Here we make use of the Q-invariant variable (6). Eq. (7) is in fact the $SU(2)$ covariant counterpart of eq. (48). The G-analyticity of Π_{12} is manifest, since only the + projections of the Grassmann variables appear. Note that the original form of the matter propagator given in (Galperin, Ivanov, Ogievetsky and Sokatchev (1985)) is different, but the equivalent form (7) is best suited for our purposes in this paper. For the gluon propagator we shall use the standard form from (Galperin, Ivanov, Ogievetsky and Sokatchev (1985)):

$$(P_{12})_{ab} = \langle V^{++}_a(1)|V^{++}_b(2)\rangle = (D^+_1)^4\left(\frac{\delta^8(\theta_1-\theta_2)}{x^2_{12}}\right)\delta^{(-2,2)}(u_1,u_2)\delta_{ab} \,. \qquad (8)$$

Its G-analyticity with respect to the first argument is manifest, since it contains the maximal number four of plus-projected spinor derivatives $D^+ = u^+_i D^i$ (just like the chiral matter propagators in $\mathcal{N}=1$ supersymmetry). G-analyticity with respect to the second argument is assured by the presence of the Grassmann and harmonic delta functions which allow us to transfer the spinor derivatives from point 1 to point 2.

Finally, the vertex describing the coupling of the gauge superfield to the hypermultiplet is shown in Fig.1. It involves a G-analytic superspace integral. Note that the harmonic integral must always be done after the Grassmann one, since the analytic Grassmann measure $d^4\theta^+$ carries a harmonic charge. The full Yang-Mills Feynman rules involve gluon vertices of arbitrary order, as well as ghosts, but none of them show up at the two-loop level.

The three-point function we want to compute involves gauge invariant composite operators of harmonic charges $\langle +2+3+3\rangle$. The simpler case $\langle +2+2+2\rangle$ turns out trivial, since the matter propagator (7) obey fermion type rules and a Furry-like theorem.

The first relevant graph is shown in Fig.2:

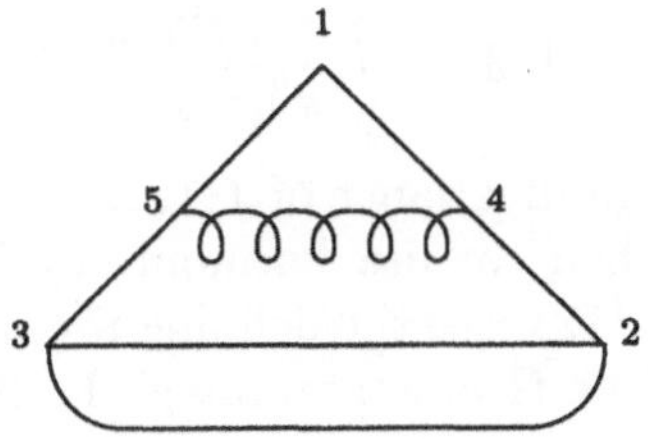

Figure 2

It should be remembered that this is a graph in x space, therefore the true loops are those involving the internal line 4-5, as opposed to the lines 2-3 which are just free propagators. Having this in mind and applying the Feynman rules above, we find the corresponding expression (the gauge group indices and factors are not shown):

$$I_1 = (\Pi_{23})^2 \int d^4x_{4,5} d^4u_{4,5} d^4\theta^+_{4,5} \times \qquad (9)$$

$$\Pi_{14}\Pi_{42}\Pi_{35}\Pi_{51}\,(D^+_4)^4 \left(\frac{\delta^8(\theta_4-\theta_5)}{x^2_{45}}\right) \delta^{(-2,2)}(u_4,u_5)\;.$$

The first step in evaluating this graph consists in using the four spinor derivatives $(D^+_4)^4$ from the gluon propagator to restore the full Grassmann integral $\int d^4\theta^+_4 (D^+_4)^4 = \int d^8\theta_4$. This is made possible by the explicit G-analyticity of the matter propagators Π_{14} and Π_{42}. Then the Grassmann $\delta^8(\theta_4-\theta_5)$ and harmonic $\delta^{(-2,2)}(u_4,u_5)$ delta functions can be used to do the integrals $\int du_4 d^8\theta_4$, thus identifying the Grassmann and harmonic points 4 and 5. In order to simplify the calculation, we shall evaluate the graph with all the external Grassmann variables put to zero, $\theta_1=\theta_2=\theta_3=0$. This corresponds to taking the lowest-order term in the θ expansion of the amplitude. This step allows us to easily deal with the hats $\hat{x}$ in the matter propagators (see (6)). For the propagators Π_{23} the choice $\theta_{1,2,3}=0$ amounts to just removing the hat, but for those involved in the vertex integrals, e.g. Π_{14}, there is still the shift due to the integration variable θ_5. Now, since the points 4 and 5 have been identified, the hats in all the propagators involve the same Grassmann structure $\theta^+_5\bar{\theta}^+_5$ but different harmonic ones. All this allows us to rewrite the amplitude (9) as follows:

$$I_1(\theta_{1,2,3}=0) = \frac{(23)^2}{x^4_{23}} \int du_5\,(15)^2(25)(35) \qquad (10)$$

$$\int d^4\theta^+_5 \exp\left\{2i\theta^+_5\bar{\theta}^+_5 \cdot \left[\frac{(15^-)}{(15)}\partial_1 + \frac{(25^-)}{(25)}\partial_2 + \frac{(35^-)}{(35)}\partial_3\right]\right\} f(1,2,1,3)\;.$$

Here $f(1,2,1,3)$ denotes the two-loop x-space integral

$$f(1,2,1,3) = \int \frac{d^4x_4 d^4x_5}{x_{14}^2 x_{24}^2 x_{15}^2 x_{35}^2 x_{45}^2} \ . \tag{11}$$

Using the translational invariance of $f(1,2,1,3)$ we can substitute $\partial_1 f = -(\partial_2 + \partial_3) f$ in (10). Then we use harmonic cyclic identities of the type $(25^-)(15) - (15^-)(25) = (12)$ (see the defining property (3)). Next we expand the exponential and do the Grassmann integral, after which (10) is reduced to (up to an overall factor)

$$I_1(0) = \frac{(23)^2}{x_{23}^4} \int du_5 \times \tag{12}$$
$$\left[\frac{(12)^2(35)}{(25)} \partial_2 \cdot \partial_2 + \frac{(13)^2(25)}{(35)} \partial_3 \cdot \partial_3 + (12)(13)\ 2\partial_2 \cdot \partial_3 \right] f(1,2,1,3).$$

The harmonic integral of the third term in (12) is trivially done ($\int du_5\ 1 = 1$), and the first two ones are computed as follows (see (Galperin, Ivanov, Ogievetsky and Sokatchev (1985)) for details):

$$\int du_5\ \frac{(35)}{(25)} = \int du_5\ \frac{\partial_5^{++}(35^-)}{(25)} = \int du_5\ (35^-)\delta^{(-1,1)}(u_2, u_5) = (32^-)\ .$$

Further, the box operators in (12) reduce the two-loop integral f to a one-loop one, e.g.

$$\partial_2 \cdot \partial_2 \int \frac{d^4x_4 d^4x_5}{x_{14}^2 x_{24}^2 x_{15}^2 x_{35}^2 x_{45}^2} = \frac{4\pi^2 i}{x_{12}^2} \int \frac{d^4x_5}{x_{15}^2 x_{25}^2 x_{35}^2} \equiv \frac{g(1,2,3)}{x_{12}^2}\ .$$

The end result of all this is:

$$I_1(0) = \frac{(23)^2}{x_{23}^4} \times \tag{13}$$
$$\left[(12)^2(32^-)\frac{g(1,2,3)}{x_{12}^2} + (13)^2(23^-)\frac{g(1,2,3)}{x_{13}^2} + (12)(13)\ 2\partial_2 \cdot \partial_3 f(1,2,1,3)\right].$$

We immediately remark the presence of negative-charged harmonics in (13) which means that the contribution of the graph in Fig.2 *is not harmonic analytic*. However, the situation changes when we take into account the two other graphs of similar type shown in Fig.3:

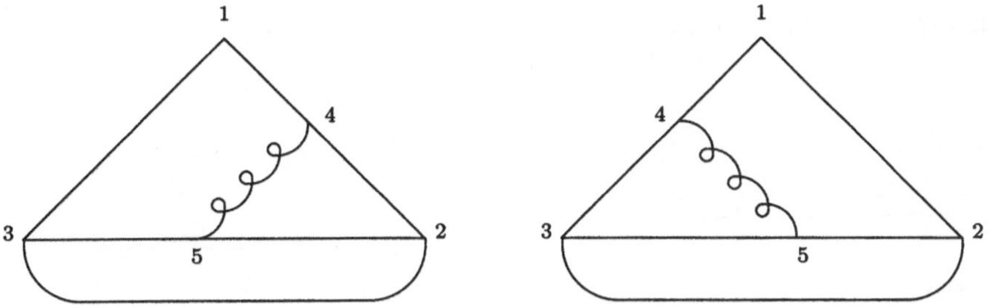

Figure 3

They can be computed in exactly the same way as the one in Fig.2. Putting all three contributions together and using harmonic cyclic identities we find that the non-analytic negative-charged harmonics disappear and we obtain the *harmonic analytic* result:

$$I(0) = I_1 + I_2 + I_3 = \frac{(12)(13)(23)^2}{x_{12}^2 x_{13}^2 x_{23}^4} a(1,2,3) \tag{14}$$

where

$$a(1,2,3) = x_{12}^2 g(1,2,3) + x_{12}^2 x_{13}^2 \, 2\partial_2 \cdot \partial_3 f(1,2,1,3) + \text{cycle} . \tag{15}$$

In (14) we observe a product of four matter propagators multiplied by the coefficient function $a(1,2,3)$. The latter, according to the general theory, must be a conformally invariant function of three points and hence can only be a constant. In fact, it can be shown to vanish. A simple argument using the Lorentz, translational and scaling properties of the integrals involved in (15) leads to the identity

$$2\partial_2 \cdot \partial_3 f(1,2,1,3) = \frac{x_{23}^2 - x_{12}^2 - x_{13}^2}{x_{12}^2 x_{13}^2} \, g(1,2,3) .$$

Substituting this in (15) gives $a(1,2,3) = 0$, so

$$I(\theta_1 = \theta_2 = \theta_3 = 0) = 0 . \tag{16}$$

In other words, the lowest-order ($\theta = 0$) term in the amplitude is zero, and this can then be generalised to the entire amplitude.

The example of a three-point correlator presented above is not unique. It is easy to construct three-point functions with higher $U(1)$ charges by simply attaching more matter propagators to the external points. This does not affect the loop structure of the graphs and thus leads to the same result as above.

In conclusion we should mention that there exist further two-loop graphs involving the following insertions:

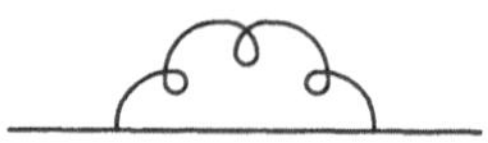

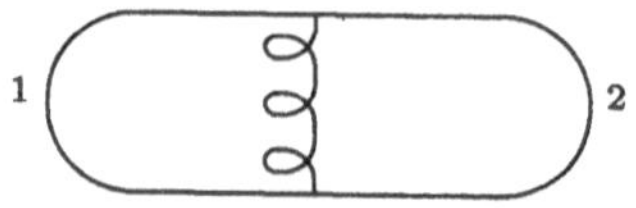

They need not be considered because both of them vanish (strictly speaking, the second one is proportional to $\delta(x_{12})$, but in our analysis we always keep the external points of an n-point function apart).

Acknowledgements This work is part of research project in which the author is collaborating with Paul Howe and Peter West from King's College, London. The author is grateful to Burkhard Eden and Christian Schubert for stimulating discussions. This work was supported in part by the EU network on Integrability, non-perturbative effects, and symmetry in quantum field theory (FMRX-CT96-0012) and by the British-French scientific programme Alliance (project 98074).

References

I.T. Todorov, M.C. Mintchev and V.B. Petkova (1978): *Conformal Invariance in Quantum Field Theory* Scuola Normale Superiore, Pisa.

M. Sohnius and P. West (1981): Phys. Lett. **B100** 45; S.Mandelstam, Nucl. Phys. **B213** 149; P.S. Howe, K.S. Stelle and P.K. Townsend, Nucl. Phys. **B214** 519; Nucl. Phys. **B236** 125; L. Brink, O. Lindgren and B. Nilsson, Nucl. Phys. **B212** 401.

A.A. Belavin, A.M. Polyakov and A.B. Zamolodchikov (1984): Nucl. Phys. **B241** (1984) 333.

P. Howe and P. West (1989): *Chiral Correlators in Landau-Ginzburg theories and $N=2$ superconformal models*, Phys. Lett. **B227** 397.

P. Howe and P. West (1995): *Non-perturbative Green's functions in theories with extended superconformal symmetry*, hep-th/9509140, Int. Journal. Mod. Phys. to be published.

B. Conlong and P. West: in PhD thesis of B. Conlong, London University.

B. Conlong and P. West (1993) J. Phys. **A26** 3325.

A. Galperin, E. Ivanov, S. Kalitzin, V. Ogievetsky and E. Sokatchev (1984,1985): Class. Quant. Grav. **1** (1984) 469; Class. Quant. Grav. **2** (1985) 155.

G. Hartwell and P. Howe (1995): Int. J. Mod. Phys **A27** 3901; Class. Quant. Grav. **12** 1823.

P. Howe and P. West (1996): *Operator product expansions in four-dimensional superconformal field theories*, Phys. Lett **B389** 273, hep-th/9607060.

P. Howe and P. West (1997): *Superconformal invariants and extended supersymmetry* Phys. Lett **B400** 307, hep-th/9611075.

P. Howe and P. West (1996): *Is $N = 4$ Yang-Mills soluble?*, in the proceedings of the 6th Quantum Gravity Seminar, Moscow, hep-th/9611074.

B. Eden, P. Howe, A. Pickering, E. Sokatchev and P. West: in preparation.

H. Liu and A. Tseytlin (1998): *$D = 4$ super Yang-Mills, $D = 5$ gauged supergravity and $D = 4$ conformal supergravity*, hep-th/9804083.

D.Z. Freedman, S. Mathur, A. Matusis and L. Rastelli (1998): *Comments on 4 point functions in the ADS/CFT correspondence*, hep-th/9808006.

J. Maldacena (1997): *The large N limit of superconformal theories and supergravity* hep-th/9711200; S.S. Gubser, I.R. Klebanov, and A.M. Polyakov (1998): Phys. Lett. **B428** 105; E.Witten (1998): *Anti-de Sitter space and holography*, hep-th/9802150.

L. Andrianopoli and S. Ferrara (1998): *KK excitations on $AdS_5 \times S^5$ and $N = 4$ primary superfields* hep-th/9803171.

S. Lee, S. Minwalla, M. Rangamani and N. Seiberg (1998): *Three-point functions of chiral operators in $D = 4, \mathcal{N} = 4$ SYN at large N*, hep-th/9806074.

D.Z. Freedman, S. Mathur, A. Matusis and L. Rastelli (1998): *Correlation functions in the CFT/AdS correspondence*, hep-th/9804058.

D. Anselmi, D.Z. Freedman, M.T. Grisaru and A.A. Johansen (1998): Nucl.Phys. **B526** 543.

E. Fradkin and A. Tseytlin (1982,1984): *One loop beta function in conformal supergravities*, Nucl. Phys. **B203** (1982) 157; *Conformal anomaly in Weyl theory and anomaly free superconformal theories*, Phys. Lett. 134B (1984) 187.

P.S. Howe, K.S. Stelle and P.K. Townsend (1984): *Miraculous cancellations in supersymmetry made manifest* Nucl. Phys. **B236** 125.

P. West (1983): *Supersymmetry and Finiteness*, Proceedings of the 1983 Shelter Island II Conference on Quantum Field theory and Fundamental Problems in Physics, edited by R. Jackiw, N Kuri, S. Weinberg and E. Witten, M.I.T. Press.

E. D'Hoker, D.Z. Freedman and W. Skiba (1998): *Field theory tests for correlators in the AdS/CFT correspondence*, hep-th/9807098.

A. Galperin, E. Ivanov, V. Ogievetsky and E. Sokatchev (1985): Class. Quantum Grav. **2** 601; 617.

Harmonic Superspaces for Three-Dimensional Theories

Boris Zupnik

Bogoliubov Laboratory of Theoretical Physics, Joint Institute for Nuclear Research, Dubna, Moscow Region, 141980, Russia

Abstract. Three-dimensional field theories with $N = 3$ and $N = 4$ supersymmetries are considered in the framework of the harmonic-superspace approach. Analytic superspaces of these supersymmetries are similar; however, the geometry of gauge theories with the manifest $N = 3$ is richer and admits construction of the topological mass term.

1 Introduction

Three-dimensional supersymmetric gauge theories have been intensively studied in the framework of new nonperturbative methods in the field theory (Seiberg and Witten (1997)). Superfield description of the simplest $D = 3$, $N = 1, 2$ theories and various applications have earlier been discussed in refs. (Schonfeld (1981), Gates et al. (1983), Zupnik and Pak (1989)). The most interesting features of $D = 3$ theories are connected with the Chern-Simons terms for gauge fields and also with duality between vector and scalar fields.

We shall discuss the harmonic-superfield formalism of three-dimensional theories with the extended supersymmetries, which reflects the intrinsic geometry of these theories relevant for quantum description. The formulation of $D = 3$, $N = 4$ superfield theories is analogous to the harmonic formalism of $D = 4$, $N = 2$ theories (Galperin et al. (1984), Galperin et al. (1985), Zupnik (1986)), although the existence of different $SU(2)$ automorphism groups allows one to choose various versions of harmonic superspaces. General $N = 3$ superfields are not covariant with respect to $N = 4$ supersymmetry, but the analytic $N = 3$ superfields are equivalent to the corresponding $N = 4$ superfields. However, interactions with the manifest $N = 3$ require a specific geometric description (Zupnik and Hetselius (1988), Kao and Lee (1992)), which does not guarantee conservation of the additional 4-th supersymmetry. The harmonic $N = 3$ superspace of ref. (Zupnik and Hetselius (1988)) has been based on the use of isovector harmonics, and now we consider the improved version of the harmonic formalism for these theories. Note that field models with $N = 3$ supersymmetry (in distinction with $N = 4$ models) are dual to the dynamics of non-orthogonal intersections of branes (Ohta and Townsend (1998)).

Our conventions for the (2,1)-dimensional γ-matrices are

$$(\gamma_m)_{\alpha\beta}(\gamma_n)^{\beta\rho} + (\gamma_n)_{\alpha\beta}(\gamma_m)^{\beta\rho} = 2\eta_{mn}\delta^\rho_\alpha\,, \quad (\gamma_m)_{\alpha\beta} \equiv \varepsilon_{\alpha\rho}(\gamma_m)^\rho_\beta\,, \tag{1}$$

where η_{mn} is the metric with signature $(1,-1,-1)$ and $\alpha, \beta \ldots$ are the $SL(2,R)$ spinor indices. We also shall use the basic notation of ref. (Galperin et al. (1984)) for the isospinor harmonics $u^\pm_i$.

2 N=4 Harmonic Superspace

Superfield models with $D=3,\ N=4$ supersymmetry can be studied via the dimensional reduction of $D=4,\ N=2$ superfield theories. We shall discuss $D=3,\ N=4$ harmonic superspace by analogy with refs. (Galperin et al. (1984), Galperin et al. (1985)). Let $z^M = (x^{\alpha\beta}, \theta^\alpha_{ka})$ be the *central* coordinates of the general $D=3,\ N=4$ superspace SS^4_3, where $k, l \ldots$ and $a, b \ldots$ are indices of the automorphism groups $SU_L(2)$ and $SU_R(2)$, respectively. In the superspace without central charges, the relations between basic spinor derivatives are

$$\{D^{ka}_\alpha, D^{lb}_\beta\} = 2i\varepsilon^{kl}\varepsilon^{ab}\partial_{\alpha\beta}\,, \tag{2}$$

where $D^{ka}_\alpha = (D^k_\alpha\,, \bar{D}^k_\alpha)$ and $\partial_{\alpha\beta} = (\gamma^m)_{\alpha\beta}\partial_m$.

The superfield constraints of $N=4$ super-Yang-Mills theory SYM^4_3 can be written as follows:

$$\{\nabla^{ka}_\alpha, \nabla^{lb}_\beta\} = 2i\varepsilon^{kl}\varepsilon^{ab}\nabla_{\alpha\beta} + \varepsilon_{\alpha\beta}\varepsilon^{kl}W^{ab}_L\,, \tag{3}$$

where ∇_M are covariant derivatives with superfield connections and W^{ab}_L is a tensor constrained superfield of the SYM^4_3 theory. Below we shall discuss the alternative convention for the SYM^4_3 constraints.

Introduce the notation $W = W^{22}_L$ and $\overline{W} = W^{11}_L$. The constraint for the Abelian gauge theory produces the following relations:

$$D^k_\alpha \overline{W} = 0\,, \quad \bar{D}^k_\alpha W = 0\,, \tag{4}$$

$$D^{k\alpha}D^l_\alpha W = \bar{D}^{k\alpha}\bar{D}^l_\alpha \overline{W}\,, \tag{5}$$

which are analogous to the constraints of $N=2,\ D=4$ vector multiplet.

One can also consider the on-shell constraints for different hypermultiplet superfields q^{kb}_L and q^{kb}_R

$$\nabla^{ia}_\alpha q^{kb}_L + \nabla^{ka}_\alpha q^{ib}_L = 0\,, \tag{6}$$

$$\nabla^{ia}_\alpha q^{kb}_R + \nabla^{ib}_\alpha q^{ka}_R = 0\,. \tag{7}$$

The coset space of the automorphism group $SU(2)$ plays an important role in the harmonic description of $N=2,\ D=4$ superfield theory. In the harmonic approach to $N=4,\ D=3$ theory we can use, alternatively, the harmonic variables for cosets spanned on the generators of the corresponding $SU(2)$ groups $L,\ R$ or $(L+R)$. Let us firstly consider the harmonics $u^\pm_i$ for the group $SU_L(2)$. Using the standard harmonic methods of ref. (Galperin et al.

(1984)) we can transform the constraints (3,6) into the following integrability (L-analyticity) condition:

$$\{\nabla^{+a}_{\alpha}, \nabla^{+b}_{\beta}\} = 0\,, \quad \nabla^{+a}_{\alpha} = u^{+}_{i}\nabla^{ia}_{\alpha}\,, \tag{8}$$

$$\nabla^{+a}_{\alpha} q^{+b} = 0, \quad q^{+a} = u^{+}_{i} q^{ia}_{L}\,. \tag{9}$$

By analogy with the $D = 4$, $N = 2$ case, one can use the L-analytic basis for the SYM^4_3 theory

$$\nabla^{+a}_{\alpha} = D^{+a}_{\alpha} = \partial/\partial\theta^{\alpha-}_{a}\,, \tag{10}$$

$$\nabla^{++} = D^{++} + V^{++}_{L}\,, \quad D^{+a}_{\alpha} V^{++}_{L} = 0\,, \tag{11}$$

where $D^{+a}_{\alpha} = (D^{+}_{\alpha},\, \bar{D}^{+}_{\alpha})$ and V^{++}_{L} is the prepotential of SYM^4_3 depending on the coordinates $\zeta_L = (x^{\alpha\beta}_{L},\, \theta^{+\alpha}_{a})$ of the L-analytic superspace LSS^4_3. In the physical gauge, it contains the components of the $D = 3$, $N = 4$ vector multiplet

$$V^{++}_{WZ} = (\theta^{+a}\theta^{+b})\Phi_{(ab)}(x_L) + i(\theta^{+a}\gamma^{m}\theta^{+}_{a})A_{m}(x_L) +$$
$$+(\theta^{+a}\theta^{+b})\theta^{+\alpha}_{b} u^{-}_{k} \lambda^{k}_{a\alpha}(x_L) + (\theta^{+})^{4} u^{-}_{k} u^{-}_{l} X^{(kl)}(x_L)\,. \tag{12}$$

The solution of the constraint (3) has the following form in the harmonic approach:

$$W^{ab}_{L} = D^{+a\alpha} D^{+b}_{\alpha} V^{--}_{L}(V^{++}_{L})\,, \tag{13}$$

where the standard solution for the 2-nd harmonic connection V^{--}_{L} (Zupnik (1986)) is considered.

The analytic superfield q^{+a} with the infinite number of auxiliary fields is the complete analog of the corresponding $D = 4$ hypermultiplet representation. An alternative form of the analytic hypermultiplet can be obtained with the help of the harmonic duality transform $q^{+a} = u^{+a}\omega + u^{-a}F^{++}$ (Galperin et al. (1985)). On mass shell, these hypermultiplets have the following components:

$$q^{+a}_{0} = u^{+}_{k} f^{ka}(x_L) + \theta^{+b\alpha}\psi^{a}_{b\alpha}(x_L)\,, \tag{14}$$

$$\omega_{0} = f(x_L) + u^{+}_{k} u^{-}_{l} f^{(kl)}(x_L) + \theta^{+b\alpha} u^{-}_{k} \psi^{k}_{b\alpha}(x_L)\,. \tag{15}$$

The holomorphic effective action of the Abelian $N = 4$ gauge theory contains the chiral superfield $W = \int du\, (\bar{D}^{-})^{2} V^{++}_{L}$ (Ivanov and Zupnik (1997)). One can consider the equivalent chiral and analytic representations of this action

$$i\int d^{3}x d^{4}\theta\, \mathcal{F}(W) + \text{c.c.} = i\int d\zeta^{(-4)}_{L} du\, V^{++}_{L}(D^{+})^{2}\frac{\mathcal{F}(W)}{W} + \text{c.c.}\,, \tag{16}$$

where $d\zeta^{(-4)}_{L}$ is the integral measure in LSS^4_3 and $d^{4}\theta$ is the spinor measure in the chiral superspace.

The u-independent chiral superfield A can be used for the construction of the complex analytic superfield $C^{++} = (D^+)^2 A$ satisfying the additional harmonic constraint $D^{++}C^{++} = 0$. The superfield A is treated as a dual variable with respect to the Abelian 'magnetic' gauge superfield V_M^{++} . The effective action of this system contains V_M^{++} as a Lagrange multiplier

$$i\int d^3x d^4\theta\, \mathcal{F}(A) + i\int d\zeta^{(-4)} du\, V_M^{++}(D^+)^2 A + \text{c.c.}\,. \tag{17}$$

One can obtain the reality constraint (5) for A and the relation between A and the magnetic prepotential varying this action with respect to the superfields V_M^{++} and A, respectively. Note that the holomorphic representation with chiral superfields breaks the $SU_R(2)$ automorphism group.

Consider the Abelian case of the constraint (7) for the R-hypermultiplet and define the harmonic projection $R^{-a} \equiv u_k^- q_R^{ka}$. In the L-analytic basis, the basic relations for this harmonic superfield are

$$D_\alpha^{+a} R^{-b} + D_\alpha^{+b} R^{-a} = 0\,, \quad (D^{++})^2 R^{-a} = 0\,, \tag{18}$$

where the 1-st relation is treated as the constraint and the 2-nd one as the equation of motion. Using the relation $\{D_\alpha^{+a}, D_\beta^{-b}\} = -2i\varepsilon^{ab}\partial_{\alpha\beta}$ one can obtain a general covariant solution of the constraint which contains the L-analytic bosonic and fermionic superfields b^{-a} and f^α

$$R^{-a} = b^{-a} + D_\alpha^{-a} f^\alpha\,. \tag{19}$$

The harmonic equation of motion is equivalent to the equation $D^{--}R^{-a} = 0$.

Thus, the hypermultiplet R^{-a} (or its derivative $R^{+a} = D^{++}R^{-a}$) is reduced to the pair of L-analytic superfields and their interactions can be described in LSS_3^4. On-shell it has the same components as the superfield q_R^{ka}.

One can identify the indices of the left and right automorphism groups and use the $SU_C(2)$-covariant spinor $N=4$ coordinates and supersymmetry parameters

$$\theta_{kl}^\alpha = \theta_{(kl)}^\alpha + \varepsilon_{kl}\,\theta^\alpha\,, \quad \epsilon_{kl}^\alpha = \epsilon_{(kl)}^\alpha + \varepsilon_{kl}\,\epsilon^\alpha\,, \tag{20}$$

where the isovector and isoscalar parts are introduced. The alternative C-form of the SYM_3^4 constraints contains the isovector superfield W^{kl}

$$\{\nabla_\alpha^{km}, \nabla_\beta^{ln}\} = 2i\varepsilon^{kl}\varepsilon^{mn}\nabla_{\alpha\beta} + \frac{1}{2}\varepsilon_{\alpha\beta}(\varepsilon^{kl}W^{mn} + \varepsilon^{mn}W^{kl})\,. \tag{21}$$

This representation allows us to separate the isoscalar covariant derivative

$$\{\nabla_\alpha, \nabla_\beta\} = i\nabla_{\alpha\beta}\,, \quad \{\nabla_\alpha, \nabla_\beta^{(kl)}\} = 0\,. \tag{22}$$

The 2-nd relation is a conventional constraint which depends on a choice of the $SU(2)$-frame. Below we shall discuss the commutation relations between the isovector $N=3$ covariant derivatives which are frame-independent.

3 New Formulation of $N = 3$ Harmonic Superspace

Let us consider now the new harmonic projections of the $N = 4$ spinor coordinates (20)

$$\theta^{\alpha\pm\pm} = u^{\pm}_{k} u^{\pm}_{l} \theta^{\alpha kl} \ , \quad \theta^{\alpha\pm\mp} = u^{\pm}_{k} u^{\mp}_{l} \theta^{\alpha kl} \ . \tag{23}$$

Coordinates of LSS^4_3 in the new representation are $\zeta_L = (x^{\alpha\beta}_L, \ \theta^{\alpha++}, \ \theta^{\alpha+-})$ where

$$x^{\alpha\beta}_L = x^{\alpha\beta} + i(\theta^{\alpha++}\theta^{\beta--} + \theta^{\beta++}\theta^{\alpha--} - \theta^{\alpha+-}\theta^{\beta-+} - \theta^{\beta+-}\theta^{\alpha-+}) \ .$$

The infinitesimal $N = 4$ spinor transformations have the following form in these coordinates:

$$\delta x^{\alpha\beta}_L = \{2iu^{-k}u^{-l}\epsilon^{\alpha}_{(kl)}\theta^{\beta++} + 2i[\epsilon^{\alpha} - u^{-k}u^{+l}\epsilon^{\alpha}_{(kl)}]\theta^{\beta+-}\} + \{\alpha \leftrightarrow \beta\} \ , \tag{24}$$

$$\delta\theta^{\alpha++} = u^{+k}u^{+l}\epsilon^{\alpha}_{(kl)} \ , \quad \delta\theta^{\alpha+-} = \epsilon^{\alpha} + u^{+k}u^{-l}\epsilon^{\alpha}_{(kl)} \ , \tag{25}$$

where ϵ^{α} is an isoscalar parameter of the 4-th supersymmetry. Using the subgroup with $\epsilon^{\alpha} = 0$ one can describe $N = 3$ supersymmetry in this L-analytic superspace.

A three-dimensional $N = 3$ supersymmetry can also be realized in the superspace SS^3_3 with the coordinates $z = (x^{\alpha\beta}, \theta^{\alpha}_{(kl)})$. The corresponding superfields do not depend on the isoscalar coordinate θ^{α} and are not covariant with respect to the 4-th supersymmetry [1]. The constraints of SYM^3_3 in this superspace are

$$\{\nabla^{(km)}_{\alpha}, \nabla^{(ln)}_{\beta}\} = i(\varepsilon^{kl}\varepsilon^{mn} + \varepsilon^{ml}\varepsilon^{kn})\nabla_{\alpha\beta} + \frac{1}{4}\varepsilon_{\alpha\beta}(\varepsilon^{kl}W^{mn} + \varepsilon^{ml}W^{kn} + \\ + \varepsilon^{mn}W^{kl} + \varepsilon^{kn}W^{ml}) \ , \tag{26}$$

where all connections do not depend on θ^{α}.

Let us introduce the alternative analytic coordinates of the $N = 3$ harmonic superspace ASS^3_3

$$x^{\alpha\beta}_A = x^{\alpha\beta} + i(\theta^{\alpha++}\theta^{\beta--} + \theta^{\beta++}\theta^{\alpha--}) \ , \tag{27}$$

$$\theta^{\alpha++} = u^{+}_{k}u^{+}_{l}\theta^{\alpha(kl)} \ , \quad \theta^{\alpha 0} = \frac{1}{2}(\theta^{\alpha+-} + \theta^{\alpha-+}) = u^{+}_{k}u^{-}_{l}\theta^{\alpha(kl)} \ . \tag{28}$$

It should be stressed that there is a one-to-one correspondence between the analytic $N = 4$ and $N = 3$ superfields.

Spinor derivatives have the following form in this $N = 3$ superspace:

$$D^{++}_{\alpha} = \partial^{++}_{\alpha} = \partial/\partial\theta^{\alpha--} \ , \quad D^{--}_{\alpha} = \partial^{--}_{\alpha} - 2i\theta^{\beta--}\partial^{A}_{\alpha\beta} \ , \tag{29}$$

$$D^{0}_{\alpha} = -\frac{1}{2}\partial^{0}_{\alpha} - i\theta^{\beta 0}\partial^{A}_{\alpha\beta} \ , \quad \partial^{\pm\pm}_{\alpha}\theta^{\beta\mp\mp} = \partial^{0}_{\alpha}\theta^{\beta 0} = \delta^{\beta}_{\alpha} \ . \tag{30}$$

---

[1] In ref. (Zupnik and Hetselius (1988)), we have used the isovector $N = 3$ spinor coordinates $\theta^{\alpha}_{B} = (1/2)(\tau_B)^{kl}\theta^{\alpha}_{(kl)}$.

The corresponding covariant harmonic derivatives are

$$D^{++} = \partial^{++} - 2i\theta^{\alpha++}\theta^{\beta 0}\partial^{A}_{\alpha\beta} + \theta^{\alpha++}\partial^{0}_{\alpha} + 2\theta^{\alpha 0}\partial^{++}_{\alpha}\,, \tag{31}$$

$$D^{--} = \partial^{--} + 2i\theta^{\alpha--}\theta^{\beta 0}\partial^{A}_{\alpha\beta} + \theta^{\alpha--}\partial^{0}_{\alpha} + 2\theta^{\alpha 0}\partial^{--}_{\alpha}\,, \tag{32}$$

$$[D^{--}, D^{++}_{\alpha}] = [D^{++}, D^{--}_{\alpha}] = 2D^{0}_{\alpha}\,, \quad [D^{\pm\pm}, D^{0}_{\alpha}] = D^{\pm\pm}_{\alpha}\,. \tag{33}$$

Analytic $N = 3$ superfields do not depend on θ^{--} and are unconstrained objects in ASS^3_3.

The $N = 3$ covariant derivatives $\nabla^{(kl)}_{\alpha}$ in the central basis can be transformed to the harmonized covariant derivatives of the SYM^3_3 theory in the basis with the analytic gauge group

$$\nabla^{++}_{\alpha} = D^{++}_{\alpha}\,, \quad \nabla^{++} = D^{++} + V^{++}\,, \tag{34}$$

$$\nabla^{--} = D^{--} + V^{--}(V^{++})\,, \quad \nabla^{0}_{\alpha} = D^{0}_{\alpha} - \frac{1}{2}D^{++}_{\alpha}V^{--}\,, \tag{35}$$

$$\nabla^{--}_{\alpha} = [\nabla^{--}, \nabla^{0}_{\alpha}]\,, \tag{36}$$

where V^{++} is the analytic gauge prepotential in the adjoint representation of the gauge group, and $V^{--}(V^{++})$ is the solution of the zero-curvature equation for harmonic connections (Zupnik (1986)). In the physical WZ-gauge the prepotential contains the components of the $N = 3$ vector supermultiplet

$$V^{++}_{WZ} = (\theta^{++})^2 u^-_k u^-_l \Phi^{(kl)}(x_A) + i(\theta^{++}\gamma^m\theta^0)A_m(x_A) + (\theta^0)^2\theta^{\alpha++}\lambda_\alpha(x_A) + \\ +(\theta^{++})^2\theta^{\alpha 0}u^-_k u^-_l \lambda^{(kl)}_{\alpha}(x_A) + (\theta^{++})^2(\theta^0)^2 u^-_k u^-_l X^{(kl)}(x_A)\,, \tag{37}$$

which are analogous to the $N = 4$ components (12) with identified L and R isospinor indices.

The basic superfield tensor of SYM^3_3 is analytic

$$W^{++} = \frac{1}{2}D^{\alpha++}D^{++}_{\alpha}V^{--} \tag{38}$$

and satisfies the additional H-constraint (Bianchi identity)

$$\nabla^{++}W^{++} \equiv 0\,. \tag{39}$$

It is a specific feature of the SYM^3_3 theory that the prepotential V^{++} and its superfield-strength W^{++} belong to the same analytic superspace ASS^3_3.

Let us define integral measures in the full and analytic $N = 3$ harmonic superspaces

$$d^9 z_A = \frac{1}{64}d^3x_A(D^{\alpha++}D^{++}_{\alpha})(D^{\alpha--}D^{--}_{\alpha})(D^{\alpha 0}D^{0}_{\alpha})\,, \tag{40}$$

$$d\zeta^{(-4)} = \frac{1}{16}d^3x_A(D^{\alpha--}D^{--}_{\alpha})(D^{\alpha 0}D^{0}_{\alpha})\,. \tag{41}$$

Note that these measures have dimensions $d = 0$ and 1, respectively.

The standard kinetic term of the SYM^3_3 action is

$$S_k = \frac{1}{g^2} \int d\zeta^{(-4)} du \, \mathrm{Tr} \, W^{++} W^{++} \,, \tag{42}$$

where g is the coupling constant with dimension $d = -1/2$.

The effective action of the Abelian $N = 3$ theory contains an arbitrary function $\mathcal{G}^{(+4)}(W^{++}, u)$ of the H-constrained superfield W^{++}

$$\int d\zeta^{(-4)} du \, [\tau (W^{++})^2 + \sum_{p=1}^{\infty} c^{l_1 \ldots l_{2p}} u^-_{l_1} \ldots u^-_{l_{2p}} (W^{++})^{p+2}] \,, \tag{43}$$

where $\tau, c^{l_1 \ldots l_{2p}}$ are constants in the decomposition of $\mathcal{G}^{(+4)}$. It is clear that only quadratic term of the general action conserves the $SU_C(2)$ symmetry. This analytic representation of the low-energy effective action is alternative to the holomorphic $N = 4$ representation (16).

The interaction of the gauge superfield $W^{++}(V^{++})$ is dual to the following interaction of the unconstrained real analytic superfields ω and A^{++}:

$$\int d\zeta^{(-4)} du \, [\mathcal{G}^{(+4)}(A^{++}, u) + A^{++} D^{++} \omega] \,. \tag{44}$$

Varying ω yields the constraint $D^{++} A^{++} = 0$. This action is the first-order form of the special interaction of ω and $D^{++}\omega$, although the elimination of the superfield A^{++} is a non-trivial algebraic problem for the general function $\mathcal{G}^{(+4)}$.

An important feature of the SYM^3_3 theory is the existence of a topological mass (Chern-Simons) term (Zupnik and Hetselius (1988)). In the improved $N = 3$ harmonic formalism this term can be constructed by the analogy with the action of SYM^2_4 (Zupnik (1986))

$$S_m = \frac{m}{g^2} \sum_{n=2}^{\infty} \frac{(-1)^{n+1}}{n} \int d^9 z du_1 \ldots du_n \frac{\mathrm{Tr} \, [V^{++}(z, u_1) \ldots V^{++}(z, u_n)]}{(u^+_1 u^+_2) \ldots (u^+_n u^+_1)} \,, \tag{45}$$

where $(u^+_1 u^+_2)^{-1}$ is the harmonic distribution (Galperin et al. (1985)). Note that the measure $d^9 z$ in this term is not covariant with respect to the 4-th supersymmetry. The analytic version of the topological mass term and the Fayet-Iliopoulos term for the Abelian theory has the following form:

$$S_m + S_{FI} = \frac{1}{g^2} \int d\zeta^{(-4)} du \, (m V^{++} W^{++} + \xi^{++} V^{++}) \,. \tag{46}$$

The action $S_k + S_m + S_{FI}$ yields the free equation of motion for the $N = 3$ Abelian gauge theory

$$[(D^{\alpha 0} D^0_\alpha) + m] \, W^{++} + \xi^{++} = 0 \,. \tag{47}$$

This equation has the following vacuum solutions:

$$W^{++} = -\frac{1}{m}\xi^{++}\,, \quad m \neq 0\,, \tag{48}$$

$$W^{++} = a^{(kl)}u_k^+u_l^+ - 2[(\theta^0)^2\xi^{++} - 2(\theta^{++}\theta^0)\xi^0 + (\theta^{++})^2\xi^{--}]\,, \quad m = 0\,, \tag{49}$$

where $\xi^{\pm\pm} = \xi^{(kl)}u_k^\pm u_l^\pm$, $\xi^0 = (1/2)D^{++}\xi^{--}$ and $a^{(kl)}$ and $\xi^{(kl)}$ are arbitrary constants. Note that the first solution does not break supersymmetry.

For the case $m = 0$, $\xi^{(kl)} = 0$ we can study the background Abelian prepotential

$$V^{++} = \frac{1}{2}a^{(kl)}[(\theta^0)^2u_k^+u_l^+ - 2(\theta^{++}\theta^0)u_k^+u_l^- + (\theta^{++})^2u_k^-u_l^-]\,, \tag{50}$$

which introduces the $N = 3$ central charges and produces masses of charged superfields.

The minimal gauge interaction of the q^+ hypermultiplet has the standard form $\int d\zeta^{(-4)}du\ \bar{q}^+(D^{++} + V^{++})q^+$. The free hypermultiplet satisfies the equation $D^{++}q^+ = 0$ and contains a finite number of complex on-shell components

$$q_0^+ = u_k^+f^k(x_A) + (\theta^{\alpha++}u_k^- - \theta^{\alpha 0}u_k^+)\psi_\alpha^k(x_A)\,. \tag{51}$$

The real $N = 3$ ω-hypermultiplet has been described in ref. (Zupnik and Hetselius (1988)).

Note that the similar harmonic methods can be used for description of two-dimensional models with (3,3) supersymmetry.

The author is grateful to E. Ivanov and N. Ohta for stimulating discussions. This work is partially supported by grants RFBR-96-02-17634, RFBR-DFG-96-02-00180, INTAS-93-127-ext and INTAS-96-0308, and by grant of Uzbek Foundation of Basic Research N 11/97.

References

Seiberg, N., Witten, E. *Mathematical Beauty of Physics*, ed. Drouffe, J.M., Zuber, J.B. (World Scientific, Singapore) pp. 333–366

Schonfeld, J.F. (1981): Nucl. Phys. **B185**, 157–171

Gates, S.J., Grisaru, M.T., Roček, M., Siegel, W. (1983): *Superspace or one thousand and one lessons in supersymmetry* (Benjamin Cummings, Massachusetts)

Zupnik, B.M., Pak, D.G. (1989): Theor. Mat. Phys. **77**, 1070–1076; Class. Quant. Grav. **6**, 723–730

Galperin, A., Ivanov, E., Kalitzin, S., Ogievetsky, V., Sokatchev, E. (1984): Class. Quant. Grav. **1**, 469–498

Galperin, A., Ivanov, E., Kalitzin, S., Ogievetsky, V., Sokatchev, E. (1985): Class. Quant. Grav. **2**, 601–630

Zupnik, B.M. (1986): Sov. J. Nucl. Phys. **44**, 512–517

Zupnik, B.M., Hetselius, D.V. Sov. J. Nucl. Phys. **47**, 730–735

Kao, H.-C., Lee, K. (1992): Phys. Rev. D **46**, 4691–4697

Ohta, N., Townsend, P.K. (1998): Phys. Lett. B **418**, 77–84

Ivanov, E.A., Zupnik, B.M. (1997): Preprint JINR E2-97-322; hep-th/9710236

Part II

Super p-Branes and M-Theory

The Neveu-Schwarz Five-Brane and Its Dual Geometries

Björn Andreas[1], Gottfried Curio[2] and Dieter Lüst[3]

[1] Humboldt-Universität, Institut für Physik, D-10115 Berlin, Germany
E-mail: andreas@physik.hu-berlin.de
[2] School of Natural Sciences, Institute for Advanced Study, Princeton, NJ 08540
E-mail: curio@ias.edu
supported by NSF grant DMS9627351
[3] Humboldt-Universität, Institut für Physik, D-10115 Berlin, Germany
E-mail: luest@physik.hu-berlin.de

Abstract. In this paper we discuss two aspects of duality transformations on the Neveu-Schwarz (NS) 5-brane solutions in type II and heterotic string theories. First we demonstrate that the non-extremal NS 5-brane background is U-dual to its CGHS limit, a two-dimensional black hole times $S^3 \times T^5$; an intermediate step is provided by the near horizon geometry which is given by the three-dimensional BTZ_3 black hole (being closely related to AdS_3) times $S^3 \times T^4$. In the second part of the paper we discuss the T-duality between k NS 5-branes and the Taub-NUT spaces respectively ALE spaces, which are related to the resolution of the A_{k-1} singularities of the non-compact orbifold $\mathbf{C}^2/\mathbf{Z}_k$. In particular in the framework of $N = 1$ supersymmetric gauge theories related to brane box constructions we give the metric dual to two sets of intersecting NS 5-branes. In this way we get a picture of a dual orbifold background $\mathbf{C}^3/\Gamma$ which is fibered together out of two $N = 2$ models ($\Gamma = \mathbf{Z}_k \times \mathbf{Z}_{k'}$). Finally we also discuss the intersection of NS 5-branes with D branes, which can serve as probes of the dual background spaces.

1 Introduction

The NS 5-brane [1] is one of the first string soliton solutions, which can be constructed both for the type IIA/B superstrings as well as for the heterotic string. In certain limits there exists a CFT description of the NS 5-brane (see also [2]). In type IIB the NS 5-brane is S-dual to the D 5-brane [3].

The NS 5-brane plays an important role in the construction of gauge theories from branes [4]. In this context it serves as a kind of background for the D-branes on which the gauge fields live, since the NS 5-branes are heavy and the D-branes are light. It is known for some time that $k+1$ parallel NS 5-branes are T-dual to the ALE space with A_k singularity [5], which is the local geometry of a K3 around a singularity. So now ALE is the background which is probed by D branes, which are then also called fractional branes [6]. If one considers intersecting NS 5-branes one gets a background for N=1 models (brane boxes) [7]. After T-duality one gets a space with a local $\mathbf{C}^3/\Gamma$, $\Gamma = \mathbf{Z}_k \times \mathbf{Z}_{k'}$ singularity. This describes the situation of a specific Calabi-Yau manifold around a singularity. The T-duality between the Hanany-Witten set up and the fractional branes was already recently discussed for $N = 2$ space-time supersymmetry in [8] and for the $N = 1$ brane box models in [9]. We will discuss several aspects of the duality among NS 5-branes and ALE.

Another aspect arises from the observation that D or M brane solutions are in fact U-dual to their own horizon geometry [10,11]. This provides a relation to supergravity on anti-de Sitter spaces and, in the gauge theory picture, to a corresponding large N limit [12]. Via S-duality one expects an analogous behaviour for NS 5-branes. In fact we will explicitly demonstrate that the NS 5-brane background is U-dual to its CGHS limit via the intermediate step of BTZ_3 [13].

This paper is organized as follows. In the next section we present a short summary on the asymptotic geometry change of string background spaces using the heterotic [14] or respectively the type IIB [3] S-duality group. Then in section 3, applying the previous discussion, we show the equivalence of the non-extremal NS 5-brane string background to its own horizon geometry, namely the BTZ black hole respectively the CGHS limit [15], via U-duality. In section 4 we will turn the discussion to the T-duality between NS 5-branes and the ALE type of spaces. Here we especially focus on the construction of the dual background spaces for intersecting NS 5-branes versus six-dimensional non-compact orbifolds $\mathbf{C}^3/\Gamma$ which describe the local neighborhood around the singularities of a certain Calabi-Yau 3-fold.

2 Asymptotic Geometry Change

Recently it has been observed [10,11] that via a sequence of duality transformations the metrics of M 2-, D 3- and M 5-branes with flat asymptotic geometry can be transformed to their own horizon geometries which correspond to the asymptotic non flat spaces $AdS_4 \times S^7$, $AdS_5 \times S^5$ and $AdS_7 \times S^4$, respectively. Here, one element within this series of duality transformations is a certain change, S, of coordinates (u, v) which has the form an $Sl(2, \mathbf{R})$ transformation:

$$S: \quad \begin{pmatrix} v \\ u \end{pmatrix} \to \begin{pmatrix} 1 & -h \\ 0 & 1 \end{pmatrix} \begin{pmatrix} v \\ u \end{pmatrix} . \tag{2.1}$$

This transformation has the effect that it shifts to zero the constant part in the harmonic functions of the above p-brane metrics.

Another way to perform the described geometry change by removing the constant part in the metric goes back to the already older work of [16], where the asymptotically flat four-dimensional Taub-NUT space of the KK monopole was transformed into the non-flat ALE manifold via a TST duality transformation. In this context, S now means a genuine strong-weak coupling duality plus an axionic shift transformation which is again an element of $Sl(2, \mathbf{R})$. This approach was recently applied to D branes [17], in particular by dualizing D 3-branes to intermediates D instantons, and also to heterotic black holes in four and two dimensions [18].

In the this chapter we like to apply analogous techniques to transform the metric of the NS 5-brane to its own horizon geometry via eliminating the constant part in the metric. For the extremal NS 5-brane this brings us to the socalled throat limit which can be described by an WZW type superconformal field theory. For the non-extremal NS 5-brane the sequence of duality transformations will bring us to the socalled CGHS limit [15].

First let us recall some facts about the strong-weak coupling S-duality group $SL(2,\mathbf{R})$ can be used to perform asymptotic geometry changes. The discussion goes in parallel both for the heterotic as well as type II NS 5-brane.

The two well-known frameworks for the non-perturbative group of S-duality symmetries are the 4-dimensional heterotic string on T^6 and the 10-dimensional type IIB string. The group $Sl(2)$ operates fractionally linear on the combination of axion and coupling constant e^{ϕ}, where in the heterotic case we have $S = a + ie^{-2\phi}$ with a being the Hodge dual of the four-dimensional $B_{\mu\nu}$-field; in the type IIB case S is given by $S = l + ie^{-\phi}$ with l being the RR scalar field.

We will see that for metrics flat in the Einstein frame the asymptotic geometry change (in the string frame) is caused, in the end, by the possible axion shift (in the heterotic string) resp. RR scalar shift in the 7-brane picture (for the type IIB string), i.e. the well-known (F-theory) monodromy in the transversal complex plane (around the "singular elliptic fibre" at the 7-brane position, cf. the stringy cosmic string [19], and [20]).

The $Sl(2)$ S-duality group has the generators $\omega_1 := \begin{pmatrix} 1 & \frac{1}{2} \\ 0 & 1 \end{pmatrix}$, $\omega_2 := \begin{pmatrix} 0 & 1 \\ -1 & 0 \end{pmatrix}$ The interpretation of these elements with respect to $S = x+iy$ is well-known:[1] ω_1 gives the shift in x[2] (axion for the heterotic string resp. RR-scalar for type IIB), ω_2 gives the coupling constant inversion $y \to -\frac{1}{y}$ (for vanishing x). One obvious question arises: besides the inversion element ω_2 and the upper triangular elements $\omega_1(B)$ one has the lower triangular elements which are conjugates of the upper triangular elements by the non-perturbative inversion element ω_2 (we write $\omega_1(B) = \left(\begin{smallmatrix} 1 & B \\ 0 & 1 \end{smallmatrix}\right)$): $\begin{pmatrix} 1 & 0 \\ C & 1 \end{pmatrix} = \omega_2 \omega_1(-C) \omega_2^{-1}$.

In the following we will elucidate their connection with the notion of *asymptotic geometry change* in both of the mentioned string theories, the heterotic string and the type IIB.

A suitably normalised element of the relevant subgroup is the shift

$$S_{\text{shift}} = \begin{pmatrix} 1 & 0 \\ -1/2 & 1 \end{pmatrix} \tag{2.2}$$

[1] In our actual computations we are going to the Euclidean space where the complex S-field is replaced by $S_{\pm} = x \pm y$.

[2] Note that we have normalized x in such a way that x can be shifted by half integers.

For reasons explained in section (4.1) we will be especially interested in the case where $x = y$. In that case S_{shift} has the effect (regard footnote 4) of mapping again to a S' of $x' = y'$ where the effect on $y = e^{-j\phi}$ ($j = 2, 1$ for the heterotic resp. IIB case) can be described as

$$S_{\text{shift}} : y^{-1} \to y'^{-1} = y^{-1} - 1 \tag{2.3}$$

(and the same for x) which reflects the nature of the involved (upper triangular=ordinary) shift element conjugated by the inversion element.

In order to eventually perform the desired geometry change the S-duality transformation has to be combined with further elements of the U-duality group. For the heterotic string the additional elements are just T-duality transformations (see Sect.(4.1) for the concrete treatment of T-duality), such that one ends up with the combined transformation of the form $TS_{\text{shift}}T$. On the other hand, for the type II geometry a more involved U-duality transformation is necessary, namely one has to consider the sequence $\mathcal{T}'S_{\text{shift}}\mathcal{T}'$ with $\mathcal{T}' := T_i S_{\text{inv}} T_{WV}$. Here T_i is a T-duality transformation along the direction i within the world volume of the 5-brane and T_{WV} denotes the T-dualisation of all world volume directions. The reason why one has to use a more involved sequence $\mathcal{T}'$ in the type II string compared to the simple T-duality in the heterotic case relies in the fact that for type II strings one is transforming the system to intermediate D instantons. Then one has a space-time interpretation for S_{shift}, acting like on the heterotic axion now on the type IIB RR-scalar field respectively, which amounts to an *asymtotic geometry change* by 'deleting the "1" in the harmonic function' (for the 4D heterotic space-time geometry resp. the 4D transversal directions to the 5-brane) (cf. [17]). Namely the background to which S_{shift} is applied to achieve this effect has to be **flat in the Einstein frame**, so that one has in the string frame

$$\begin{aligned} ds^2 &= V(\xi) d\xi^2 \\ e^{j\phi} &= V(\xi) \end{aligned} \tag{2.4}$$

As $y^{-1} = V$ the desired effect in the metric follows.

Let us describe briefly the process of asymtotic geometry change from the extremal type II NS 5-brane to its horizon geometry, i.e. to the socalled throat limit. As said before, the throat limit corresponds to 'deleting the "1" in the harmonic function' H_5 (this time in the transversal geometry of the 5-brane) and is achieved by the element of the U-duality group $\mathcal{T}'S_{shift}\mathcal{T}'$ where $\mathcal{T}' := TS_{\text{inv}}T_{WV}$ which changes the type IIA NS 5-brane to a type IIB D(-1) brane. Note that the (-1) brane carries the electric charge (measured by an integral over the S^9 of its transversal space, i.e. all of space-time) for the RR-scalar, whereas the Hodge-dual 7-brane carries the magnetic charge (measured by the l_{RR} upper triangular monodromy in $\mathbf{C}_{\text{transversal}}$).

In the sequence $\mathcal{T}'$ the NS-NS B field is by S-duality of type IIB mapped to a RR B field the Hodge-dual 6-form of which becomes after T_{WV} the RR scalar. As the 5-brane metric is *flat in the Einstein frame*, which means essentially that $e^{2\phi} = H_5$, the shift in the inverse of $y = e^{-\phi} = H_5^{-1}$ leads to the desired effect in the metric (which in the string frame is just given by the harmonic function times the flat metric). So the metric is indeed dually mapped to some 'subsector' of itself, the near-horizon geometry. So we see that, if we include the conjugating inversion element ω_2 of S_{shift} in the conjugation process $\mathcal{T}'$, that it is in the end really the (upper triangular) monodromy of the RR-scalar in the transversal complex plane of the dual 7-brane which causes the asymptotic geometry change. The 7-brane is magnetically charged for the RR-scalar (which is $\mathcal{T}'$-dual to the magnetic NSNS B-field of the NS 5-brane in type IIA we started with), which is detected by the mentioned monodromy related to the stringy cosmoc string resp. the singular elliptic fibre of the associated F-theory situation.

3 U-duality of the NS Five-Brane with Its CGHS Limit

Now we want to describe in more detail the geometry change from the non-extremal type NS 5-brane to the socalled CGHS limit [15]. For this below a two step process is described which shows that the CGHS-limit,[3] $BH_2 \times S^3$, (which can be interpreted as an α' exact solution) of the near extremal NS 5-brane (cf. for example [21]),

$$NS_5 \longrightarrow BH_2 \times S^3_{Q_5}, \tag{3.5}$$

is actually U-dual to it via the combination of the U-duality [11]

$$NS_5 \simeq BTZ_3 \times S^3 \times T^4 \tag{3.6}$$

with the T-duality [24]

$$BTZ_3 \times S^3 \times T^4 \simeq BH_2 \times S^3 \times (S^1 \times T^4). \tag{3.7}$$

As this comes down to 'deleting the additive constant "1" in the harmonic function', i.e. a *change of asymptotic geometry*, this compares nicely with a corresponding duality (TST, the classical Ehlers-Geroch transform) between two purely gravitational backgrounds [16], which one also gets by 'deleting the additive constant "1" in the harmonic function', namely between the ALE instanton and the multi Taub-NUT space. This is a transformation operating purely in the 4-dimensional transversal space whereas the dualities shown above make use of transformations in the world-volume sector, too. Nevertheless the comparison matches nicely as the 5-brane is well known to be T-dual to the ALE-space [5] - but on a transversal direction of the 5-brane compactified on a circle (corresponding with the S^1_τ-fibration of the ALE-space), which on closer inspection has some quite non-trivial subtleties [25]. These issues are described in a Sect.3.

[3] This limit is relevant for the new QFT's in D=6 and D=5 [22].

3.1 The CGHS limit NS 5 $\to BH_2 \times S^3$

For the near-extremal NS 5-brane in type IIA with its world-volume compactified on $S^1_1 \times T^4_{2345}$ (we suppress the flat spatial world-volume directions) one has [21]

$$ds^2 = -(1 - \frac{r_0^2}{r^2})dt^2 + (1 + \frac{Q_5\alpha'}{r^2})(\frac{dr^2}{1 - \frac{r_0^2}{r^2}} + r^2 d\Omega_3^2),$$
$$e^{2\phi} = e^{2\phi_\infty}(1 + \frac{Q_5\alpha'}{r^2}), \tag{3.8}$$

with $H = Q_5\epsilon_3$ (cf. sect. (3.2)). Let us introduce the near-horizon coordinate σ with

$$r = r_0 \cosh\sigma, \tag{3.9}$$

the non-extremality-parameter α_5

$$\sqrt{Q_5} = \frac{r_0}{\sqrt{\alpha'}} \sinh\alpha_5, \tag{3.10}$$

and the energy density parameter

$$\mu = \frac{r_0^2}{g^2\alpha'}, \tag{3.11}$$

(here $g := e^{\phi_\infty}$) which occurs [21] in the (string-frame) energy per unit 5-volume[4] $M_5 := \frac{M}{V_5} = \frac{1}{\alpha'^3(2\pi)^5}(\frac{Q_5}{g^2} + \mu) = \frac{1}{\alpha'^3(2\pi)^5}\mu\cosh^2\alpha_5$. Then one gets

$$ds^2 = -\tanh^2\sigma dt^2 + (\mu g^2\cosh^2\sigma + Q_5)\alpha'(d\sigma^2 + d\Omega_3^2),$$
$$e^{2\phi} = g^2 + \frac{Q_5}{\mu\cosh^2\sigma}. \tag{3.12}$$

One sees that making the $g \to 0$ limit (CGHS-limit [15]), while keeping the energy density parameter μ at order one, corresponds to 'deleting the additive constant "1" in the harmonic function' $1 + \frac{Q_5\alpha'}{r^2}$; at the same time this causes the decoupling of the S^3 sector with $ds^2_{S^3} = Q_5 d\Omega_3^2$ and $H = Q_5\epsilon_3$, leading to the 2-dimensional black hole times the $SU(2)$ WZW model:

$$ds^2 = -\tanh^2\sigma dt^2 + Q_5\alpha' d\sigma^2,$$
$$e^{2\phi} = \frac{Q_5}{\mu\cosh^2\sigma}. \tag{3.13}$$

For the sake of later comparison let us transform the coordinates back by $\bar{r} := \sqrt{\frac{\mu}{Q_5}}\cosh\sigma = \bar{r_0}\cosh\sigma$:

[4] Here $V_5 = (2\pi)^5 R_1 R_2 R_3 R_4 R_5$.

$$ds^2 = -(1 - \frac{\bar{r}_0^2}{\bar{r}^2})dt^2 + \frac{Q_5\alpha'}{\bar{r}^2}\frac{d\bar{r}^2}{1 - \frac{\bar{r}_0^2}{\bar{r}^2}},$$
$$e^{2\phi} = \frac{1}{\bar{r}^2}. \tag{3.14}$$

Let us remark that $g \to 0$ implied here that $r_0 \to 0$ as α' is kept fixed here. [12] makes $\alpha' \to 0$ instead of $g \to 0$, apart from that similar reduction to the near-horizon region.

3.2 The Duality NS 5 $\simeq BTZ_3 \times S^3 \times T^4$

Let us start again with the NS 5-brane in type IIA with its world-volume compactified on $S_1^1 \times T_{2345}^4$ in the notation of [11] (set $H_1 = 1$ for now; let $dy^2 := dx_2^2 + \cdots + dx_5^2$):

$$ds^2 = \frac{1}{H_1}[-(1 - \frac{r_0^2}{r^2})dt^2 + dx_1^2] + dy^2 + (1 + \frac{Q_5\alpha'}{r^2})(\frac{dr^2}{1 - \frac{r_0^2}{r^2}} + r^2 d\Omega_3^2),$$
$$e^{2\phi} = \frac{1}{H_1}(1 + \frac{Q_5\alpha'}{r^2}),$$
$$H_{ijk} = \frac{1}{2}\epsilon_{ijkl}\partial_l(1 + \coth\alpha_5 \frac{Q_5\alpha'}{r^2}). \tag{3.15}$$

Note that ϕ is shifted by ϕ_∞ compared to the previous equations.

So we are in the case[5] $Q_1 = \alpha_1 = 0$ of [11] where $H_i = 1 + \frac{Q_i\alpha'}{r^2}$ and $Q_i = \frac{r_0^2}{\alpha'}\sinh^2\alpha_i$ with $i = 1, 5$. Note that because of the $\coth\alpha_5 = \sqrt{1 + \frac{r_0^2}{Q_5\alpha'}}$ in H only for the extremal case of α_5 very large one has an axionic instanton. (Note that $g \to 0$ implies (by the μ condition) $r_0 \to 0$ and so $\alpha_5 \to \infty$.)

This is U-dual via $\mathcal{T}S_{\text{shift}}\mathcal{T}$ (with $\mathcal{T} := T_1ST_{1234}ST_5$, cf. [11])[6] to a configuration with $H_1 = \frac{r_0^2}{r^2} = H_5$

$$ds^2 = \frac{r^2}{r_0^2}[-(1 - \frac{r_0^2}{r^2})dt^2 + dx_1^2] + dy^2 + \frac{r_0^2}{r^2}(\frac{dr^2}{1 - \frac{r_0^2}{r^2}} + r^2 d\Omega_3^2) \tag{3.16}$$

---

[5] Also $\alpha_K = Q_K = 0, H_K = 1$.

[6] We use here as a technical device the shift interpretation as coordinate change in the fundamental wave (reached from the IIB D-string, which we got from the IIA NS 5 brane after the part T_1ST_{1234} of $\mathcal{T}$, by the remaninig part ST_5); this is a technical alternative to the interpretation of the the wave in a 12D sense (the D-(-1) brane [28]) which one gets after part T_{05} which follows in the construction of $\mathcal{T}' = TST_{123405}$ described in the introduction. This gives the shift (deleting the "1") for H_5; it has actually to be coupled with a similar procedure for H_1 (cf. [11]). (Operator products to be read from the left.)

and $e^{2\phi} = 1, B_{01} = \frac{r^2}{r_0^2} - 1, H_{ijk} = \frac{1}{2}\epsilon_{ijkl}\partial_l(\frac{r_0^2}{r^2} - 1)$.

Note that the Q_5-dependence, which seems to be lost, is still kept as the rescaling $R_5 \to R_5 \cosh \alpha_5$ has happened. Because of the R_5-rescaling the 3-dimensional Newton constant is not[7] $\mathcal{G}_N^{(3)} = \frac{G_N^{(10)}}{V_{T^4}\cdot(r_0^3 \Omega_3)}$ but $G_N^{(3)} = \frac{\mathcal{G}_N^{(3)}}{\cosh \alpha_5}$; so $G_N^{(3)}$ is here a function of r_0 and Q_5.

Now by effectively 'deleting the additive constant "1" in the harmonic function' H_5 (besides changing $Q_5\alpha'$ to r_0^2) the S^3 sector has decoupled where one has an S^3 of radius r_0 with $ds_{S^3}^2 = r_0^2 d\Omega_3^2$ and $H = r_0^2 \epsilon_3$. This leads to the structure $BTZ_3 \times S^3 \times T_y^4$ with the metric (besides $e^{2\phi} = 1, B_{t\varphi} = r_0(\frac{r^2}{r_0^2} - 1)$; $\varphi := x_1/r_0$)

$$ds_{BTZ}^2 = -(\frac{r^2}{r_0^2} - 1)dt^2 + r^2 d\varphi^2 + \frac{dr^2}{\frac{r^2}{r_0^2} - 1}. \tag{3.17}$$

After the rescaling $t \to ct$, $\varphi \to c\varphi$, $r \to c^{-1}r$, $r_0 \to c^{-1}r_0 = \sqrt{Q_5\alpha'}$ of the metric by $c = \frac{r_0}{\sqrt{Q_5\alpha'}}$ it takes with $M_3 = c^2 = r_0^2/(Q_5\alpha')$ the form

$$ds^2 = -M_3(\frac{r^2}{r_0^2} - 1)dt^2 + r^2 d\varphi^2 + \frac{1}{M_3}\frac{dr^2}{\frac{r^2}{r_0^2} - 1}. \tag{3.18}$$

3.3 The Non-Compact Untwisting $BTZ_3 \simeq BH_2 \times S^1$

To describe the 3-dimensional BTZ black hole [13] in its relation to anti-de Sitter space note first that AdS_3 is

$$-x_0^2 - x_1^2 + x_2^2 + x_3^2 = -l^2 \tag{3.19}$$

in the flat space of signature $(- - ++)$

$$ds^2 = -dx_0^2 - dx_1^2 + dx_2^2 + dx_3^2. \tag{3.20}$$

To get the physical coordinates for the black hole of mass M_3 with horizon at r_0 one makes first the coordinate change to r, φ and t (with $l^2 = \frac{r_0^2}{M_3}$, $\frac{r^2}{M_3} = x_1^2 - x_2^2, e^{\varphi\sqrt{M_3}} = \frac{\sqrt{M_3}}{r}(x_1 + x_2)$)

$$\begin{aligned}
x_0 &= \frac{r_0}{\sqrt{M_3}}\sqrt{1 - \frac{r^2}{r_0^2}}\cosh(t\frac{M_3}{r_0}), \\
x_1 &= \frac{r}{\sqrt{M_3}}\cosh \varphi\sqrt{M_3}, \\
x_2 &= \frac{r}{\sqrt{M_3}}\sinh \varphi\sqrt{M_3}, \\
x_3 &= \frac{r_0}{\sqrt{M_3}}\sqrt{1 - \frac{r^2}{r_0^2}}\sinh(t\frac{M_3}{r_0}),
\end{aligned} \tag{3.21}$$

[7] Here $V_{T^4} = (2\pi)^4 R_2 R_3 R_4 R_5$ is the volume in the beginning.

and then identifies φ with period 2π to get BTZ_3 from AdS_3 [13],[24] leading to the metric (besides $e^{2\phi} = 1, B_{\varphi t} = \sqrt{M_3} r^2/r_0$)

$$ds^2 = -M_3\left(\frac{r^2}{r_0^2} - 1\right)dt^2 + \frac{1}{M_3}\frac{dr^2}{\frac{r^2}{r_0^2} - 1} + r^2 d\varphi^2. \tag{3.22}$$

If one makes a T-duality along the φ-direction, where one has a translational symmetry, one gets (after the further coordinate change[8] $\tilde{t} = \varphi/r_0, \tilde{\varphi} = \sqrt{M_3}t + \tilde{t}$) [24]

$$\tilde{ds}^2 = -\left(1 - \frac{r_0^2}{r^2}\right)d\tilde{t}^2 + \frac{r_0^2/M_3}{r^2}\frac{dr^2}{1 - \frac{r_0^2}{r^2}} + d\tilde{\varphi}^2,$$

$$e^{2\phi} = \frac{1}{r^2}, \tag{3.23}$$

with $B_{\tilde{\varphi}\tilde{t}} = 0$. This is the 2-dimensional black hole times S^1. As AdS_3 is[9] (up to signature) $SL(2,\mathbf{R})$, and the 2-dimensional black hole is the $SL(2,\mathbf{R})$ WZW model with a $U(1)$ gauged [23], we see that the T-duality above is just a non-compact version of the T-duality-'untwisting' of $S^3 = SU(2)$ to $S^2 \times S^1 = SU(2)/U(1) \times U(1)$.

As $r_0^2/M_3 = Q_5\alpha'$ we find coincidence with the crucial prefactor of the $d\bar{r}^2$ term in the CGHS limit.

4 Some Aspects of the T-Duality of the Taub-NUT Spaces with the Five-Brane

We consider the T-duality for the ALE spaces with the five-brane. This was first made plausible by the argument of [5] that under a fibrewise T-duality for an elliptically fibered $K3$ with A_{k-1} singularity the monodromy for the complex structure parameter of the elliptic fibre (caused by the singularity) goes over to the monodromy for the Kähler parameter which gives the crucial H charge k. On closer inspection [25] this has some non-trivial points described below. (For a treatment from an other perspective cf. [26].)[10] The

[8] As φ is periodic, so are $\tilde{t}$ and $\tilde{\varphi}$; one actually works then on the covering space, to avoid CTC's.

[9] The $\det g = 1$ condition for $g = \begin{pmatrix} a & b \\ c & d \end{pmatrix} \in Sl_2(\mathbf{R})$ with $x_{0/3} = \frac{b \pm c}{2}, x_{2/1} = \frac{a \mp d}{2}$ translates to $l = i$ causing the signature change.

[10] [26] argues that at a point in moduli space, where the center positions of the Taub-NUT metric (=KK monopole of IIa resp. M) merge (the critical point for gauge symmetry enhancement), and near the singularity (where the membranes wrapping the vanishing S^2 become massless giving the new non-abalian gauge bosons) the pol terms dominate so one can effectively neglect the "1" in the harmonic function leading to the ALE situation; then a TST transformation is made to the system of coalescing D6 branes in IIA (=KK monopole of M=Taub-NUT) where the mentioned membranes become the stretched strings between the D6 branes.

T-duality between ALE spaces and axionic instantons was also discussed in [27].

4.1 Case of One Isometric Direction

We will consider the case where we perform the T-duality with respect to one isometric direction. The spaces we are going to start with are the purely gravitational backgrounds given by the ALE and Taub-NUT spaces. Of these the ALE spaces describing the resolutions of the A_{k-1} singularities are complex two-dimensional non-compact relatives of $K3$, i.e. non-compact Ricci-flat hyperkaehler manifolds. The ALE manifold of the A_{k-1} series corresponds to the metric given by the Gibbons-Hawking multi-center ansatz

$$ds^2 = V(\mathbf{x})d\mathbf{x}^2 + V^{-1}(\mathbf{x})(d\tau + \boldsymbol{\omega} \cdot d\mathbf{x})^2 \tag{4.1}$$

with the self-duality condition $\boldsymbol{\nabla} V = \boldsymbol{\nabla} \times \boldsymbol{\omega}$, where we are in the case $\epsilon = 0$ of

$$V = \epsilon + \sum_{i=1}^{k} \frac{1}{|\mathbf{x} - \mathbf{x_i}|} \tag{4.2}$$

This space M_{k-1} is the smooth resolution of the singular variety $xy = z^k$ in $\mathbf{C}^3$ of type A_{k-1} with $\partial M_{k-1} = S^3/\mathbf{Z}_k$. The singular situation corresponds to the pol-terms coalescing: $V = \frac{k}{|\mathbf{x}|}$. The case $\epsilon = 1$ corresponds to the Taub-NUT spaces.

Now T-duality with respect to the $U(1)$-isometry generated by the Killing vector $\partial/\partial\tau$ gives with the well-known Buscher formula the conformal flat metric of the extremal NS 5-brane (cfr. eq.(3.8) with $r_0 = 0$)

$$\begin{aligned} ds^2 &= V(\mathbf{y})(d\tau^2 + d\mathbf{y}^2), \\ B_{0i} &= \omega_i, \\ e^{2\phi} &= V(\mathbf{y}), \end{aligned} \tag{4.3}$$

where the self-duality condition for the original metric is now, in the new axion-dilaton sector, assuring the condition for an axionic instanton

$$H_{\mu\nu\rho} = \sqrt{g}\epsilon_{\mu\nu\rho}{}^{\sigma}\partial_\sigma\phi. \tag{4.4}$$

The H charge is from $H_{\mu\nu\rho} = \sqrt{g}\epsilon_{\mu\nu\rho}{}^{\sigma}\partial_\sigma\phi$ with $e^{2\phi} = V = \frac{k}{|\mathbf{x}|}$ easily seen to be $\frac{1}{2\pi^2}\int_{S^3} H = k$. This shows the appearence of the required H charge k.

The heterotic[11] axion a in the dual geometry is defined by dualizing the dual B field

$$\partial a = \pm e^{-2\phi} H_D^* = \frac{1}{2} V^{-2} \partial\omega^* \tag{4.5}$$

[11] A analogous argument can be made for the type II RR-scalar after dualizing down to the D-instantons.

The axion charge of the T-dual solution is called the nut charge of the original gravitational solution. One can say that the S-duality group is related to a duality between the electric aspects of the original gravity background (characterised by the Maxwell field $A = V^{-1}(d\tau + \boldsymbol{\omega} \cdot \mathbf{x})$) and the magnetic aspects (characterised by the nut potential a). Note that the the isometry we are using is called 'translational' (the main importance of this is keeping the SUSY manifest after dualisation), i.e. the covariant derivative of the Killing vector field is self-dual which means $(\partial S_-)^2 = 0$. So from $\partial S_- = 0$ one gets $\partial V = \partial \omega^*$ so that V satisfies the 3D Laplace equation and one has $S_- = 0$, i.e. $a = V^{-1}$. We see that the Taub-NUT metric with can be mapped $S_\pm = a \pm e^{-2\phi} = a \pm V^{-1}$ to the ALE space via the shift $V \to V - 1$.

Note however that in $e^{2\phi} = \frac{k}{|\mathbf{y}|}$ the 3-dimensional harmonic function $\frac{1}{r_3}$ occurs whereas in the five-brane the 4-dimensional harmonic function $\frac{1}{r_4^2}$ occurs. The reason is of course that by doing T-duality in the periodic τ-direction in Taub-NUT space one arrives at the five-brane with one of its four transverse directions compactified on a circle. In other words, since the harmonic function of the original extremal 5-brane metric also depends on τ, before the T-duality from the 5-brane to the ALE space one has to *enforce* an isometric direction by taking the transversal space to be $R^3_{\mathbf{x}} \times S^1_\tau$ and requiring H_5 to be a 3D harmonic function $H_5 = V = 1 + \frac{Q_5}{r_3}$ (here $r_3 := |\mathbf{x}|$) in R^3 and independent of the S^1_τ direction. More precisely if the original 5-brane metric is scaled so that the τ-direction is very large and the $\mathbf{x}$-space looks correspondingly contracted, then in the dual space the direction of the corresponding little circle is 'suppressed', i.e. the ansatz there leads to a harmonic function in only three variables.
But note that this is only *one* possibility to realise the duality. One could also tune directly the merging without making the radius R of the S^1_τ large; then the dual circle (of radius $\tilde{R} = \alpha'/R$) is not 'suppressed'.
More concretely one has in the dual picture the Fourier decomposition along the dual circle [25]

$$e^{2\phi}(\mathbf{x}, \tau) = \sum_{n \in \mathbf{Z}} e^{in\tau/\tilde{R}} \Psi_n(\mathbf{x}) \tag{4.6}$$

where (the Ψ_n are no longer suppressed for $\tilde{R}$ being no longer small)

$$\Psi_0 = e^{2\phi_0} + \frac{k\alpha'}{2r\tilde{R}}, \quad \Psi_n = \frac{\alpha'}{2r\tilde{R}} e^{-|n|r/\tilde{R}} e^{-in\tau_0/\tilde{R}} \quad (n \neq 0) \tag{4.7}$$

One can see that the Ψ_n interpolate between the 3-dimensional and the 4-dimensional harmonic function according to $\tilde{R}$ being very small or very large [29]. Furthermore the occurrence of these momentum modes in the dual picture leads to the idea that in the original picture winding modes have to be included in the description.
This can also be understood from the following perspective. In the original

picture of the ALE space one has besides the degree of freedom $\mathbf{x}_i$ (positions of the centers) also to take into account the parameters $\int_{\Sigma_i} B$ (the Σ_i being the non-trivial 2-cycles 'between the centers')[12]; that they play an important role in the game was the insight of [30], who showed that actual gauge symmetry enhancement occurs only, if not only the positions of the centers merge (giving the A_{k-1} singularity), but also the B-field parameters have the value zero (and not the CFT orbifold value π; this leads to the breakdown of the CFT reasoning, necessary for the non-perturbative gauge symmetry enhancement). Now each of these 4 real parameters (for $i = 1 \cdots, k-1$), the center-distance and the B-field parameter, constitute a hypermultiplet. In the dual picture this corresponds to the positions of the 5-branes in the transversal space $R^3 \times S^1$. But the position parameter in the (dual) S^1-drection breaks the expected isometry in this direction.

This leads to the alternative view on the necessity of including winding modes in the original Taub-NUT picture. Above we saw this was caused by the actual 'occurence' (being no longer suppressed) of the S^1 in the 5-brane picture which lead to the momentum modes there and so to the winding modes in the original picture. Here we see that the S^1-position degree of freedom in the 5-brane picture corresponds in the original picture to the $\int_{\Sigma_i} B$ degree of freedom; but one gets the necessary compact S^2-cycles Σ_i 'between the centers' exactly beacuse the S^1-fibration in the τ variable in the Taub-NUT space collapses to circles of zero radius at the center points; and this means that the corresponding winding modes there become light and so winding modes should be included in the description.

At the end of this section we briefly give the duality transformation for the non-extremal 5-brane. Its metric becomes

$$ds^2 = -(1 - \frac{r_0}{r_3})dt^2 + H_5(\frac{dr_3^2}{1 - \frac{r_0}{r_3}} + r_3^2 d\Omega_2^2 + d\tau^2) \tag{4.8}$$

with the H-monopole magnetic field $H_{\tau\theta\varphi} = -\partial_\theta B_{\tau\varphi} = r_3^2 \sin\theta \partial_{r_3}(1 + \coth\alpha_5 \frac{Q_5}{r_3})$.

Then the non-extremal KK-monopole dual to non-extremal 5-brane is

$$ds^2 = -(1 - \frac{r_0}{r_3})dt^2 + V^{-1}(d\tau + \omega \cdot d\mathbf{x})^2 + V(\frac{dr_3^2}{1 - \frac{r_0}{r_3}} + r_3^2 d\Omega_2^2) \tag{4.9}$$

with KK magnetic field $F = \partial\omega$. More precisely if $\omega = A_\varphi$ then $F_{\theta\varphi} = -\partial_\theta A_\varphi = r_3^2 \sin\theta \partial_{r_3}(1 + \coth\alpha_5 \frac{Q_5}{r_3})$ with $Q_5 = r_0 \sinh^2 \alpha_5$.

In the case of two isometries one will see an effective reduction by two dimensions down to a function harmonic in two dimensions, the logarithm. As there is an effective 2+2 split of the coordinates and in view of the hyper-Kähler nature of the relevant background, it is appropriate to describe the

[12] being given by the S^1_τ-fibration over the line in $R^3_{\mathbf{x}}$ connecting two centers; as the S^1 shrinks at the centers this is an S^2

situation in an complex superfield formalism, the general features of which we describe first (cf. for ex. [31]).

4.2 Superfield Formalism of Buscher Duality and Two Isometries

The $N = 2$ superspace action for one chiral ($\bar{D}_{\pm}U = 0$) superfield U and one twisted chiral ($\bar{D}_{+}V = D_{-}V = 0$) superfield V is determined by the real potential function $K(U, \bar{U}, V, \bar{V})$

$$S = \frac{1}{2\pi\alpha'} \int d^2x D_+ D_- \bar{D}_+ \bar{D}_- K(U, \bar{U}, V, \bar{V})$$

with the target space interpretation ($K_u := \frac{\partial K}{\partial u}$)

$$S_{bos} = -\frac{1}{2\pi\alpha'} \int d^2x (K_{u\bar{u}} \partial^a u \partial_a \bar{u} - K_{v\bar{v}} \partial^a v \partial_a \bar{v} + \epsilon_{ab}(K_{u\bar{v}} \partial^a u \partial^b \bar{v} + K_{v\bar{u}} \partial^a v \partial^b \bar{u}))$$

which shows the $G_{\mu\nu}$ and the $B_{\mu\nu}$ part; so, for example, the H field components become

$$\begin{aligned} H_{u\bar{u}v} &= K_{u\bar{u}v} \; , \; H_{v\bar{v}u} = K_{v\bar{v}u} \\ H_{u\bar{u}\bar{v}} &= -K_{u\bar{u}\bar{v}} \; , \; H_{v\bar{v}\bar{u}} = -K_{v\bar{v}\bar{u}}. \end{aligned} \tag{4.10}$$

Furthermore the string equations of motion have to be satisfied (vanishing β-function equations). If the central charge deficit (determined by the dilaton β-function) vanishes, one actually has $N = 4$ supersymmetry in two dimensions.

In general one gets $N = 4$ supersymmetry for a potential K satisfying the Laplace equation $K_{u\bar{u}} + K_{v\bar{v}} = 0$ (this is the generalization of the hyper-Kähler condition for backgrounds including a B field; it is only a sufficient condition in case of non-trivial dilaton). From the string equations of motion one has then

$$\begin{aligned} \partial_u \log K_{v\bar{v}} &= 2\partial_u \phi, \\ \partial_v \log K_{u\bar{u}} &= 2\partial_v \phi, \end{aligned} \tag{4.11}$$

giving $e^{2\phi} \sim K_{u\bar{u}}$ so that the metric is flat in the Einstein metric $G^{\rm Einst}_{\mu\nu} = e^{-2\phi} G^{\sigma}_{\mu\nu}$, i.e. only the axion-dilaton sector is non-trivial and one has

$$H_{u\bar{u}v} = K_{u\bar{u}v} = 2e^{2\phi} \partial_v \phi \tag{4.12}$$

and $d\phi = \pm\frac{1}{2} e^{-2\phi} H^*$, the self-duality condition for the axion-dilaton sector.

For T-duality one has to assume the existence of (at least) one $U(1)$-isometry; this cooresponds to a Killing symmetry of the potential K

$$K = K(u + \bar{u}, v, \bar{v}).$$

The duality will trade in a twisted field w for the untwisted field u by a Legendre transformation leading to the dual potential ($r := u + \bar{u}$) [32,31]

$$\tilde{K}(r, w + \bar{w}, v, \bar{v}) = K(u + \bar{u}, v, \bar{v}) - r(w + \bar{w})$$

i.e. after the variation w.r.t. u: $\frac{\delta S}{\delta u} = 0 \Rightarrow w + \bar{w} = K_r = K_u$ the independent variables for $\tilde{K}$ are $w, \bar{w}, v, \bar{v}$. As these are now only twisted fields (a set containig only untwisted fields would of course do it equally well) $\tilde{K}$ is a true Kähler potential providing a Kähler metric with Ricci-tensor

$$\tilde{R}_{i\bar{j}} = -\partial_i \partial_{\bar{j}} \log \det \tilde{G}_{i\bar{j}} = -\partial_i \partial_{\bar{j}} \log - \frac{K_{v\bar{v}}}{K_{u\bar{u}}}$$

i.e. the dual background is Ricci-flat for $K_{v\bar{v}} \sim K_{u\bar{u}}$.

In the case of 2 translational $U(1)$ Killing symmetries

$$K = K(u + \bar{u}, v + \bar{v})$$

the Laplace equation is solved by ($u := r_1 + i\theta, v := r_2 + i\phi, z := r_1 + ir_2$)

$$K(r_1, r_2) = iT(r_1 + ir_2) - i\bar{T}(r_1 - ir_2) = -2\mathrm{Im}T(z)$$

where $T(z)$ is an arbitrary holomorphic function and the associated metric is

$$ds^2 = -4\mathrm{Im}T_{zz}(dud\bar{u} + dvd\bar{v}). \tag{4.13}$$

For our axionic instanton background consisting of the 5-brane with two isometric directions the relevant harmonic function is now just $H_5 = k \log |z|$.

Then the dual metric becomes (with $w = \frac{1}{2}K_u + i\theta$) [31]

$$\begin{aligned} ds^2 &= \frac{1}{K_{u\bar{u}}}(dw - K_{uv}dv)(d\bar{w} - K_{u\bar{v}}d\bar{v}) - K_{v\bar{v}}dvd\bar{v} \\ &= \mathrm{Im}S dz d\bar{z} + \frac{1}{\mathrm{Im}S}(d\theta - Sd\phi)(d\theta - \bar{S}d\phi) \end{aligned} \tag{4.14}$$

with $S(z) := -\frac{1}{2}T_{zz}(z)$.

We interpret this as a part, local in the base (z-variable), of an elliptic fibration (cf. the discussion of the stringy cosmic string [19]), which is thus dual to the axionic instanton background we started with. In [19] also global issues in the base were treated making S a true well-defined modular invariant by multyplying it by an η-function term, i.e. $S(z)$ is just a local version of $\tau(z)$. If one specialises to an A_{k-1} singularity one has k cosmic strings at $z = 0$, i.e. $j(\tau(z)) = \frac{1}{z^k}$. Now at $\tau \approx i\infty, j \approx \infty$ one has $j(\tau) \sim e^{-2\pi\tau(z)}$ or $\tau(z) = -\frac{k}{2\pi}i \log z$, so

$$\mathrm{Im}S(z) = -\frac{k}{2\pi} \log |z|. \tag{4.15}$$

Note that the H charge of the original axionic instanton background is from $H_{u\bar{u}v} = K_{u\bar{u}v} = 2e^{2\phi}\partial_v\phi$ found to be proportional to n as the dilaton was $e^{2\phi} \sim K_{u\bar{u}} = -2\text{Im}T_{zz} = 4\text{ImS}$, i.e.

$$e^{2\phi} \sim k \log|z| \tag{4.16}$$

which shows the consistency of the interpretation.

Note that the ALE description (which is local around the (resolved) singular point) is related to the resolution of the singularity, whereas the description given here in the fibration picture (which is local in the base around the (desingularised) fibre, but global in the fibre) is related to the deformation of the singularity to an elliptic fibration of smooth total space.

4.3 Intersection of Two NS 5-branes – $N = 1$ Brane Boxes

In the gauge-theory-from-branes setup the NS 5-branes are considered to be heavy relative to the D4-branes whose world-volume gives the gauge theory. So the NS 5-branes (later, after T-duality, the KK-monopoles, resp., if one is interested in the singular situation at the neighborhood of the singularity, the ALE spaces in the transversal dimensions) constitute the 'background', the D4-branes (later, after T-duality, the fractional D3-branes) are the 'probes'.

So there are really two levels of consideration here: first the gauge theory where gravity is turned off and the light D3-branes; second there is a background, probed by the D3-branes, of ALE/KK-monopoles, the T-dual of the NS 5-branes which are considered to be heavy. In the case of $N = 2$ space-time supersymmetry with parallel NS 5-branes the background is just the well-known ALE/KK-monopole space; in the $N = 1$ case it is a background of two ALE/KK-monopole spaces fibered together over a common R^2 direction. This back ground arises as the T-dual of intersecting NS and NS' 5-branes which build the socalled brane boxes of [7], as we will discuss in the following.

If on the other hand the backreaction of the D3-branes on the background is included, the former NS 5-branes (resp. their T-duals) become dynamical and it is appropriate to give a common metric for the total brane system. This will be the topic of the next subsection.

Let us describe the metric for the $N = 1$ situation of a $\mathbf{C}^3/\Gamma$ singularity, $\Gamma = \mathbf{Z}_k \times \mathbf{Z}_{k'}$, (probed by the D3-branes) [9] in the case of "adding up" two ALE spaces. Let us forget about the D-branes and just concentrate on the two now non-parallel NS 5-branes. Specifically to compute the metric of the NS-NS' 5-brane system one starts one step earlier with k D5-branes in 012345 and k' D5-branes in 012367 with compact directions 4 and 6. This has the metric [33] (with $e^{-2\phi} = H_5 H_{5'}$)

$$ds^2 = \frac{1}{\sqrt{H_5 H_{5'}}} ds^2_{0123} + \sqrt{\frac{H_{5'}}{H_5}} ds^2_{45} + \sqrt{\frac{H_5}{H_{5'}}} ds^2_{67} + \sqrt{H_5 H_{5'}} ds^2_{89} \tag{4.17}$$

Then this is S-dualised to k NS 5-branes in 012345 and k' NS' 5-branes in 012367 giving

$$ds^2 = ds^2_{0123} + H_{5'} ds^2_{45} + H_5 ds^2_{67} + H_5 H_{5'} ds^2_{89} \tag{4.18}$$

with $e^{2\phi} = H_5 H_{5'}$.

Then one makes T-dualities in the compact directions 4 and 6 giving (at the singularity) an A_{k-1} in 6789 and an $A_{k'-1}$ in 4589.
This "adding up" of the two ALE spaces is difficult to perform in the usual representationfor the ALE metric which has a 3+1 split in the coordinates. Instead one would like to have a representation which isolates the singularity in a $2 + 2_{89}$ description. This is provided by the stringy cosmic string description of Sect.(4.2) which gives the A_{k-1} singularity in an elliptic fibration over $\mathbf{C} = R_{89}$ (so here the directions 5 and 7 are considered to be compact too; the respective "τ-circles" of the ALE spaces A_{k-1} in 6789 and an $A_{k'-1}$ in 4589 are well known to shrink to zero radius at the, in the merging limit common, center; they are the vanishing S^1's in the elliptic fibration).

So one assumes that H_5 (and correspondingly for $H_{5'}$), which - to make T_6 - was assumed to be independent of x_6 and just living as a harmonic function in 789, is now actually independent also of x_7 and so lives just as a logarithmic function in 89 (cf. eqn. (4.16) above: $e^{2\phi} \sim k \log |z_{89}|$).

Because of the fibration structure now this type of representation of the ALE metric is - in contrast to the 3+1 representation - easily "added up". So this extends the $N = 2$ supersymmetric case with the (local in the base) description of an A_{k-1} singularity of an elliptic fibration (over $\mathbf{C}_{89}$) to the $N = 1$ supersymmetric case of a description[13] of the singularities of the doubly elliptically fibered Calabi-Yau space[14] $CY^{19,19} = \left[\begin{array}{c|cc} P^2_{z_{45}} & 3 & 0 \\ P^1_{z_{89}} & 1 & 1 \\ P^2_{z_{67}} & 0 & 3 \end{array}\right] = dP_9 \times_{P^1} dP_9$. This Calabi-Yau space can be constructed as a $T^6/\mathbf{Z}_k \times \mathbf{Z}_{k'}$ orbifold, where we have identified $z_{45} = x_4 + ix_5$, $z_{67} = x_6 + ix_7$, $z_{89} = x_8 + ix_9$. Locally around the singularities we therefore consider the non-compact space $\mathbf{C}^3/\mathbf{Z}_k \times \mathbf{Z}_{k'}$, where the $\mathbf{Z}_k \times \mathbf{Z}_{k'}$ orbifold action, giving a genuine $\Gamma \subset SU(3)$, is fibered together by corresponding actions giving an A_{k-1} resp. $A_{k'-1}$ singularity in 6789 resp. 4589. Concretely the corresponding total metric has the following explicit form:

$$\begin{aligned} ds^2 = ds^2_{0123} &+ \frac{1}{\mathrm{Im} S_{k'}(z_{89})} (d\theta_4 - S_{k'}(z_{89}) d\phi_5)(d\theta_4 - \bar{S}_{k'}(\bar{z}_{89}) d\phi_5) \\ &+ \frac{1}{\mathrm{Im} S_k(z_{89})} (d\theta_6 - S_k(z_{89}) d\phi_7)(d\theta_6 - \bar{S}_k(\bar{z}_{89}) d\phi_7) \\ &+ \mathrm{Im} S_k(z_{89}) \mathrm{Im} S_{k'}(z_{89}) dz_{89} d\bar{z}_{89}. \end{aligned} \tag{4.19}$$

---

[13] G. C. thanks A. Uranga for discussion on this point

[14] $dP_9 = \frac{1}{2} K3$ is the elliptically fibered surface $dP_9 = \left[\begin{array}{c|c} P^2 & 3 \\ P^1 & 1 \end{array}\right]$.

Here $S_k(z_{89}) = \frac{k}{2\pi i} \log z_{89}$ and analogously for $S_{k'}$.

This metric descibes a fibration of a Γ-singularity in the z_{4567} respectively z_{67} directions over a singular point in the common base space with coordiates z_{89}, just like the $CY^{19,19}$ is a $T^2 \times T^2$ fibration over the common base P^1. The A_{k-1} singular point of one dP_9 direction times the S^1 of the remaining z-plane gives the complex curve of singularities $z_{67} = z_{89} = 0$ resp. $z_{45} = z_{89} = 0$ intersecting the S^5 relevant to the $AdS_5 \times S^5/(\mathbf{Z}_k \times \mathbf{Z}_{k'})$ in an S^1 of singularities given by the unit circle in z_{45} resp. z_{67} (cf. [9]; if k and k' are not coprime there exists a third curve of singularities).

4.4 Intersection of D 4-Brane with NS 5-Branes

Finally we want to describe the metric for the configuration of n D4-branes between k parallel NS 5-branes in type IIA, corresponding to the $N = 2$ supersymmetric $SU(n)^{k-1}$ gauge theory [34]. Let us start two steps earlier with D5-branes in type IIB, from which we get the NS 5-branes by type IIB S-duality, and D3-branes, being invariant under the S-duality, from which we get the D4-branes in type IIA by T-duality.

Starting with the intersection of D5-branes with D3-branes we obtain after a S-duality transformation the following metric which describes the intersection of NS 5-branes with D3-branes ($e^{2\phi} = H_5$):

$$\mathrm{NS5} \perp \mathrm{D3} : ds^2 = \frac{1}{\sqrt{H_3}} ds^2_{012} + \sqrt{H_3} s^2_{345} + \frac{H_5}{\sqrt{H_3}} ds^2_6 + H_5 \sqrt{H_3} ds^2_{789} \quad (4.20)$$

Next let us perform a T-duality transformation with respect to the x_3 direction. Under this duality transformation the type IIB D3-brane turns into a type IIA D4-brane, i.e. one of the transverse directions of the D3-brane becomes a worldvolume direction of the D4-brane. Here we have to assume that H_3 is independent of x_3 and following [33] we may write $\tilde{H}_3 = H_4$. Finally let us apply a T-duality with respect to x_6 which leads to a multi Taub-NUT configuration and a fractional D3-brane (where $\omega_i = B_{6i}$ for $i = 7, 8, 9$ and $e^{2\phi} = 1$)

$$ds^2 = \frac{1}{\sqrt{H_3}} ds^2_{0123} + \sqrt{H_3} [ds^2_{45} + \frac{1}{H_5}(dx^2_6 + \omega d\mathbf{x}_{789})^2 + H_5 ds^2_{789}] . \quad (4.21)$$

Inspection of this metric clearly exhibits the D3-brane with world volume along the (0123)-directions as well as the Taub-NUT space in the transversal directions (6789). Now one can proceed as before and 'delete' the constant in the harmonic function H_5 via a U-duality transformation. In this way the D3-brane is localized at the A_{k-1} singularity in the transversal space (6789). In addition one can also consider the limit where the constant part in the harmonic function H_3 can be neglected. In this case the geometry becomes equivalent to $AdS_5 \times S^5/\mathbf{Z}_k$. This space describes $N = 2$ supersymmetric gauge theories in the large N-limit, where the theories are supposed to be superconformal. Finally let us remark that we can also consider the combined system of D-branes which are positioned in the brane boxes of intersecting NS 5-branes, as described in section (4.3). The dual geometry of this set up is then given by D3-branes plus a Γ singularity, which extends into the full transversal space with directions (456789). At the horizon of the D3-branes the large N-limit of $N = 1$ supersymmetric gauge theories, based on the space $AdS_5 \times S^5/(\mathbf{Z}_k \times \mathbf{Z}_{k'})$, is obtained.

Acknowledgements

We thank K. Behrndt, K. Sfetsos and A. Uranga for discussion.

References

1. A. Strominger, Nucl. Phys. **B343** (1990) 167;
   C.G. Callan, J.A. Harvey and A. Strominger, Nucl. Phys. **B359** (1991) 611, Nucl. Phys. **B367** (1991) 60.
2. S. Rey, *Axionic String instantons and their low energy implications*, published in proceedings to Tuscaloosa Workshop 1989; *On string theory and axionic strings and instantons*, published in DPF Conf. 1991.
3. J.H. Schwarz, Phys. Lett. **B360** (1995) 13, hep-th/9508143.
4. A. Hanany and E. Witten, *Type IIB superstrings, BPS monopoles and three-dimensional gauge dynamics*, Nucl. Phys. **B492** (1997) 152, hep-th/9611230.
5. H. Ooguri and C. Vafa, *Two-dimensional black hole and singularities of CY manifolds*, Nucl. Phys. **B463** (1996) 55, hep-th/9511164.
6. M.R. Douglas and G. Moore, *D-branes, quivers and ALE instantons*, hep-th/9603167.
7. A. Hanany and A. Zaffaroni, *On the realization of chiral four-dimensional gauge theories using branes*, J. High Energy Phys. **5** (1998) 1, hep-th/9801134.
8. A. Karch, D. Lüst and D. J. Smith, *Equivalence of Geometric Engineering and Hanany-Witten via Fractional Branes*, hep-th/9803232.
9. A. Hanany and A. Uranga, *Brane Boxes and Branes on Singularities*, hep-th/9805139.
10. H. J. Boonstra, B. Peeters and K. Skenderis, *Duality and asymptotic geometries*, Phys. Lett. **B411** (1997) 59, hep-th/9706192.
11. K. Sfetsos and K. Skenderis, *Microscopic derivation of the Bekenstein-Hawking entropy formula for non-extremal black holes*, Nucl. Phys. **B 517** (1998) 179, hep-th/9711138.
12. J. Maldacena, *The Large N Limit of Superconformal field theories and supergravity*, hep-th/9711200.
13. M. Banados, M. Henneaux, C. Teitelboim and J. Zanelli, *Geometry of the* $2+1$ *Black Hole* Phys. Rev. **D 48** (1993) 1506, gr-qc/9302012.

14. A. Font, L. Ibanez, D. Lüst and F. Quevedo, Phys. Lett **B249** (1990) 35;
S. Rey, Phys. Rev. **D43** (1991) 526;
A. Sen, Phys. Lett. **B303** (1993) 22;
J. Schwarz and A. Sen, Nucl. Phys. **B411** (1994) 35.
15. C. Callan, S. Giddings, J. Harvey and A. Strominger, *Evanescent Black Holes*, Phys. Rev. **D 45** (1992) 1005, hep-th/9111056.
16. I. Bakas, *Space Time Interpretation of S-Duality and Supersymmetry Violations of T-Duality*, Phys. Lett. B343 (1995) 103, hep-th/9410104.
17. E. Bergshoeff and K. Behrndt, *D-Instantons and asymptotic geometries*, hep-th/9803090.
18. G. Lopes Cardoso and T. Mohaupt, *Dual heterotic black-holes in four and two dimensions*, hep-th/9806036.
19. B. Greene, A. Shapere, C. Vafa and S.T. Yau, Nucl. Phys **B 337** (1990) 1.
20. C. Vafa, *Evidence for F-Theory*, Nucl. Phys. **B 469** (1996) 403, hep-th/9602022.
21. J.M. Maldacena and A. Strominger, *Semiclassical decay of near extremal fivebranes*, hep-th/9710014.
22. N. Seiberg, hep-th/9608111; hep-th/9609161; hep-th/9705221.
23. E. Witten, Phys. Rev 44 (1991) 314.
24. G.T. Horowitz and D.L. Welch, *Exact Three Dimensional Black Holes in String Theory*, Phys. Rev. Lett. **71** (1993) 328, hep-th/9302126.
25. R. Gregory, J.A. Harvey and G. Moore, hep-th/9708086.
26. A. Sen, hep-th/9707042; hep-th/9707123.
27. M. Bianchi, F. Fucito, G.C. Rossi and M. Martinelli, *ALE Instantons in string effective theory*, Nucl. Phys. **B 440** (1995) 129, hep-th/9409037.
28. A.A. Tseytlin, *Type IIB instanton as a wave in twelve dimensions*, Phys. Rev. Lett. **78** (1997) 1864, hep-th/9612164.
29. D.-E. Diaconescu and N. Seiberg, hep-th/9707158.
30. P.S. Aspinwall, Phys. Lett. **B 357** (1995) 329.
31. E. Kiritsis, C. Kounnas and D. Lüst, Journ. of Mod. Phys. **A9** (1994) 1361, hep-th/9308124 and hep-th/9312143;
C. Kounnas, hep-th/9402080.
32. S. Gates, C. Hull and M. Rocek, Nucl. Phys. **B258** (1984) 157.
33. K. Behrndt, E. Bergshoeff and B. Janssen, *Intersecting D-branes in Ten and Six Dimensions*, hep-th/9604168.
34. E. Witten *Solutions Of Four-Dimensional Field Theories Via M Theory*, Nucl. Phys. **B 500** (1997) 3, hep-th/9703166.

Superembedding Approach and Generalized Action in String/M-Theory

Igor Bandos

[1] Institute for Theoretical Physics, NSC Kharkov Institute of Physics and Technology, 310108 Kharkiv, Ukraine
[2] ICTP, Trieste, Italy

Abstract. A brief introduction to superembedding approach (SEA) in its variant based on the generalized action principle (GAP) for super-p-branes is given. A role of harmonic variables for Lorentz group is stressed. A relation of the GAP with complete superfield actions is noted. Recent applications in studying of Dirichlet branes (super–Dp–branes) and M-branes are discussed.

1996 was the hard year for our Science. In January of this year we had lost our teacher Dmitrij V. Volkov and, in March, Victor I. Ogievetsky left us.

In this contribution I present a brief description of generalized action principle (GAP) and superembedding approach (SEA) for supersymmetric extended objects. We have proposed them and elaborated for $D = 10$ superstrings (called now fundamental strings) and $D = 11$ supermembrane (called now M2-brane) in collaboration with D.V. Volkov (BPSTV (1995), Bandos, Sorokin and Volkov (1995), Volkov (1995)). On the other side, they are based on the works of D.V. Volkov (STVZ (1988)) on doubly supersymmetric twistor-like approach, unify the latter with the Lorentz harmonic approach (Bandos and Zheltukhin (1991)) and, thus, uses essentially the concept of harmonic variables, developed by V.I. Ogievetsky with collaborators (GIKOS (1984)). So the subject of my talk originates from the work of both these great scientists.

Recently the superembedding approach has been applied for investigation of super-D-branes (Howe and Sezgin (1996), Bandos, Sorokin and Tonin (1997), Akulov et al. (1998)), super-M5-brane (Howe and Sezgin (1997), Howe et al (1998)) as well as intersecting branes (Chu et. al. (1998)) and brane models of gauge theories (Howe, Lambert, West et al. (1997)). A derivation of brane action from superembedding equations, which can be regarded as an inversion of the line of the GAP approach, has been proposed in (Howe, Raetzel and Sezgin (1998)). The GAP for $D = 11$ supermembrane (Bandos, Sorokin and Volkov (1995)) in $AdS_4 \times M_7$ background has been used to obtain the supersymmetric $Osp(8|4)$ singleton action (Dall'Agata et. al. (1998)).

In this talk I review the main ingredients of the SEA and the GAP using relatively simple example of $D = 10$ heterotic string and describe briefly achievements and problems of the SEA in studying String/M-theory.

1 Generalized Action for $D = 10$ Heterotic String

1.1. One of the main ingredients of the generalized action (GAP) is the **Lagrangian form** defined on the *world volume superspace* $\Sigma^{(p+1|n)} = \{(\zeta^M)\} = \{(\xi^m, \eta^q)\}$ of the super-p-brane. For $D = 10$ heterotic string ($\underline{m} = 0, \ldots, 9$; $p = 1$; $m = 0, 1$; $q = 1, \ldots 8$) the Lagrangian two- form [1]

$$\mathcal{L}_2 = \frac{1}{2} E^{++} \wedge E^{--} - i\Pi^{\underline{m}} \wedge d\Theta \Gamma_{\underline{m}} \Theta + \frac{1}{2} E^{--} \wedge \Psi^I_- d\Psi^I_-, \qquad I = 1, \ldots, 32 \quad (1)$$

is constructed from the pull-backs of the basic one forms of target superspace $\Pi^{\underline{m}} = dX^{\underline{m}} - id\Theta\Gamma^{\underline{m}}\Theta$, $d\Theta^{\underline{\mu}}$ onto the world sheet superspace $\Sigma^{(1+1|n)} = \{(\zeta^M)\} = \{(\xi^{(\pm\pm)}, \eta^q)\}$ $\Pi^{\underline{m}} = d\zeta^M \Pi_M^{\underline{m}} \equiv d\zeta^M (\partial_M X^{\underline{m}}(\xi^{(\pm\pm)}, \eta^q) - i\partial_M \Theta \Gamma^{\underline{m}} \Theta)$, $d\Theta^{\underline{\mu}} = d\zeta^M \partial_M \Theta^{\underline{\mu}}(\xi^{(\pm\pm)}, \eta^q)$ and heterotic fermion 1-form $\Psi^I_- d\Psi^I_- = -d\zeta^M \, (\partial_M \Psi^I_-) \Psi^I_-(\xi^{(\pm\pm)}, \eta^q)$ *by the use of the external product of the superforms only*. Supervielbeine of flat target superspace $E^{\underline{A}} = (E^{\underline{a}}, E^{\underline{\alpha}})$

$$E^{\underline{a}} = (E^{\pm\pm}, E^i) = \Pi^{\underline{m}} u_{\underline{m}}^{\underline{a}} = (\Pi^{\underline{m}} u_{\underline{m}}^{\pm\pm}, \Pi^{\underline{m}} u_{\underline{m}}^i) \quad (2)$$
$$E^{\underline{\alpha}} = (E^{+q}, E^{-\dot{q}}) = d\Theta^{\underline{\mu}} v_{\underline{\mu}}^{\underline{\alpha}} = (d\Theta^{\underline{\mu}} v_{\underline{\mu}q}^{\;+}, d\Theta^{\underline{\mu}} v_{\underline{\mu}\dot{q}}^{\;-})$$

distinct from the standard one $(\Pi^{\underline{m}}, d\Theta^{\underline{\mu}})$ by a Lorentz rotation. The vector and spinor representations of this $SO(1, D-1)$ transformation are given by the matrices

$$u_{\underline{m}}^{\underline{a}} = (u_{\underline{m}}^{\pm\pm}, u_{\underline{m}}^{\,i}) \in SO(1,9) \qquad v_{\underline{\mu}}^{\underline{\alpha}} = (v_{\underline{\mu}q}^{\;+}, v_{\underline{\mu}\dot{q}}^{\;-}) \in Spin(1,9)$$

$$\Leftrightarrow \quad u_{\underline{m}}^{\underline{a}} \eta^{\underline{mn}} u_{\underline{n}}^{\underline{b}} = \eta^{\underline{ab}} \quad \Leftrightarrow \quad \begin{cases} u_{\underline{m}}^{++} u^{++\underline{m}} = 0, \quad u_{\underline{m}}^{--} u^{--\underline{m}} = 0, \\ u_{\underline{m}}^{--} u^{++\underline{m}} = 2, \\ u_{\underline{m}}^{i} u^{j\underline{m}} = -\delta^{ij}, \quad u_{\underline{m}}^{\pm\pm} u^{i\underline{m}} = 0, \end{cases} \quad (3)$$

(vector and spinor $\frac{SO(1,9)}{SO(1,1)\times SO(8)}$ *Lorentz harmonics*, see BPSTV (1995) and refs. therein). They are related by $u_{\underline{m}}^{\underline{a}} \Gamma^{\underline{m}}_{\underline{\mu\nu}} = v_{\underline{\mu}}^{\underline{\alpha}} \Gamma^{\underline{a}}_{\underline{\alpha\beta}} v_{\underline{\nu}}^{\underline{\beta}}$, $u_{\underline{m}}^{\underline{a}} \Gamma_{\underline{a}}^{\underline{\alpha\beta}} = v_{\underline{\mu}}^{\underline{\alpha}} \Gamma_{\underline{m}}^{\underline{\mu\nu}} v_{\underline{\nu}}^{\underline{\beta}}$,

$$\Leftrightarrow \quad \delta_{qp} u_{\underline{m}}^{++} = v_q^+ \Gamma_{\underline{m}} v_p^+, \qquad \delta_{\dot{q}\dot{p}} u_{\underline{m}}^{--} = v_{\dot{q}}^- \Gamma_{\underline{m}} v_{\dot{p}}^-, \qquad \gamma^i_{q\dot{p}} u_{\underline{m}}^i = v_q^+ \Gamma_{\underline{m}} v_{\dot{p}}^-, \quad (4)$$

(where $\Gamma_{\underline{m}}$ and $\gamma^i_{q\dot{p}}$ are the $SO(1,9)$ and $SO(8)$ γ–matrices). Their differentials

$$du_{\underline{m}}^{\underline{a}} = u_{\underline{m}}^{\underline{b}} \Omega_{\underline{b}}^{\;\underline{a}}(d) \quad \Leftrightarrow \quad \begin{cases} du_{\underline{m}}^{\pm\pm} = \pm u_{\underline{m}}^{\pm\pm} \Omega^{(0)}(d) + u_{\underline{m}}^{\,i} \Omega^{\pm\pm i}(d), \\ du_{\underline{m}}^{i} = -u_{\underline{m}}^{j} \Omega^{ji} + \frac{1}{2} u_{\underline{m}}^{\pm\pm} \Omega^{\mp\mp i}(d) \end{cases} \quad (5)$$

$$dv_{\underline{\mu}}^{\underline{\alpha}} = 1/4 v_{\underline{\mu}}^{\underline{\beta}} (\Gamma_{\underline{ab}})_{\underline{\beta}}^{\;\underline{\alpha}} \Omega^{\;\underline{ab}}(d) \quad (6)$$

---

[1] For simplicity we restrict ourself by the case of flat target superspace. The generalization for supergravity background is straightforward (Bandos, Sorokin and Tonin (1997)).

are expressed in terms of the $so(1,D-1)$ valued Cartan 1–forms

$$\Omega^{\underline{ab}} = -\Omega^{\underline{ba}} = \begin{pmatrix} \Omega^{ab} & \Omega^{aj} \\ -\Omega^{bi} & \Omega^{ij} \end{pmatrix} = u_{\underline{m}}^{\underline{a}} du^{\underline{bm}} = \frac{1}{16} dv_{\underline{\mu}}^{\underline{\alpha}} (\Gamma^{\underline{ab}})_{\underline{\alpha}}^{\underline{\beta}} v_{\underline{\beta}}^{\underline{\mu}} \tag{7}$$

Integrating the Lagrangian form (1) over *pure bosonic* world sheet

$$S_0 = \int_{\mathcal{M}_0^{1+1}} \mathcal{L}_2 \equiv \int d^2\xi \ \epsilon^{nm} \ (\mathcal{L}_2)_{mn}|_{\eta^q=0} \tag{8}$$

gives the usual (component) superstring action in the first order form (Bandos and Zheltukhin (1991), BPSTV (1995)). The equations of motion following from the functional (8)[2]

$$E^i \equiv \Pi^{\underline{m}} u_{\underline{m}}^i = 0. \tag{9}$$

$$E^{--} \wedge \mathcal{D}\Psi_-^I = 0 \quad \Rightarrow \quad \mathcal{D}\Psi_-^I = E^{--}\mathcal{D}_{--}\Psi_-^I \tag{10}$$

$$E_{\dot q}^- = (E^{++} - \frac{1}{2}\Psi_-^I d\Psi_-^I)\psi_{++\dot q}^{\ -} = (E^{++} - E^{--}\frac{1}{2}\Psi_-^I \mathcal{D}_{--}\Psi_-^I)\psi_{++\dot q}^{\ -} \tag{11}$$

$$E^{--} \wedge \Omega^{++i} - (E^{++} - \Psi_-^I d\Psi_-^I) \wedge \Omega^{++i} + 4iE_q^+ \wedge E_{\dot q}^- \gamma_{q\dot q}^i = 0 \tag{12}$$

can be reduced to the standard superstring equations upon eliminating the auxiliary fields (harmonics). A part of variations does not produce independent equations. They, hence, can be identified with the parameters of the gauge symmetries: $i_\delta \Omega^{(0)} = \frac{1}{2} u^{--}\cdot \delta u^{++} \ \rightarrow SO(1,1)$, $i_\delta \Omega^{ij} = u^i \cdot \delta u^j \ \rightarrow$ $SO(8)$, $i_\delta E^{\pm\pm} \rightarrow$ 'b–symmetry' or reparametrization, $\kappa_q^+ = i_\delta E_q^+ \rightarrow$ κ–symmetry.

The *generalized action* for $D = 10$ heterotic string

$$S = \int_{\mathcal{M}^{1+1}} \mathcal{L}_2 \equiv \int_{\mathcal{M}_0^{1+1}} \mathcal{L}_2|_{\eta^q=\eta^q(\xi)} \tag{13}$$

(Bandos, Sorokin and Volkov (1995), see Nieman and Regge (1978), D'Auria, Frè and Regge (1980), Castellani, D'Auria and Frè (1991) for supergravity) is given by the integral of the Lagrangian 2–form (1) **over arbitrary 2 – dimensional bosonic surface** $\mathcal{M}^{1+1} = \{(\xi^{(\pm\pm)}, \eta^{(+)q}) : \eta^{(+)q} = \eta^{(+)q}(\xi)\}$ **in the world volume superspace** $\Sigma^{(1+1|n)} = \{(\xi^{(\pm\pm)}, \eta^{(+)q})\}$. Hence, in the functional (13) all the variables shall be considered as world volume superfields but, taken on the surface $\mathcal{M}^{1+1}$ (i.e. with $\eta^q = \eta^q(\xi)$):
$X^{\underline{m}} = X^{\underline{m}}(\xi,\eta(\xi)), \Theta^{\underline{\mu}} = \Theta^{\underline{\mu}}(\xi,\eta(\xi)), u_{\underline{m}}^{\underline{a}} = u_{\underline{m}}^{\underline{a}}(\xi,\eta(\xi)), \Psi_-^I = \Psi_-^I(\xi,\eta(\xi))$.

The variation of the GAP (13) should vanishes for arbitrary variations of the (super)fields involved *as well as for arbitrary variations of the surface* $\mathcal{M}^{1+1}$ (i.e. $\delta S/\delta\eta(\xi) = 0$). For the Lagrangian form under consideration it can be proved that the latter variation does not lead to new equations of motion. Instead, the arbitrariness of the surface $\mathcal{M}^{1+1}$ provides the possibility to regard the 'field' equations (9), (10), (11), (12), **as equations for forms and superfields defined on the whole world volume superspace** $\Sigma^{(1+1|8)}$.

[2] These nontrivial equations are produced by variations of Lorentz harmonic variables $\delta S/i_\delta \Omega^{\pm\pm i} \equiv u^{\underline{m}i}\delta S/\delta u^{\underline{m}\mp\mp} = 0$ (9), heterotic fermions $\delta S/\delta \Psi_-^I = 0$ (10), and superspace coordinate fields $\delta S/i_\delta E^i \equiv u^{\underline{m}i}\delta S/\delta X^{\underline{m}} = 0$ (11), $\delta S/i_\delta E_{\dot q}^- \equiv v_{\dot q}^{+\underline{\mu}}\delta S/\delta\Theta^{\underline{\mu}} = 0$ (12) respectively.

2 Superembedding Equations of Heterotic String and Complete Superfield Action

Thus the GAP produces *formally the same equations* (9), (10), (11), (12). However, now they *can be considered as equations for superfields and differential forms on the world volume superspace* $\Sigma^{(1+1|8)}$.

A pragmatic way to find that a Lagrangian form $\mathcal{L}$ is proper for constructing the GAP is to prove that corresponding equations are selfconsistent as equations on the world volume superspace $\Sigma^{(p+1|n)}$ of the super-p-brane, i.e. form *the free differential algebra on this superspace* (Nieman and Regge (1978), D'Auria, Frè and Regge (1980), Castellani, D'Auria and Frè (1991)).

The Eqs. (9), (10), (11), (12) are selfconsistent on $\Sigma^{(1+1|8)}$ and describe the **minimal (on-shell) embedding** of the heterotic string world volume superspace $\Sigma^{(1+1|8)}$ into the flat $D = 10, N = 1$ superspace. Not all of them are independent. Eqs. (11), and the bosonic component of (10) $D_{+q}\Psi^{I}_{-} = 0$ are *independent dynamical superfield equations* while Eq. (12) and Grassmann component of (10) $D_{++}\Psi^{I}_{-} = 0$ appear as their consequences.

The **off–shell superembedding** $\Sigma^{(1+1|8)}$ is specified by the equation (9) only. The Eq. (9) can be recognized as the *geometrodynamic equation* (STVZ (1988), DGHS 1992, BPSTV (1995) and refs.therein) or superembedding condition (Howe and Sezgin (1996), Howe and Sezgin (1997)). To justify the equivalence of (9) with the standard form of superembedding equations

$$E^{i} \equiv \Pi^{\underline{m}} u^{i}_{\underline{m}} = 0 \qquad \Leftrightarrow \qquad \Pi^{\underline{m}}_{+q} \equiv \mathcal{D}_{+q} X^{\underline{m}} - i\mathcal{D}_{+q}\Theta\Gamma^{\underline{m}}\Theta, \tag{14}$$

we have to take into account the freedom in choice of the intrinsic supervielbein of the world volume superspace $e^{A} = (e^{\pm\pm}, e^{+q}) = d\xi^{(\pm\pm)}e^{A}_{(\pm\pm)} + d\eta^{(+)q}e^{A}_{(+)q}$, $(d = e^{A}\mathcal{D}_{A} = e^{+q}\mathcal{D}_{+q} + e^{\pm\pm}\mathcal{D}_{\pm\pm})$ as well as a freedom to adapt the moving frame (Lorentz harmonics) to the world volume (see BPSTV (1995), Howe, Raetzel and Sezgin (1998)) [3].

As it is known (DGHS 1992), *for heterotic string the geometrodynamic condition (14) does not contain equations of motion among its consequences.* Indeed, from the integrability condition for Eq. (14) one obtains

$$E^{-}_{\dot{q}} = E^{++}\psi_{++\dot{q}}^{\;-} + E^{--}\psi_{--\dot{q}}^{\;-} \tag{15}$$

while the superfield equations of motion (11) specifies

$$\psi_{--\dot{q}}^{\;-} = -\frac{1}{2}\Psi^{I}_{-}\mathcal{D}_{--}\Psi^{I}_{-}\psi_{++\dot{q}}^{\;-}.$$

[3] In the presence of heterotic fermions the most convenient choice of the world volume supervielbein is $e^{--} = E^{--} \equiv \Pi^{\underline{m}}u^{--}_{\underline{m}}$, $e^{++} = E^{++} - a\Psi^{I}_{-}d\Psi^{I}_{-}$, $e^{+q} = E^{+q} \equiv d\Theta^{\underline{\mu}}v^{+}_{\underline{\mu}q}$ with $a = 1/2$. The harmonic frame can be chosen in such a way that $\Pi^{\underline{m}}_{a}u^{i}_{\underline{m}} = 0$ holds.

In the absence of heterotic fermions the external derivative of the Lagrangian form (1)

$$d\mathcal{L}_2|_{\Psi^I_-=0} = -2iE^-_{\dot{q}} \wedge E^-_{\dot{q}} \wedge E^{++} + \frac{1}{2}E^i \wedge (E^{\mp\mp} \wedge \Omega^{\pm\pm i} - 4iE^+_q \wedge E^-_{\dot{q}} \gamma^i_{q\dot{q}}) \quad (16)$$

evidently vanishes as a consequence of the geometrodynamic condition (14) only. As the latter has no dynamical content, this statement means the *off-shall superdiffeomorphism invariance* of the GAP (in the *rheonomic sense*).

This provides the possibility to construct the **complete superfield action** using the GAP Lagrangian form $\mathcal{L}_2 \equiv 1/2 d\zeta^M \wedge d\zeta^N \mathcal{L}_{NM}$

$$S_{superfield} = \int d^2\xi d^8\eta \; (P^{+q}_{\underline{m}} \Pi^{\underline{m}}_{+q} + P^{MN} \; (\mathcal{L}_2 - dY)_{NM}). \quad (17)$$

Here $P^{+q}_{\underline{m}}$, P^{MN} are Lagrangian multipliers and $Y = d\zeta^M Y_M$ is an auxiliary 1–form superfield. The functional (17) is just the superfield action for heterotic string discovered in (DGHS 1992), where the GAP Lagrangian form $\mathcal{L}_2$ was called 'Wess–Zumino 2–form'. It is worth mentioned that in all the superfield functionals for superbranes (Berkovits (1989), Ivanov and Kapustnikov (1991), Pasti and Tonin (1993), Bergshoeffand Sezgin (1993) and refs. in BPSTV (1995)) the so-called **'Wess-Zumino form' is** nothing else then **the GAP Lagrangian form** for corresponding brane.

When *the heterotic fermions are present*, the derivative of the Lagrangian form is

$$d\mathcal{L}_2 = d\mathcal{L}_2|_{\Psi^I_-=0} + iE^-_{\dot{q}} \wedge E^-_{\dot{q}} \wedge \Psi^I_- d\Psi^I_- + \frac{1}{2}E^{--} \wedge \mathcal{D}\Psi^I_- \wedge \mathcal{D}\Psi^I_-$$

To write the complete superfield action, the inputs from heterotic fermions has to be separated and considered with some care (Sorokin and Tonin (1993), Ivanov and Sokatchev (1994), Howe (1994), BCSV (1994)).

Thus the GAP produces superembedding equations together with their consequences (including proper equations of motion). This can simplify essentially the studying of brane superembeddings. On the other hand, when the Lagrangian form is closed on the surface of nondynamical equations, GAP can be used for construction of the complete superfield action.

3 Generalized Action and Superembedding Approach to D-Branes and M-Branes

The superembedding approach (SEA) demonstrated its strength in studying the new objects of String/M-theory: $D = 10$ Dirichlet superbranes (super-Dp-branes) and M5-brane. The equations of motion for these objects had been obtained in the frame of superembedding approach (Howe and Sezgin (1996), Howe and Sezgin (1997)) before the covariant action functionals were constructed ((Cederwall et. al. (1996), Aganagic, Popescu and Schwarz

(1996), Bergshoeff and Townsend (1996), BLNPST (1997), Aganagic et al. (1997)).

The complete bridge between the covariant actions and the SEA description can be build by constructing the GAP.

The GAP Lagrangian form in all the cases includes *the same Wess–Zumino term* $\mathcal{L}^{WZ}_{p+1}$ as the 'standard' (component) action does. So the only problem is to write down the 'kinetic' part (in terms of differential forms, without any use of the Hodge operation). This goal is achieved *using Lorentz harmonics* $u^{\underline{a}}_{\underline{m}} = (u^{a}_{\underline{m}}, u^{i}_{\underline{m}}) \in SO(1, D-1)$ $(a = 0, \ldots p,\ i = 1, \ldots (D-p-1))$ providing the possibility to adapt the bosonic component of target superspace supervielbein $E^{\underline{a}}$ to the superembedding of the world volume superspace
$E^{\underline{a}} = (E^{a}, E^{i})$ $E^{a} \equiv \Pi^{\underline{m}} u^{a}_{\underline{m}}$, $E^{i} \equiv \Pi^{\underline{m}} u^{i}_{\underline{m}}$

The GAP for **M2-brane (supermembrane)** is known since 1995 (Bandos, Sorokin and Volkov (1995)) and is based on the Lagrangian form

$$\mathcal{L}^{M2}_{3} = E^{\wedge 3} + \mathcal{L}^{WZ\ M2}_{3}, \qquad with \qquad E^{\wedge 3} \equiv \frac{1}{3!}\epsilon_{abc} E^{a} \wedge E^{b} \wedge E^{c} \qquad (18)$$

It produces the superembedding equation $E^{i} = 0$, its consequence $E^{\alpha}_{\dot{q}} \equiv d\Theta^{\underline{\mu}} v_{\underline{\mu}\dot{q}}^{\ \alpha} = E^{a} \psi_{a\dot{q}}^{\ \alpha}$ as well as the superfield equations of motion $\psi_{a\dot{q}}^{\ \alpha} \gamma^{a}_{\alpha\beta} = 0$.

The Lagrangian form for $D = 10$ **super-Dp-branes**

$$\mathcal{L}^{Dp}_{p+1} = E^{\wedge(p+1)} \sqrt{-det(\eta_{ab} + F_{ab})} + Q_{p-1} \wedge (dA - B_{2} - \frac{1}{2} E^{b} \wedge E^{c} F_{cb}) + \mathcal{L}^{WZ\ Dp}_{p+1} \qquad (19)$$

(Bandos, Sorokin and Tonin (1997)) includes the auxiliary tensor field $F_{ab} = -F_{ba}$ as well as $(p-1)$–form Lagrange multiplier Q_{p-1}. It produces the generalized **gauge field constraints** $dA - B_{2} = \frac{1}{2} E^{b} \wedge E^{c} F_{cb}$ which complete the geometrodynamic condition $E^{i} = 0$ up to the complete set of superfield equations for *any Dp – brane* (cf. with Howe and Sezgin (1996), Howe, Lambert, West et al. (1997)). The geometrodynamic equation $E^{i} = 0$ as well as its consequence (*fermionic superembedding condition*) $E^{\alpha 2}_{q} = E^{\beta 1}_{q} h_{\beta}^{\ \alpha} + E^{a} \psi^{\alpha}_{aq}$ and equations of motion $\psi^{\ \alpha}_{aq} (\eta + F)^{-1\ ab} (\gamma_{b})_{\alpha\beta} = 0$ follow from the GAP (19) as well.

It is worth mentioned that the **spin–tensor field** $h_{\beta}^{\ \alpha}$ involved into fermionic superembedding conditions $E^{\alpha 2}_{q} = E^{\beta 1}_{q} h_{\beta}^{\ \alpha} + E^{a} \psi^{\alpha}_{aq}$ (Howe and Sezgin (1996),(1997)) is *Lorentz group valued* $h_{\beta}^{\ \alpha} \in Spin(1,p)$ and provides the *spinor representation for the Cayley image*

$$k_{b}^{\ a} \equiv (\eta - F)_{bc} (\eta + F)^{-1\ ca} \in SO(1,p)$$

of the gauge field strength F_{ab} (Akulov et al. (1998)) $h_{\beta}^{\ \beta'} \gamma^{a}_{\beta'\alpha'} h_{\alpha}^{\ \alpha'} = \gamma^{b}_{\beta\alpha} k_{b}^{\ a}$.

Returning to M-branes, note that the GAP for $D = 11$ massless superparticle or **M0-brane** (Bergshoeff and Townsend (1996)) can be constructed on the basis of Lagrangian 1–form $\mathcal{L}_{1} = P_{--} v^{-}_{\underline{\mu}A} v^{-}_{\underline{\nu}A} \Gamma^{\underline{\mu\nu}}_{\underline{m}} \Pi^{\underline{m}}$,

($A = 1, ..., 16$) (Bandos and Lukierski (1998)) where (cf. Bandos and Nurmagambetov (1997)) P_{--} is Lagrange multiplier, and $v_{\underline{\mu}}^{\underline{\alpha}} = (v_{\mu A}^{-}, v_{\mu A}^{+}) \in Spin(1,10)$ are spinor Lorentz harmonics parametrizing the coset $SO(1,10)/[SO(1,1) \otimes SO(9) \otimes K_9]$

One of the **open problems** is the construction of the GAP for M5-brane. The superembedding equations for the M5-brane are known (Howe and Sezgin (1997)), produces the same field equations as the covariant action (BLNPST (1997), Aganagic et al. (1997)) and found many physical applications (Howe, Lambert, West et al. (1997)).

However, the natural candidate for the GAP Lagrangian form $\mathcal{L}_6^{M5}$ (see BPST (1997)), whose integration over the bosonic world volume provides the first order form for the action (BLNPST (1997)), produces (in addition to the superembedding equations $E^i = 0$, $E_{\alpha q} = E_q^{\beta} h_{\beta\alpha} + E^a \psi_{a\alpha q}$ (Howe and Sezgin (1997))) the equation $da = E^a u_a$, whose superspace extension has only trivial solutions. As the presence of the closed form da is characteristic for the PST approach, developed for covariant Lagrangian description of self–dual (chiral) fields (Pasti, Sorokin and Tonin (1995-97)) and used in (BLNPST (1997)), the problem can be formulated in more general way as looking for a consistent unification of the GAP with the PST approach. Another possible way consists in searching for reformulation of the M5–brane action in terms of the spin–tensor field $h_{\alpha\beta} \equiv h^{abc} \tilde{\gamma}_{abc}^{\alpha\beta}$ (involved into the fermionic superembedding condition $E_{\alpha q} = E_q^{\beta} h_{\beta\alpha} + E^a \psi_{a\alpha q}$) instead of $H_3 = dB_2 - A_3$ and the PST scalar a. Similar reformulation of D3-brane action was found recently (Bandos and Kummer (1997)).

Another relevant problem for further study consists in searching for SEA and GAP description for $D = 11$ KK monopole (Bergshoeff, Jassen and Ortin (1997)) and M–brane (Bergshoeff and Van der Schaar (1998)).

Acknowledgements. Recent results reviewed in this talk were obtained in fruitful collaboration with V. Akulov, W. Kummer, K. Lechner, A. Nurmagambetov, P.Pasti, D. Sorokin, M. Tonin, V. Zima. Author is thankful to D. Sorokin, M.Tonin and E. Sezgin for useful discussions. Author is grateful to Prof. M. Virasoro and Prof. R. Randjbar-Daemi for hospitality at the ICTP were this work was completed.

References

APPS (1997): Aganagic, M., Park, J., Popescu, C. Schwarz, J.H., *Nucl.Phys.* **B496** (1997) 191-214, hep-th/9701166.

Aganagic, M., Popescu, C., Schwarz, J.H., *Phys.Lett.* **B393** (1997) 311–315, hep-th/9610249;*Nucl.Phys.* **B490** (1997) 178, hep-th/9612080.

Akulov, V., Bandos, I., Kummer, W., Zima, V. (1998): *D=10 super-D9-brane*, hep-th/9802032, *Nucl.Phys.* **B** (in press).

BCSV (1994): Bandos, I., Cederwall, M.. Sorokin, D., Volkov, D.V., *Mod. Phys. Lett.* **A9** (1994) 2987-2998, hep-th/9403181.

Bandos, I., Kummer, W. (1997): *Phys.Lett.* **B413** 311-321; Err. **B420** (1998) 405.

BLNPST (1997): Bandos, I., Lechner, K., Nurmagambetov, A., Pasti, P., Sorokin, D., Tonin, M., *Phys.Rev.Lett.* **78** (1997) 4332-4334, **hep-th/9701149**; *Phys.Lett.* **B408** (1997) 135-141 hep-th/9703127.

BPSTV(1995): Bandos,I., Pasti,P., Sorokin,D., Tonin,M., Volkov,D.V., *Nucl.Phys.* **B446** (1995) 79–119, hep-th/9501113.

Bandos, I., Sorokin, D., Volkov, D.V. (1995): *Phys.Lett.* **B 352**269, hep-th/9502141

BPST (1997): Bandos,I., Pasti,P., Sorokin,D., Tonin,M., *Lect.Not.Phys.* **509** (1998) 79-91, hep-th/9705064.

Bandos, I., Sorokin, D., Tonin, M. (1997) *Nucl.Phys.* **B497** 275-296, hep-th/9701127.

Bandos, I. Zheltukhin, A. (1991): *JETP Lett.* **54** (1991) 421-424; *Phys.Lett.* **B288** (1992) 77-84; **Class.Quant.Grav. 12**(1995)609-626 *and refs. therin*

Bandos, I., Lukierski, J. (1998): paper in preparation.

Bandos, I., Nurmagambetov, A. (1997): *Class.Quant.Grav.* **14** 1597-1621.

Bergshoeff, E., Jassen, B., Ortin, T. (1997): *Phys.Lett.* **B410** 379, hep-th/9706117.

Bergshoeff, E., Van der Schaar J.P. (1998), *On M-9-branes*, hep-th/9806069.

Bergshoeff, E., Sezgin , E. (1993): *Nucl. Phys.* **422** (1994) 329

Bergshoeff, E., Townsend, P.K. (1996): *Nucl.Phys.* **B490** (1997) 145–162, hep-th/9611173.

Berkovits, N. (1989): *Phys. Lett.* **232B** (1989) 184; **241B** (1990) 497.

Castellani, L., D'Auria, R., Frè, P. (1991): *Supergravity and superstrings, a geometric perspective*, World Scientific, Singapore, 1991 *and references therein*

Cederwall, M., Von Gussich, A. Nilsson, B.E.W., Sundell, P., Westerberg, A. *Nucl.Phys.* **B490** (1997) 163–178, hep-th/9610148; **B490** (1997) 179–201, hep-th/9611159.

Chu, C.S., Howe, P.S., Sezgin, E., West, P.C. (1998): *Mod. Phys. Lett.* **A13** (1998) 1407, hep-th/9803041.
Chu, C.S., Sezgin, E. (1997): *High Energy Phys.* **12** 001. hep-th/9801202.
Chu, C.S., Howe, P.S., Sezgin, E. (1998): *Phys.Lett.* **B429** 273-280, hep-th/9710223.

Dall'Agata, G., Fabbri, D., Fraser, C., Fre', P., Termonia, P., Trigiante, M., *The* $Osp(8|4)$ *singleton action from the supermembrane*, hep-th/9807115.

D'Auria, R., Frè, P., Regge, T. (1980): *Revista del Nuovo Cim.* 3 1980 1.

GIKOS (1984): Galperin, A., Ivanov,E., Kalitzin,S., Ogievetsky,V., Sokatchev,E., *Class.Quantum Grav.* **1** 1984 498; **2** (1985) 155.

DGHS 1992: Delduc,. F., Galperin, A., Howe, P., Sokatchev, E., *Phys. Rev.* **D47** (1993) 578-593; Galperin, A., Sokatchev, E., *Phys. Rev.* **D46** (1992) 714-725; **D48** (1993) 4810-4820 (hep-th/9203051, hep-th/9304046).

Howe, P.S. (1994): *Phys. Lett.* **B332**, 61-65, hep-th/9403177

Howe, P.S., Sezgin, E. (1996): *Phys.Lett.* **B390** (1997) 133-142, hep-th/9607227.

Howe, P.S., Sezgin, E. (1997), *Phys.Lett.* **B394** (1997) 62-66, hep-th/9611008

Howe, P.S., Sezgin, E., West, P.C. (1997) *Phys.Lett.* **B399** 49-59; **B400** 255-259.

Howe, P.S., Lambert, N.D., West, P.C. et. al. (1997): *Phys.Lett.* **B418** (1998) 85-90 hep-th/9710033; Phys.Lett. **B419** (1998) 79-83, hep-th/9710034;
Lambert, N.D., West, P.C., Phys.Lett. B424 (1998) 281-287, hep-th/9712040; hep-th/9801104 ;
J. P. Gauntlett, N. D. Lambert, P. C. West, Branes and Calibrated Geometries, hep-th/9803216 *and refs. therin*

Howe, P.S., Raetzel, O., Sezgin, E. (1998): *On Brane Actions and Superembeddings* , hep-th/9804051.

Ivanov, E.A., Kapustnikov, A. (1991): *Phys. Lett.* **B267** (1991) 175.

Ivanov, E. and Sokatchev, E. (1994): Chiral fermion action with (8,0) worldsheet supersymmetry, hep-th/9406071.

Nieman, Y., and Regge, T., *Phys.Lett.* **B74** (1978) 31, *Revista del Nuovo Cim.* 1 1978 1.

Pasti, P., Sorokin, D., Tonin, M., *Phys.Lett.* **B352** (1995) 269, **B398** (1997) 41-46.

Pasti, P. and Tonin, M. (1993): *Nucl. Phys.* **418** (1994) 337.

STVZ(1988): Sorokin, D., Tkach, V., Volkov, D.V., *Mod.Phys.Lett.* **A4** (9189) 901-908; Volkov, D.V., Zheltukhin, A.A., *Lett.Math.Phys.* **17** (1989) 141; *Nucl.Phys.* **B335** (1990) 723; Sorokin, D., Tkach, V., Volkov, D.V., Zheltukhin, A.A., *Phys.Lett.* **B216** (1989) 302-306.

Sorokin, D., Tonin, M. (1993): *Phys. Lett.***B326**(1994)84-88, hep-th/9307195.

Volkov, D.V., *Generalized Action Principle for superstrings and supermembranes* , ""SUSY-95". Proc. Int. Conf., Paris, 1995, May 15-19, hep-th/9512103.

Domain Walls and Spacetime-Filling Branes

Eric Bergshoeff

Institute for Theoretical Physics, University of Groningen, Nijenborgh 4,
9747 AG Groningen, The Netherlands

Abstract. We discuss branes with one transversal direction (domain walls) and no transversal direction (spacetime-filling branes). In particular, we briefly discuss a relationship between spacetime-filling branes and superstring theories with sixteen supercharges.

1 Introduction

It is a pleasure for me to write a contribution dedicated to the memory of Victor I. Ogievetsky. Given the many important contributions of Ogievetsky in the field of supersymmetry and supergravity, it seems appropriate to discuss, as a further development along these lines, some properties of branes which are a generalization of strings. In particular I would like to discuss two special kinds of branes: those with just one transversal direction (domain walls) and those with no transversal direction at all (spacetime-filling branes). As we will see, there turns out to be an intriguing relationship between spacetime-filling branes and superstring theories with sixteen supercharges. The work I describe is based on a recent paper I wrote with J.P. van der Schaar (Bergshoeff and van de Schaar (1998)).

It it well known that the existence of branes can be understood from the occurrence of corresponding central charges in the supersymmetry algebra (Townsend (1997) and Hull (1997)). For instance, the ten–dimensional IIA supersymmetry algebra with central charges is given by ($\alpha = 1, \cdots, 32$; $M = 0, \cdots 9$))

$$\begin{aligned}\{Q_\alpha, Q_\beta\} = &\left(\Gamma^M C\right)_{\alpha\beta} P_M + \left(\Gamma_{11} C\right)_{\alpha\beta} Z + \left(\Gamma^M \Gamma_{11} C\right)_{\alpha\beta} Z_M \\ &+ \tfrac{1}{2!}\left(\Gamma^{MN} C\right)_{\alpha\beta} Z_{MN} + \tfrac{1}{4!}\left(\Gamma^{MNPQ}\Gamma_{11} C\right)_{\alpha\beta} Z_{MNPQ} \\ &+ \tfrac{1}{5!}\left(\Gamma^{MNPQR}\Gamma_{11} C\right)_{\alpha\beta} Z_{MNPQR}\,.\end{aligned} \tag{1}$$

Note that the right-hand-side contains the maximum number of allowed central charges:

$$\tfrac{1}{2} \times 32 \times 33 = 1 + 10 + 10 + 45 + 210 + 252\,. \tag{2}$$

Every central charge (except scalar central charges) gives rise to two branes. Scanning the known IIA branes we find the following correspondences[1] :

$$
\begin{aligned}
P_M &\to \text{IIA wave}\,,\\
Z &\to \text{D0}\,,\\
Z_M &\to \text{Type IIA F1} - \text{string and } \underline{\text{NS} - \text{9A brane}}\,,\\
Z_{MN} &\to \text{D2 and D8}\,,\\
Z_{MNPQ} &\to \text{D4 and D6}\,,\\
Z_{MNPQR} &\to \text{NS} - \text{5A brane and Type IIA KK monopole}\,.
\end{aligned}
\tag{3}
$$

All cases are well understood except for the underlined case. This case suggests the existence of a hypothetical spacetime-filling brane, called the NS-9A brane, which is the subject of this work. Note that there is also a domain wall: the D8-brane.

The other case to be considered is the ten-dimensional IIB supersymmetry algebra with central charges: in this case there are two Majorana-Weyl charges Q^i_α $(i = 1, 2)$ with the same chirality. The algebra is given by

$$
\begin{aligned}
\{Q^i_\alpha, Q^j_\beta\} = {}& \delta^{ij}\left(\mathcal{P}\Gamma^M C\right)_{\alpha\beta} P_M + \left(\mathcal{P}\Gamma^M C\right)_{\alpha\beta} Z^{ij}_M \\
&+\tfrac{1}{3!}\epsilon^{ij}\left(\mathcal{P}\Gamma^{MNP} C\right)_{\alpha\beta} Z_{MNP} + \tfrac{1}{5!}\delta^{ij}\left(\mathcal{P}\Gamma^{MNPQR} C\right)_{\alpha\beta} Z^{+}_{MNPQR} \\
&+\tfrac{1}{5!}\left(\mathcal{P}\Gamma^{MNPQR} C\right)_{\alpha\beta} Z^{+,ij}_{MNPQR}\,.
\end{aligned}
\tag{4}
$$

Here $\mathcal{P}$ is a chiral projection operator and $Z^{ij}_M\,, Z^{+,ij}_{MNPQR}$ are doublets of SO(2) (symmetric traceless representations). The super index + indicates that the corresponding 5-form central charge is self-dual in the M indices. Like in the IIA case we have the maximum number of central charges:

$$
\tfrac{1}{2} \times 32 \times 33 = 10 + 20 + 120 + 126 + 252\,. \tag{5}
$$

Scanning the known IIB branes we find the following correspondences[2]:

$$
\begin{aligned}
P_M &\to \text{IIB wave}\,,\\
Z^{ij}_M &\to \text{D1 and Type IIB F1} - \text{string}\,,
\end{aligned}
$$

[1] The time component of the translation generator P_M is identified as the Hamiltonian and does not give rise to a brane.

[2] The self-dual 5-form central charges only give rise to a single brane.

$$
\begin{aligned}
&\underline{\text{D9}-\text{brane}} \text{ and } \underline{\text{NS}-9\text{B brane}}\,, \\
Z_{MNP} &\to \text{D3 and D7}\,, \qquad (6)\\
Z^{+,ij}_{MNPQR} &\to \text{D5 and NS}-5\text{B brane}\,, \\
Z^{+}_{MNPQR} &\to \text{Type IIB KK monopole}\,.
\end{aligned}
$$

We see that the IIB central charges suggest the existence of two further spacetime-filling branes: the D9-brane and the NS-9B brane.

We conclude that the central charges of the IIA and IIB supersymmetry algebra suggest the existence of one domain wall: the D8-brane and three hypothetical spacetime-filling branes: the NS-9A brane, the D9-brane and the NS-9B brane. It is the purpose of this work to discuss the role played by these domain walls and spacetime-filling branes. The reason that we include a discussion of domain walls is that upon direct dimensional reduction they become spacetime-filling branes in one dimension lower. It turns out that in this sense they naturally occur in the discussion of nine-dimensional superstring theories with sixteen supercharges.

The organization of this contribution is as follows. In section 2 we dicuss the domain walls and in section 3 the spacetime-filling branes. In the same section we will combine these results and discuss the relation between spacetime-filling branes and superstring theories with sixteen supecharges.

2 Domain Walls

The ten-dimensional domain wall is identified as the D8-brane. It is a domain wall solution of massive IIA supergravity (Romans (1986)). When the cosmological constant term is dualized to a 9-form potential one can write down a domain wall solution with different values of the cosmological constant at different sides of the domain wall (Bergshoeff et al. (1996)). This solution is coordinate equivalent to the conformally flat solution given in Polchinski and Witten (1996).

The relevant part of the action to consider is given by

$$
S_{\text{massive IIA}}(m) \sim \int d^{10}x\sqrt{|g|}\,\{e^{-2\phi}\left[R-4(\partial\phi)^2\right]+\tfrac{1}{2}m^2+\cdots\}\,, \qquad (7)
$$

where m is a parameter with the dimension of a mass[3]. For the construction of the domain wall solution it is convenient to use an alternative formulation where the mass parameter m is replaced by a 9-form potential $C^{(9)}$ with curvature $G^{(10)}$ (Bergshoeff et al. (1996)):

[3] The parameter m can be positive or negative. The sign of the cosmological constant is determined by T-duality (see below) and is such that for constant dilaton $\phi = \phi_0$ the vacuum solution to the Einstein equations is *anti*-de Sitter spacetime.

$$S_{\text{massive IIA}}(C^{(9)}) \sim \int d^{10}x\sqrt{|g|}\left\{e^{-2\phi}\left[R - 4(\partial\phi)^2\right] - \tfrac{1}{2\times 10!}(G^{(10)})^2 + \cdots\right\}. \tag{8}$$

This 9-form potential naturally appears in the low-energy limit of Type IIA superstring theory (Polchinski (1995)). The sign of the kinetic term for $C^{(9)}$ follows, via T-duality, from the (standard) sign of the kinetic term for the other Ramond-Ramond p-form potentials $C^{(p)}$ $(p < 9)$. This in turn fixes the sign of the cosmological constant in (7). More precisely, the equation of motion of $C^{(9)}$

$$d\,{}^*G^{(10)} = 0 \tag{9}$$

is solved for by

$$G^{\mu_1\cdots\mu_{10}}(C) = \frac{1}{\sqrt{|g|}}\epsilon^{\mu_1\cdots\mu_{10}}c\,, \tag{10}$$

where c is an integration constant. Comparing with (7) we find that $c = \pm m$.

The expression of $G^{(10)}$ given in (10) is not the most general solution of (9). In the presence of a domain wall, the integration constant c can be *piece-wise constant* (Bergshoeff et al. (1996)). This possibility is excluded in the formulation without the 9-form potential where the parameter m is constant everywhere.

We consider the following Ansatz for the extreme D8-brane solution (in string frame metric):

$$\begin{aligned} ds^2_{10} &= H^\alpha(dt^2 - dx^2_{(8)}) - H^\beta dy^2\,, \\ e^{2\phi} &= H^\gamma\,, \qquad C^{(9)}_{012345678} = H^\epsilon\,, \end{aligned} \tag{11}$$

where $H = H(y)$ is a harmonic function over the single transverse direction y whose form, in a local neighbourhood, is given by

$$H(y) = c + Q|y| \tag{12}$$

in terms of two constants c and Q. In order to avoid a singularity at $H = 0$ and to obtain a real dilaton, we use the absolute value of y in the harmonic function and take $c > 0$ and $Q > 0$. The Ansatz (11) describes a domain wall positioned at $y = 0$.

It turns out that the parameter ϵ can not be determined by the equations of motion obtained by minimizing the action (8). This is in contradistinction to the Dp-branes with $p < 8$ where one finds, for all $p < 8$, that $\epsilon = -1$. Solving the equations of motion obtained from the action (8) we find the following expressions for α, β, γ:

$$\alpha = \tfrac{1}{2}\epsilon\,, \qquad \beta = -\tfrac{5}{2}\epsilon - 2\,, \qquad \gamma = \tfrac{5}{2}\epsilon\,. \tag{13}$$

Furthermore, substituting the solution (13) into (10), we find the following relation between m and Q:

$$m = \pm\epsilon Q\,. \tag{14}$$

Notice that for any fixed m and Q we find two solutions corresponding to ϵ and $-\epsilon$. For $m = 0$ there is a single solution with $\epsilon = 0$ or, equivalently, $Q = 0$. In this case the metric reduces to that of a flat Minkowski spacetime.

The value of the cosmological constant differs at the two sides of the domain wall whenever we make different choices for the constant Q at the left and right of the domain wall:

$$\begin{aligned} H(y) &= c + Q_L|y|\,, \qquad & y < 0,\\ H(y) &= c + Q_R|y|\,, \qquad & y > 0\,. \end{aligned} \tag{15}$$

The free parameter ϵ labelling the above D8-brane solutions is related to the freedom to perform a coordinate transformation in y keeping the solution within the ansatz (11). To show this we first note that at one side of the domain wall one can always use coordinates such that $y \geq 0$. By performing a suitable shift transformation, $y' = y + c/Q$, the harmonic function can always be written as $H(y') = Qy'$, where $y' \in (c/Q, \infty)$ and $Q > 0$. We only consider coordinate transformations that keep the transversal coordinate y within this (positive and infinite) range. Consider a D8-brane for a given negative value of ϵ, say $\epsilon = -1$ or $m = \pm Q$. We perform the following coordinate transformation labelled by ϵ (we omit the prime on the shifted coordinate):

$$y \to y' = f(\epsilon)y^{-\frac{1}{\epsilon}}\,, \tag{16}$$

with the function $f(\epsilon)$ given by

$$f(\epsilon) = -\epsilon Q^{-\frac{1+\epsilon}{\epsilon}}\,. \tag{17}$$

We restrict ourselves to negative ϵ (for positive ϵ the range of y' would become negative and finite). Under the above coordinate transformation the harmonic function H transforms as

$$H_{(\epsilon=-1)}(y) = -my = \left(H_{(\epsilon)}(y')\right)^{-\epsilon} = (Q'y')^{-\epsilon}\,, \tag{18}$$

with $Q' = -Q/\epsilon$. Therefore all solutions with negative ϵ, defined at positive y, are related to each other by the coordinate transformations (16). The same holds for all positive ϵ solutions. Starting with a solution for a given positive value of ϵ, say $\epsilon = 1$, we obtain all other positive ϵ solutions by performing the transformation (16) with ϵ replaced by $-\epsilon$.

To relate solutions with positive and negative values of ϵ one must perform a coordinate transformation of the form

$$y \to 1/y\,. \tag{19}$$

This transformation is included in (16) if we also allow positive values of ϵ. However, under the transformation (19) the infinite positive domain of y gets mapped to a finite negative domain in y' and Q becomes negative to keep $H(y)$ positive. This means that in the new coordinate system $y = 0$ acts as infinity. In any case, we see that for all values of ϵ, positive and negative, the domain wall solutions are coordinate equivalent to each other. We often concentrate on the $\epsilon < 0$ solutions because in that case the position of the domain wall occurs for a finite value of y and the asymptotic region far away from the domain wall is reached by taking $y \to \infty$.

3 Spacetime-Filling Branes

We have seen in the Introduction that there are three kinds of spacetime-filling branes: the D9-brane, the NS-9B brane and the NS-9A brane. At the level of target space solutions all these spacetime-filling branes are represented by a d=10 Minkowski spacetime which has unbroken supersymmetry. However, since the lines of forces cannot escape to infinity (there is no non-compact transversal direction) spacetime itself cannot be chargeless. Consider for instance the D9-brane. In order to obtain a spacetime with zero charge one must introduce an orientifold O9 that carries a negative Ramond-Ramond charge[4]. At the same time the orientifold breaks half of the supersymmetry so that effectively the spacetime filling brane is a natural place in wich N=1 open superstrings (with the end points living in the d=10 worldvolume of the spacetime-filling brane) live. At this point it is natural to suggest that the effective worldvolume action of *each* spacetime-filling brane must correspond to the effective action of one of the N=1 superstring theories. The known superstring theories with sixteen supercharges and the dualities between them suggest the following relationships:

$$\begin{aligned} \mathrm{NS}-9\mathrm{A} &\longleftrightarrow \text{heterotic } E_8 \times E_8\,, \\ \mathrm{NS}-9\mathrm{B} &\longleftrightarrow \text{heterotic } SO(32)\,, \\ \mathrm{D}9 &\longleftrightarrow \text{Type I } SO(32)\,. \end{aligned} \tag{20}$$

The following S- and T-dualities relate these spacetime-filling branes/N=1 superstrings to each other:

[4] The nonzero Ramond-Ramond charge is due to the fact that the $\mathbb{Z}_2$-projection of the orientifold not only acts on spacetime but also on the worldsheet.

$$\mathrm{NS}-9\mathrm{A} \stackrel{\mathrm{T}}{\longleftrightarrow} \mathrm{NS}-9\mathrm{B} \stackrel{\mathrm{S}}{\longleftrightarrow} \mathrm{D}9\,. \tag{21}$$

It can be shown that these dualities require the following effective tension in front of the leading Nambu-Goto term of the worldvolume actions (Bergshoeff and van de Schaar (1998)) (we take the string-length $l_s = \sqrt{\alpha'} = 1$):

$$T_{\mathrm{D9}} = e^{-\phi}\,, \quad T_{\mathrm{NS-9B}} = e^{-4\phi}\,, \quad T_{\mathrm{NS-9A}} = |k|^3 e^{-4\phi}\,. \tag{22}$$

Here ϕ is the dilaton and $|k|^2 = k^\mu k^\nu g_{\mu\nu}$. Note that the existence of the Killing vector k in the case of the NS-9A brane requires an isometry direction for the background geometry effectively reducing the theory to one in d=9 dimensions.

A few remarks are in order. First of all, from (21) we see that an S-duality transformation relates the NS-9B brane, or (1,0)-brane, to the D9-brane, or (0,1)-brane[5]. We have used here the S-duality rules (using string-frame metric)

$$\begin{aligned} \phi &\longrightarrow -\phi\,, \\ g_{\mu\nu} &\longrightarrow e^{-\phi} g_{\mu\nu}\,. \end{aligned} \tag{23}$$

In calculating the S-duality rules of the effective tensions (22), the latter rule can be taken care of by re-introducing the string length l_s in the effective tension with the S-duality rule

$$l_s \longrightarrow g_s^{1/2} l_s\,. \tag{24}$$

The inverse stringth length occurs with a power that counts the $g^{1/2}$ metric factors in the worldvolume action. This power is given by p+1 (coming from the metric factors inside the square root of the determinant) plus the power of $|k| = R$ in front of the action[6]. The effective tensions (22) with the appropriate powers of l_s inserted are given by (Bergshoeff et al. (1998))

$$T_{\mathrm{D9}} = \frac{1}{g_s l_s^{10}}\,, \quad T_{\mathrm{NS-9B}} = \frac{1}{g_s^4 l_s^{10}}\,, \quad T_{\mathrm{NS-9A}} = \frac{R^3}{g_s^4 l_s^{12}}\,. \tag{25}$$

Secondly, in order to realize the T-duality relation given in (21) between the NS-9A brane and the NS-9B brane we must assume that the NS-9A brane is a *wrapped* one. To see this, consider the effective tension of the unwrapped NS-9B brane (see (25)). Performing a T-duality in the isometry direction of the NS-9B brane:

---

[5] The (1,0)-brane and (0,1)-brane are two special cases of a whole family of (p, q) 9-branes (Hull (1997)).

[6] This is due to the fact that $|k|^2 \equiv g_{\mu\nu} k^\mu k^\nu$ contains a single metric factor.

$$R \to l_s^2/R\,, \qquad g_s \to g_s l_s/R\,, \qquad l_s \to l_s \tag{26}$$

we find that the effective tension of the NS-9A brane is proportional to R^3 as given in (25). This shows that the full action contains a Killing vector and therefore represents a *wrapped* NS-9A brane. Note that the T-duality between the wrapped NS-9A brane and the unwrapped NS-9B brane is similar to the T-duality between the wrapped Type IIA KK monopole and the unwrapped NS-5B brane. In both cases one brane has a special Killing vector (Type IIA KK monopole, NS-9A brane) whereas the T-dual brane has no such Killing vector (NS-5B brane, NS-9B brane).

Finally, one cannot only consider the T-dual of the NS-9B brane (leading to the NS-9A brane/Heterotic SO(32) string) but also the T-dual of the D9-brane. This leads to the D8 brane/Type I′ superstring. This superstring has a Killing vector and is effectively a d=9 string. The corresponding spacetime-filling brane is the direct reduction of the D8-brane which we call a nine-dimensional RR-8 brane. We thus find the relation:

$$RR - 8 \leftrightarrow \text{Type I}'\ SO(16) \times SO(16)\,. \tag{27}$$

It remains to be further investigated in which sense the spacetime-filling branes discussed in this work can be used to describe the different superstring theories with sixteen supercharges (Bergshoeff et al. (1998)).

Acknowledgements

The work described here was done in collaboration with Jan Pieter van der Schaar. I also report about work in preparation done in collaboration with Eduardo Eyras, Rein Halbersma, Chris Hull, Yolanda Lozano and Jan Pieter van der Schaar. This work is supported by the European Commission TMR programme ERBFMRX-CT96-0045, in which I am associated to the University of Utrecht.

References

Bergshoeff, E.A., van de Schaar, J.P. (1998): *On M-9-Branes*, `hep-th/9806069`, submitted to Class. & Quant. Grav.

Townsend, P.K. (1997): *M-Theory from its Superalgebra*, `hep-th/9712004`, Cargèse Lectures 1997.

Hull, C.M. (1997): *Gravitational Duality, Branes and Charges*, `hep-th/9705162`, Nucl. Phys. **B509** (1998) 216.

Romans, L.J. (1986): *Massive N=2a Supergravity in Ten Dimensions*, Phys. Lett. **169B** (1986) 374.

Bergshoeff, E.A., de Roo, M., Green, M.B., Papadopoulos, G. and Townsend, P.K. (1996): *Duality of type-II 7-branes and 8-branes*, Nucl. Phys. **B470** (1996) 113, `hep-th/9601150`.

Polchinski, J. and Witten, E. (1996): *Evidence for Heterotic - Type I String Duality*, Nucl. Phys. **B460** (1996) 525, `hep-th/9510169`.

Polchinski, J. (1995): *Dirichlet-Branes and Ramond-Ramond Charges*, Phys. Rev. Lett. **75** (1995) 4724, `hep-th/9510017`.

Lü, H., Pope, C.N. and Townsend, P.K. (1997): *Domain Walls from Anti-de Sitter Spacetime*, Phys. Lett. **B391** (1997) 39, `hep-th/9607164`.

Bergshoeff, E., Eyras, E., Halbersma, R, Hull, C.M., Lozano, Y. and van de Schaar, J.P. (1998): *in preparation*.

The Quantum Geometry of Branes

C. M. Hull

Physics Department, Queen Mary and Westfield College,
Mile End Road, London E1 4NS, U. K.

1 Introduction

It is a pleasure to contribute to this volume in honour of Viktor Ogievetsky. In this article, I shall describe some work done in collaboration with Mohab Abou Zeid (Abou Zeid and Hull (1997), Abou Zeid and Hull (1998), Maldacena(1997)) on the geometry of the non-linear actions that describe the dynamics of the p-dimensional extended objects known as p-branes that arise in string theories and M-theory.

The Nambu-Goto action for a string can be generalised to an action for a p-brane, which is proportional to the $p+1$-volume swept out by the brane as it evolves in time. Writing $p = n - 1$, the action is

$$S_{NG} = -T_p \int d^n\sigma \sqrt{-\det(G_{\mu\nu})}, \tag{1}$$

where T_p is the p-brane tension and

$$G_{\mu\nu} = G_{ij}\partial_\mu X^i \partial_\nu X^j \tag{2}$$

is the world-volume metric induced by the spacetime metric G_{ij}. The non-linear form of the action (1) is inconvenient for many purposes. However, introducing an intrinsic worldvolume metric $g_{\mu\nu}$ allows one to write down the equivalent action (Polyakov (1981), Brink, Di Vecchia and Howe. (1976), Howe and Tucker (1977))

$$S_P = -\frac{1}{2}T'_p \int d^n\sigma \sqrt{-g}\,[g^{\mu\nu}G_{\mu\nu} - (n-2)\Lambda]\,, \tag{3}$$

where $g \equiv \det(g_{\mu\nu})$ and Λ is a constant. The metric $g_{\mu\nu}$ is an auxiliary field which can be eliminated using its equation of motion to recover action (1). The constants T_p and T'_p are related by

$$T'_p = \Lambda^{\frac{n}{2}-1} T_p. \tag{4}$$

This form of the action is much more convenient for many purposes, such as quantiztion and the study of T-dualities as it is quadratic in ∂X. For the string, this form of the action (with $n = 2$) allows the covariant quantization of the string (Polyakov (1981)).

The Born-Infeld action for a vector field A_μ in an n-dimensional space-time with metric $G_{\mu\nu}$ is

$$S_{BI} = -T_p \int d^n\sigma \sqrt{-\det\left(G_{\mu\nu} + F_{\mu\nu}\right)}, \tag{5}$$

where $F = dA$ is the Maxwell field strength. A related $(n-1)$-brane action is

$$S_{DBI} = -T_p \int d^n\sigma \sqrt{-\det\left(G_{\mu\nu} + \mathcal{F}_{\mu\nu}\right)}, \tag{6}$$

where $G_{\mu\nu}$ is the induced metric (2) and $\mathcal{F}_{\mu\nu}$ is the antisymmetric tensor field

$$\mathcal{F}_{\mu\nu} \equiv F_{\mu\nu} - B_{\mu\nu} \tag{7}$$

with $B_{\mu\nu}$ the pull-back of a space-time 2-form gauge field B,

$$B_{\mu\nu} = B_{ij}\partial_\mu X^i \partial_\nu X^j. \tag{8}$$

The action (6) is closely related to the D-brane action, which has been the subject of much recent work (Schmidhuber (1996), de Alwis and Sato (1996), Douglas (1995), Townsend (1996), Tseytlin (1996a), Cederwall, von Gussich, Nilsson and Westerberg (1996), Cederwall, von Gussich, Nilsson, Sundell and Westerberg (1996), Cederwall (1996), Aganagic, and Schwarz (1996a), Aganagic, and Schwarz (1996b), Green, Hull and Townsend (1996), Bergshoeff and Townsend (1996)) and the Born-Infeld action (5) can be thought of as a special case of this, but with a different interpretation of $G_{\mu\nu}$. However, just as in the case of the action (5), the non-linearity of (6) makes it rather difficult to study. In particular, dualising the action (6) has proved rather difficult in this approach, and has only been achieved for $n \leq 5$ (Townsend (1996), Tseytlin (1996a), Aganagic, Park, Popescu and Schwarz (1997)). Clearly, an action analogous to (3) for this case would be very useful, and here we will discuss two such actions, the first of which is linear in $\mathcal{F}_{\mu\nu}$ and the second of which is quadratic.

2 Action with Non-symmetric 'Metric'

For the first action (Lindström (1988), Abou Zeid and Hull (1997), Maldacena(1997)), the key is to introduce a 'non-symmetric metric', in the form of an auxiliary world-volume tensor field

$$k_{\mu\nu} \equiv g_{\mu\nu} + b_{\mu\nu} \tag{9}$$

with both a symmetric part $g_{\mu\nu}$ and an antisymmetric part $b_{\mu\nu}$.[1] The action which is classically equivalent to (6) is

[1] Such 'metrics' have been used in alternative theories of gravitation; see e.g. (Einstein (1925,1945,1948), Moffat (1995)).

$$S = -\frac{1}{2}T'_p \int d^n\sigma \sqrt{-k}\left[(k^{-1})^{\mu\nu}\left(G_{\mu\nu} + \mathcal{F}_{\mu\nu}\right) - (n-2)\Lambda\right], \tag{10}$$

where $k \equiv \det(k_{\mu\nu})$; the inverse metric $(k^{-1})^{\mu\nu}$ satisfies

$$(k^{-1})^{\mu\nu} k_{\nu\rho} = \delta^\mu{}_\rho. \tag{11}$$

Such an action was first proposed for Born-Infeld theory in (Lindström (1988)). For $n \neq 2$, the $k_{\mu\nu}$ field equation implies

$$G_{\mu\nu} + \mathcal{F}_{\mu\nu} = \Lambda k_{\nu\mu} \tag{12}$$

and substituting back
into (10) yields the Born-Infeld-type action (6) where the constants T_p, T'_p are related as in eq. (4). For $n = 2$, the action (10) is invariant under the generalised Weyl transformation

$$k_{\mu\nu} \to \omega(\sigma) k_{\mu\nu} \tag{13}$$

and the $k_{\mu\nu}$ field equation implies

$$G_{\mu\nu} + \mathcal{F}_{\mu\nu} = \Omega k_{\nu\mu} \tag{14}$$

for some conformal factor Ω.

The action (10) is *linear* in $(G_{\mu\nu} + \mathcal{F}_{\mu\nu})$ and so is much easier to analyse than (6). In particular, it is linear in F, so that A_μ is a Lagrange multiplier imposing the constraint

$$\partial_\mu \left(\sqrt{-k}(k^{-1})^{[\mu\nu]}\right) = 0. \tag{15}$$

The general solution of this is $\sqrt{-k}(k^{-1})^{[\mu\nu]} = \tilde{H}^{\mu\nu}$, where

$$\tilde{H}^{\mu\nu} \equiv \frac{1}{(n-2)!}\epsilon^{\mu\nu\rho\gamma_1\ldots\gamma_{n-3}}\partial_{[\rho}\tilde{A}_{\gamma_1\ldots\gamma_{n-3}]}, \tag{16}$$

$\tilde{A}$ is an $n-3$ form and $\epsilon^{\mu\nu\rho\ldots}$ is the alternating tensor density. The anti-symmetric part of $k_{\mu\nu}$ can then in principle be solved for in terms of $\tilde{A}$, leaving a dual form of the action involving only the symmetric part of $k_{\mu\nu}$ and the dual potential $\tilde{A}$. To do this explicitly requires a judicious choice of variables, as we will show in Sect.4.

3 Action of the Second Type

We will refer to the actions of the previous section as being of the first type, and those introduced below as of the second type. We begin with the Born-Infeld action in $p+1$ dimensions (Born and Infeld (1935))

$$S = -T_p \int d^{p+1}\sigma \sqrt{-\det\left(g_{\mu\nu} + F_{\mu\nu}\right)} \tag{17}$$

where

$$F_{\mu\nu} = \partial_\mu A_\nu - \partial_\nu A_\mu \tag{18}$$

is the field strength of a $U(1)$ gauge field A_μ, $\mu, \nu = 0, \ldots, p$ are space-time indices and $g_{\mu\nu}$ is the space-time metric. We now show that the action (17) can be rewritten in a form which is quadratic in the field strength F, and is therefore simpler to analyse and quantise. The key is to use the fact that

$$\det\left(g_{\mu\nu} + F_{\mu\nu}\right) = \det\left(g_{\mu\nu} - F_{\mu\nu}\right) \tag{19}$$

to write the integrand in (17) in the form

$$\begin{aligned} \left[-\det\left(g_{\mu\nu} + F_{\mu\nu}\right)\right]^{\frac{1}{2}} &= \left[\det\left(g_{\mu\nu} + F_{\mu\nu}\right)\right]^{\frac{1}{4}} \left[\det\left(g_{\mu\nu} - F_{\mu\nu}\right)\right]^{\frac{1}{4}} \\ &= (-g)^{\frac{1}{4}} \left\{-\det\left[\left(g_{\mu\nu} + F_{\mu\nu}\right) g^{\nu\rho} \left(g_{\rho\sigma} - F_{\rho\sigma}\right)\right]\right\}^{\frac{1}{4}} \\ &= (-g)^{\frac{1}{4}} \left[-\det\left(g_{\mu\sigma} - g^{\nu\rho} F_{\mu\nu} F_{\rho\sigma}\right)\right]^{\frac{1}{4}}, \end{aligned} \tag{20}$$

where $g \equiv \det\left(g_{\mu\nu}\right)$. The action (17) can thus be rewritten as

$$S' = -T_p \int d^{p+1}\sigma (-g)^{\frac{1}{4}} (-\mathcal{G})^{\frac{1}{4}}, \tag{21}$$

where

$$\mathcal{G}_{\mu\nu} = g_{\mu\nu} - g^{\rho\sigma} F_{\mu\rho} F_{\sigma\nu} \tag{22}$$

and $\mathcal{G} \equiv \det\left[\mathcal{G}_{\mu\nu}\right]$. Introducing an intrinsic metric $\gamma_{\mu\nu}$ allows us to rewrite (21) in the following classically equivalent form which is quadratic in the gauge field strength $F_{\mu\nu}$

$$\begin{aligned} S' &= -T'_p \int d^{p+1}\sigma (-g)^{\frac{1}{4}} (-\gamma)^{\frac{1}{4}} \left[\gamma^{\mu\nu} \mathcal{G}_{\mu\nu} - (p-3)\Lambda\right] \\ &= -T'_p \int d^{p+1}\sigma (-g)^{\frac{1}{4}} (-\gamma)^{\frac{1}{4}} \\ &\quad \left[\gamma^{\mu\nu}\left(g_{\mu\nu} - g^{\rho\sigma} F_{\mu\rho} F_{\sigma\nu}\right) - (p-3)\Lambda\right], \end{aligned} \tag{23}$$

where $\gamma \equiv \det\left(\gamma_{\mu\nu}\right)$ and Λ is a constant. For $p \neq 3$, the $\gamma_{\mu\nu}$ field equation implies

$$\gamma_{\mu\nu} = \frac{1}{\Lambda}\left(g_{\mu\nu} - g^{\rho\sigma} F_{\mu\rho} F_{\sigma\nu}\right) \tag{24}$$

and substituting back into (23) yields the action (21), which is identical to the Born-Infeld action (17). The constants T_p, T'_p are related by

$$T'_p = \frac{1}{4} \Lambda^{\frac{p-3}{4}} T_p. \tag{25}$$

For $p = 3$, the four-dimensional action (23) is invariant under the Weyl transformation

$$\gamma_{\mu\nu} \to \omega(\sigma) \gamma_{\mu\nu} \tag{26}$$

and the $\gamma_{\mu\nu}$ field equation implies

$$\gamma_{\mu\nu} = \Omega \mathcal{G}_{\mu\nu} \tag{27}$$

for some Ω, which is found by taking traces of both sides; this gives

$$\gamma^{\rho\sigma} \left(g_{\rho\sigma} - g^{\kappa\delta} F_{\rho\kappa} F_{\delta\sigma} \right) \gamma_{\mu\nu} = 4 \left(g_{\mu\nu} - g^{\rho\sigma} F_{\mu\rho} F_{\sigma\nu} \right). \tag{28}$$

Substituting this back into (23) gives (21), which is identical to the Born-Infeld action (17), so that (17) and (23) are classically equivalent.

This can be generalised to the D-brane kinetic term

$$S = -T_p \int d^{p+1}\sigma e^{-\phi} \sqrt{-\det \left(g_{\mu\nu} + \mathcal{F}_{\mu\nu} \right)} \tag{29}$$

where

$$\mathcal{F}_{\mu\nu} \equiv F_{\mu\nu} - B_{\mu\nu}, \tag{30}$$

ϕ, $g_{\mu\nu}$ and $B_{\mu\nu}$ are the pullbacks to the worldvolume of the background dilaton, metric and NS antisymmetric two-form fields and $F = dA$, with A the $U(1)$ world-volume gauge field. This action gives the effective dynamics of the zero-modes of the open strings with ends tethered on a D-brane when F is slowly varying, so that corrections involving ∇F can be ignored, and has therefore played a central role in recent studies of D-brane dynamics and string theory duality (Polchinski (1996)). However, the non-linearity of (29) makes it rather difficult to study. In particular, the action (29) is inconvenient for the purpose of quantisation, and its dualisation has proved rather difficult (Townsend (1996), Schmidhuber (1996), de Alwis and Sato (1996), Tseytlin (1996b), Lozano (1997), Aganagic, Park, Popescu and Schwarz (1997)). It is therefore useful to know classically equivalent, alternative forms of this action which have a more tractable dependence on the spacetime coordinates X or on the field strength F. In the previous section, we obtained an alternative form of (29) which is *linear* in F and *quadratic* in derivatives of X by introducing an auxiliary worldvolume tensor with both symmetric and antisymmetric parts. Here, we give an alternative form of (29) that is *quadratic* in F.

As before, introducing an intrinsic metric $\gamma_{\mu\nu}$ allows us to rewrite (29) in the classically equivalent form

$$\begin{aligned}S' &= -T'_p \int d^{p+1}\sigma e^{-\phi}(-g)^{\frac{1}{4}}(-\gamma)^{\frac{1}{4}}\left[\gamma^{\mu\nu}\mathcal{G}_{\mu\nu} - (p-3)\Lambda\right] \\ &= -T'_p \int d^{p+1}\sigma e^{-\phi}(-g)^{\frac{1}{4}}(-\gamma)^{\frac{1}{4}} \\ &\quad \left[\gamma^{\mu\nu}(g_{\mu\nu} - g^{\rho\sigma}\mathcal{F}_{\mu\rho}\mathcal{F}_{\sigma\nu}) - (p-3)\Lambda\right],\end{aligned} \tag{31}$$

where the tensions T_p, T'_p are related as in eq. (25).

The energy-momentum tensor $T_{\mu\nu}$ can be defined from the form (31) of the D-brane kinetic term by

$$T_{\mu\nu} \equiv -\frac{1}{T'_p}\frac{1}{(-\gamma)^{\frac{1}{4}}}\frac{\delta S}{\delta\gamma^{\mu\nu}} \tag{32}$$

and we find

$$\begin{aligned}T_{\mu\nu} = (-g)^{\frac{1}{4}}\Big\{&-\frac{1}{4}\gamma_{\mu\nu}\left[\gamma^{\rho\sigma}\left(g_{\rho\sigma} - g^{\kappa\tau}F_{\rho\kappa}F_{\tau\sigma}\right) - (p-3)\Lambda\right] + \\ &g_{\mu\nu} - g^{\rho\sigma}F_{\mu\rho}F_{\sigma\nu}\Big\}.\end{aligned} \tag{33}$$

This is traceless (i. e. $\gamma^{\mu\nu}T_{\mu\nu} = 0$) if $p = 3$ as a result of the Weyl invariance (26), and the equation $T_{\mu\nu} = 0$ implies the field equation of the metric (24) or (27).

The low-energy effective action for an open type I string includes the terms given by (29) with $p = 9$, but with $g_{\mu\nu}, B_{\mu\nu}$ the space-time metric and anti-symmetric tensor gauge field (rather than their pull-backs) (Tseytlin (1996b)), and can be rewritten in the equivalent form (31) with $p = 9$. The dimensional reduction of the type I string action (29) to $p+1$ dimensions gives the action for a D-p-brane in static gauge (and with vanishing RR gauge fields), with the 9+1 vector field A giving rise to a vector and $9-p$ scalars X_i on reduction. The reduction of the form (31) of the action then gives a useful form of the static-gauge D-p-brane action which is quadratic in A, X.

We now turn to the reduction of (31) from 9+1 to p+1 dimensions. We use the notation that hatted quantities are ten-dimensional, so $\hat{\mu} = 0, \ldots, 9$, while $\mu = 0, \ldots, p$ and $i = p+1, \ldots, 9$. Then the vector field $A_{\hat{\mu}} = (A_\mu, X_i)$ gives a vector and $9-p$ scalars X_i. We choose (for simplicity) a flat space-time metric $\hat{g}_{\hat{\mu}\hat{\nu}} = \eta_{\hat{\mu}\hat{\nu}}$ and vanishing 2-form $\hat{B}_{\hat{\mu}\hat{\nu}}$, and make the following Ansatz for the metric $\hat{\gamma}_{\hat{\mu}\hat{\nu}}$:

$$\hat{\gamma}_{\hat{\mu}\hat{\nu}} = \begin{pmatrix}\gamma_{\mu\nu} + C^i{}_\mu C^j{}_\nu\gamma_{ij} & C^j{}_\mu\gamma_{ij} \\ C^k{}_\nu\gamma_{kj} & \gamma_{ij}\end{pmatrix}. \tag{34}$$

Then the metric $\hat{\gamma}_{\hat{\mu}\hat{\nu}}$ gives, as usual, a $p+1$-dimensional metric $\gamma_{\mu\nu}$, $9-p$ vector fields $C^i{}_\mu$ and $(9-p)(10-p)/2$ scalar fields taking values in the coset $GL(9-p,R)/SO(9-p)$. The inverse of (34) is

$$\hat{\gamma}^{\hat{\mu}\hat{\nu}} = \begin{pmatrix}\gamma^{\mu\nu} & -C^{\mu i} \\ -C^{\mu j} & \gamma^{ij} + C^i{}_\rho\gamma^{\rho\sigma}C^j{}_\sigma\end{pmatrix} \tag{35}$$

and its determinant is

$$\det \hat{\gamma}_{\hat{\mu}\hat{\nu}} = \det \gamma_{\mu\nu} \det \gamma_{ij} . \tag{36}$$

Setting $F_{ij} \equiv 0$ and $F_{\mu i} \equiv \partial_\mu X_i$, this gives the following static gauge D-p-brane action which is quadratic in both F and ∂X^i:

$$\begin{aligned} S' = -T'_p \int d^{p+1}\sigma e^{-\phi} \left[-\det \gamma_{\mu\nu} \det \gamma_{ij}\right]^{\frac{1}{4}} \\ \left[\gamma^{\mu\nu}(\eta_{\mu\nu} + \eta^{ij}\partial_\mu X_i \partial_\nu X_j + \eta^{\rho\sigma} F_{\mu\rho} F_{\nu\sigma}) \right. \\ +2C^{\mu i} F_{\mu\rho}\partial_\sigma X_i \eta^{\rho\sigma} + (\eta^{ij} + C^i{}_\upsilon \gamma^{\upsilon\tau} C^j{}_\tau)(\eta_{ij} + \eta^{\rho\sigma}\partial_\rho X_i \partial_\sigma X_j) \\ \left. -(p-3)\Lambda\right] . \end{aligned} \tag{37}$$

This quadratic action should be a convenient starting point for the study of D-p-brane dynamics, taking into account the Born-Infeld corrections. These methods can also be applied to the M-theory five-brane action (Bandos, Lechner, Nurmagambetov, Pasti, Sorokin and Tonin (1997), Aganagic, Park, Popescu and Schwarz (1997)), as shown in (Abou Zeid and Hull (1998)).

4 Duals of Actions of the First Type

Instead of introducing a tensor $k_{\mu\nu}$, we introduce a tensor density $\tilde{k}^{\mu\nu}$ with $\tilde{k} \equiv \det(\tilde{k}^{\mu\nu})$. For $n \neq 2$, the action

$$\tilde{S} = -\frac{1}{2}T'_p \int d^n\sigma \left[\tilde{k}^{\mu\nu}(G_{\mu\nu} - B_{\mu\nu} + F_{\mu\nu}) - (n-2)(-\tilde{k})^{\frac{1}{n-2}}\Lambda\right] \tag{38}$$

is equivalent to (10), as can be seen by defining a tensor $k_{\mu\nu}$ by $(k^{-1})^{\mu\nu} = (-\tilde{k})^{-\frac{1}{n-2}}\tilde{k}^{\mu\nu}$, so that

$$\tilde{k}^{\mu\nu} \equiv \sqrt{-k}(k^{-1})^{\mu\nu} . \tag{39}$$

Integrating out $\tilde{k}^{\mu\nu}$ yields the action (6) as before.

Integrating out the world-volume vector field A_μ from (38) gives $\partial_\mu \tilde{k}^{[\mu\nu]} = 0$ which is solved by $\tilde{k}^{[\mu\nu]} = \tilde{H}^{\mu\nu}$ where $\tilde{H}^{\mu\nu}$ is given in terms of an unconstrained $n-3$ form $\tilde{A}$ by (16), so that

$$\tilde{k}^{\mu\nu} = \tilde{g}^{\mu\nu} + \tilde{H}^{\mu\nu}, \tag{40}$$

where the symmetric tensor density $\tilde{g}^{\mu\nu}$ is defined by $\tilde{g}^{\mu\nu} \equiv \tilde{k}^{(\mu\nu)}$. The action (38) then becomes

$$\begin{aligned} S' = -\frac{1}{2}T'_p \int d^n\sigma \left[\left(\tilde{g}^{\mu\nu} + \tilde{H}^{\mu\nu}\right)(G_{\mu\nu} - B_{\mu\nu}) - \right. \\ \left. (n-2)\Lambda\left(-\det\left[\tilde{g}^{\mu\nu} + \tilde{H}^{\mu\nu}\right]\right)^{\frac{1}{n-2}}\right] . \end{aligned} \tag{41}$$

This is a dual form of the action in which A_μ has been replaced by $\tilde{A}$. It contains the auxiliary symmetric tensor density $\tilde{g}^{\mu\nu}$ which can in principle be integrated out; this can be done explicitly for low values of n, but is harder for general n.

We define a symmetric metric tensor $g_{\mu\nu}$ with inverse $g^{\mu\nu}$ by $g^{\mu\nu} = (-\tilde{g})^{-\frac{1}{n-2}}\tilde{g}^{\mu\nu}$ where $\tilde{g} = \det(\tilde{g}^{\mu\nu})$, so that

$$\tilde{g}^{\mu\nu} = \sqrt{-g}g^{\mu\nu}, \tag{42}$$

where $g = \det(g_{\mu\nu})$, and an anti-symmetric tensor by

$$H^{\mu\nu} = \frac{1}{\sqrt{-g}}\tilde{H}^{\mu\nu}, \tag{43}$$

so that

$$\tilde{k}^{\mu\nu} = \sqrt{-g}\,(g^{\mu\nu} + H^{\mu\nu}) \tag{44}$$

and

$$\tilde{k} \equiv \det(\tilde{k}^{\mu\nu}) = -(-g)^{\frac{n}{2}-1}\Delta, \tag{45}$$

where

$$\Delta(g, H) \equiv \det(\delta_\mu{}^\nu + H_\mu{}^\nu) \tag{46}$$

and $H_\mu{}^\nu = g_{\mu\rho}H^{\rho\nu}$. Then the action (41) can be rewritten as

$$\tilde{S}_D = -\frac{1}{2}T_p' \int d^n\sigma\sqrt{-g}\,(g^{\mu\nu}G_{\mu\nu} + \Sigma)\,, \tag{47}$$

where

$$\Sigma \equiv -(n-2)\Lambda\Delta^{\frac{1}{n-2}} - H^{\mu\nu}B_{\mu\nu}. \tag{48}$$

The action (47) is the dual form of action (38). Unfortunately, the metric dependence of Δ makes it hard to eliminate $g_{\mu\nu}$ from this action explicitly.

For $n = 2$, the action (10) has the Weyl symmetry (13) and can be rewritten using a tensor density $\tilde{k}^{\mu\nu}$ as

$$\tilde{S}^2 = -\frac{1}{2}T_1 \int d^2\sigma \left\{\tilde{k}^{\mu\nu}(G_{\mu\nu} + \mathcal{F}_{\mu\nu}) + \lambda\left[\det(\tilde{k}^{\mu\nu}) + 1\right]\right\}. \tag{49}$$

Integrating out λ yields the constraint

$$\tilde{k} = -1 \tag{50}$$

which is solved in $n = 2$ dimensions by

$$\tilde{k}^{\mu\nu} \equiv \sqrt{-k}(k^{-1})^{\mu\nu}, \tag{51}$$

so that one recovers the original action (10). If instead one keeps the Lagrange multiplier and integrates out the world-volume vector A, one finds the constraint eq. (15) again. For $n = 2$, this is solved by

$$\tilde{k}^{[\mu\nu]} = \epsilon^{\mu\nu}\Lambda, \tag{52}$$

where Λ is a constant. The dual action for $n = 2$ is then

$$\tilde{S}_D^2 = -\frac{1}{2}T_1 \int d^2\sigma \left\{ [\tilde{g}^{\mu\nu} + \Lambda\epsilon^{\mu\nu}][G_{\mu\nu} - B_{\mu\nu}] + \lambda \left[\det(\tilde{g}^{\mu\nu}) + 1 + \Lambda^2 \right] \right\} \tag{53}$$

where $\tilde{g}^{\mu\nu} = \tilde{k}^{(\mu\nu)}$ and we have used the identity

$$\det(\tilde{g}^{\mu\nu} + \epsilon^{\mu\nu}\Lambda) = \det(\tilde{g}^{\mu\nu}) + \Lambda^2. \tag{54}$$

Integrating out λ gives $\det(\tilde{g}^{\mu\nu}) = -1 - \Lambda^2$, which is solved in terms of an unconstrained metric $g_{\mu\nu}$ by

$$\tilde{g}^{\mu\nu} = \sqrt{1+\Lambda^2}\sqrt{-g}g^{\mu\nu} \tag{55}$$

so that the action becomes

$$S_D^2 = -\frac{1}{2}T_1 \int d^2\sigma \left(\sqrt{1+\Lambda^2}\sqrt{-g}g^{\mu\nu}G_{\mu\nu} + \Lambda\epsilon^{\mu\nu}B_{\mu\nu} \right). \tag{56}$$

The metric can be eliminated from this to give the dual action of ref. (Schmidhuber (1996), de Alwis and Sato (1996), Tseytlin (1996a))

$$S_D^2 = -T_1 \int d^2\sigma \left(\sqrt{1+\Lambda^2}\sqrt{-\det(G_{\mu\nu})} + \frac{1}{2}\Lambda\epsilon^{\mu\nu}B_{\mu\nu} \right). \tag{57}$$

5 More Duals of Actions of the First Type

Consider actions given by the sum of (6) and some action $S_F = \int d^n\sigma f(F)$ which is algebraic in F; in the next section we will be interested in the example of D-brane actions which are of this form. Defining

$$N_{\mu\nu} \equiv G_{\mu\nu} - B_{\mu\nu}, \tag{58}$$

the action can be rewritten in first order form as

$$-\frac{1}{2}T_p' \int d^n\sigma \Big\{ \sqrt{-k}\left[(k^{-1})^{\mu\nu}(N_{\mu\nu} + F_{\mu\nu}) - (n-2)\Lambda \right] + \frac{1}{2}\tilde{H}^{\mu\nu}\left(F_{\mu\nu} - 2\partial_{[\mu}A_{\nu]}\right) + f(F) \Big\}. \tag{59}$$

Here the anti-symmetric tensor density $\tilde{H}^{\mu\nu}$ is a Lagrange multiplier imposing $F = dA$ and can be integrated out to regain the original action. Alternatively, integrating over A_μ imposes

$$\partial_\mu \tilde{H}^{\mu\nu} = 0, \tag{60}$$

which can be solved in terms of an $n-3$ form $\tilde{A}$ as before:

$$\tilde{H}^{\mu\nu} = \frac{1}{(n-2)!}\epsilon^{\mu\nu\rho\gamma_1\ldots\gamma_{n-3}}\partial_{[\rho}\tilde{A}_{\gamma_1\ldots\gamma_{n-3}]}. \tag{61}$$

Now F is an auxiliary 2-form occuring algebraically; we emphasize this by rewriting $F \to L$ so that the action is

$$-\frac{1}{2}T'_p \int d^n\sigma \left\{ \sqrt{-k}\left[(k^{-1})^{\mu\nu}(N_{\mu\nu}+L_{\mu\nu}) - (n-2)\Lambda \right] + \frac{1}{2}\tilde{H}^{\mu\nu}L_{\mu\nu} + f(L) \right\}. \tag{62}$$

The field equation for $L_{\mu\nu}$ is

$$\sqrt{-k}(k^{-1})^{[\mu\nu]} + \frac{1}{2}\tilde{H}^{\mu\nu} + \frac{\delta f}{\delta L_{\mu\nu}} = 0. \tag{63}$$

If $f = 0$, this can be used to recover the dual action (47) of the last section. More generally, if $f(L)$ is at most quartic in L, this can be solved to give an expression for $L_{\mu\nu}$ which can then be re-substituted in (62) to give a dual action analogous to (47). This is applicable to the D-brane actions considered in the next two sections, in which f is at most quartic for $p < 9$ branes.

Integrating out $k_{\mu\nu}$ from (62) gives

$$-T_p \int d^n\sigma \left\{ \sqrt{-\det(N_{\mu\nu}+L_{\mu\nu})} + \frac{1}{2}\tilde{H}^{\mu\nu}L_{\mu\nu} + \frac{T'_p}{T_p}f(L) \right\}. \tag{64}$$

If $f = 0$ and $n \leq 5$, the equation of motion for L can be solved explicitly and the solution substituted in (62) to get the dual action (Tseytlin (1996a), Aganagic, Park, Popescu and Schwarz (1997))

$$S_D = -T_p \int d^n\sigma \left\{ \sqrt{-\det(G_{\mu\nu}+iK_{\mu\nu})} + \frac{1}{2}\tilde{H}^{\mu\nu}B_{\mu\nu} \right\}, \tag{65}$$

where

$$K_{\mu\nu} \equiv \frac{1}{\sqrt{-\det(G_{\mu\nu})}} G_{\mu\rho}G_{\nu\lambda}\tilde{H}^{\rho\lambda}. \tag{66}$$

This can in turn be linearised to give the equivalent action

$$-\frac{1}{2}T'_p \int d^n\sigma \left\{ \sqrt{-k}\left[(k^{-1})^{\mu\nu}(G_{\mu\nu}+iK_{\mu\nu}) - (n-2)\Lambda \right] + \frac{1}{2}\tilde{H}^{\mu\nu}B_{\mu\nu} \right\}. \tag{67}$$

6 D-Brane Actions of the First Type

The bosonic part of the effective world-volume action for a D-brane in a type II supergravity background is (Schmidhuber (1996), Douglas (1995), Green, Hull and Townsend (1996))

$$S_1 = -T_p \int d^n\sigma e^{-\phi}\sqrt{-\det(G_{\mu\nu}+\mathcal{F}_{\mu\nu})} + T_p \int_{W_n} Ce^{\mathcal{F}}, \tag{68}$$

The first term is of the form (6) with an extra dependence on the dilaton field ϕ. The second term is a Wess-Zumino term and gives the coupling to the background Ramond-Ramond r-form gauge fields $C^{(r)}$ (where r is odd for type IIA and even for type IIB). The potentials $C^{(r)}$ for $r > 4$ are the duals of the potentials $C^{(8-r)}$. In (68), C is the formal sum (Green, Hull and Townsend (1996))

$$C \equiv \sum_{r=0}^{9} C^{(r)}, \tag{69}$$

all forms in space-time are pulled back to the worldvolume of the brane W_n and it is understood that the n-form part of $Ce^{\mathcal{F}}$, which is $C^{(n)} + C^{(n-2)}\mathcal{F} + \frac{1}{2}C^{(n-4)}\mathcal{F}^2 + \ldots$, is selected. The case of the 9-form potential $C^{(9)}$ is special because its equation of motion implies that the dual of its field strength is a constant m. This constant will be taken to be zero here, so that $C^{(9)} = 0$.

Introducing $k_{\mu\nu}$, we obtain the classically equivalent D-brane action

$$S_1' = -\frac{1}{2}T_p' \int d^n\sigma \sqrt{-k} e^{-\phi} \left[(k^{-1})^{\mu\nu} \left(G_{\mu\nu} + \mathcal{F}_{\mu\nu} \right) - (n-2)\Lambda \right] + T_p \int_{W_n} Ce^{\mathcal{F}}. \tag{70}$$

The field equation for $k_{\mu\nu}$ is given in (12); substituting back into (70) yields (68). The action is of the form (59) (apart from the introduction of the dilaton) so that the dual action is (cf. (62))

$$-\frac{1}{2}T_p' \int d^n\sigma \left\{ \sqrt{-k} e^{-\phi} \left[(k^{-1})^{\mu\nu} \left(N_{\mu\nu} + L_{\mu\nu} \right) - (n-2)\Lambda \right] + \frac{1}{2}\tilde{H}^{\mu\nu} L_{\mu\nu} \right\}$$
$$+ T_p \int_{W_n} Ce^{L-B}. \tag{71}$$

The potential $f(L) \sim Ce^{L-B}$ is a polynomial of order $[n/2]$ in L (i.e. the integer part of $n/2$), so that the field equation for $L_{\mu\nu}$ (63) is of order $[n/2]-1$ in L and so should be soluble explicitly for all $n \leq 10$. In particular, it is quadratic for $n \leq 8$, so that the dual action for p-branes with $p \leq 7$ can be obtained straightforwardly. Here we will consider only the case in which $C^{(n-4)} = C^{(n-6)} = C^{(n-8)} = 0$ so that the action is linear in F. Then A is a Lagrange multiplier imposing the constraint

$$\partial_\mu \left(\sqrt{-k}(k^{-1})^{[\mu\nu]} e^{-\phi} - \frac{2T_p}{T_p'} \frac{1}{(n-2)!} \epsilon^{\mu\nu\gamma_1\ldots\gamma_{n-2}} (C^{(n-2)})_{\gamma_1\ldots\gamma_{n-2}} \right) = 0. \tag{72}$$

The general solution of this constraint is

$$\sqrt{-k}(k^{-1})^{[\mu\nu]} e^{-\phi} = \tilde{H}^{\mu\nu} + \frac{2T_p}{T_p'} \frac{1}{(n-2)!} \epsilon^{\mu\nu\gamma_1\ldots\gamma_{n-2}} (C^{(n-2)})_{\gamma_1\ldots\gamma_{n-2}} \equiv \tilde{\mathcal{H}}^{\mu\nu}, \tag{73}$$

where $\tilde{H}^{\mu\nu}$ is given in terms of $\tilde{A}$ by (61).

To obtain the dual action, we first express (70) in terms of a density $\tilde{k}^{\mu\nu}$. For $n \neq 2$, this gives the equivalent action

$$\tilde{S}_1 = -\frac{1}{2}T'_p \int d^n\sigma e^{-\phi} \left[\tilde{k}^{\mu\nu}\left(G_{\mu\nu} + \mathcal{F}_{\mu\nu}\right) - (n-2)(-\tilde{k})^{\frac{1}{n-2}} \Lambda\right]$$
$$+T_p \int_{W_n} C^{(n)} + C^{(n-2)}\mathcal{F}. \tag{74}$$

Integrating out $\tilde{k}^{\mu\nu}$ yields the action (68), while integrating out A gives

$$\tilde{k}^{\mu\nu} = \tilde{g}^{\mu\nu} + \tilde{\mathcal{H}}^{\mu\nu}, \tag{75}$$

where $\tilde{\mathcal{H}}^{\mu\nu}$ is given in terms of $\tilde{A}$ and $C^{(n-2)}$ by eq. (73), and $\tilde{g}^{\mu\nu}$ is a symmetric tensor density. The action (74) can be written in terms of tensors $g^{\mu\nu} = (-\tilde{g})^{-\frac{1}{n-2}}\tilde{g}^{\mu\nu}$ (with $\tilde{g} = \det(\tilde{g}^{\mu\nu})$) and $\mathcal{H}^{\mu\nu} = (-g)^{-\frac{1}{2}}\tilde{\mathcal{H}}^{\mu\nu}$ (with $g = \det(g_{\mu\nu})$) as

$$\tilde{S}_{1D} = -\frac{1}{2}T'_p \int d^n\sigma e^{-\phi}\sqrt{-g}\left[\left(g^{\mu\nu} + \mathcal{H}^{\mu\nu}\right)N_{\mu\nu} - (n-2)\Omega^{\frac{1}{n-2}}\Lambda\right]$$
$$+T_p \int_{W_n} C^{(n)} - C^{(n-2)}B, \tag{76}$$

where

$$\Omega \equiv \det\left(\delta_\mu{}^\nu + \mathcal{H}_\mu{}^\nu\right) \tag{77}$$

and $\mathcal{H}_\mu{}^\nu = (\tilde{g}^{-1})_{\mu\rho}\mathcal{H}^{\rho\nu}$. The action (76) is the dual form of action (70). Again, the dependence of Ω on the metric $g_{\mu\nu}$ makes the elimination of the latter from the action difficult, although possible in principle.

For $n = 2$, the action (70) still possesses the generalised Weyl symmetry (13) and can be rewritten in terms of a tensor density $\tilde{k}^{\mu\nu}$ as

$$\tilde{S}_1^2 = -\frac{1}{2}T_1 \int d^2\sigma e^{-\phi}\left[\tilde{k}^{\mu\nu}\left(N_{\mu\nu} + F_{\mu\nu}\right) + \lambda\left(\tilde{k}+1\right)\right] + T_1 \int_{W_2} C^{(2)} + C^{(0)}\mathcal{F}. \tag{78}$$

Integrating out λ yields the constraint $\tilde{k} = -1$, which is solved by eq. (51), so that one recovers the original action. Keeping the Lagrange multiplier and integrating out A one finds the constraint

$$\partial_\mu\left(e^{-\phi}\tilde{k}^{[\mu\nu]} - 2\epsilon^{\mu\nu}C^{(0)}\right) = 0, \tag{79}$$

which for $n = 2$ is solved by

$$e^{-\phi}\tilde{k}^{[\mu\nu]} = \epsilon^{\mu\nu}\mathcal{E}, \tag{80}$$

where

$$\mathcal{E} \equiv \tilde{\Lambda} + 2C^{(0)} \tag{81}$$

and $\tilde{\Lambda}$ is a constant. The dual action for $n = 2$ is then

$$\tilde{S}_{1D}^2 = -\frac{1}{2}T_1 \int d^2\sigma e^{-\phi}\left[\left(\tilde{g}^{\mu\nu} + e^{\phi}\epsilon^{\mu\nu}\mathcal{E}\right)N_{\mu\nu} + \lambda\left(\det(\tilde{g}^{\mu\nu}) + 1 + e^{2\phi}\mathcal{E}^2\right)\right]$$
$$+T_1 \int_{W_2} C^{(2)} - C^{(0)}B, \tag{82}$$

where $\tilde{g}^{\mu\nu} = \tilde{k}^{(\mu\nu)}$. Integrating out λ gives the dual action in the form

$$S_1^2 = -\frac{1}{2}T_1 \int d^2\sigma \sqrt{e^{-2\phi} + \mathcal{E}^2}\sqrt{-g}g^{\mu\nu}G_{\mu\nu} + T_1 \int_{W_2} C^{(2)} + (\mathcal{E} - C^{(0)})B. \quad (83)$$

Finally, integrating out $g_{\mu\nu}$ gives the action (Tseytlin (1996a))

$$S_1^2 = -T_1 \int d^2\sigma \sqrt{e^{-2\phi} + \mathcal{E}^2}\sqrt{-\det(G_{\mu\nu})} + T_1 \int_{W_2} C^{(2)} + (\mathcal{E} - C^{(0)})B. \quad (84)$$

7 Supersymmetric D-Brane Actions of the First Type

The new actions discussed above can be extended to supersymmetric D-brane actions with local kappa symmetry equivalent to those presented in refs. (Cederwall, von Gussich, Nilsson and Westerberg (1996), Cederwall, von Gussich, Nilsson, Sundell and Westerberg (1996), Cederwall (1996), Bergshoeff and Townsend (1996), Aganagic, and Schwarz (1996a), Aganagic, and Schwarz (1996b)) at the classical level.

The (flat) superspace coordinates are the $D = 10$ space-time coordinates X^i and the Grassmann coordinates θ, which are space-time spinors and world-volume scalars. For the type IIA superstring (even p), θ is Majorana but not Weyl while in the IIB superstring there are two Majorana-Weyl spinors θ_α ($\alpha = 1, 2$) of the same chirality. The superspace (global) supersymmetry transformations are

$$\delta_\epsilon \theta = \epsilon \ , \quad \delta_\epsilon X^i = \bar{\epsilon}\Gamma^i \theta. \quad (85)$$

The world-volume theory has global type IIA or type IIB super-Poincaré symmetry and is constructed using the supersymmetric one-forms $\partial_\mu \theta$ and

$$\Pi^i_\mu = \partial_\mu X^i - \bar{\theta}\Gamma^i \partial_\mu \theta. \quad (86)$$

The induced world-volume metric is

$$G_{\mu\nu} = G_{ij}\Pi^i_\mu \Pi^j_\nu. \quad (87)$$

The supersymmetric world-volume gauge field-strength two form $\mathcal{F}_{\mu\nu}$ is given by (7) for the following choice of the two form B (Townsend (1996))

$$B = -\bar{\theta}\Gamma_{11}\Gamma_i d\theta \left(dX^i + \frac{1}{2}\bar{\theta}\Gamma^i d\theta \right) \quad (88)$$

when p is even or the same formula with Γ_{11} replaced with the Pauli matrix τ_3 when p is odd. With the choice (88), $\delta_\epsilon B$ is an exact two-form and $\mathcal{F}$ is supersymmetric for an appropriate choice of $\delta_\epsilon A$ (Aganagic, and Schwarz (1996b)).

The effective world-volume action for a D-brane in flat superspace with constant dilaton is

$$S_1^{DBI} = -T_p \int d^n\sigma e^{-\phi}\sqrt{-\det(G_{\mu\nu} + \mathcal{F}_{\mu\nu})} + T_p \int_{W_n} Ce^{\mathcal{F}}, \tag{89}$$

which is formally of the same form as (68). Again C represents a complex of differential forms $C^{(r)}$, as in (69), but now the r-forms $C^{(r)}$ are the pull-backs of superspace forms $C^{(r)} = d\bar{\theta}T^{(r-2)}d\theta$ for certain $r-2$ forms $T^{(r-2)}$ given explicitly in (Cederwall, von Gussich, Nilsson and Westerberg (1996), Cederwall, von Gussich, Nilsson, Sundell and Westerberg (1996), Cederwall (1996), Bergshoeff and Townsend (1996), Aganagic, and Schwarz (1996a), Aganagic, and Schwarz (1996b)). This action is supersymmetric and invariant under local kappa symmetry (Cederwall, von Gussich, Nilsson and Westerberg (1996), Cederwall, von Gussich, Nilsson, Sundell and Westerberg (1996), Cederwall (1996), Bergshoeff and Townsend (1996), Aganagic, and Schwarz (1996a), Aganagic, and Schwarz (1996b)).

A classically equivalent form of the D-brane action is given by

$$S_1^P = -\frac{1}{2}T_p' \int d^n\sigma e^{-\phi}\sqrt{-k}\left[(k^{-1})^{\mu\nu}(G_{\mu\nu} + \mathcal{F}_{\mu\nu}) - (n-2)\Lambda\right] + T_p \int_{W_n} Ce^{\mathcal{F}} \tag{90}$$

The action is of the same form as (70) and can be dualised to give (71) for $n \neq 2$ or (82) for $n = 2$.

8 Duals of Actions of the Second Type

The dualisation of the form (23) of the Born-Infeld action can be achieved via the addition of a Lagrange multiplier term imposing eq. (18). Consider the action

$$S = -T_p' \int d^{p+1}\sigma \Big\{(-g)^{\frac{1}{4}}(-\gamma)^{\frac{1}{4}}\left[\gamma^{\mu\nu}(g_{\mu\nu} - g^{\rho\sigma}F_{\mu\rho}F_{\sigma\nu}) - (p-3)\Lambda\right] + 2\tilde{H}^{\mu\nu}\left(F_{\mu\nu} - \partial_{[\mu}A_{\nu]}\right)\Big\}, \tag{91}$$

where $\tilde{H}^{\mu\nu}$ is a tensor density and F is regarded as an independent field. Integrating out $\tilde{H}^{\mu\nu}$ sets $F = dA$ and yields the original action (23). Alternatively, integrating out A_μ imposes the constraint

$$\partial_\mu \tilde{H}^{\mu\nu} = 0 \tag{92}$$

which can be solved in terms of a $(p-2)$-form $\tilde{A}$,

$$\tilde{H}^{\mu\nu} = \frac{1}{(p-1)!}\epsilon^{\mu\nu\rho\gamma_1\ldots\gamma_{p-2}}\partial_{[\rho}\tilde{A}_{\gamma_1\ldots\gamma_{p-2}]}, \tag{93}$$

where $\epsilon^{\mu\nu\rho\ldots}$ is the alternating tensor density. Now F is an auxiliary two-form occuring quadratically in the action and can be integrated out. The field equation for $F_{\mu\nu}$ is

$$(-g)^{\frac{1}{4}}(-\gamma)^{\frac{1}{4}}\left(\gamma^{\mu\rho}g^{\nu\sigma}+\gamma^{\nu\sigma}g^{\mu\rho}\right)F_{\sigma\rho}=2\tilde{H}^{\mu\nu}, \tag{94}$$

where $\tilde{H}^{\mu\nu}$ is given by the solution (93), and the Gaussian integration amounts to solving this for $F_{\mu\nu}$ and substituting the solution $F[g_{\mu\nu},\gamma_{\mu\nu},\tilde{H}^{\mu\nu}]$ in the action (31). This gives the dual action $S[g_{\mu\nu},\gamma_{\mu\nu},\tilde{H}^{\mu\nu}]$. In principle, an equivalent dual action $S_D[g_{\mu\nu},\tilde{H}^{\mu\nu}]$ can then be obtained by integrating out the auxiliary metric $\gamma_{\mu\nu}$, but in practice this procedure is difficult to carry out explicitly because of the non-linearity in the worldvolume metric of eq. (94) and of the action $S[g_{\mu\nu},\gamma_{\mu\nu},\tilde{H}^{\mu\nu}]$.

Defining the matrices

$$f_\mu{}^\nu=F_{\mu\rho}g^{\rho\nu},\qquad h_\mu{}^\nu=(-g)^{-\frac{1}{4}}(-\gamma)^{-\frac{1}{4}}g_{\mu\rho}\tilde{H}^{\nu\rho},\qquad \beta_\mu{}^\nu=2\left(g_{\mu\rho}\gamma^{\rho\nu}-\delta_\mu{}^\nu\right), \tag{95}$$

the equation (94) can be written as

$$h=f+\{\beta,f\}\equiv(1+L_\beta)f \tag{96}$$

where for any matrices X,Y, the operator L_X is defined by

$$L_XY\equiv\{X,Y\}. \tag{97}$$

Then (96) can be inverted to give

$$\begin{aligned}f&=(1+L_\beta)^{-1}h=\left(1-L_\beta+L_\beta^2-L_\beta^3+\ldots\right)h\\&=h-\{\beta,h\}+\{\beta,\{\beta,h\}\}-\{\beta,\{\beta,\{\beta,h\}\}\}+\ldots.\end{aligned} \tag{98}$$

Substituting this solution for F back in (91) gives

$$\begin{aligned}S=&-T_p'\int d^{p+1}\sigma\left\{(-g)^{\frac{1}{4}}(-\gamma)^{\frac{1}{4}}\left[\gamma^{\mu\nu}g_{\mu\nu}-(p-3)\Lambda\right]\right.\\&\left.+2(-g)^{\frac{1}{4}}(-\gamma)^{\frac{1}{4}}\mathrm{tr}\left[h(1+L_\beta)^{-1}h\right]\right\}\\=&-T_p'\int d^{p+1}\sigma\left\{(-g)^{\frac{1}{4}}(-\gamma)^{\frac{1}{4}}\left[\gamma^{\mu\nu}g_{\mu\nu}-(p-3)\Lambda\right]\right.\\&\left.+2(-g)^{-\frac{1}{4}}(-\gamma)^{-\frac{1}{4}}\tilde{H}^{\mu\sigma}M_{\mu\rho\nu\sigma}\tilde{H}^{\nu\rho}\right\},\end{aligned} \tag{99}$$

where the tensor $M_{\mu\nu\rho\sigma}$ is defined by

$$\mathrm{tr}\left[h(1+L_\beta)^{-1}h\right]=h^{\mu\sigma}M_{\mu\rho\nu\sigma}h^{\nu\rho} \tag{100}$$

(where $h^{\mu\sigma}=g^{\mu\tau}h_\tau{}^\sigma$) and is given to lowest orders by

$$\begin{aligned}M_{\mu\nu\rho\sigma}=\gamma_{\mu\kappa}g_{\nu\tau}&\left[\delta_\rho^\kappa\delta_\sigma^\tau-\Sigma^\tau{}_\rho\Sigma_\sigma{}^\kappa+\Sigma^\alpha{}_\rho\Sigma^\tau{}_\alpha\Sigma_\sigma{}^\beta\Sigma_\beta{}^\kappa-\right.\\&\left.-\Sigma^\alpha{}_\rho\Sigma^\beta{}_\alpha\Sigma^\tau{}_\beta\Sigma_\sigma{}^\delta\Sigma_\delta{}^\kappa+\Sigma^\alpha{}_\rho\Sigma^\beta{}_\alpha\Sigma^\delta{}_\beta\Sigma^\tau{}_\delta\Sigma_\sigma{}^\epsilon\Sigma_\epsilon{}^\lambda\Sigma_\lambda{}^\kappa+\ldots\right],\end{aligned} \tag{101}$$

where

$$\Sigma^\mu{}_\nu\equiv g_{\nu\rho}\gamma^{\rho\mu} \tag{102}$$

and $\Sigma_\mu{}^\rho$ denotes the inverse of the matrix $\Sigma^\mu{}_\rho$. The auxiliary metric $\gamma_{\mu\nu}$ occurs algebraically and can in principle be eliminated using its equation of motion, giving $\gamma_{\mu\nu}$ as a function of $g_{\nu\rho}$ and $\tilde{H}^{\mu\nu}$. Although this is hard to do explicitly, it can be done perturbatively, giving $\gamma_{\mu\nu}$ to any desired order in $\tilde{H}^{\mu\nu}$.

The dualisation of the action (31), which is classically equivalent to the D-brane kinetic term (29), proceeds in a similar way. Consider the action

$$S = -T_p' \int d^{p+1}\sigma \left\{ e^{-\phi}(-g)^{\frac{1}{4}}(-\gamma)^{\frac{1}{4}} \left[\gamma^{\mu\nu}\left(g_{\mu\nu} - g^{\rho\sigma}\mathcal{F}_{\mu\rho}\mathcal{F}_{\sigma\nu}\right) - (p-3)\Lambda\right] \right.$$
$$\left. +2\tilde{H}^{\mu\nu}\left(F_{\mu\nu} - 2\partial_{[\mu}A_{\nu]}\right)\right\}. \tag{103}$$

Integrating out $\tilde{H}^{\mu\nu}$ yields the original action (31). Alternatively, integrating over A_μ imposes the constraint (92), which is solved in terms of a $(p-2)$ form $\tilde{A}$ as in (93). Now F is an auxiliary two-form occuring algebraically. The field equation for $F_{\mu\nu}$ is

$$(-g)^{\frac{1}{4}}(-\gamma)^{\frac{1}{4}}\left(\gamma^{\mu\rho}g^{\nu\sigma} + \gamma^{\nu\sigma}g^{\mu\rho}\right)\left(F_{\sigma\rho} - B_{\sigma\rho}\right) = 2\tilde{H}^{\mu\nu}, \tag{104}$$

where $\tilde{H}^{\mu\nu}$ is given by the solution (93).

Defining the matrix

$$\tilde{f}_\mu{}^\nu \equiv \left(F_{\mu\rho} - B_{\mu\rho}\right) g^{\rho\sigma} \tag{105}$$

the equation (104) can be written as

$$h = \tilde{f} + \{\beta, \tilde{f}\} = (1 + L_\beta)\tilde{f} \tag{106}$$

where the matrices h, β and the operator L_β are defined as in (95) and (97). This can be inverted to give

$$\tilde{f} = (1 + L_\beta)^{-1}h = h - \{\beta, h\} + \{\beta, \{\beta, h\}\} - \{\beta, \{\beta, \{\beta, h\}\}\} + \ldots. \tag{107}$$

Substituting this solution for $\mathcal{F}$ back in (91) gives

$$S = -T_p' \int d^{p+1}\sigma \left\{(-g)^{\frac{1}{4}}(-\gamma)^{\frac{1}{4}}\left[\gamma^{\mu\nu}g_{\mu\nu} - (p-3)\Lambda\right] + 2\tilde{H}^{\mu\nu}B_{\mu\nu}\right.$$
$$\left. +2(-g)^{-\frac{1}{4}}(-\gamma)^{-\frac{1}{2}}\tilde{H}^{\mu\sigma}M_{\mu\rho\nu\sigma}\tilde{H}^{\nu\rho}\right\}, \tag{108}$$

with the tensor $M_{\mu\nu\rho\sigma}$ defined as in (101).

References

M. Abou Zeid and C. M. Hull. (1997): *Intrinsic Geometry of D-Branes*, Phys.Lett. **B404** (1997) 264-270, [hep-th/9704021].

M. Abou Zeid and C. M. Hull. (1998): *Geometric Actions for D-Branes and M-Branes*, Phys.Lett. **B428** 277-283, [hep-th/9802179].

M. Abou Zeid. (1997): *D-Brane Actions, Intrinsic Geometry and Duality*, [hep-th/9712163].

A. M. Polyakov. (1981): *Quantum Geometry of Bosonic Strings*, Phys. Lett. **B103** (1981) 207

L. Brink, P. Di Vecchia and P. S. Howe. (1976): *A Locally Supersymmetric and Reparametrization Invariant Action for the Spinning String*, Phys. Lett. **B65** 471

P. S. Howe and R. W. Tucker. (1997): *A Locally Supersymmetric and Reparametrization Invariant Action for a Spinning Membrane*, J. Phys. **A10** L155

U. Lindström. (1988): *First Order Actions for Gravitational Systems, Strings and Membranes*, Int. J. Mod. Phys. **A3** 2401

C. Schmidhuber. (1996): *D-Brane Actions*, Nucl. Phys. **B467** 146

S. P. de Alwis and K. Sato. (1996): *D-Strings and F-Strings from String Loops*, Phys. Rev. **D53** 7187

M. Douglas. (1995): *Branes Within Branes*, hep-th/9512077

P. K. Townsend. (1996): *D-Branes from M-Branes*, Phys. Lett. **B373** 68

A. A. Tseytlin. (1996): *Self-Duality of Born-Infeld Action and Dirichlet 3-Brane of Type IIB Superstring Theory*, Nucl. Phys. **B469** 51

M. Cederwall, A. von Gussich, B. E. W. Nilsson and A. Westerberg. (1996): *The Dirichlet Super-Three-Brane in Ten-Dimensional Type IIB Supergravity*, hep-th/9610148

M. Cederwall, A. von Gussich, B. E. W. Nilsson, P. Sundell and A. Westerberg. (1996): *The Dirichlet Super-p-Branes in Ten-Dimensional Type IIA and IIB Supergravity*, hep-th/9611159

M. Cederwall. (1996) *Aspects of D-Brane Actions*, hep-th/9612153

M. Aganagic, C. Popescu and J. H. Schwarz. (1996): *D-Brane Actions with Local Kappa Symmetry*, hep-th/9610249

M. Aganagic, C. Popescu and J. H. Schwarz. (1996): *Gauge-Invariant and Gauge-Fixed D-Brane Actions*, hep-th/9612080

M. Aganagic, J. Park, C. Popescu and J. Schwarz. (1997): *Dual D-Brane Actions*, hep-th/9702133

A. Einstein. : Sitzungsberichte der Preussische Akademie der Wissenschaften (1925) 414; Ann. Math. **46** (1945) 578; Rev. Mod. Phys. **20** (1948) 35

J. W. Moffat. (1995): *Nonsymmetric Gravitational Theory*, J. Math. Phys. **36** 3722

M. B. Green, C. M. Hull and P. K. Townsend. (1996): *D-Brane Wess-Zumino Actions, T-Duality and the Cosmological Constant*, Phys. Lett. **B382** 65

E. Bergshoeff and P. K. Townsend. (1996): *Super D-Branes*, hep-th/9611173

M. Born and L. Infeld. (1935): *Foundations of the New Field Theory*, Proc. Roc. Soc. **A144** 425.

J. Polchinski. (1996): *TASI Lectures on D-Branes*, [hep-th/9611050]

A. A. Tseytlin. (1996): *Self-Duality of Born-Infeld Action and Dirichlet 3-Brane of Type IIB Superstring Theory*, Nucl. Phys. **B469** 51, [hep-th/9602064].

Y. Lozano. (1997): *D-Brane Dualities as Canonical Transformations*, Phys. Lett. **B399** 233, [hep-th/9701186].

I. Bandos, K. Lechner, A. Nurmagambetov, P. Pasti, D. Sorokin and M. Tonin. (1997): *Covariant Action for the Super-Five-Brane of M-Theory*, Phys. Rev. Lett. **78** 4332, [hep-th/9701149]

M. Aganagic, J. Park, C. Popescu and J. H. Schwarz. (1997): *World Volume Action of the M Theory Five-Brane*, Nucl. Phys. **B496** 191, [hep-th/9701166]

String Action on adS Space

Renata Kallosh

Physics Department, Stanford University,
Stanford CA 94305

Abstract. String theory in the background of the near horizon D3 brane can be described in terms of the Green-Schwarz type action with manifest space-time supersymmetry. This allows to overcome one of the basic drawbacks of the standard NS string action: the fact that it does not couple to Ramond-Ramond forms. The five-form of the supersymmetric adS geometry does couple to fermions in GS formulation. We describe the classical string action in this background which depends on up to 32+2 powers of fermions as well as the one with the gauge-fixed κ-symmetry which has only up to 2+2 powers of fermions. Finally we also give a T-dual action with the simple quadratic dependence on fermions.

1 An Appreciation of Victor Isaakovich Ogievetsky

Victor Ogievetsky had a great influence on the development of new concepts in high-energy physics which were associated with supersymmetry and geometry. His way of thinking about quantum theories and gravity became a dominating point of view in attempts to understand the issues in quantum gravity. With deep gratitude and sadness I am writing this note dedicated to the memory of Victor Isaakovich Ogievetsky. I am also happy to present the recent development in string theory and supergravity on which I was working with J. Rahmfeld, A. Rajaraman and A. Tseytlin. This development as the rest of my work to large extent was based on the early influence of Victor Isaakovich Ogievetsky on my choice of the problems in physics.

2 Classical IIB String Action in $adS_5 \times S^5$ + 5-Form Background

Manifestly supersymmetric in space-time approach to IIB string theory starts with the superspace of supergravity with the coordinates (X, Θ), where there are 10 bosonic coordinates X and 32 fermionic ones Θ. These coordinates which from the point of view of supergravity theory were just labels of the superspace become functions of the 2-dimensional world-sheet of the string: $(X(\sigma), \Theta(X(\sigma), \sigma))$. The 5-form of the background $F_{a_1a_2a_3a_4a_5}$ interacts with the string in presence of fermions Θ, Metsaev and Tseytlin (1998a, 1998b)

$$\partial X^m \partial X^n \bar{\Theta} \Gamma_m \Gamma^{a_1a_2a_3a_4a_5} \Gamma_n \Theta F_{a_1a_2a_3a_4a_5} \tag{1}$$

It is therefore important to have a complete dependence of the string action on space-time fermions Θ.

The classical superstring action in manifestly supersymmetric form was constructed by Green and Schwarz (1984) in a flat superspace. This action was generalized to an arbitrary background of supergravity by Grisaru, Howe, Mezincescu, Nilsson and Townsend (1985). More recently the string action was found in a maximally supersymmetric $D = 10$ type IIB supergravity vacuum, which is also the near horizon space of the D3 brane Gibbons and Townsend (1993), i.e. in the $AdS_5 \times S^5$ background. The string action in supercoset approach was constructed by Metsaev and Tseytlin (1998a, 1998b) and shown to be equivalent to using a supergravity superspace by Kallosh, Rahmfeld and Rajaraman (1998) where also the closed form of the superspace vielbeins and forms was found. This action has local κ-Symmetry and 2-d reparametrization symmetry. The classical $AdS_5 \times S^5$ action is $(2\pi\alpha' = 1)$

$$S = -\frac{1}{2}\int d^2\sigma \left(\sqrt{-g}\, g^{ij} L_i^{\hat{a}} L_j^{\hat{a}} + 4i\epsilon^{ij}\int_0^1 ds\; L_{is}^{\hat{a}} \mathcal{K}^{IJ} \bar{\Theta}^I \Gamma^{\hat{a}} L_{is}^J\right) . \quad (2)$$

Here $\mathcal{K}^{IJ} \equiv \mathrm{diag}(1,-1)$, $I, J = 1,2$ and $\hat{a} = (a, a') = (0, ..., 4, 5, ..., 9)$. The invariant 1-forms $L^I = L^I_{s=1}$, $L^{\hat{a}} = L^{\hat{a}}_{s=1}$ are given by

$$L_s^I = \left(\frac{\sinh(s\mathcal{M})}{\mathcal{M}} D\Theta\right)^I , \quad (3)$$

and

$$L_s^{\hat{a}} = e_{\hat{m}}^{\hat{a}}(X) dX^{\hat{m}} - 4i\bar{\Theta}^I \Gamma^{\hat{a}} \left(\frac{\sinh^2(\frac{1}{2}s\mathcal{M})}{\mathcal{M}^2} D\Theta\right)^I , \quad (4)$$

where $X^{\hat{m}}$ and Θ^I are the bosonic and fermionic superstring coordinates and

$$\begin{aligned}(\mathcal{M}^2)^{IL} &= \epsilon^{IJ}(-\gamma^a \Theta^J \bar{\Theta}^L \gamma^a + \gamma^{a'} \Theta^J \bar{\Theta}^L \gamma^{a'}) \\ &\quad + \frac{1}{2}\epsilon^{KL}(\gamma^{ab}\Theta^I \bar{\Theta}^K \gamma^{ab} - \gamma^{a'b'}\Theta^I \bar{\Theta}^K \gamma^{a'b'}) , \end{aligned} \quad (5)$$

$$(D\Theta)^I = \left[d + \frac{1}{4}(\omega^{ab}\gamma_{ab} + \omega^{a'b'}\gamma_{a'b'})\right]\Theta^I - \frac{1}{2} i\epsilon^{IJ}(e^a\gamma_a + ie^{a'}\gamma_{a'})\Theta^J . \quad (6)$$

The Dirac matrices are split in the '5+5' way, $\Gamma^a = \gamma^a \times 1 \times \sigma_1$, $\Gamma^{a'} = 1 \times \gamma^{a'} \times \sigma_2$, where σ_k are Pauli matrices.

Thus the classical action has all terms of the form

$$\Theta^{2n} d\Theta d\Theta, \quad n = 0, 1, \ldots 16. \quad (7)$$

and one would like to gauge-fix κ-Symmetry to minimize the dependence on fermions in the gauge-fixed action. We will show that by using a proper gauge one can reduce this dependence up to

$$\Theta^2 d\Theta d\Theta. \quad (8)$$

Furthermore by performing a 2d T-duality we will find only terms of the form

$$d\Theta d\Theta. \quad (9)$$

3 κ-Symmetry Gauge-Fixed Action

The κ-Symmetry gauge fixing of the classical string action was performed in Kallosh and Rahmfeld (1998). We shall use the 'D3-brane adapted' or '4+6' bosonic coordinates $X^{\hat{m}} = (x^p, y^t)$ in which the $AdS_5 \times S^5$ metric is split into the parts parallel and transverse to the D3-brane directions (we take the radius parameter to be $R = 1$)

$$ds^2 = y^2 dx^p dx^p + \frac{1}{y^2} dy^t dy^t \ , \qquad y^2 \equiv y^t y^t \ , \tag{10}$$

where $p = 0, ..., 3$, $t = 4, ..., 9$. In what follows the contractions of the indices p is understood with Minkowski metric and indices t – with Euclidean metric. The κ-Symmetry gauge is fixed using the 'parallel to D3-brane' Γ-matrix projector

$$\Theta_-^I = 0 \ , \qquad \Theta_\pm^I \equiv \mathcal{P}_\pm^{IJ} \Theta^J \ , \qquad \mathcal{P}_\pm^{IJ} = \frac{1}{2}\left(\delta^{IJ} \pm \Gamma_{0123}\epsilon^{IJ}\right) \ , \qquad \mathcal{P}_+ \mathcal{P}_- = 0 \ . \tag{11}$$

In '5+5' coordinates ($x^a = (x^p, x^4 = y)$ and $\xi^{a'}$ coordinates on S^5) one finds that ($\Gamma_{0123} = i\gamma_4 \times 1 \times 1$, $\omega^{p4} = e^p$)

$$(D\Theta)^I = \left[\delta^{IJ}(d + \frac{1}{4}\omega^{a'b'}\gamma_{a'b'}) + \frac{1}{2}\epsilon^{IJ}(e^{a'}\gamma_{a'} - ie^4\gamma_4) + \frac{1}{2}e^p\gamma_p\gamma_4\mathcal{P}_-^{IJ}\right]\Theta^J \ . \tag{12}$$

Using that the S^5 part of the covariant derivative satisfies $D_5^{IJ} \equiv \delta^{IJ}(d + \frac{1}{4}\omega^{a'b'}\gamma_{a'b'}) + \frac{1}{2}\epsilon^{IJ}e^{a'}\gamma_{a'} = (\Lambda d\Lambda^{-1})^{IJ}$, $(D_5)^2 = 0$, where the spinor matrix $\Lambda^{IJ} = \Lambda^{IJ}(\xi)$ is a function of the S^5 coordinates, one finds that $(D\Theta_+)^I$ can be written as

$$D\Theta_+ = (d - \frac{1}{2}d\log y + \Lambda d\Lambda^{-1})\Theta_+^I = y^{1/2}\Lambda \ d \ (y^{-1/2}\Lambda^{-1}\Theta_+) \ . \tag{13}$$

Eq. (13) suggests to make the change of the fermionic variable $\Theta \to \theta$

$$\Theta_+^I = y^{1/2}\Lambda^{IJ}(\xi) \ \theta_+^J \ , \qquad \mathcal{P}_-^{IJ}\theta_+^J = 0 \ , \qquad D\Theta_+^I = y^{1/2}\Lambda d\theta_+^I \ . \tag{14}$$

If we further transform from the coordinates $(y, \xi^{a'})$ to the 6-d Cartesian coordinates y^t in (10), $y^t = \frac{y}{\sqrt{1+\xi^2}}(1, \xi^{a'})$, that would effectively absorb the matrix Λ into an $SO(6)$ spinor rotation. This simplification is suggested, as shown in Kallosh (1998), by the form of the Killing spinors in $AdS_5 \times S^5$ space viewed as the near-horizon D3-brane background. In particular, writing the 10-d covariant derivative (including the Lorentz connection and 5-form terms) in the '4+6' coordinates in (10) one learns that when acting on the constrained spinor Θ_+ it becomes simply $D\Theta_+ = y^{1/2}d\theta_+$, $\theta_+ \equiv y^{-1/2}\Theta_+$.

As a result, $\mathcal{M}^2 D\Theta_+ = 0$ as found in Kallosh and Rahmfeld (1998) and the fermionic sector of the action reduces only to terms quadratic and quartic in θ_+^I. Using $\mathcal{P}_-^{IJ}\theta_+^J = 0$ to eliminate θ_+^2 in favour of

$$\theta_+^1 \equiv \vartheta \tag{15}$$

one finds that the κ-Symmetry gauge-fixed string action in $AdS_5 \times S^5$ background (2) expressed in terms of the bosonic coordinates $X^{\hat{m}} = (x^p, y^t)$ and the *single* $D = 10$ Majorana-Weyl spinor ϑ takes the following simple form [1]

$$S = -\frac{1}{2}\int d^2\sigma \left[\sqrt{-g}\, g^{ij}\ \ y^2(\partial_i x^p - 2i\bar{\vartheta}\Gamma^p\partial_i\vartheta)(\partial_j x^p - 2i\bar{\vartheta}\Gamma^p\partial_j\vartheta) \right.$$
$$\left. +\sqrt{-g}\, g^{ij}\frac{1}{y^2}\partial_i y^t \partial_j y^t \ + 4i\epsilon^{ij}\partial_i y^t \bar{\vartheta}\Gamma^t\partial_j\vartheta \right] \tag{16}$$

The $\Theta\Theta\partial X\partial X$ terms representing the coupling to the RR background which is shown to be present in Metsaev and Tseytlin (1998a, 1998b) in the original action (2) are now 'hidden' in the $\bar{\vartheta}\partial\vartheta\partial X$ terms because of the redefinition made in (14).

The same action but without y^2 and $1/y^2$ factors is found by fixing the κ-Symmetry gauge (11) in the flat-space type IIB Green-Schwarz action. This 'D3-brane' gauge breaks the $SO(1,9)$ Lorentz invariance of the action to $SO(1,3) \times SO(6)$, i.e. distinguishes between the 4 'parallel' and 6 'transverse' coordinates. In particular, only the latter ones (y^t) survive in the WZ term. This special property of the gauge (11) turns out to be crucial for the observation below that the fermionic terms in the action can be put in a much simpler *quadratic* form by making 2-d duality transformation of the 'parallel' coordinates x^p.

4 T-Dualized Action

Let us now perform the 2-d duality transformation of the four x^p coordinates as suggested in Kallosh and Tseytlin (1998). As usual, this is done by putting the action in the first-order form by introducing the 'momenta' (Lagrange multipliers) P_i^p

$$S = -\frac{1}{2}\int d^2\sigma \left[\sqrt{-g}\Big(g^{ij}\frac{1}{y^2}P_i^p P_j^p + 2P_i^p(\partial_j x^p - 2i\bar{\vartheta}\Gamma^p\partial_j\vartheta) + \frac{1}{y^2}\partial_i y^t\partial_j y^t\Big)\right.$$
$$\left. + \ 4i\epsilon^{ij}\partial_i y^t\bar{\vartheta}\Gamma^t\partial_j\vartheta\right] . \tag{17}$$

Integrating out x^p and solving the resulting constraint on P_i^p as

$$\sqrt{-g}g^{ij}P_j^p = \epsilon^{ij}\partial_j\tilde{x}^p \ , \tag{18}$$

we finish with the dual action

[1] The same action was obtained in Pesando (1998a, 1998b) using the Supersolvable algebra approach and a change of variables which brings the action to the simple form (16).

$$\tilde{S} = -\frac{1}{2}\int d^2\sigma \left[\sqrt{-g}g^{ij}\ \frac{1}{y^2}(\partial_i\tilde{x}^p\partial_j\tilde{x}^p + \partial_i y^t\partial_j y^t)\right.$$

$$\left. + 4i\epsilon^{ij}\bar{\vartheta}(\partial_i\tilde{x}^p\Gamma^p + \partial_i y^t\Gamma^t)\partial_j\vartheta\right] . \quad (19)$$

At the quantum level, the 2-d duality is accompanied (Buscher (1988)) by the dilaton term which should be added to the dual action (19) to preserve its conformal invariance,

$$\Delta\tilde{S} = \frac{1}{4\pi}\int d^2\sigma\sqrt{-g}R^{(2)}\phi(X) , \qquad \phi = \phi_0 - 4\ln y . \quad (20)$$

A remarkable property of the action (19) is not only that its fermionic part is quadratic in θ but also that it does not depend on 2-d metric, i.e. is given by a WZ type term. This WZ term is linear in the bosonic coordinates, i.e. has formally the same form as in flat target space. A somewhat surprising conclusion is that adding this fermionic term to the bosonic symmetric space $AdS_5 \times S^5$ sigma model action (with dilaton term (20) also included) should give a conformally invariant 2-d theory!

Let us note that if we would start with the '5+5' form of the action in which the WZ term has a more complicated structure depending on the S^5 spinor matrix $\Lambda(\xi)$ in (14) (which drops out of the 'kinetic' term in (16)), that would not affect the 2-d duality transformation step, and the resulting dual action will still be quadratic in fermions.

The duality has partially restored a 'symmetry' between 'parallel' and 'transverse' coordinates. Starting with the flat-space analogue of (16) and performing the same duality transformation one would get (19) without the $1/y^2$ factor, i.e. obtain indeed the $SO(1,9)$ invariant action for $\tilde{X}^{\hat{a}} = (\tilde{x}^p, y^t)$ and ϑ

$$\tilde{S}_{flat} = -\frac{1}{2}\int d^2\sigma \left(\sqrt{-g}g^{ij}\partial_i\tilde{X}^{\hat{a}}\partial_j\tilde{X}^{\hat{a}}\ +\ 4i\epsilon^{ij}\partial_i\tilde{X}^{\hat{a}}\ \bar{\vartheta}\Gamma^{\hat{a}}\partial_j\vartheta\right) . \quad (21)$$

This looks like the type I (or heterotic) flat-space string action but without the standard fermionic terms complementing ∂X in the 'kinetic' part of the action (as a result, the κ-symmetry is broken, as, of course, should be in the present type IIB gauge-fixed theory).We would arrive at exactly the same flat-space dual action (21) had we started with the flat-space IIB action, used the 'Dq-brane' combination $\Gamma_{0...q}$ (q =odd) instead of Γ_{0123} in (11) and dualized the 'parallel' coordinates x^p, $p = 0, ..., q$. However, that procedure would no longer generalize to curved $AdS_5 \times S^5$ space unless $q = 3$: the form of the $AdS_5 \times S^5$ background (10) prefers the 'D3-brane' gauge choice (11).

As was already mentioned above, the fact that 2-d duality simplified the structure of the fermionic terms is related to the key property of the gauge-fixed action (16) or its flat-space counterpart: only part of the bosonic coordinates ('transverse' ones) appear in the WZ term. For example, the standard type I superstring action which has similar form of a sum of a 'kinetic' (2-d metric dependent) and a WZ term, i.e. $\int (\partial X - \bar{\theta}\partial\theta)^2 + idX \wedge \bar{\theta}d\theta$, *preserves* its form under 2-d duality applied to any of the coordinates $X^{\hat{a}}$, as all of them enter both the first and the second term in that action.

The dual action (19) can be interpreted as describing the fundamental string propagating in the background representing the near-core region of the D-instanton smeared in the 4 directions $\tilde{x}^p$. This background is T-dual to the original D3-brane background and has the form

$$ds^2 = H^{1/2}(d\tilde{x}^p d\tilde{x}^p + dy^t dy^t) , \quad e^\phi = H , \quad C = H^{-1} , \quad H = \frac{\tilde{R}^4}{y^4} . \tag{22}$$

Note that this conformally flat $D = 10$ string-frame metric is actually equivalent to the $AdS_5 \times S^5$ metric (10): changing the coordinates y^t so that the radial coordinate gets inverted, $y = 1/y'$, we get (we set $\tilde{R} = 1$ as above)

$$ds^2 = y'^2 d\tilde{x}^p d\tilde{x}^p + \frac{1}{y'^2} dy'^t dy'^t , \qquad e^\phi = C^{-1} = y'^4 . \tag{23}$$

Thus, like the action (16), the dual action (19) can also be *directly* interpreted as describing a superstring propagating in $AdS_5 \times S^5$ space, now supplemented not by the 4-form background but by the dilaton and 0-form backgrounds.

Since the fermionic term in (19) does not depend on the 2-d metric, the semiclassically equivalent Nambu-type action obtained by solving for g_{ij} is thus *also* quadratic in θ:

$$\tilde{S} = -\frac{1}{2} \int d^2\sigma \Big[\sqrt{-\det \big[\frac{1}{y^2}(\partial_i \tilde{x}^p \partial_j \tilde{x}^p + \partial_i y^t \partial_j y^t)\big]} + 4i\epsilon^{ij} \bar{\vartheta} (\partial_i \tilde{x}^p \Gamma^p + \partial_i y^t \Gamma^t) \partial_j \vartheta \Big] . \tag{24}$$

A possible reparametrization gauge choice here may be the static gauge: $\tilde{x}^i = \sigma^i$, $i = 0, 1$, leading to the free 'kinetic' $\bar{\vartheta}\partial\vartheta$ term in the action ($\pi = 2, 3$)

$$\tilde{S} = -\frac{1}{2} \int d^2\sigma \Big[\frac{1}{y^2} \sqrt{-\det (\eta_{ij} + \partial_i \tilde{x}^\pi \partial_j \tilde{x}^\pi + \partial_i y^t \partial_j y^t)} + 4i\epsilon^{ij} \bar{\vartheta} \Gamma_i \partial_j \vartheta + 4i\epsilon^{ij} \bar{\vartheta} (\partial_i \tilde{x}^\pi \Gamma^\pi + \partial_i y^t \Gamma^t) \partial_j \vartheta \Big] . \tag{25}$$

The semiclassical expansion is then developed by starting with a particular solution for the string coordinates and integrating over small fluctuations near it.

Let us note that in fixing the static gauge one assumes that the ground state of the string is massive, i.e. this gauge is appropriate for a solitonic string or wound string state but cannot be used to describe a spectrum of a fundamental string in the zero-winding sector. One can find more details in

Kallosh and Tseytlin (1998) about the conditions on the background which make the gauge-fixing in the action (16) as well as in (19) acceptable.

To conclude, we have shown that choosing the 'D3-brane' or '4-d space-time' adapted κ-Symmetry gauge in the $AdS_5 \times S^5$ superstring action and duality-rotating the four isometric space-time coordinates one obtains an action in which the fermionic term is quadratic and does not depend on the world-sheet metric. The '4+6' Cartesian parametrization of the 10-d space thus led to a substantial simplification of the fermionic sector of the theory. This should hopefully allow one to make progress towards extracting more non-trivial information about the $AdS_5 \times S^5$ string theory and thus about its dual (Maldacena(1997)) – $N = 4$ super Yang-Mills theory.

Thus an essential progress in understanding string theory in a curved space came out from implementing the geometry concepts like Killing vectors and Killing spinors into the quantum theory of strings and by keeping supersymmetry manifest. This all is in a spirit of Victor Isaakovich vision about how to approach quantum gravity.

References

M.B. Green and J.H. Schwarz (1984) : Covariant description of superstrings. Phys. Lett. **B136**, 367.

M.T. Grisaru, P. Howe, L. Mezincescu, B. Nilsson and P.K. Townsend (1985) : N=2 superstrings in a supergravity background. Phys. Lett. **B162**, 116.

G.W. Gibbons and P.K. Townsend (1993) : Vacuum interpolation in supergravity via super p-branes, Phys. Rev. Lett. 71 (1993) 5223, hep-th/9307049.

R.R. Metsaev and A.A. Tseytlin (1998a) : Type IIB superstring action in $AdS_5 \times S^5$ background, hep-th/9805028; (1998b) Supersymmetric D3-brane action in $AdS_5 \times S^5$, hep-th/9806095

R. Kallosh, J. Rahmfeld and A. Rajaraman (1998) : Near horizon superspace, hep-th/9805217.

R. Kallosh and J. Rahmfeld: The GS string action on $AdS_5 \times S^5$, hep-th/9808038.

R. Kallosh (1998) : Superconformal actions in Killing gauge, hep-th/9807206.

I. Pesando (1998a) : A kappa gauge fixed type IIB superstring action on $AdS_5 \times S^5$, hep-th/9808020; (1998b) All roads lead to Rome: Supersolvable and Supercosets, hep-th/9808146.

R. Kallosh and A. Tseytlin (1998): Simplifying Superstring Action on $AdS_5 \times S^5$, hep-th/9808088.

T.H. Buscher (1988) : Path integral derivation of quantum duality in non-linear sigma models, Phys. Lett. B201 466.

J. Maldacena (1997) : The large N limit of superconformal field theories and supergravity, hep-th/9711200.

Making Manifest the Symmetry Enhancement for Coinciding BPS Branes

Sergei V. Ketov[1]

Institut für Theoretische Physik, Universität Hannover, Appelstraße 2, D-30167 Hannover, Germany

Abstract. We consider $g = N - 1$ *coinciding* M-5-branes on top of each other, in the multiple KK monopole background Q. The worldvolume of an M-5-brane is the local product of the four-dimensional spacetime $R^{1,3}$ and an elliptic curve. Taken together, these genus-one Riemann surfaces are supposed to give a single (Seiberg-Witten) hyperelliptic curve Σ_g in the coincidence limit, where the gauge symmetry is known to be enhanced to $SU(N)$. We make this gauge symmetry enhancement manifest by analyzing the corresponding *hypermultiplet* LEEA which is given by the N=2 non-linear sigma-model having Q as the target space. The hyper-Kähler manifold Q is known to be given by the multicentre Taub-NUT space, which in the coincidence limit amounts to the multiple Eguchi-Hanson (ALE) space $Q_{\rm mEH}$. The latter is most naturally described by using the hyper-Kähler coset construction in harmonic superspace, in terms of the auxiliary N=2 vector superfields as Lagrange multipliers in the presence of FI terms. The Maldacena limit, in which the LEEA is given by the N=4 SYM with the gauge group $SU(N)$, corresponds to sending all the FI terms to zero at large N, when all the auxiliary N=2 vector superfields become dynamical.

1 Brane Technology and KK Monopoles

The (Seiberg-Witten-type) exact solution to the N=2 super-QCD can be identified with the LEEA of the effective (called N=2 MQCD) field theory defined in the worldvolume of the *single* M-5-brane, locally given by the product of the uncompacftified four-dimensional spacetime $R^{1,3}$ and the hyperelliptic (Seiberg-Witten) curve Σ_g of genus $g = N_c - 1$: (Witten (1997)) (see, e.g. (Karch et al. (1998), Ketov (1997), Ketov (1998)) for a review or an introduction). The hyperelliptic curve Σ_g is supposed to be holomorphically embedded into the hyper-Kähler four-dimensional multiple Taub-NUT space $Q_{\rm mTN}$ associated with the multiple KK-monopole. The identification of the *low-energy effective actions* (LEEA) in the two apparently different field theories (the $N = 2$ super-QCD in the Coulomb branch on the one side, and the $N = 2$ MQCD defined in the M-5-brane world-volume, on the other side) is highly non-trivial, since the former is defined as the leading contribution to the *quantum* LEEA in quantum field theory, whereas the latter is determined by the *classical* M-5-brane dynamics or the $D = 11$ supergravity equations of motion, whose BPS solutions preserving some part of supersymmetry are the M-theory branes under consideration.

1.1 Multiple KK Monopole

The multiple KK monopole is a non-singular BPS solution to the eleven-dimensional supergravity equations of motion, given by the product of the seven-dimensional (flat) Minkowski spacetime $R^{1,6}$ and the four-dimensional Euclidean multicentre Taub-NUT space $Q_{\rm mTN}$ (Townsend (1995)):

$$\begin{aligned} ds^2_{[11]} &= dx^\mu dx^\nu \eta_{\mu\nu} + H(d\mathbf{y})^2 + H^{-1}(d\varrho + \mathbf{C}\cdot d\mathbf{y})^2 \ , \\ \nabla\times\mathbf{C} &= \nabla H \ , \qquad F_{(4)} \equiv dA_{(3)} = 0 \ , \end{aligned} \tag{1}$$

where $\mu = 0,1,2,3,7,8,9$, $\mathbf{y} = \{y_i\}$, $i = 4,5,6$, the eleventh coordinate ϱ is supposed to be periodic (with the period $2\pi k$ – this is just necessary to avoid conical singularities of the metric), while the harmonic function $H(\mathbf{y})$ is given by

$$H(\mathbf{y}) = 1 + \sum_{p=1}^{N} \frac{|k|}{2\,|\mathbf{y}-\mathbf{y}_p|} \ . \tag{2}$$

The moduli $(k,\mathbf{y}_i)$ can be interpreted as (equal) charges and the locations of the KK monopoles, respectively. The multiple Taub-NUT space can be thought of as a non-trivial bundle (Hopf fibration) with the base R^3 and the fiber S^1 of magnetic charge k. There exist N linearly independent normalizable self-dual harmonic 2-forms ω_p in $Q_{\rm mTN}$, which satisfy the orthogonality condition (see (Gibbons at al. (1988)), and references therein)

$$\frac{1}{(2\pi k)^2}\int_{Q_{\rm mTN}} \omega_p \wedge \omega_q = \delta_{pq} \ . \tag{3}$$

As is clear from eq. (1), two adjacent KK monopoles are connected by a homology 2-sphere having poles at the positions of the two monopoles. Near a singularity of H, the KK circle S^1 contracts to a point. A *holomorphic* embedding of the SW curve Σ_g into the hyper-Kähler manifold $Q_{\rm mTN}$ is the consequence of the *BPS condition* (Mikhailov (1997))

$$\text{Area}(\Sigma) = \left| \int_\Sigma \Omega \right| \ , \tag{4}$$

where Ω is the Kähler form in $Q_{\rm mTN}$.

1.2 $N = 2$ QCD LEEA in Coulomb Branch from Brane Dynamics

The geometrical origin and the physical interpretation of the hyperelliptic curve Σ_g, parameterizing the exact Seiberg-Witten solution to the LEEA of $N = 2$ supersymmetric QCD in the Coulomb branch, becomes transparent when using brane technology after M-theory resolution of UV singularities (Witten (1997)), where Σ_g appears to be the part of the M-5-brane worldvolume in eleven dimensions. The Nambu-Goto (NG) term (proportional to the

M-5-brane worldvolume) of the effective M-5-brane action in the low-energy approximation gives rise to the scalar *non-linear sigma model* (NLSM) having the special geometry after the KK compactification of the six-dimensional NG action on the Seiberg-Witten curve Σ_g down to four spacetime dimensions. This is enough to unambiguously restore the full $N = 2$ supersymmetric Seiberg-Witten LEEA, by the use of $N = 2$ supersymmetrization of the special bosonic NLSM, when considering its complex scalars as the leading components of abelian $N = 2$ vector multiplets in four spacetime dimensions and then deducing the Seiberg-Witten holomorphic potential out of the already derived special Kähler NLSM potential (see (Ketov (1998)) for a review).

Being applied to a derivation of the *hypermultiplet* LEEA of $N = 2$ super-QCD in the Coulomb branch, brane technology suggests to dimensionally reduce the effective action of a D-6-brane (to be described in M-theory by a KK-monopole) down to four spacetime dimensions (Ketov (1998)). In a static gauge for the D-6-brane, the indiced metric in the brane worldvolume is given by

$$g_{\mu\nu} = \eta_{\mu\nu} + G_{mn}\partial_\mu y^m \partial_\nu y^n \ , \tag{5}$$

where $\mu, \nu = 0, 1, 2, 3, 7, 8, 9$, $m, n = 4, 5, 6, 10$, and G_{mn} is the multicentre ETN metric. After expanding the NG-part of the D-6-brane effective action

$$S_{\rm NG} = \int d^7\xi \sqrt{-\det(g_{\mu\nu})} \tag{6}$$

up to the second-order in the spacetime derivatives, and performing plain dimensional reduction down to four dimensions, one arrives at the hyper-Kähler NLSM

$$S[y] = -\frac{1}{2}\int d^4x \, G_{mn}(y)\partial_{\underline{\mu}} y^m \partial^{\underline{\mu}} y^n \ , \qquad \underline{\mu} = 0, 1, 2, 3 \ , \tag{7}$$

whose $N = 2$ supersymmetrization yields the full hypermultiplet LEEA, in precise agreement with the $N = 2$ supersymmetric quantum field theory calculations in harmonic superspace (Ivanov et al. (1997)).

2 Symmetry Enhancement for 2 Coinciding $D-6$-Branes

As is well known, the isolated singularities of the harmonic function (2) are just the coordinate singularities of the *eleven*-dimensional metric (1), though they are truly singular with respect to the (dimensionally reduced) *ten*-dimensional metric to be associated with the D-6-branes in the type-IIA picture. The physical significance of these ten-dimensional singularities can therefore be understood due to the illegitimate neglect of the KK modes related to the compactification circle S^1 in ten dimensions, since the KK particles (also called D-0-branes) become massless near the D-6-brane core

(Townsend (1995)). Their inclusion is equivalent to accounting instanton corrections in the four-dimensional N=2 supersymmetric gauge field theory.

When some parallel and similarly oriented D-branes coincide, the symmetry enhancement happens (Kovner and Rosenstein (1990), Witten (1996)). Since the brane singularities become non-isolated in the coincidence limit, they first have to be resolved by considering the branes separated by some distance r and then taking the limit $r \to 0$. In the simplest non-trivial case of two D-6-branes, one considers the harmonic function

$$H(\mathbf{y}) = \lambda + \frac{1}{2}\left\{ \frac{1}{|\mathbf{y}-\xi\mathbf{e}|} + \frac{1}{|\mathbf{y}+\xi\mathbf{e}|} \right\} , \qquad r = 2\xi , \tag{8}$$

describing the double Taub-NUT metric in (1) with non-vanishing constant potential at infinity, and both centers on the line $\mathbf{e}$ in 6-th direction, $\mathbf{e}^2 = 1$. After being substituted into eq. (1), eq. (8) describes two parallel and similarly oriented KK-monopoles in M-theory, which dimensionally reduce to the double D-6-brane configuration in ten dimensions. The homology 2-sphere connecting the KK monopoles contracts to a point in this limit, which gives rise to a curvature singularity of the dimensionally reduced metric in ten dimensions.

The BPS states in M-theory, whose zero modes appear in the effective $D = 4$ field theory defined in the M-5-brane worldvolume, correspond to the minimal area M-2-branes ending on the M-5-brane. The topology of an M-2-brane determines the type of the corresponding $N = 2$ supermultiplet in $D = 4$: a cylinder leads to an N=2 vector multiplet, whereas a disc gives rise to a hypermultiplet (Mikhailov (1997)). The M-2-branes can wrap about the 2-sphere connecting the KK monopoles, while the energy of the wrapped M-2-brane is proportional to the area of the sphere (Hull and Townsend (1995)). When the sphere collapses, its area vanishes and, hence, an additional massless vector state appears due to the zero mode of the wrapped M-2-brane. One thus expects the gauge symmetry enhancement from $U(1) \times U(1)$ to $U(2)$ assiciated with the A_1-type singularity (Ooguri and Vafa (1996)). From the ten-dimensional perspective, the wrapped M-2-branes are just the (6-6) superstrings stretched between the D-6-branes, so that it is the massless zero modes of these 6-6 superstrings that become massless in the coincidence limit.

In order to make this symmetry enhancement manifest, let's start with the hypermultiplet low-energy effective action to be obtained by $N = 2$ supersymmetrization of the bosonic NLSM (7) in four spacetime dimensions, whose hyper-Kähler (double Taub-NUT) metric is determined by the harmonic function (8). In terms of this NLSM metric, the symmetry enhancement amounts to the appearance of gauged isometries in the NLSM target space, while the latter can be made manifest in the $N = 2$ harmonic superspace, as we are now going to demonstrate. First, let's note that the $N = 2$ supersymmetric double Taub-NUT NLSM is known to be equivalent to the one with the *mixed* (=Eguchi-Hanson-Taub-NUT) metric (Gibbons at al. (1988)). The mixed NLSM is described by the following $N = 2$ harmonic superspace action over the analytic subspace:

$$S_{\text{mixed}}[q, V] = \int_{\text{analytic}} \{q_A^{a+} D^{++} q_{aA}^{+} + V_{\text{L}}^{++}(\frac{1}{2}\varepsilon^{AB} q_A^{a+} q_{aB}^{+} + \xi^{++}) + \frac{1}{4}\lambda(q_A^{a+} q_{aA}^{+})^2\}, \tag{9}$$

which is written down in terms of a gauged $O(2)$ analytic hypermultiplet superfield q_A^+, $A = 1, 2$, and the auxiliary $O(2)$ vector gauge analytic superfield (Lagrange multiplier) V_{L}^{++} having no kinetic term. The parameters λ in eqs. (8) and (9) can be identified, whereas the parameter $\xi^{++} = \xi^{ij} u_i^+ u_j^+$ in eq. (9) is simply related to the constant ξ appearing in eq. (8) after choosing the coordinate frame in which $\xi^{12} = 2i\xi$ and $\xi^{11} = \xi^{22} = 0$. The hyper-Kähler NLSM metric, which is deduced out of the superspace action (9) after eliminating all the auxiliary fields in components, yields the double Taub-NUT metric, as can be verified by explicit calculation (Gibbons at al. (1988)). This is, in fact, ensured by the manifest $U(1)_A \times U(1)_{\text{PG}}$ symmetry of the superspace action (9), where the first $U(1)_A$ factor is the unbroken part of the $SU(2)_A$ automorphisms of the $N = 2$ supersymmetry algebra rotating two supercharges, whereas the second $U(1)_{\text{PG}}$ symmetry only acts on the pseudoreal indices $a = 1, 2$ in $q^{a+} = (\overset{*}{\bar{q}}{}^{+}, q^+)$ and implies an abelian isometry of the NLSM metric. Any four-dimensional hyper-Kähler metric having the $U(1)_{\text{PG}}$ isometry is known to be a multicentre Taub-NUT metric (see (Gibbons at al. (1988)) and references therein).

The mixed four-dimensional hyper-Kähler metric of the $N = 2$ supersymmetric NLSM (9) interpolates between the *Eguchi-Hanson* (EH) metric ($\lambda = 0$) and the Taub-NUT ($\xi = 0$), both having thr maximal isometry group $U(2)$. The action of the $U(2)$ isometry is linear in both limiting cases, while it is even holomorphic in the second case. In the harmonic superspace approach, this symmetry enhancement can be simply understood either as the restoration of the $SU(2)_A$ automorphism symmetry in the Taub-NUT limit, or as the restoration of the $SU(2)_{\text{PG}}$ symmetry in the Eguchi-Hanson limit.

3 Symmetry Enhancement for N Coinciding $D-6$-Branes

The $D = 11$ supergravity approximation to M-theory is only valid for well-separated KK monopoles. When the KK monopoles coincide, their low-energy dynamics can be approximated by weakly coupled superstrings propagating in the multi-Eguchi-Hanson (ALE) background (Sen (1997)). This background naturally originates from the multi-Taub-NUT space. Indeed, when all the D-6-branes coincide, they can be described in M-theory by sending all the moduli $\mathbf{y}_p$ in the harmonic function (2) to zero, so that the additive constant (asymptotic potential) 1 in eq. (2) can be ignored near the core of N D-6-branes on top of each other. The multi-Eguchi-Hanson space thus possesses A_{N-1} simple singularity which implies the enhanced non-abelian gauge symmetry $SU(N)$ in the effective supersymmetric field theory defined in the common worldvolume of the coinciding D-6-branes.

The effective gauge field theory is supposed to be defined in the limit where the gravity decouples. The $D = 11$ supergravity has a 3-form $A^{[11]}_{(3)}$ which is decomposed in the full spacetime given by the product of the D-6-brane worldvolume $R^{1,6}$ and the multi-Taub-NUT space $Q_{\rm mTN}$ as

$$A^{[11]}_{(3)} = \sum_{p=1}^{N} A^{[7]}_{p(1)} \wedge \omega^{[4]}_{p(2)} \ , \tag{10}$$

where the 2-forms ω_p in $Q_{\rm mTN}$ have been introduced in subsect. 1.1, whereas A_p are N massless vectors (1-forms) in $R^{1,6}$. In addition, there are $3N$ scalar fields associated with the translational zero modes (or moduli) $\mathbf{y}_p$. All these vectors and scalars are the bosonic components of N massless vector supermultiplets in $1+6$ dimensions, each having $8_{\rm B} + 8_{\rm F}$ on-shell components. The gauge group of the effective field theory (in the case of separated KK monopoles) in the Coulomb branch is therefore given by $U(1)^N$. Since the intersection matrix of 2-cycles in $Q_{\rm mTN}$ is just given by the Cartan matrix of A_{N-1}, the abelian gauge symmetry $U(1)^N$ is to be enhanced to $U(N)$ (the non-abelian Coulomb branch) in the coincidence limit (Sen (1997)). The area of the 2-cycles vanishes in this limit, so that the M-2-branes wrapped around these 2-cycles give rise to the additional massless vectors as their zero modes. In the type-IIA picture, the 6-6 strings stretched between separated D-6-branes do not contribute to the effective LEEA in the abelian Coulomb branch. However, since the zero modes of 6-6 strings become massless when the brane separation vanishes, they do contribute to the LEEA in the non-abelian Coulomb branch. After plain dimensional reduction from $R^{1,6}$ to $R^{1,3}$ the effective $N = 1$ super-Yang-Mills theory in $1+6$ dimensions yields the $N = 4$ super-Yang-Mills theory in $1+3$ dimensions, which has the same number of on-shell components, the same number of conserved supercharges and the same gauge group.

The hypermultiplet LEEA in the coincidence limit is described by the N=2 NLSM in the four-dimensional spacetime, with the NLSM target space being given by the ALE space $Q_{\rm mEH}$. A proper generalization of eq. (9) reads

$$S_{\rm mEH} = {\rm tr} \int_{\rm analytic} \left\{ \overset{*}{\bar{q}}{}^{+} \mathcal{D}^{++} q^{+} + V^{++} \xi^{++} \right\} \ , \quad \mathcal{D}^{++} = D^{++} + iV^{++} \ , \tag{11}$$

in terms of the N=2 vector multiplet V^{++} valued in the Lie algebra of $SU(N)$, the hypermultiplet q^{+} valued in the Lie algebra of $U(N)$, and the Fayet-

Iliopoluols terms $\xi^{++} = \xi^{(ij)}u_i^+u_j^+$ valued in the Cartan algebra of $SU(N)$. Note that the FI terms explicitly resolve the A_{N-1} singularity in this action.

If the $(1+6)$-dimensional $N = 1$ supersymmetric effective field theory were compactified on the circle S^1, this would yield the gauge field theory in $(1+5)$ dimensions, whose T-dual is an $(2,0)$ supersymmetric gauge field theory with N tensor multiplets. Therefore, in the type-IIB picture, we do not get the enhanced gauge symmetry but N tensor multiplets instead.

Yet another gauge symmetry enhancement pattern is known when N of the D-6-branes come on top of an *orientifold* six-plane, which leads to the $SO(2N)$ gauge symmetry (Ooguri and Vafa (1996)). In M-theory, the orientifold six-plane is to be represented by the *Atiyah-Hitchin* space (Atiyah and Hitchin (1988)) instead of a KK monopole (Sen (1997)). Indeed, far from the origin the Atiyah-Hitchin space has the topology $R^3 \times S^1/\mathcal{T}_4$, i.e. it looks like $Q_{\rm mTN}$ whose points are supposed to be identified under the action of the discrete symmetry $\mathcal{T}_4$ reversing signs of all four coordinates of $Q_{\rm mTN}$, which matches with the definition of the orientifold six-plane (Sen (1997)).

4 Large N Limit

To reproduce the Seiberg-Witten-type solution to $N = 2$ super-QCD from M-theory, merely a *single* and smooth M-5-brane in a KK-monopole background is needed. The M-5-brane worldvolume should just be compactified on the SW curve Σ_g down to $(1+3)$ dimensions i.e. to the worldvolume of a D-3-brane. Taking N M-5-branes (whose worldvolume is now locally given by the product $R^{1,3} \times \Sigma_1$, with Σ_1 being an elliptic curve) and allowing them to coincide in the background of multi-KK monopole yields (at a generic point in the moduli space) an $N = 2$ supersymmetric gauge field theory with the non-abelian gauge group $U(N)$ as the LEEA in the common (macroscopically $(1+3)$-dimensional) brane worldvolume. The KK monopoles are supposed to merge too, which also implies the non-abelian gauge group and extra massless supermultiplets in the LEEA. Indeed, the configuration of N parallel M-5-branes can support M-2-branes ending on different M-5-branes. When some or all of these M-5-branes coincide, the zero modes of the M-2-branes stretched between them give rise to the additional massless $N = 2$ multiplets in the effective field theory defined in the common worldvolume. The type of a supermultiplet depends upon the topology of M-2-brane: a cylinder yields an $N = 2$ vector multiplet, while a disc yields a hypermultiplet. The M-2-branes wrapped about S^1 and connecting different M-5-branes are strings in ten dimensions, which become *tensionless* in the M-5-brane coincidence limit.

In the coincidence limit for N KK monopoles, near the A_{N-1} singularity, all FI parameters ξ in the hypermultiplet LEEA (11) vanish, so that this action formally exhibits the genuine non-abelian gauge symmetry $SU(N)$. At large N, the auxiliary N=2 vector gauge multiplet V^{++} valued in the Lie algebra of $SU(N)$ becomes *propagating* via the well-known mechanism of the dynamical generation of massless vector bosons in non-compact NLSM (Polyakov (1987)). This N=2 vector multiplet and the hypermultiplet q^+, both in the adjoint of the gauge group, constitute an $N = 4$ super-Yang-Mills multiplet, whose action is a straightforward $N = 4$ extension of eq. (11) !

Our result is closely related to the recent conjecture of Maldacena (1997). He discussed 'simple' M-5-branes, having no Riemann surface in their world-volumes, in the particular limit (down the 'throat') given by the product $AdS_7 \times S^4$ whose both radii are proportional to $N^{1/3}$. At large N, the Maldacena LEEA is given by a $(2,0)$ superconformally invariant gauge field theory in six dimensions, which is supposed to be dual to M-theory compactified on $AdS_7 \times S_4$ (Maldacena 1997). Our results imply that the Maldacena limit can also be approached from the hypermultiplet part of the LEEA near the singularity, after taking into account the dynamical generation of an $SU(N)$ massless N=2 vector maultiplet at large N. Our approach can be easily generalized to the other simply-laced gauge groups.

References

Atiyah, M.F., Hitchin, N.J. (1988): *The Geometry and Dynamics of Magnetic Monopoles.* (Princeton University Press, Princeton) pp. 1–134

Gibbons, G.W., Olivier, D., Ruback, P.J., Valent, G., (1988): Multicentre metrics and harmonic superspace. Nucl. Phys. **B296**, 679–696

Hull, C.M., Townsend, P.K. (1995): Enhanced gauge symmetries in superstring theories. Nucl. Phys. **B451**, 525–546

Ivanov, E.A., Ketov, S.V., Zupnik, B.M. (1997): Induced hypermultiplet selfinteractions in N=2 gauge theories. Nucl. Phys. **B509**, 53–82

Karch, A., Lüst, D., Smith, D.J. (1998): Equivalence of geometric engineering and Hanany-Witten via fractional branes. Berlin preprint HUB–EP 98/22, 31 pages; hep-th/9803232

Ketov, S.V., (1997): Solitons, Monopoles and Duality. Fortschr. Phys. **45**, 237–292

Ketov, S.V., (1998): Analytic tools to brane technology in N=2 gauge theories with matter. DESY and Hannover preprint, DESY 98–059 and ITP–UH–12/98, 80 pages; hep-th/9806009

Klemm, A., Lerche, P., Mayr, P., Vafa, C., Warner, N. (1996): Self-dual strings and N=2 supersymmetric field theory. Nucl. Phys. **B477**, 746–766

Maldacena, J. (1997): The large N limit of superconformal field theories and supergravity. Harvard preprint HUTP–97/AO97, 21 pages; hep-th/9711200

Mikhailov, A., (1997): BPS states and minimal surfaces. Moscow and Princeton preprint, ITEP–TH–33/97 and PUTP–1714, 31 pages; hep-th/9708068

Polyakov, A.M. (1987): *Gauge Fields and Strings* (Harwood Academic Publishers, Chur, Switzerland) pp. 1–301

Ooguri, H., Vafa, C. (1996): Two-dimensional black holes and singularities of CY manifolds. Nucl. Phys. **B463**, 55–72

Sen, A. (1997): Dynamics of multiple KK monopoles in M- and string theory. Adv. Theor. Math. Phys. **1**, 115–126;

A note on enhanced gauge symmetries in M- and string theory. Electronically published in JHEP **9** 001

Townsend, P.K., (1995): The 11-dimensional supermembrane revisited. Phys. Lett. **350B**, 184–188

Witten, E., (1996): Bound states of strings and p-branes. Nucl. Phys. **B460**, 335–350

Witten, E., (1997): Solutions of four-dimensional field theory via M theory. Nucl. Phys. **B500**, 3–42.

On Some Stability Properties of Compactified $D = 11$ Supermembranes

I. Martin and A. Restuccia

Universidad Simon Bolivar, Departamento de Fisica
Caracas 89000, Venezuela.
e-mail:isbeliam@usb.ve, arestu@usb.ve

Abstract. We desribe the minimal configurations of the bosonic membrane potential, when the membrane wraps up in an irreducible way over $S^1 \times S^1$. The membrane 2-dimensional spatial world volume is taken as a Riemann Surface of genus g with an arbitrary metric over it. All the minimal solutions are obtained and described in terms of 1-forms over an associated $U(1)$ fiber bundle. It it shown that there are no infinite dimensional valleys at the minima.

1 Introduction

The Minkowski $D = 11$ Supermembrane, when the $SU(N)$, $N \to \infty$, regularization is used, was shown to have a continuous spectrum from zero to infinity (de Wit , Lüscher and Nicolai(1989)). The instability problem may be understood as a consequence of the existence of string like configuration with the same energy. The configurations which give the minimum of the potential consists of infinite dimensional functional subspaces. The potential as a functional may be described around those subspaces as having infinite dimensional valleys. This property appears already in the bosonic membrane, however because of the quantum zero point energy of the oscillators transversal to the valleys stability is attained. In the supermembrane, because of the property of the $SUSY$ harmonic oscillators to have no zero point energy the theory is unstable. Because of duality arguments and the relation between the supermembrane and its dual D-brane, we are really more interested in studying the Hamiltonian of the compactified supermembranes where at least one dimension in the target space is compactified to S^1 (Duff et. al. (1988)). The spectrum of the Hamiltonian of the compactified supermembrane has been recently studied by several authors without a conclusive result (Russo (1997))(de Wit, Peeters and Plefka (1997)).

In a recent paper (Martin, Restuccia ant Torrealba (1998))we showed that the Hamiltonian of the membrane wrapped up in an irreducible way over $S^1 \times S^1$ has no infinite dimensional valleys. Moreover we found all the minimal configurations, when the metric over Σ the 2-dimensional spatial world volume of the membrane was the canonical generalization of the induced metric over S^2 from R^3. We considered the general situation where Σ was a Riemann Surface of genus g. The construction was performed in terms of intrinsic harmonic coordinates. The minimal configuration was found to be unique up to closed forms over Σ. We expect that this properties of the bosonic Hamiltonian will also be valid for the complete supermembrane Hamiltonian. That is, that there are no infinite dimensional valleys at the minimal configurations of the supermembrane wrapped up in an irreducible way over $S^1 \times S^1$. We will discuss this problem in a forthcoming paper. In the present work we would like to give the general solution for the minimal configurations when other metrics, not necessarily the canonical one, is assumed over Σ.

2 Minimal Configurations of the Membrane Potential

We will analyze in this section minimal configurations of the Hamiltonian of the bosonic membrane. To each configuration of the membrane, determined by $dx(\sigma)$ we will associate a connection 1-form on a $U(1)$ bundle over $\sum$, a compact Riemann surface of genus g. We will then show that the minimal configurations correspond exactly to the magnetic monopoles over Riemann surfaces found in (Martin and Restuccia (1997))(Ferrari (1993)). Given a set of maps X_a, $a = 1, \ldots, 9$, from $\sum$ to the target space T we define

$$F_{ab} \equiv d(X_a dX_b), \ a, b = 1, \ldots, 9. \tag{2.1}$$

as the $U(1)$ curvatures associated to the connection 1-forms

$$A_{ab} \equiv X_{[a} dX_{b]} \tag{2.2}$$

The potential of the bosonic membrane may then be rewritten as

$$< V >_\sigma \equiv \int_\Sigma d^2\sigma \sqrt{g} V(\sigma) = \int_\Sigma {}^*F_{ab} F_{ab} \tag{2.3}$$

where ${}^*F_{ab}$ is the Hodge dual of the curvature 2-form F_{ab} and g is the determinant of the metric over the Riemann surface Σ. We normalize the metric by the condition $Vol\ \Sigma = 2\pi$.

It is clear from (2.3) that the infinite dimensional configuration

$$X_a(\sigma) = \lambda_a X(\sigma), \ a = 1, \ldots, 9. \tag{2.4}$$

where λ_a are arbitrary parameters, has zero potential.

In fact $F_{ab} = 0$ for all a and b. The configuration (2.4) is infinite dimensional since $X(\sigma)$ is an arbitrary map, with the only restriction to have a well defined potential (2.3). The space of functions over Σ satisfying that requirement is infinite dimensional. These are the valleys which give rise to the continuous spectrum from zero to infinite for the non-compactified supermembrane.

The existence of the valleys is, of course, not restricted to the minima, as emphasized in (de Wit, Peeters and Plefka (1997)).

We are interested in the analysis of the supermembrane in the case in which the target space T is compactified. Let $X(\sigma)$ denote a compactified coordinate over T. Let us assume X is a map from

$$\Sigma \to S^1 \tag{2.5}$$

It is then straightforward to see that dx satisfies the following conditions for a 1-form L

$$dL = 0 \tag{2.6}$$
$$\oint_{C_i} L = 2\pi n^i$$

where C_i denotes a basis of the integral homology of dimension one over Σ. It is interesting to note that the converse to (2.7) is also valid. That is, given a globally defined 1-form L over Σ satisfying (2.7) there exists a map X from

$$\Sigma \to S^1 \tag{2.7}$$

for which $L = dX$.

We will say the supermembrane wraps up in a non-trivial way when at least one of the n^i is different from zero.

We also say the supermembrane wraps up in an irreducible way over $S^1 \times S^1$ when

$$\int_\Sigma dX \wedge dY \neq 0 \tag{2.8}$$

where X and Y are two maps. That is, the irreducibility condition requires a compactification of at least two coordinates on the target T. If the membrane wraps up in an irreducible way over $S^1 \times S^1$ it does it in a nontrivial way over each of the S^1. Moreover

$$\frac{1}{2\pi} \int_\Sigma dX \wedge dY = 2\pi N \neq 0, \tag{2.9}$$

N being an integer.

(2.8) may be interpreted in terms of the $U(1)$ bundle associated to

the membrane configuration, using (2.1) and (2.2). It tell us that the corresponding Chern class is non-trivial. The integral number N in (2.9) determines the $U(1)$ bundle over which the connection (2.2) is defined. We notice that the existence of infinite dimensional valleys still persist when the target space T is of the form

$$T = M_{10} \times S^1. \tag{2.10}$$

In fact the coordinates mapping $\Sigma \to M_{10}$, 10-dimensional Minkowski space, are single valued over Σ.

We may always take the compactified coordinate, say X_1 to be

$$X_1 = \phi \tag{2.11}$$

where ϕ is the angular coordinate of S^1 in (2.10). An admissible configuration is then given by

$$X_2 = X_3 = \ldots = X_9 = \frac{dX(\phi)}{d\phi}$$

$$\phi = \phi(\sigma), \tag{2.12}$$

where $X(\phi)$ is a differentiable single valued function on ϕ, that is a regular periodic function, and $\phi(\sigma)$ is a map from Σ to S^1. It then follows that the curvature (2.1) is zero for all a and b. The subspace (2.12) is still infinite dimensional.

We look now for the stationary points of (2.3) over the space of maps defining supermembranes with irreducible wrapping over $S^1 \times S^1$. It is straightforward to see in this case that the minimal configurations occur when all but X, Y maps are zero. Associated to this space we may introduce an $U(1)$ principle bundle. We proceed by noting that

$$F = \frac{1}{2\pi} dX \wedge dY \tag{2.13}$$

is a closed 2-form globally defined over Σ satisfying (2.9). By Weil's Theorem (Weil (1957)), (Caicedo, Martin and Restuccia (1997), there exists a $U(1)$ principle bundle and a connection over it such that its pull back by sections over Σ are 1-form connections with curvatures given by (2.13).

The stationary points of the potential satisfy

$$\begin{aligned} \delta X dY \wedge d^* F &= 0 \\ \delta Y dX \wedge d^* F &= 0 \end{aligned} \tag{2.14}$$

which imply

$$d^* F = 0. \tag{2.15}$$

Now, since $*F$ is a 0-form we get

$$^*F = constant.$$

over Σ, and using (2.9) we finally obtain

$$^*F = \frac{2\pi N}{Vol \Sigma} = N. \tag{2.16}$$

We will now show that the configurations with X and Y satisfying (2.16) and all other X_a maps to zero are minima of the potential (2.3) within the space of configurations with irreducible winding (2.9).

For any connection on the $U(1)$ principle bundle over Σ determined by N the associated curvature 2-form satisfies

$$dF = 0 \tag{2.17}$$
$$\int_{\Sigma} F = 2\pi N$$

Let A_o be a connection 1-form satisfying (2.16), and A_1 any other connection 1-form on the sample principle bundle. Then, using (2.18) for $F(A_1)$, we obtain

$$\int_{\Sigma} {}^*F(A_1 - A_0)F(A_1 - A_0) = \int_{\Sigma} [{}^*F(A_1)F(A_1) -{}^* F(A_0)F(A_0)]. \tag{2.18}$$

The left hand member of (2.18) is greater or equal to zero, we then have

$$\int_{\Sigma} {}^*F(A_1)F(A_1) \geq \int_{\Sigma} {}^*F(A_0)F(A_0) \tag{2.19}$$

The equality in (2.19) is obtained when the left hand member of (2.18) is zero. This implies

$${}^*F(A_1 - A_0) ={}^* F(A_1) -{}^* F(A_0) = 0. \tag{2.20}$$

That is, the equality in (2.19) is obtained if and only if

$$A_1 = A_0 + d\Lambda \tag{2.21}$$

where $d\Lambda$ is a closed 1-form globally defined over Σ.

The space of regular closed 1-forms, modulo exact 1-forms, over a compact (closed) Riemann Surface is finite dimensional. The exact 1-forms correspond to gauge transformations on the $U(1)$ bundle. We will show in the Section 4 that they are generated by the area preserving transformations on the membrane maps. We will discuss there the non-existence of infinite dimensional valleys at the minima for the membranes wrapping up in an irreducible way onto $S^1 \times S^1$.

3 Minimal Connections: Magnetic Monopoles over Riemann Surfaces of genus g

We will show in this section how to construct all the minimal connections over S^2 and over all topologically non-trivial Riemann Surfaces. To do so we will construct one minimal connection for each N. All others are obtained from (2.21). The space of closed 1-forms modulo exact forms is the space of harmonic 1-forms over Σ. It has been extensively studied in the literature, so it is not necessary to discuss it there. Our problem reduces then to find one minimal connection for each N, That is for each principle bundle over Σ. We will describe now that construction.

The explicit expression of the monopole connections is obtained in terms of the abelian differential $d\tilde{\Phi}$ of the third kind over the compact Riemann surface Σ of genus g. $d\tilde{\Phi}$ is a meromorphic 1-form with poles of residue +1 and -1 at points a and b, with real normalization. $\tilde{\Phi}$ is the abelian integral, its real part $G(z, \bar{z}, a, b, t)$ is a harmonic univalent function over Σ with logarithmic behavior around a and b

$$\begin{aligned} &\ln\left(\frac{1}{|z+a|}\right) + \text{regular terms} , \\ &\ln|z-b| + \text{regular terms} , \end{aligned} \tag{3.1}$$

It is a conformally invariant geometrical object. z denotes the local coordinate over Σ and t the set of $3g-3$ parameters describing the moduli space of Riemann surfaces. We are considering maps from $\Sigma \mapsto S^1 \times S^1 \times M^7$ for a given Σ, so the parameters t are kept fixed. They show however that the construction of minimal connections is a conformally invariant one.

Let a_i, $i = 1, ..., m$ be m points over the compact Riemann surface. We associate to them integer weights α_i, $i = 1, ..., m$, satisfying

$$\sum_{i=1}^{m} \alpha_i = 0 \tag{3.2}$$

We define

$$\phi = \sum_{i=1}^{m} \alpha_i G(z, \bar{z}, a_i, b, t). \tag{3.3}$$

and have

$$\begin{aligned} &\phi \to -\infty \text{ at } a_i \text{ with negative weights} \\ &\phi \to +\infty \text{ at } a_i \text{ with positive weights.} \end{aligned} \tag{3.4}$$

α_i are integers in order to have univalent transition functions over the non-trivial fiber bundle that we consider.

We denote $\tilde{\Phi}$ the abelian integral with real part ϕ. Its imaginary part φ is also harmonic but multivalued over Σ,

$$\tilde{\Phi} = \phi + i\varphi. \tag{3.5}$$

Let us consider the curve C over Σ defined by

$$\phi = \textit{constant}.$$

It is a closed curve homologous to zero. It divides the Riemann surface into two regions U_+ and U_-, where U_+ contains all the points a_i with negative weights and U_- the ones with positive weights.

We define over U_+ and U_- the connection 1-forms

$$\begin{aligned} A_+ &= \frac{1}{2}(1 + \tanh(\phi))d\varphi \\ A_- &= \frac{1}{2}(-1 + \tanh(\phi))d\varphi \end{aligned} \tag{3.6}$$

respectively. A_+ is regular in U_+ and A_- in U_-. In the overlapping $U_+ \bigcap U_-$ we have

$$A_+ = A_- + d\varphi \tag{3.7}$$

$g = exp(i\varphi)$ defines the transition function on the overlapping $U_+ \bigcap U_-$, and because of the integer weights it is univalued over $U_+ \bigcap U_-$.

The base manifold Σ, the transition function g and the structure group $U(1)$ have a unique class of equivalent $U(1)$ principle bundles over Σ associated to them. (3.6) defines a 1-form connection over Σ with curvature

$$F = \frac{1}{2}\frac{1}{\cosh^2\phi} d\phi \wedge d\varphi \tag{3.8}$$

The $U(1)$ principle bundles are classified by the sum of the positive integer weights α_i

$$N = \sum_i \alpha_i^+ \ , \alpha_i^+ > 0. \tag{3.9}$$

which is the only integer determining the number of times φ wraps around $\mathcal{C}$. All the bundles with the same N are equivalent. (3.8) satisfies (2.17), moreover it also satisfies (2.16). In fact, since φ and ϕ are harmonic over Σ, the metric is

$$d^2s = \frac{1}{\cosh^2\phi}((d\varphi)^2 + (d\phi)^2), \tag{3.10}$$

and then (2.16) follows directly.

We have then found for each $U(1)$ principle bundle over Σ, a connection 1-form (3.6) with curvature 2-form (3.8) satisfying (2.16)for the metric (3.10) on the Riemann Surface. We have obtained several expressions (3.6),(3.8) and (3.10), since we are allowed to consider different harmonic coordinates ϕ and φ. In fact, for any set of a_i with total positive weight N, we may define coordinates ϕ and φ away from the points a_1 and b. By so doing we are only using different coordinates over Σ to describe the same connection over the $U(1)$ principle bundle. It is, as if in the expressions of the 1-form connection describing the Dirac monopole we use different coordinates (θ, φ) with different North and South poles to describe the magnetic field of the monopole. Since ${}^*F$ is scalar field over Σ then it is always equal to N.

4 The General Solution

We have thus obtained all the solutions satisfying ${}^*F = N$ for the metric (3.10). The question arises then, what happens when we consider, instead of (3.10), the metric

$$|\lambda(\phi,\varphi)|^2 \frac{1}{\cosh^2(\phi)}[(d\varphi)^2 + (d\phi)^2] \tag{4.1}$$

that is, an element of the conformal class of (3.10). The questions is relevant since (2.16) is not conformal invariant. It depends on the metric through the factor $(\sqrt{g})^{-1}$. We may then ask for the solutions of (2.16) when the new metric (4.1) is considered over Σ. We will answer the question starting with the case in which Σ is the sphere S^2, and giving afterwards the solutions for topologically non-trivial Riemann Surfaces.

We consider the Hopf fiber bundle over S_2. The three dimensional sphere S_3 may be defined by $z_0, z_1 \in C$, the complex numbers, satisfying

$$z_0\bar{z}_0 + z_1\bar{z}_1 = 1. \tag{4.2}$$

The group $U(1)$ acts on S_3 by

$$(z_0, z_1) \to (z_0 u, z_1 u) \tag{4.3}$$

where $u\bar{u} = 1$, $\bar{u}$ being the complex conjugate to $u \in C$. The projection $S_3 \to S_2$ is defined by the composition of

$$(z_0, z_1) \to \begin{cases} \frac{z_1}{z_0} & z_0 \neq 0 \\ \frac{z_0}{z_1} & z_1 \neq 0 \end{cases} \tag{4.4}$$

with the stereographic projection

$$C \to S_2$$

defined by

$$\rho = \frac{\sin(\theta)}{1 - \cos(\theta)} \tag{4.5}$$

where $z = \rho e^{i\phi} \in C$ and (θ, ϕ) are the coordinates S_2.

There is a natural connection over the Hopf fiber bundle which may be obtained from the line element of S_3,

$$ds^2 = 4\left(d\bar{z}_0 dz_0 + d\bar{z}_1 dz_1\right) \tag{4.6}$$

where z_0, z_1 satisfy (4.2). We will use spherical coordinates (χ, θ, ϕ) over S_3, defined in the following way

$$\begin{aligned} z_0 &= \exp\left[\frac{1}{2}i(\chi + \varphi)\right] C(\theta, \varphi) \\ z_1 &= \exp\left[\frac{1}{2}i(\chi - \varphi)\right] S(\theta, \varphi) \end{aligned} \tag{4.7}$$

where

$$C^2 + S^2 = 1. \tag{4.8}$$

We then obtain the line element of S_3 as

$$\frac{1}{4}ds^2 = (dc)^2 + (ds)^2 + \frac{1}{4}\left[(d\chi)^2 + (d\varphi)^2 + 2(c^2 - s^2)d\chi d\phi\right].$$

We denote

$$c^2 - s^2 \equiv g(u, \varphi)$$
$$u \equiv \cos(\theta)$$

We then get

$$ds^2 = (d\chi + g(u, \varphi)d\phi)^2 + \frac{(dg)^2}{1 - g^2} + (1 - g^2)(d\varphi)^2. \tag{4.9}$$

We notice that in the particular case

$$C(\theta, \varphi) = \cos(\frac{\theta}{2}) \tag{4.10}$$
$$S(\theta, \varphi = \sin(\frac{\theta}{2}))$$

(4.9) yields

$$ds^2 = (d\chi + \cos(\theta)d\varphi)^2 + (d\theta)^2 + (\sin(\theta)d\varphi)^2. \tag{4.11}$$

Coming back to the general case (4.9), the line element of S_3 decomposes into the line element of S_2

$$\frac{(dg)^2}{1 - g^2} + (1 - g^2)(d\varphi)^2 \tag{4.12}$$

and the tensorial square of the 1-form

$$\omega = d\chi + g(u, \varphi)d\varphi. \tag{4.13}$$

The above decomposition allows to determine a 1-form (4.13) over the fiber bundle S_3. Notice that from (4.3) and (4.7) the group acts on χ as follows

$$\chi \to \chi + \lambda \tag{4.14}$$

where $u = \exp(i\lambda)$. We are then interested in considering

$$\frac{1}{2}\omega$$

as a connection over the fiber bundle S_3.

To obtain the $U(1)$ connection 1-form over S_2, one may consider the local section

$$\tilde{z}_0 = \exp(i\varphi)C(\theta, \varphi) \tag{4.15}$$
$$\tilde{z}_1 = S(\theta, \varphi)$$

over S_2 with the point $\theta = 0$ removed, which we denote U_+. The $U(1)$ connection over U_+ is then

$$A_+ = \frac{1}{2}(1 + g(u,\varphi))d\varphi. \tag{4.16}$$

To give a covering of S_2 we define U_-, another local section, by

$$\begin{aligned} \hat{z}_0 &= C(\theta,\varphi) \\ \hat{z}_1 &= e^{-i\varphi}S(\theta,\varphi) \end{aligned} \tag{4.17}$$

over S_2 with the point $\theta = \pi$ removed. We have assumed in (4.15) and (4.17) that

$$\begin{aligned} C|_{\theta=\pi} &= 0 \\ S|_{\theta=0} &= 0. \end{aligned} \tag{4.18}$$

Over U_- the connection 1-form is then given by

$$A_- = \frac{1}{2}(g(u,\varphi) - 1)d\varphi. \tag{4.19}$$

In the overlapping region $U_+ \cap U_-$ we have

$$A_+ - A_- = d\varphi. \tag{4.20}$$

The curvature 2-form F is then given by

$$F = \frac{1}{2}\partial_\mu g(u,\varphi)\sin(\theta) d\varphi \wedge d\theta. \tag{4.21}$$

In the particular case (4.11), it reduces to

$$F = \frac{1}{2}\sin(\theta) d\varphi \wedge d\theta.$$

In order to obtain the general solution satisfying (2.16) we may proceed in two ways. We may extend the Hopf fibring

$$S_3 \rightarrow S_2$$

to

$$S_{2n+1} \rightarrow CP_n$$

as considered in (Trautman (1977)), and repeat the procedure. This approach yields the explicit expressions of the connection 1-form over CP_n and then over S_2. Otherwise we may consider

$$F = N\frac{1}{2}\partial_\mu g(u,\varphi)\sin(\theta) d\varphi \wedge d\theta \tag{4.22}$$

and check (2.17). Weil's theorem ensures the existence of the connection over a $U(1)$ fiber bundle over Σ, with a curvature 2-form given by (4.22). This second procedure, although more direct, does not provide the explicit expression of the connection 1-form as in (4.16) and (4.19).

If is straightforward to check that (4.22) satisfies (2.17). In fact

$$C^2 = \frac{1+g}{2} \qquad (4.23)$$
$$S^2 = \frac{1-g}{2}$$

and hence at $\theta = \pi$, g=-1 and at $\theta = 0$, $g = 1$. We may now evaluate ${}^*F$ for our general solution (4.22) and the metric (4.12).

If we use our normalization $Vol\Sigma = 2\pi$, we obtain

$$\sqrt{g} = \frac{1}{2}\partial_\mu g \sin(\theta),$$

and

$${}^*F = \frac{2}{\sqrt{g}} F_{\varphi\theta} = N \qquad (4.24)$$

as required for the minima of the membrane potential. The above construction may be extended to obtain all the minimal connection 1-form over topologically non-trivial Riemann surfaces. Using the global coordinates introduced in Sect.3, the connection 1-form over U_+ and U_- may be expressed as

$$A_+ = \frac{N}{2}\left(1 + g(u,\varphi)\right) d\varphi, \qquad (4.25)$$
$$A_- = \frac{N}{2}\left(-1 + g(u,\varphi)\right) d\varphi,$$

respectively, where $u = \tanh(\phi)$ and $g(u,\varphi)$ is a single valued function over Σ satisfying

$$u \to +1 \quad g \to +1 \qquad (4.26)$$
$$u \to -1 \quad g \to -1.$$

The curvature 2-form has then the form

$$F = N\frac{1}{2}\partial_\mu g(u,\varphi)\frac{1}{\cosh^2(\theta)} d\phi \wedge d\varphi. \qquad (4.27)$$

Its Hodge dual, over the metric (4.12), is consequently

$${}^*F = N. \qquad (4.28)$$

(4.25) gives then the general solution for the minimization problem of the membrane potential over topologically non-trivial Riemann Surface.

Having constructed all minimal configurations of the membrane potential, in terms of connection 1-forms over $U(1)$ fiber bundles, one has to determine the configurations maps in terms of them. If $\hat{A}$ is a minimal connection, then under an area preserving diffeomorphism

$$\delta\hat{A}_r = \partial_r(-\epsilon^{st}\partial_t\xi\hat{A}_s - \frac{1}{2}\xi^*\hat{F})$$

where ξ is the infinitesimal parameter of the transformation. It is then equivalent to a gauge transformation on the $U(1)$ fiber bundle. Using this property it was shown in (Martin, Restuccia ant Torrealba (1998)) that the space of all minimal connections may be generated by considering a particular minimal connection in terms of the minimal maps plus a representative of each real cohomology class of 1-form over Σ. The space of maps given rise to the particular minimal connection being finite dimensional.

References

B. de Wit, M. Lüscher and H. Nicolai(1989), Nucl. Phys.**B320** 135

M.J.Duff, T. Inami, C.N. Pope, E. Sezgin and K.S.Stelle (1988), Nucl. Phys. **B297** 515.

J.G. Russo (1997), Nucl. Phys. **B492**205.

B. de Wit, K. Peeters and J. Plefka (1997), hep-th/9705225.

I. Martin, A. Restuccia ant R. Torrealba (1998), Nucl. Phys. **B521** 117.

M. Caicedo, I. Martin and A. Restuccia (1997),hep-th/9701010; Proceedings of I SILAFAE, Yucatan, Mexico November 1996.

A. Weil (1957), *Varits Kaehlriennes*, Hermann .

I. Martin and A. Restuccia (1997), Lett. Math. Phys. **39** (4).

F. Ferrari (1993), hep-th/9310024.

A. Trautman (1977), Internat. J. Theoret. Phys. **16**,561

U-Duality and M-Theory Cosmology

Burt A. Ovrut

Abstract. A manifestly U–duality covariant approach to M–theory cosmology is discussed and applied to cosmologies in dimension $D = 5$.

1 Introduction

In this talk, we will discuss a manifestly U–duality covariant formulation of M–theory cosmology first presented in (Lukas and Ovrut (1997)). We are going to reduce 11–dimensional supergravity on a Ricci–flat manifold to D dimensions, thereby keeping the breathing mode of the internal space as the only modulus (Maeda (1986)) and focusing on the D–dimensional "external" part of U–duality. As an explicit example, we will study the case $D = 5$, corresponding to the U–duality group $G = SL(5)$. This example is motivated by the Horava–Witten construction of M–theory on S^1/Z_2 (Horava and Witten (1996) Witten (1996)) which represents the effective theory of strongly coupled heterotic string theory .

2 Compactification of 11-Dimensional Supergravity

The relevant bosonic part of the 11–dimensional supergravity Lagrangian reads

$$\mathcal{L} = \sqrt{-\hat{g}}\left[\hat{R} - \frac{1}{2\cdot 4!}\hat{F}_{MNPQ}\hat{F}^{MNPQ}\right] . \tag{1}$$

We are using the conventions of ref. (Cremmer, Julia and Scherk (1978), Cremmer and Julia (1978,1979), Cremmer (1981)). The 4–form field strength $\hat{F}_{MNPQ}$ is expressed in terms of the 3–form gauge field $\hat{A}_{NPQ}$ as $\hat{F}_{MNPQ} = 4\,\partial_{[M}\hat{A}_{NPQ]}$. We have omitted the Chern–Simons term in eq. (1) since this vanishes for the class of compactifications we will be interested in. Our main purpose is to investigate the relation of cosmological solutions and U–duality symmetries for the action (1). In our reduction to D dimensions we will keep a minimal moduli content only, that is, the breathing mode of the Ricci–flat manifold.

We are using indices $\mu, \nu ... = 0, ..., d \equiv D - 1$ for the external space–time, indices $m, n, ... = 1, ..., d$ for the external spatial directions and indices $a, b, ... = d+1, ..., 10$ for the δ–dimensional internal space, where $\delta = 11 - D$. Our Ansatz for the 11–dimensional fields is as follows

$$\begin{aligned}\hat{g}_{\mu\nu} &= g_{\mu\nu}(x^\rho)\\ \hat{g}_{\mu b} &= 0\\ \hat{g}_{ab} &= \bar{b}^2(x^\rho)\Omega_{ab}(x^c)\\ \hat{A}_{\mu\nu\rho} &= B_{\mu\nu\rho}(x^\sigma)\,.\end{aligned} \tag{2}$$

All other components of $\hat{A}_{NPQ}$ are set to zero. Here Ω_{ab} is the metric of a δ–dimensional Ricci–flat manifold and $\bar{b}$ is its breathing mode. Depending on the dimension, this manifold can be a Calabi–Yau space, a torus or even a product of both.

Using the truncation (2) the action (1) turns into

$$\mathcal{L} = \sqrt{-\bar{g}}\left[\bar{R} - k^2\,\bar{b}^{-2}(\partial_\mu \bar{b})^2 - \bar{b}^{\frac{6(11-D)}{D-2}}\frac{1}{2\cdot 4!}F_{\mu\nu\rho\sigma}F^{\mu\nu\rho\sigma}\right] \tag{3}$$

where $F_{\mu\nu\rho\sigma} = 4\,\partial_{[\mu}B_{\nu\rho\sigma]}$, $k^2 = \frac{D-1}{D-2}\delta^2 - \delta(\delta-1)$ and to get a canonical curvature term we have performed a Weyl rotation to the Einstein frame metric $\bar{g}_{\mu\nu}$. For a study of cosmological solutions of this Lagrangian we consider the Ansatz

$$\begin{aligned}\bar{g}_{\mu\nu} &= \begin{pmatrix} -\bar{N}^2(\tau) & 0\\ 0 & \bar{G}_{mn}(\tau)\end{pmatrix}\\ B_{mnr} &= B_{mnr}(\tau)\\ \bar{b} &= \bar{b}(\tau)\,.\end{aligned} \tag{4}$$

Here, time has been denoted by τ. The equations of motion with these specialized fields inserted can be derived from the 1–dimensional Lagrangian

$$\begin{aligned}\mathcal{L} = \bar{N}^{-1}\sqrt{\bar{\Phi}}\Big[&k^2\,\bar{b}^{-2}\dot{\bar{b}}^2 - \frac{1}{4}\bar{\Phi}^2\dot{\bar{\Phi}}^2 - \frac{1}{4}\dot{\bar{G}}_{mn}\dot{\bar{G}}^{mn}\\ &+\bar{b}^{\frac{6(11-d)}{D-2}}\bar{G}^{mm'}\bar{G}^{nn'}\bar{G}^{rr'}\dot{B}_{mnr}\dot{B}_{m'n'r'}\Big]\end{aligned} \tag{5}$$

where $\bar{\Phi} = \det(\bar{G})$. The dot denotes the derivative with respect to the time τ.

One should expect the U–duality group of $(11-d)$–dimensional maximal supergravity as a symmetry group of the Lagrangian (5). For example, for $D=5$ $(d=4)$ one expects $G = SL(5)$, the U–duality group of 7–dimensional supergravity. It is, however, hard to see directly from the Lagrangian in the form (5). One can perform nonlinear field redefinitions which express the physical Einstein frame fields in terms of the new fields G_{mn}, N, b and $\Phi = \det(G)$. Written in terms of these new variables the Lagrangian (5) finally reads

$$\begin{aligned}\mathcal{L} = N^{-1}\,b^\delta\Big[&-\delta(\delta-1)\,b^{-2}\dot{b}^2 + \frac{1}{4(10-D)}\Phi^{-2}\dot{\Phi}^2 - \frac{1}{4}\dot{G}_{mn}\dot{G}^{mn}\\ &+\frac{1}{2\cdot 3!}G^{mm'}G^{nn'}G^{rr'}\dot{B}_{mnr}\dot{B}_{m'n'r'}\Big]\,.\end{aligned} \tag{6}$$

3 U-Duality Covariant Formulation

For arbitrary D, all cases can be treated uniformly by considering an $SL(n)/SO(n)$ sigma model (where $n = 5$ for $D = 5$, for example) written in terms of the coset parameterization $M \in SL(n)/SO(n)$. The coset M can be characterized by the conditions $\det(M) = 1$ and $M = M^T$. We are therefore considering the Lagrangian

$$\mathcal{L}_1 = N^{-1}\, b^\delta \left[-\delta(\delta-1)\, b^{-2}\dot{b}^2 + \frac{1}{4}\mathrm{tr}\left(M^{-1}\dot{M}M^{-1}\dot{M}\right)\right] + \\ \lambda\,(\det(M) - 1) + \mathrm{tr}\left(\gamma(M - M^T)\right) \tag{7}$$

with the Lagrange multipliers λ, γ. The $SL(n)$ symmetry transformations are given by

$$b \to b\,,\ \lambda \to \lambda \\ M \to PMP^T\,,\ \gamma \to P^{T^{-1}}\gamma P^{-1}\,, \tag{8}$$

where $P \in SL(n)$. After eliminating the Lagrange multipliers, we find as the $SL(n)$ covariant equations of motion for M, b and N (in the gauge $N = 1$)

$$\frac{d}{d\tau}\left(M^{-1}\dot{M}\right) + \delta H M^{-1}\dot{M} = 0 \\ (\delta-1)\dot{H} + \frac{1}{2}\delta(\delta-1)H^2 = -\rho \\ \frac{1}{2}\delta(\delta-1)H^2 = \rho\,, \tag{9}$$

respectively. The Hubble constant H and the energy density ρ are defined by

$$H = \frac{\dot{b}}{b}\,, \qquad \rho = \frac{1}{8}\mathrm{tr}\left(M^{-1}\dot{M}M^{-1}\dot{M}\right)\,. \tag{10}$$

The second and third equation in (9) can be combined give

$$H = \frac{1}{\delta\tau}\,, \qquad b = b_0\,|\tau|^{1/\delta}\,, \tag{11}$$

where b_0 is an arbitrary constant. Inserting this into the first equation (9) we find for the coset M

$$M = M_0\, e^{I \ln|\tau|}\,, \tag{12}$$

where

$$\det(M_0) = 1\,, \quad \mathrm{tr}(I) = 0 \\ M_0 = M_0^T\,, \quad M_0 I = I^T M_0\,. \tag{13}$$

Furthermore, from the last equation (9) one obtains the zero energy constraint

$$\mathrm{tr}(I^2) = 4\frac{\delta-1}{\delta}\,. \tag{14}$$

Eqs. (11)–(14) represent the complete solution of the Lagrangian (7) written in a manifestly $SL(n)$ covariant form. The $SL(n)$ transformation (8) on the coset M acts as

$$\begin{aligned} M_0 &\to P M_0 P^T \\ I &\to P^{T^{-1}} I P^T\,. \end{aligned} \tag{15}$$

The matrices M_0, I satisfying the constraints (13) contain n^2+n-2 independent parameters. The zero energy condition (14) eliminates one of them so that n^2+n-3 remain. On the other hand, the group $SL(n)$ consists of n^2-1 parameters which implies that for $n>2$ not all solutions can be connected to each other by $SL(n)$ transformations. More precisely, the total n^2+n-3–dimensional solution space splits into n^2-1–dimensional equivalence classes, each consisting of solutions related to each other by $SL(n)$ transformations via eq. (15). The remaining $n-2$ integration constants label different equivalence classes. It is useful in the following to make this structure more explicit. Diagonalizing M_0 and I using eq. (15) with appropriate matrices P, eqs. (12), (13), (14) can be written in the form

$$M = P\,\mathrm{diag}(|\tau|^{p_1},\ldots,|\tau|^{p_n})\,P^T \tag{16}$$

with

$$\sum_{i=1}^{n} p_i = 0\,,\qquad \sum_{i=1}^{n} p_i^2 = 4\frac{\delta-1}{\delta} \tag{17}$$

and $P \in SL(n)$. The equivalence classes of $SL(n)$ unrelated solutions are parameterized by the $n-2$ constants $\{p_i\}$ subject to the constraints (17). On the other hand, a specific class, characterized by a fixed set $\{p_i\}$, is generated by the matrices $P \in SL(n)$ in eq. (16).

4 The Example $D = 5$

The U–duality group in the $D=5$ case is $G = SL(5)$. Let us define the vector $\mathbf{B} = (B^s)$ by $B_{mnr} = \frac{1}{\Phi}\epsilon_{mnrs}B^s$. Then the $SL(5)/SO(5)$ coset $\mathcal{M}$ can be parameterized by (Duff and Lu. (1990))

$$\mathcal{M} = \Phi^{-2/5}\begin{pmatrix} G & -G\mathbf{B} \\ -\mathbf{B}^T G & \Phi + \mathbf{B}^T G \mathbf{B} \end{pmatrix}, \tag{18}$$

where we have used a matrix notation $G = (G_{mn})$ for the metric. With the internal dimension $\delta = 6$, Lagrangian (6) can then be put into the form

$$\mathcal{L} = N^{-1} b^6 \left[-30\, b^{-2}\dot{b}^2 - \frac{1}{4}\mathrm{tr}\left(\dot{\mathcal{M}}\dot{\mathcal{M}}^{-1}\right)\right], \tag{19}$$

which has manifest $SL(5)$ invariance. The explicit transformations are given by

$$\begin{aligned} b &\to b \\ \mathcal{M} &\to P\mathcal{M}P^T \end{aligned} \tag{20}$$

with $P \in SL(5)$.

The equations of motion for the Lagrangian (19) are given by the general expressions (9), (10) with $\delta = 6$ and $M = \mathcal{M}$ inserted. Here $\mathcal{M}$ is the $SL(5)/SO(5)$ coset explicitly given in terms of the metric and the 3–form in eq. (18). The solution for the breathing mode can be read off from eq. (11)

$$H = \frac{1}{6\tau}\,, \qquad b = b_0\,|\tau|^{1/6}\,. \tag{21}$$

For the coset $\mathcal{M}$ we have from eq. (16) and (17)

$$\mathcal{M} = P\,\mathrm{diag}(|\tau|^{p_1}, ..., |\tau|^{p_5})P^T\,, \tag{22}$$

with

$$\sum_{i=1}^{5} p_i = 0\,, \quad \sum_{i=1}^{5} p_i^2 = \frac{10}{3} \tag{23}$$

and $P \in SL(5)$. We require $p_i \geq p_j$ for $i < j$. The solution (22) contains 27 integration constants. Three of them are given by the parameters $\{p_i\}$ subject to the constraints (23), labeling the $SL(5)$ equivalence classes. The remaining 24 integration constants parameterize the $SL(5)$ matrix P in eq. (22).

What is the general physical picture emerging from the solution (22)? One expects that in the asymptotic regions, the evolution rates are controlled by the $\{p_i\}$ and are therefore $SL(5)$ invariant. The $SL(5)$ parameters, however, determine the precise time range for the asymptotic regions and the time of transition.

We will concentrate on the physically interesting case of FRW universes in the following. For the metric and the form field this implies

$$G = \begin{pmatrix} c\mathbf{1}_3 & 0 \\ 0 & \phi \end{pmatrix}\,, \quad \mathbf{B} = \begin{pmatrix} 0_3 \\ B \end{pmatrix}\,, \tag{24}$$

where c, ϕ, B are time dependent scalars. Inserting (24) into the coset parameterization (18) for $\mathcal{M}$ results in

$$\mathcal{M} = \begin{pmatrix} \mathcal{M}_3 & 0 \\ 0 & \mathcal{M}_2 \end{pmatrix}\,, \quad \mathcal{M}_3 = c\Phi^{-2/5}\mathbf{1}_3\,, \quad \mathcal{M}_2 = \Phi^{-2/5}\begin{pmatrix} \phi & -\phi B \\ -\phi B & \Phi + B^2\phi \end{pmatrix}\,, \tag{25}$$

with $\Phi = c^3\phi$. Eq. (25) shows that FRW universes can be mapped into anisotropic solutions and vice versa using appropriate $SL(5)$ transformations. Since we wish to stay within the class of FRW solutions, we should restrict ourselves to the subgroup $H \equiv SL(3) \times SL(2) \times U(1) \subset SL(5)$ which leaves the structure of $\mathcal{M}$ in eq. (25) invariant. The FRW universes are specified by

$$p \equiv p_1 = p_2 = p_3 \tag{26}$$

in eq. (22). From eq. (23) we derive

$$p_{4,5} = -\frac{3p}{2} \pm \sqrt{\frac{5}{3} - \frac{15}{4}p^2}\,, \quad |p| \leq \frac{2}{3}\,. \tag{27}$$

We have therefore found a one parameter set (with parameter p) of H-inequivalent classes of FRW universes, each equivalence class for a fixed p spanned by the action of the group H. How does H act explicitly? First of all, $GL(3) \subset H$ is again part of the global coordinate transformations and therefore trivial. We concentrate on the $SL(2)$ part and write

$$P_2 = \begin{pmatrix} \alpha\, \beta \\ \gamma\, \delta \end{pmatrix}, \quad \alpha\delta - \beta\gamma = 1\,. \tag{28}$$

Then, $\mathcal{M}_2$, $\mathcal{M}_3$ take the form

$$\mathcal{M}_2 = \begin{pmatrix} \alpha^2|\tau|^{p_4} + \beta^2|\tau|^{p_5} & \alpha\gamma|\tau|^{p_4} + \beta\delta|\tau|^{p_5} \\ \alpha\gamma|\tau|^{p_4} + \beta\delta|\tau|^{p_5} & \gamma^2|\tau|^{p_4} + \delta^2|\tau|^{p_5} \end{pmatrix}, \quad \mathcal{M}_3 = |\tau|^p \mathbf{1}_3\,, \tag{29}$$

where $p_{4,5}$ are given by (27) in terms of the free parameter p. By comparison with eq. (25) we can read off the expressions for c, Φ, B, ϕ and convert them to the physical fields. We can then solve for the comoving time t in two asymptotic regions, leading to

$$\begin{aligned} t &= \frac{2b_0^2\,|\alpha|}{3p + p_4 + 8/3}|\tau|^{3p/2+p_4/2+4/3}\,\mathrm{sgn}(\tau)\,, \quad \text{for} \quad |\tau| \gg \tau_{\mathrm{form}} \\ t &= b_0^2\,|\beta|\frac{2}{3p + p_5 + 8/3}|\tau|^{3p/2+p_5/2+4/3}\,\mathrm{sgn}(\tau)\,, \quad \text{for} \quad |\tau| \ll \tau_{\mathrm{form}}\,, \end{aligned} \tag{30}$$

where

$$\tau_{\mathrm{form}} = \left(\frac{\beta}{\alpha}\right)^{\frac{2}{p_4 - p_5}}\,. \tag{31}$$

In these regions, we find for the Hubble parameter H_a

$$H_a = \frac{P(p)}{t}\,, \quad P(p) = \begin{cases} \frac{2}{3}\frac{3p+p_4+1}{3p+p_4+8/3} & \text{for } |\tau| \gg \tau_{\mathrm{form}} \\ \frac{2}{3}\frac{3p+p_5+1}{3p+p_5+8/3} & \text{for } |\tau| \ll \tau_{\mathrm{form}} \end{cases}\,. \tag{32}$$

The expansion coefficient $P(p)$ depends on the free parameter p and is generically different in the two asymptotic regions. We can show that it is always positive for large $|\tau| \gg \tau_{\mathrm{form}}$ and can have both signs for $|\tau| \ll \tau_{\mathrm{form}}$. For the positive branch $t > 0$ this implies a universe which is expanding or contracting at early *time and* is turned into an expanding universe at late time. The situation for the negative branch is reversed; the universe is always contracting at early time ($|t|$ large and $t < 0$) and can be contracting or expanding later. The Hubble parameter (32) and hence the aforementioned properties are $SL(2)$ invariant. The transition time given by

$$|\tau| \sim \tau_{\rm form} = \left(\frac{\beta}{\alpha}\right)^{\frac{2}{p_4 - p_5}}, \qquad (33)$$

on the other hand, depends on $SL(2)$ parameters along with the details of the transition.

Suppose now, we choose a solution in the positive branch which is contracting for a short period of time and then turns into expansion. By applying appropriate $SL(2)$ transformations to this solution the contraction period can be made arbitrarily long. An additional time reversal $t \to -t$ leads to a negative branch solution with an expansion period that can be arbitrarily long. The extreme limits are possible. By choosing $P_2 = \mathbf{1}$ ($\beta = 0$ in particular) we have a positive branch solution which is always expanding. As the other extreme we may set

$$P_2 = \begin{pmatrix} 0 & 1 \\ -1 & 0 \end{pmatrix}$$

($\alpha = 0$ in particular) which generates an expanding negative time branch solution. We have therefore shown that a combination of U–duality and time reversal can map expanding negative and positive time branch solutions into each other.

Acknowledgments Supported in part by DOE under contract No. DE-AC02-76-ER-03071.

References

A. Lukas, B. A. Ovrut. (1997): hep-th/9709030, to appear Phys. Lett. B.
K. Maeda. (1986): *Phys. Lett.* **B166** 59
P. Horava and E. Witten. (1996): *Nucl. Phys.* **B475** 94; *Nucl. Phys.* **B460** 506
E. Witten. (1996): *Nucl. Phys.* **B471** (1996) 135
E. Cremmer, B. Julia and J. Scherk. (1978): *Phys. Lett.* **B76** (1978) 409
E. Cremmer and B. Julia. (1978,1979): *Phys. Lett.* **B80** (1978) 48; *Nucl. Phys.* **B159** (1979) 141
E. Cremmer. (1981): LPTENS 81/18, Lectures given at ICTP Spring School Supergravity, Trieste 1981, published in Trieste Supergrav.School 1981:313
M. J. Duff and J. X. Lu. (1990): *Nucl. Phys.* **B347** (1990) 394

Brane Scattering and Supersymmetric σ-Model Geometries

K.S. Stelle

The Blackett Laboratory, Imperial College,
Prince Consort Road, London SW7 2BZ, UK
and
TH Division, CERN
CH-1211 Geneva 23, Switzerland

Abstract. The study of low-velocity scattering between components of a multi p-brane solitonic supergravity solution uncovers various possibilities of supersymmetric nonlinear σ models for the relative moduli. Specifically, a study of the scattering of two supersymmetric black holes in $D = 9$ spacetime leads to an unanticipated $d = 1$, $N = 8\text{b}$ supersymmetry, with geometry corresponding to an octonionic Kähler σ-model with torsion.

1 An Appreciation of Victor Isaakovich Ogievetsky

I met Victor Ogievetsky at a historic moment in the development of quantum gravity physics, namely the first Moscow Quantum Gravity Seminar, organized by Professor Moisei Alexandrovich Markov in 1978. At that moment, Russia and the West were opening up to each other, and the pioneering work of a number of our Soviet colleagues in the fields of quantum gravity, supersymmetry and supergravity was on its way to being fully, although belatedly, recognized. Victor and I immediately became good friends, and this friendship continued throughout the rest of his life. I am indebted to Victor for the many insights he gave me into the secrets of supersymmetric theories, and equally for providing me with an entrée into the intense world of physics in Russia. Victor's passing is a great scientific loss to many of us who have explored this subject, and to me, personally, it represents the loss of a great friend.

2 Intersecting Branes

In the following, we shall study the interaction between geometry and supersymmetry, a subject well frequented by Victor Ogievetsky in many of his pioneering works. This brief overview is derived from Gibbons, Papadopoulos and Stelle (1997). One of the remarkable features of the set of extended-object solutions to supergravity theories in dimensions $D \leq 11$ is that, for theories descending from the maximal $D = 11$ supergravity *via* dimensional reduction, all the static BPS p-brane solutions may be considered to be composed of various "intersecting" combinations (Papadopoulos and Townsend (1996); Tseytlin (1996); Gauntlett, Kastor and Traschen (1996)) of four basic "elemental" solutions to the $D = 11$ theory. These four elements are the pp wave and its "dual" Taub-NUT solution, together with the 2-brane (Duff and Stelle (1991)) and its dual 5-brane solution (Güven (1992)). As an example of an intersecting solution, one may consider a pp wave on a 2-brane,

$$\begin{aligned} ds_{11}^2 &= H_1^{\frac{1}{3}}(y)\left[H_1^{-1}(y)\{-dt^2 + d\rho^2 + d\sigma^2 + (H_2(y) - 1)(dt + d\rho)^2\} + \right. \\ &\qquad \left. ds^2(\mathbb{E}^8)\right] \\ A_{[3]} &= H_1^{-1}(y) dt \wedge d\rho \wedge d\sigma \qquad \nabla^2 H_1 = \nabla^2 H_2 = 0 \,, \end{aligned} \tag{1}$$

where the transverse-space harmonic function $H_1(y)$ is associated to the 2-brane while $H_2(y)$ is associated to the wave (y^m are coordinates on the 8-dimensional transverse space). Since waves are purely gravitational solutions, the harmonic function $H_2(y)$ does not appear in the 3-form gauge potential $A_{[3]}$.

Note that the solution (1) is independent of time and also of the two spacelike coordinates σ, ρ. This permits one to make a standard Kaluza-Klein dimensional reduction down to lower dimensions, and since this reduction removes simultaneously a translation-symmetric dimension of the "extended object" as well as a spacetime dimension, this falls into the class of "diagonal" dimensional reductions (Duff, Howe, Inami and Stelle (1987); Lü, Pope and Stelle (1996)). Making first a reduction from 11 to 10 spacetime dimensions on the coordinate σ, the solution (1) becomes a "wave-on-a-string." Analyzing the conditions for supersymmetry preservation in this spacetime background, one finds that the geometrical conditions required of the background are properly satisfied, while the supersymmetry parameter ϵ is required to satisfy simultaneously the projection conditions

$$\Gamma_{01\sigma}\epsilon = -\epsilon \tag{2a}$$

$$\Gamma_{01}\epsilon = -\epsilon \,, \tag{2b}$$

which may be imposed consistently because $[\Gamma_{01\sigma}, \Gamma_{01}] = 0$. Each of these two conditions cuts down the surviving supersymmetry by a factor of $\frac{1}{2}$, so that in the end the solution preserves $\frac{1}{4}$ of the maximum possible 32 rigid supersymmetry components (the latter would correspond, *e.g.*, to $D = 11$ flat space). As one can see from (2a), this supersymmetry is *chiral* on the $d = 1+1$ string worldsheet. In $d = 1 + 1$ dimensions, there are two available types of supersymmetry that correspond to $\frac{1}{4}$ supersymmetry preservation: (4,4) and (8,0). Only the latter is chiral, so this appears to be the type required to describe the $D = 10$ wave-on-a-string solution descending from (1).

After one more dimensional reduction, on the ρ coordinate this time, the solution now falls properly into the family of intersecting p-branes. There is actually not much left to intersect in this case, because we are left with a $D = 9$ solution comprising just two black holes, but the solution still falls into the family of intersecting p-brane solutions because the two elements still share a common time direction. Since the two harmonic functions H_1 and H_2 can be chosen independently, with unrelated charge centers, the solution now looks like a configuration of two BPS black holes, each supported by its own independent 1-form gauge potential. Since this solution exists for arbitrary choices of the integration constants determining the charge center locations, it is clear that this static configuration satisfies a zero-force condition, *i.e.* the resulting mutual potential energy must be flat. In the present case, this occurs as a result of a competition between gravitational and dilatonic forces, since the two components couple to different 1-form gauge potentials and so do not feel each other's electromagnetic forces.

3 Scattering Branes

Although the interaction between the two elements in the solution (1) vanishes for static configurations, this does not mean that the forces will cancel for time-dependent generalizations of (1). In particular, one can expect velocity-dependent forces to arise. The main question then is at what order in $(\text{velocity})^2$ do these forces arise. For the well-known case of parallel and similarly-oriented $D = 11$ membranes (Duff and Stelle (1991)), these forces continue to vanish at order $(\text{velocity})^2$, and thus can occur for the first time only at order $(\text{velocity})^4$. The reason for this continued cancellation of forces may be understood by considering the relative motion problem for the two membranes to be one of finding the corresponding nonlinear σ-model for the relative moduli. In other words, one promotes the relative-position integration constants of the solution to scalar fields on the worldvolume and then works out the lowest-order effective action for these modulus fields. This is necessarily some kind of nonlinear σ-model. Now, for the two-membrane scattering problem, the fraction of residual supersymmetry is $\frac{1}{2}$, so in $d = 2+1$ this corresponds to $N = 8$ supersymmetry, with 16 independent preserved supersymmetry components. This is too much, however, to permit a non-trivial σ-model to exist: the only option for the corresponding σ-model metric is a flat space, yielding free modulus fields, *i.e.* vanishing $(\text{velocity})^2$ forces. Consequently, the lowest order at which velocity-dependent forces can occur in this case is $(\text{velocity})^4$ (odd powers being ruled out by time-reversal invariance).

The case of wave-membrane scattering generalizing the solution (1), or, equivalently, after dimensional reduction, the scattering of two $D = 9$ black holes, looks at first blush like it should go similarly to that of membrane-membrane scattering in $D = 11$, yielding a flat sigma model metric. This expectation arises because the corresponding $d = 1+1$ (8,0) supersymmetry is still very strong, even though it has only half as many surviving components as in the two-membrane scattering problem. However, this $d = 1+1$ p-brane *worldvolume* analysis is still not completely appropriate, because in the corresponding $D = 10$ theory one is still dealing with a wave-string scattering problem, and the analysis of gravitational wave scattering is more involved than that of first-quantized brane-brane scattering.

Let us now examine more directly the scattering problem corresponding to (1) by the standard approximation technique of considering a brane-probe source in a fixed background. We shall consider that one of the two elements in (1) is heavy and situated at the origin. Since it is taken to be heavy, the motion of the other one in its field will not cause it to move from the origin. The motion of the other, "light," brane may be discussed by considering a brane-source action coupled to the background fields of the heavy brane. The brane-source action should properly be added to the supergravity action in order to provide δ-function sources on the right-hand sides of the supergravity equations; these sharply defined sources should then coincide with the singularities at the charge centers occurring in the harmonic functions in (1). The general bosonic p-brane action that provides the supergravity-equation sources is

$$I_{\rm probe} = -T_\alpha \int d^{p+1}\xi \,(-\det(\partial_\mu x^m \partial_\nu x^n g_{mn}(x))^{\frac{1}{2}} e^{\frac{1}{2}\varsigma^{\rm pr}\mathbf{a}_\alpha\cdot\boldsymbol{\phi}} + Q_\alpha \int \tilde{A}^\alpha_{[p+1]} \tag{3a}$$

$$\tilde{A}^\alpha_{[p+1]} = (p+1)^{-1}\partial_{\mu_1}x^{m_1}\cdots\partial_{\mu_{p+1}}x^{m_{p+1}}A^\alpha_{m_1\cdots m_{p+1}}d\xi^{\mu_1}\wedge\cdots\wedge d\xi^{\mu_{p+1}} \tag{3b}$$

The dilaton coupling in (3a) occurs because one needs to have the correct source for the Einstein frame, *i.e.* the conformal frame in which the Einstein-Hilbert action is free from dilatonic scalar factors. Requiring that the source match correctly to the known p-brane solution demands the presence of the dilaton coupling $e^{\frac{1}{2}\varsigma^{\rm pr}\mathbf{a}_\alpha\cdot\boldsymbol{\phi}}$, where $\varsigma^{\rm pr} = \pm 1$ according to whether the p-brane probe is of electric or magnetic type, and where $\mathbf{a}_\alpha$ is the dilaton vector controlling the dilaton coupling in the kinetic term for the gauge potential $A^\alpha_{[p+1]}$.

Now apply the brane-probe discussion to the $D = 9$ case at hand, which is a standard black hole – black hole scattering problem. One firstly notes that, owing to the fact that the light probe and the heavy background brane couple to different 1-form gauge potentials $A^\alpha_{[1]}$, any potential energy must come purely from the kinetic term in (3). Choosing the "static gauge," which for general p-branes is $\xi^\mu = x^\mu$ and in the present case is just $\xi^0 = t$, one may expand the determinant of the induced metric in (3). In general, for a background metric $ds^2 = e^{2A(y)}dx^\mu dx^\nu\eta_{\mu\nu} + e^{2B(y)}dy^m dy^m$, one thus obtains $\det(e^{2A(y)}\eta_{\mu\nu} + e^{2B(y)}\partial_\mu y^m\partial_\nu y^m)$. Then, expanding the determinant and the square root in (3), one obtains the potential energy term from the ∂y-independent part of the resulting expansion. In the present case, this gives $V_{\rm probe} = e^A e^{-\frac{3}{2\sqrt{7}}\phi}$, but this potential is a constant here because the heavy-brane background satisfies $A = \frac{3}{2\sqrt{7}}\phi$. Thus, we confirm the expected static zero-force condition for (1).

Continuing on to the (velocity)^2 order, one now obtains a non-trivial nonlinear sigma model with a modulus metric

$$\gamma^{mn} = H_{\text{back}}(y)\delta^{mn} , \tag{4}$$

where H_{back} is the harmonic function that determines the heavy brane's background field configuration. For the case of two black holes in $D = 9$, the harmonic function H_{back} has the structure $(1 + 1/r^6)$. But in the face of this patently non-flat modulus metric, one must ask what has gone wrong with the "folk-theorem" expectation that (8,0) σ-models must be flat.

The above test-brane analysis is confirmed by a more detailed study of the low-velocity scattering of supersymmetric black holes given by Shiraishi (1993). The procedure here is a standard one in soliton physics: one promotes the moduli of a static solution to time-dependent functions and then substitutes the resulting generalized field configuration back into the original field equations. This leads to a set of differential equations on the modulus variables which may then be viewed as the effective equations for the moduli. In the general case of multiple black hole scattering, the resulting system of differential equations is generally quite complicated. This system, however, simplifies dramatically in cases corresponding to the scattering of supersymmetric black holes, including the pair of $D = 9$ black holes in the present case, where the result involves only 2-body forces. These two-body forces may be derived from an effective action involving the position vectors of the two black holes. Separating the center-of-mass motion from the relative motion, one obtains the same modulus metric (4) as that found in the brane-probe analysis, aside from a rescaling that incorporates a replacement of the brane-probe mass by the reduced mass of the two-black-hole system.

4 Supersymmetric σ-Models in $d = 1$ Dimension

Now we should resolve the puzzle of how this non-trivial scattering modulus σ-model turns out to be consistent with supersymmetry. The modulus variables of the two-black-hole system are fields in one dimension, *i.e.* time. The N-extended supersymmetry algebra in $d = 1$ is

$$\{Q^I, Q^J\} = 2\delta^{IJ}\hat{H} \qquad I = 1, \ldots, N , \tag{5}$$

where $\hat{H}$ is the Hamiltonian. A $d = 1$, $N = 1$ σ-model is specified by a triple $(\mathcal{M}, \gamma, C_{[3]})$, where $\mathcal{M}$ is the Riemannian σ-model manifold, γ is the metric on $\mathcal{M}$ and $C_{[3]}$ is a 3-form on $\mathcal{M}$ playing the rôle of torsion in the derivative operator acting on fermions, $\nabla_t^{(+)} = \partial_t x^i \nabla_i^{(+)}$, where $\nabla_i^{(+)}\lambda^j = \nabla_i\lambda^j + \frac{1}{2}C^j{}_{ik}\lambda^k$. The σ-model action may be written using $N = 1$ superfields $x^i(t,\theta)$ (where $x^i(t) = x^i\big|_{\theta=0}$, $\lambda^i(t) = Dx^i\big|_{\theta=0}$) as

$$I = -\tfrac{1}{2}\int dtd\theta(\mathrm{i}\gamma_{ij}Dx^i\frac{d}{dt}x^j + \frac{1}{3!}C_{ijk}Dx^iDx^jDx^k) . \tag{6}$$

One may additionally have a set of spinorial $N = 1$ superfields ψ^a, with Lagrangian $-\frac{1}{2}h_{ab}\psi^a\nabla_t\psi^b$, where h_{ab} is a fibre metric and ∇_t is constructed using an appropriate connection for the fibre corresponding to the ψ^a (Coles and Papadopoulos (1990)). However, in the present case we shall not include this extra superfield. In order to have extended supersymmetry in (6), one starts by positing a second set of supersymmetry transformations of the form $\delta x^i = \eta I^i{}_j Dx^j$, and then requires these transformations to close and form the $N = 2$ algebra (5); finally, one also requires that the action (6) be invariant. In this way, one obtains the equations

$$I^2 = -\mathbf{1} \tag{7a}$$

$$N^i_{jk} \equiv I^i_{[j,k]} = 0 \tag{7b}$$

$$\gamma_{kl}I^k{}_iI^l{}_j = \gamma_{ij} \tag{7c}$$

$$\nabla^{(+)}_{(i}I^k{}_{j)} = 0 \tag{7d}$$

$$\partial_{[i}(I^m{}_jC_{|m|kl]}) - 2I^m{}_{[i}\partial_{[m}C_{jkl]]}) = 0\ , \tag{7e}$$

where (7a,b) follow from requiring the closure of the algebra (5) and (7c-e) follow from requiring invariance of the action (6). Conditions (7a,b) imply that $\mathcal{M}$ is a complex manifold, with $I^i{}_j$ as its complex structure.

What may seem surprising about the conditions (7) is that their structure is more complicated than might have been expected. Experience with $d = 1+1$ extended supersymmetry might lead one to expect, by dimensional reduction, the condition $\nabla^{(+)}_i I^j{}_k = 0$ (Coles and Papadopoulos (1990)). Certainly, solutions of this condition also satisfy (7c–e), but the converse is not true. Thus, the $d = 1$ extended supersymmetry conditions are "weaker" than those expected by dimensional reduction from $d = 1 + 1$, even though the $d = 1+1$ minimal spinors are, as in $d = 1$, just real single-component objects. Thus, the $d = 1 + 1$ theory implies a "stronger" condition; the difference is explained by $d = 1+1$ Lorentz invariance: not all $d = 1$ theories can be "oxidized" up to Lorentz-invariant $d = 1 + 1$ theories. Note also that the $d = 1$ "torsion" $C_{[3]}$ is not required to be closed in (7). $d = 1$ supersymmetric theories satisfying (7) are *analogous* to the (2,0) chiral supersymmetric theories in $d = 1 + 1$, but the weaker conditions (7) warrant a different notation for this wider class of models; one may call them 2b supersymmetric σ-models (Gibbons, Papadopoulos and Stelle (1997)). Such models are characterized by a Kähler geometry with torsion.

Continuing on to $N = 8$ supersymmetry, one finds an 8b generalization of the conditions (7), with 7 independent complex structures built using the octonionic structure constants $\varphi_{ab}{}^c$: $\delta x^i = \eta^a I_a{}^i{}_j Dx^j$, $a = 1, \ldots 7$, with $(I_a)^0{}_b = \delta_{ab}$, $(I_a)^b{}_0 = -\delta^b{}_a$, $(I_a)^b{}_c = \varphi_a{}^b{}_c$, where the octonion multiplication rule is $e_ae_b = -\delta_{ab} + \varphi_{ab}{}^ce_c$. Models satisfying such conditions have an "octonionic Kähler geometry with torsion," and are called OKT models (Gibbons, Papadopoulos and Stelle (1997)).

Now, are there any non-trivial solutions to these conditions? Evidently, from the brane-probe and Shiraishi analyses, there must be. For our $D = 9$ black holes with a $D = 8$ transverse space, one may start from the ansatz $ds^2 = H(y)ds^2(\mathbb{E}^8)$, $C_{\mu\nu\rho} = \Omega_{\mu\nu\rho}{}^{\lambda}\partial_{\lambda}H$, where Ω is a 4-form on $\mathbb{E}^8$. Then, from the 8b generalization of condition (7d) one learns $\Omega_{0abc} = \varphi_{abc}$ and $\Omega_{abcd} = -{}^*\varphi_{abcd}$; from the 8b generalization of condition (7e) one learns $\delta^{\mu\nu}\partial_{\mu}\partial_{\nu}H = 0$. Thus we recover the familiar dependence of p-brane solutions on transverse-space harmonic functions, and so one reobtains the brane-probe or Shiraishi structure of the black-hole modulus scattering metric with

$$H_{\text{relative}} = 1 + \frac{1}{|y_1 - y_2|^6} \; . \tag{8}$$

This story of the relation between supersymmetry and brane scattering is offered in the memory of my friend, Victor Isaakovich Ogievetsky, in the hope that he might have been amused by it. The interplay between gravity, supersymmetry and nonlinear σ-models was one of his best-loved subjects. As a friend, I will miss him. As a physicist, I will miss his guidance.

References

Gibbons, G.W., Papadopoulos, G., Stelle, K.S. (1997): HKT and OKT geometries on soliton black hole moduli spaces. Nucl. Phys. **B508**, 623; hep-th/9706207.

Papadopoulos, G., Townsend, P.K. (1996): Intersecting *M*-branes. Phys. Lett. **B380**, 273; hep-th/9603087.

Tseytlin, A.A. (1996): Harmonic superpositions of *M*-branes. Nucl. Phys. **B475**, 149; hep-th/9604035.

Gauntlett, J.P., Kastor, D.A., Traschen, J. (1996): Overlapping branes in *M* theory. Nucl. Phys. **B478**, 544; hep-th/9604179.

Duff, M.J., Stelle, K.S. (1991): Multi-membrane solutions of $D = 11$ supergravity. Phys. Lett. **B253**, 113.

Güven, R. (1992): Black *p*-brane solitons of $D = 11$ supergravity theory. Phys. Lett. **B276**, 49.

Duff, M.J., Howe, P.S., Inami, T., Stelle, K.S. (1987): Superstrings in D=10 from Supermembranes in D=11. Phys. Lett. **B191**, 70.

Lü, H., Pope, C.N., Stelle, K.S. (1996): Vertical versus diagonal dimensional reduction for *p*-branes. Nucl. Phys. **B481**, 313; hep-th/9605082.

Shiraishi, K. (1993): Nucl. Phys. **B402**, 399.

Coles, R., Papadopoulos, G. (1990): The geometry of one-dimensional supersymmetric non-linear sigma models. Class. Quantum Grav. **7**, 427.

Graviton Scattering in Eleven-Dimensional Supergravity

Arkady A. Tseytlin

[1] Imperial College, London SW7 2BZ, U.K.
[2] Lebedev Physics Institute, Moscow, Russia

Abstract. We find explicit expression for the one-loop four-graviton amplitude in eleven-dimensional supergravity compactified on a circle. Represented in terms of the string coupling (proportional to the compactification radius) it takes the form of an infinite sum of perturbative string loop corrections. We discuss the structure of quantum corrections in eleven-dimensional theory and their relation to string theory.

1 Introduction

Recent suggestions indicate that $D = 11$ supergravity is a low-energy effective field theory of a more fundamental M-theory. One expects that various properties of ten-dimensional string theories may be understood from eleven-dimensional perspective. Most of known relations between type IIA string theory and M-theory, viewed as its strong-coupling limit, are restricted to BPS states. A surprising observation [1] is that the *tree-level* type II string correction $\zeta(3)\alpha'^3\mathcal{R}^4$ can be interpreted as originating from a *one-loop* $D = 11$ supergravity contribution. Below we review the computation [2] of the one-loop four-graviton amplitude in $D = 11$ supergravity compactified on a circle. We shall demonstrate that it has the structure of an infinite sum of perturbative higher-loop string corrections. This suggests that the one-loop quantum $D = 11$ theory (with a proper UV cutoff implied by string theory) may contain information about certain string corrections to all orders in string coupling.

The reason why the $D = 11$ amplitude has this form may be understood as follows. The one-loop contribution to the effective action of $D = 11$ supergravity compactified on a circle of radius R_{11} can be represented as the one-loop correction in type IIA $D = 10$ supergravity plus an infinite sum of one-loop contributions of massive Kaluza-Klein modes (0-brane supermultiplets). That sum may be represented as a local series using inverse mass expansion, $\sum M^{-2n}C_n$. Since the masses of Kaluza-Klein modes are proportional to inverse string coupling $M \sim R_{11}^{-1} \sim g_s^{-1}$, the contribution of Kaluza-Klein modes has the structure of a sum of perturbative higher-loop closed string corrections, $\sum_n g_s^{2n}C'_n$. This suggests that some perturbative string theory results may be reproduced in the 'dual' formulation of the theory, in which certain solitons (0-branes) play a central role.

The scattering amplitude computed below corresponds to external gravitons with vanishing values of the 11-th component of momentum p_{11}. Using $D = 11$ Lorentz invariance it is, in principle, straightforward to generalise the final expression for the amplitude to the case when external momenta are arbitrary, subject only to the zero-mass on-shell condition in $D = 11$. The resulting amplitude with p_{11} = fixed may then be interpreted as a one-loop correction to the scattering of 0-branes in $D = 10$ and may be of interest from the point of view of testing Matrix theory. In particular, one should be able to analise the one-loop $D = 11$ supergravity contribution to the phase shift, which was previously obtained only in a semiclassical (eikonal) approximation (see [3] and refs. there).

We shall first make some general remarks on cutoff dependence of the $D = 11$ supergravity effective action, suggesting that certain curvature invariants should play a special role in both $D = 11$ and $D = 10$ theories. Then we shall present the result for the one-loop four-graviton amplitude in $D = 11$ supergravity on a circle. Finally, we shall discuss possible relation of these amplitudes to perturbative and non-perturbative contributions in string theory.

2 Loop Corrections in $D = 11$ Theory and String Theory

Let us start with some comments on the structure of higher-loop terms in low-energy $D = 11$ supergravity effective action and their relation to string theory. We shall consider the $D = 11$ theory compactified on a circle of radius R_{11} with the action

$$S = -\frac{1}{2\kappa^2}\int d^{11}x\sqrt{-g}\,\mathcal{R} + \dots , \qquad \kappa^2 = 16\pi^5\lambda^9 , \tag{1}$$

where λ is the $D = 11$ Planck scale. The two parameters of the compactified $D = 11$ theory R_{11} and κ are related to the string scale $l_{10} = \sqrt{\alpha'}$ and the string coupling g_s by

$$\lambda = (2\pi g_s)^{1/3}\sqrt{\alpha'} , \quad R_{11} = g_s\sqrt{\alpha'} , \quad \kappa_{10}^2 = \frac{\kappa^2}{2\pi R_{11}} = 64\pi^7 g_s^2 \alpha'^4 , \tag{2}$$

$$\alpha' = \frac{\lambda^3}{2\pi R_{11}}, \qquad g_s^2 = \frac{2\pi R_{11}^3}{\lambda^3} .$$

The $D = 11$ supergravity is UV divergent, so one needs to introduce a cutoff Λ_{11}. Since the $D = 11$ and $D = 10$ supergravities are related by dimensional reduction, Λ_{11} should be proportional to a cutoff Λ_{10} in type IIA $D = 10$ supergravity. The two cutoffs may be related, e.g., by comparing the divergent terms in the one-loop effective actions in $D = 11$ and $D = 10$ supergravities. The $D = 10$ supergravity is a low-energy limit of type IIA string theory, so its effective cutoff is $\Lambda_{10} \sim \frac{1}{\sqrt{\alpha'}}$. Expressed in terms of the (proper-time) cutoff Λ_{10}, the cutoff Λ_{11} is given by

$$R_{11}\Lambda_{11}^3 = a\Lambda_{10}^2 , \qquad \Lambda_{10}^2 = \frac{1}{2\pi\alpha'} , \tag{3}$$

where a is a numerical constant. Then $\Lambda_{11} = a^{1/3}\lambda^{-1} \sim \kappa^{-2/9}$, i.e. that Λ_{11} depends only on κ and not on R_{11}. This has a natural 'membrane-theory' interpretation: just as the $D = 10$ cutoff Λ_{10} is proportional to the square root of the string tension $T_1 = \frac{1}{2\pi\alpha'} = \frac{1}{2\pi l_{10}^2}$, the $D = 11$ cutoff Λ_{11} is proportional to the cubic root of the membrane tension

$$\Lambda_{11} = (2\pi a T_2)^{1/3} , \qquad T_2 = \frac{1}{2\pi l_{11}^3} = \frac{1}{4\pi^2 g_s \alpha'^{3/2}} . \tag{4}$$

The general structure of the cutoff-dependent part of the effective action of $D = 11$ supergravity at the L-loop level is

$$S_{L\infty} = \kappa^{2(L-1)} \sum \Lambda_{11}^n (\ln \Lambda_{11})^l \int d^{11}x\sqrt{-g}\, \mathcal{R}^m , \tag{5}$$

where $\mathcal{R}^m$ stands for all possible scalars built out of curvature and its covariant derivatives which have length dimension $-2m$. On dimensional grounds, $n + 2m = 9(L-1) + 11$. Note that purely logarithmic divergences ($n = 0$) may appear only at even loop orders and have $m = 10, 19, \ldots$.

At the one-loop order, the leading $\mathcal{R}^m$ ($m = 0, 1, 2, 3$) divergences cancel out [4], so that

$$S_{1\infty} \propto \int d^{11}x\sqrt{-g}\, \Lambda_{11}^3 \mathcal{R}^4 .$$

The presence of the cubic $\mathcal{R}^4$ divergence in $D = 11$ supergravity is implied [4] by the presence of quadratic $\mathcal{R}^4$ divergence in the $D = 10$ supergravity, which, in turn, can be found as the $\alpha' \to 0$ limit [5], [6] of the one-loop string-theory contribution $\frac{1}{\alpha'} \int d^{10}x\sqrt{-g}\mathcal{R}^4$.

An uncompactified $D = 11$ M-theory (having $D = 11$ supergravity as its low-energy approximation) is suggested to be a strong-coupling limit of type IIA string theory [7]. Let us suppose that there are special terms $f(g_s)\mathcal{R}^m$ in the string theory effective action which do not receive corrections beyond certain order L in string loop expansion. Then their coefficients will have simple power-like (or 'perturbative') dependence on g_s in the limit of $g_s \gg 1$, i.e. $f(g_s) \sim g_s^{2(L-1)}$. Such terms must then have a natural $D = 11$ theory interpretation. Using this logic, one may be able to obtain certain constraints on possible terms in the effective action of M-theory. As we will argue below, such special terms in the string-theory action may correspond to covariant $\mathcal{R}^m$ terms in the uncompactified $D = 11$ theory only if $m = 3k + 1$, $k = 0, 1, 2, \ldots$.

Using

$$\int d^{11}x \to 2\pi R_{11} \int d^{10}x \,, \quad \kappa^2 \sim g_s^3 \,, \qquad R_{11} \sim g_s \,, \quad \Lambda_{11} \sim g_s^{-1/3} \,,$$

and $ds_{11}^2 = dx_{11}^2 + g_{\mu\nu}dx^\mu dx^\nu$ one finds

$$\kappa^{2(L-1)}\Lambda_{11}^n \int d^{11}x\sqrt{-g}\,\mathcal{R}^m \quad \to \quad g_s^{\frac{2}{3}(m-4)} \int d^{10}x\sqrt{-g}\,\mathcal{R}^m \,.$$

The condition $\frac{1}{3}(m-4) = k-1$ where k is an integer (effective loop order in string theory) implies

$$m = 3k+1 \,, \qquad n = 9L - 6k \,, \qquad k = 0,1,2,\ldots \,. \tag{6}$$

Thus the terms in the $D = 11$ action related to the special string-theory terms with coefficients which have 'perturbative' dependence on $g_s \gg 1$ are

$$\kappa^{2(L-1)}\Lambda_{11}^{9L-6k} \int d^{11}x\sqrt{-g}\,\mathcal{R}^{3k+1} \;\sim\; l_{11}^{6k-9} \int d^{11}x\sqrt{-g}\,\mathcal{R}^{3k+1} \,. \tag{7}$$

Let us note that supersymmetry may also impose certain constraints on possible $\mathcal{R}^m$ curvature invariants. The $\mathcal{R}^m$ invariants that originate from the *full* (on-shell) superspace integral [8], [9], $\int d^{11}x\, d^{32}\theta\, \mathcal{D}^{2p}\, W^m, \quad W \sim \theta^2\mathcal{R} + \ldots$, have $m = 16+p$ (combined with above relation with $n=0$, this gives further restriction on possible purely-logarithmic counterterms: $m+p = 9L-14$ [10]). This condition includes $m = 3k+1 \geq 16$ for $p = 3k-15$. The terms with $m = 3k+1 < 16$ (i.e. $\mathcal{R}^4$, $\mathcal{R}^7$, etc.) should correspond to super-invariants constructed as integrals over parts of superspace.

One may arrive at the same restriction on powers of curvature invariants in the $D = 11$ theory (i.e. $m = 1,4,7,10,\ldots$) by an independent argument. In general, local perturbative contributions to the string-theory effective action are given by series of terms in expansion in string coupling and inverse string tension, $\sum \kappa_{10}^{2(L-1)} T_1^{-n} \int d^{10}x\sqrt{-g}\mathcal{R}^m$, where on dimensional grounds, $m = 2(L-1) + n + 5$. The natural parameter in M-theory has dimension $(\text{length})^{-3}$, which may be interpreted as the membrane tension T_2. If we assume that the M-theory effective action should similarly contain only terms which may appear in expansion in integer powers of inverse membrane tension, then the only possible curvature invariants will be those given above. Indeed, $T_2^{-n} \int d^{11}x\sqrt{-g}\mathcal{R}^m \sim l_{11}^{3n} \int d^{11}x\sqrt{-g}\mathcal{R}^m$, so that $2m = 3n+11$. Since m is a positive integer, n must be an odd number of the form $n = 2k-3$, $k = 0,1,2,\ldots$, and hence $m = 3k+1$.

To summarise, a term $f(g_s)\mathcal{R}^{3k+1}$ in type IIA superstring theory corresponds to a covariant term in the eleven-dimensional Lagrangian only if it scales like g_s^{2k-2} in the limit $g_s^2 \gg 1$. Although it is not excluded that the sum of an infinite number of string loop corrections may behave like g_s^{2k-2} at strong coupling, the non-renormalization of the $\mathcal{R}^{3k+1}$ terms seems a natural generalization of the suggestion about the non-renormalisation of $\mathcal{R}^4$ term made in [1], [11] to the case of $k > 1$. Thus we conjecture that all $\mathcal{R}^{3k+1}$

terms should not receive contributions beyond the k-th loop order in type IIA string perturbation theory. The existence of terms in uncompactified type II string theory action which receive corrections only at one specific loop order was conjectured in [12]. Examples of such terms are known in the case of $N=2, D=4$ supersymmetric compactifications of type II string theory [13].

At the same time, contributions to $\mathcal{R}^{3k+1}$ terms at *lower* loop orders in string perturbation theory are not excluded (as they will be subleading in the $g_s \to \infty$ limit). Their $D=11$ origin should be in the finite 'Casimir-type' R_{11}^{-n} terms, which accompany Λ_{11}^n-terms when the $D=11$ effective action is computed in the space with one circular dimension. For example, the one-loop $\Lambda_{11}^3 \mathcal{R}^4$ term in the case of finite radius R_{11} is replaced by $(\Lambda_{11}^3 + c_1 R_{11}^{-3})\mathcal{R}^4$ [1]. In general,

$$\kappa^{2(L-1)} R_{11}^{-n} \int d^{11}x\sqrt{-g}\, \mathcal{R}^m \quad \to \quad g_s^{2q} \int d^{10}x\sqrt{-g}\, \mathcal{R}^m \,, \qquad q = m - 3L - 2 \,.$$

Remarkably, if $m = 3L+1$ then $q=-1$, i.e. we conclude that the term $\kappa^{2(L-1)}(\Lambda_{11}^n + cR_{11}^{-n}) \int d^{11}x\sqrt{-g}\mathcal{R}^m$ in the $D=11$ effective action corresponds to a sum of L-loop and tree-level $\mathcal{R}^{3L+1}$ terms in the $D=10$ string theory effective action. For example, like the one-loop $D=11$ terms $(\Lambda_{11}^3 + c_1 R_{11}^{-3})\mathcal{R}^4$, which correspond to a sum of one-loop and tree-level terms in $D=10$ [1], the two-loop terms $\kappa^2(\Lambda_{11}^6 + c_2 R_{11}^{-6})\mathcal{R}^7$ should correspond to a sum of two-loop $(\frac{\kappa_{10}^2}{\alpha'^2}\mathcal{R}^7)$ and tree-level $(\frac{\alpha'^6}{\kappa_{10}^2}\mathcal{R}^7)$ terms in string theory.

3 One-Loop Four-Graviton Amplitude in $D=11$ Supergravity

Deriving the one-loop four-graviton amplitude directly from the component formulation of $D=11$ supergravity would be quite complicated. Fortunately, there is a short-cut way using the known expression for the one-loop $D=10$ closed superstring 4-point amplitude. It was shown in [5] that the one-loop graviton scattering amplitude in $D \le 8$ maximal supergravities can be obtained as a limit ($\alpha' \to 0, R \to 0, \kappa_D =$fixed) of the amplitude of $D=10$ string theory compactified on a torus. To find the amplitude in $D=10$ type II supergravity theory one should take $\alpha' \to 0$ limit treating $1/\alpha'$ as a proper-time UV cutoff [6]. The resulting expression is formally the same as for $D<8$ (where the $\alpha' \to 0$ limit is regular), but it still depends on α' via the cutoff (and it is quadratically divergent for $\alpha' \to 0$). According to [5]

$$A_4^{(D)} = \left(\frac{2\pi R}{\sqrt{\alpha'}}\right)^{D-10} \frac{\kappa_{10}^2}{\alpha'} \int_1^\infty d\tau_2\, \tau_2^{\frac{6-D}{2}}\, F(s,t;\tau_2)$$

$$= \kappa_D^2 \int_{\alpha'}^\infty d\tau\, \tau^{\frac{6-D}{2}}\, F(s,t;\tau) \,, \tag{8}$$

$$F(s,t;\tau) = \int [d\rho]\, e^{-\tau M(s,t;\rho)} \equiv \int_0^1 d\rho_3 \int_0^{\rho_3} d\rho_2 \int_0^{\rho_2} d\rho_1\, e^{-\tau M(s,t;\rho)} + \text{perms.}$$

$$M(s,t;\rho) \equiv s\rho_1\rho_2 + t\rho_3\rho_2 + u\rho_1\rho_3 + t(\rho_1 - \rho_2) \ , \qquad s+t+u=0 \ .$$

The dependence on the cutoff $\alpha' \to 0$ disappears in $D < 8$ (where maximal supergravities are one-loop finite), but remains in $D = 10$

$$A_4^{(10)} = \kappa_{10}^2 \int_{\epsilon_{10}}^{\infty} \frac{d\tau}{\tau^2}\, F(s,t;\tau) \ \sim \ \Lambda_{10}^2 \ + \ \text{finite part} \ . \tag{9}$$

Here $\tau \equiv \alpha'\tau_2 = \frac{\mathrm{t}}{2\pi}$ is related to the standard proper-time parameter t so that the effective proper-time cutoff is $\Lambda_{10} = \frac{1}{\sqrt{2\pi\epsilon_{10}}} = \frac{1}{\sqrt{2\pi\alpha'}}$.

The four-graviton amplitude in $D = 11$ supergravity compactified on a circle (with all external particles having ten-dimensional polarisations and $p_{11} = 0$) is thus given by above equation with an extra Kaluza-Klein factor, i.e.

$$A_4^{(11)} = \kappa_{11}^2 (2\pi R_{11})^{-1} A_4(s,t) \ , \tag{10}$$

$$A_4 = \sum_{m=-\infty}^{\infty} \int_{\epsilon_{11}}^{\infty} \frac{d\tau}{\tau^2}\, e^{-\frac{\pi\tau m^2}{R_{11}^2}}\, F(s,t;\tau) \ , \qquad \epsilon_{11} = \Lambda_{11}^{-2} \ .$$

Because of the sum over the Kaluza-Klein modes, the τ-integral here has a stronger (cubic instead of quadratic) divergence, as appropriate to the $D = 11$ theory.

The resulting amplitude is in agreement with the general expression for the $D = 11$ supergravity four-graviton amplitude suggested (on the basis of a somewhat different reasoning) in [1].

The integrand in the amplitude can be expanded in powers of M

$$A_4(s,t) = \sum_{m=-\infty}^{\infty} \int_{\epsilon_{11}}^{\infty} \frac{d\tau}{\tau^2}\, e^{-\frac{\pi\tau m^2}{R_{11}^2}} \int [d\rho] \sum_{k=0}^{\infty} \frac{(-1)^k}{k!} \tau^k M^k(s,t) \ .$$

Let us separate the first ($k = 0$) term $A_4^{(a)}$ in the expansion,

$$A_4(s,t) = A_4^{(a)} + A_4^{(b)}(s,t) \ , \qquad A_4^{(a)} = \sum_{m=-\infty}^{\infty} \int_{\epsilon_{11}}^{\infty} \frac{d\tau}{\tau^2}\, e^{-\frac{\pi\tau m^2}{R_{11}^2}} \ .$$

One can show that

$$A_4^{(a)} = \frac{2}{3} R_{11} \Lambda_{11}^3 \ + \ \frac{\zeta(3)}{\pi R_{11}^2} \ .$$

Defining where

$$\mathcal{H}(s,t) \equiv \int [d\rho]\, M(s,t;\rho) \ln M(s,t;\rho) = s\bar{\mathcal{H}}\left(\frac{s}{t}\right) \ , \tag{11}$$

$$c_k = \frac{2(-1)^k}{\pi^{k-1} k(k-1)}\, \zeta(2k-2) \ ,$$

$$H_k(s,t) \equiv \int [d\rho]\ M^k(s,t;\rho) = s^k \bar{H}_k\left(\frac{s}{t}\right) ,$$

where $\bar{H}_k$ is a polynomial of order k, we find

$$A_4(s,t) = \frac{2}{3} R_{11} \Lambda_{11}^3 + \frac{\zeta(3)}{\pi R_{11}^2} + s\bar{\mathcal{H}}\left(\frac{s}{t}\right) + \sum_{k=2}^{\infty} c_k R_{11}^{2k-2} s^k \bar{H}_k\left(\frac{s}{t}\right) . \qquad (12)$$

The third term is the finite part of the contribution of the massless $D = 10$ supergravity fields. In the case when all 11 dimensions are non-compact, the amplitude is given by the same universal expression (with $D = 11$ cutoff)

$$A_4^{(11)} = \kappa_{11}^2 \int_{\epsilon_{11}}^{\infty} \frac{d\tau}{\tau^{5/2}}\ F(s,t;\tau) = \kappa_{11}^2 \left[\frac{2}{3} \Lambda_{11}^3 + \frac{4}{3} \sqrt{\pi} s^{3/2} \bar{H}_{3/2}\left(\frac{s}{t}\right)\right] .$$

4 Remarks on Relation to String Theory

Expressing the $D = 11$ supergravity amplitude in terms of the string coupling g_s and the string scale $\sqrt{\alpha'}$ we find

$$A_4(s,t) = \frac{a}{3\pi\alpha'} + \frac{\zeta(3)}{\pi\alpha' g_s^2} + s\bar{\mathcal{H}}\left(\frac{s}{t}\right) + \sum_{k=2}^{\infty} c_k g_s^{2k-2} \alpha'^{k-1} s^k \bar{H}_k\left(\frac{s}{t}\right) . \qquad (13)$$

The first two constant terms in this amplitude (multiplied by the kinematic factor) correspond to the one-loop and tree-level $\mathcal{R}^4$ terms in the type II string effective action. That the one-loop amplitude in $D = 11$ supergravity effectively includes [1] the tree-level $\zeta(3)\mathcal{R}^4$ term of string theory may look miraculous: while in string theory this term is produced by exchanges of massive string modes, in $D = 11$ expression it originates from the loop of the Kaluza-Klein modes which are 0-brane solitons of string theory. This fact is suggesting that the uncompactified type IIA string theory ('dual' to $D = 11$ theory) may have a reformulation in terms of solitonic objects. The relation between tree-level $\mathcal{R}^4$ term in type IIA theory and one-loop $\mathcal{R}^4$ term in $D = 11$ theory is reminiscent of the relation between tree-level F^4 term in type I theory and one-loop F^4 term in heterotic theory [12].

In general, one expects the $\mathcal{R}^4$ terms in $D = 10$ string theory to be a linear combination of the $D = 10$ terms $\mathcal{J}_1 \equiv t_8 t_8 R^4$ and $\mathcal{J}_2 \equiv \frac{1}{8}\epsilon_{10}\epsilon_{10} R^4$. $\mathcal{J}_2$ is the higher-dimensional extension of the Gauss-Bonnett invariant in 8 dimensions ($\epsilon_8\epsilon_8 \rightarrow -\frac{1}{2}\epsilon_{10}\epsilon_{10}$). Its expansion near flat space ($g_{mn} = \eta_{mn} + h_{mn}$) starts with h^5 terms and thus its coefficient cannot be determined from consideration of the on-shell 4-graviton amplitude only. The sigma-model approach implies [14], [15] that (up to the usual field redefinition ambiguities) the tree-level type II string term is $L_0 \sim \zeta(3) J_0$, $J_0 = \mathcal{J}_1 + \mathcal{J}_2$. The structure of the kinematic factor in the one-loop type IIA 4-point amplitude with transverse polarisations and momenta $(t_8 + \frac{1}{2}\epsilon_8)(t_8 + \frac{1}{2}\epsilon_8)$ hints that the one-loop $\mathcal{R}^4$ terms in $D = 10$ type IIA theory should be proportional to the opposite-sign combination $\mathcal{J}_1 - \mathcal{J}_2$. Combining with the $B\mathcal{R}^4$ term, we get $L_{1A} = \mathcal{J}_1 - \mathcal{J}_2 + b_1\epsilon_{10}B[\mathrm{tr}R^4 - \frac{1}{4}(\mathrm{tr}R^2)^2]$. This can be rewritten (using $\mathcal{J}_1 = 24[t_8\mathrm{tr}R^4 - \frac{1}{4}t_8(\mathrm{tr}R^2)^2]$) as a combination of the bosonic parts of the *three* $N = 1$ super-invariants $I_3 = t_8\mathrm{tr}R^4 - \frac{1}{4}\epsilon_{10}B\mathrm{tr}R^4$, $I_4 = t_8\mathrm{tr}R^2\mathrm{tr}R^2 - \frac{1}{4}\epsilon_{10}B(\mathrm{tr}R^2)^2$ and J_0 *provided* $b_1 = -12$. Then $L_{1A} = -J_0 + 48(I_3 - \frac{1}{4}I_4)$. While the coefficients of I_3 and I_4 are expected not to be renormalised, there is no reason for this to be true for the coefficient of J_0 and thus of the $\mathcal{J}_2$ term. This may preclude one from identifying the $D = 11$ counterpart of this term as $\frac{1}{24}\epsilon_{11}\epsilon_{11}R^4$.

We observe that not only the two constant terms but also all *momentum-dependent* terms in the $D = 11$ amplitude have 'perturbative' dependence on the type IIA string coupling. It appears as if the *one-loop* four-graviton amplitude in $D = 11$ supergravity represents a sum of certain perturbative string corrections, containing contributions of *all orders* in the string loop expansion. It is not clear, however, which regions of the moduli spaces of higher genus Riemann surfaces this expression is accounting for. Moreover, while the first two terms (or $\mathcal{R}^4$ terms in the type II string effective action) are expected to be unchanged by both $D = 11$ supergravity and type IIA string higher-loop corrections [11], [1], this may not be true for other s, t-dependent terms. If this is the case, one may be unable to relate the $D = 10$ and $D = 11$ expressions in a simple way.

I would like to thank the organisers for the invitation and acknowledge the support of PPARC and the European Commission TMR programme grant ERBFMRX-CT96-0045.

References

[1] M.B. Green, M. Gutperle and P. Vanhove, hep-th/9706175.
[2] J. Russo and A.A. Tseytlin, Nucl. Phys. B508 (1997) 245, hep-th/9707134.
[3] K. Becker, M. Becker, J. Polchinski and A.A. Tseytlin, hep-th/9706072.
[4] E.S. Fradkin and A.A. Tseytlin, Nucl. Phys. B227 (1983) 252.
[5] M.B. Green, J.H. Schwarz and L. Brink, Nucl. Phys. B198 (1982) 474.
[6] R.R. Metsaev and A.A. Tseytlin, Nucl. Phys. B298 (1988) 109.
[7] E. Witten, Nucl. Phys. B443 (1995) 85, hep-th/9503124.
[8] R. Kallosh, unpublished; P. Howe, unpublished.
[9] M. Grisaru and W. Siegel, Nucl. Phys. B201 (1982) 292.
[10] M.J. Duff and D.J. Toms, in: *Unification of Fundamental Particle Interactions II*, Proceedings of the Europhysics Study Conference, Erice, 4-16 October 1981, ed. by J. Ellis and S. Ferrara (Plenum, 1983).
[11] M.B. Green and P. Vanhove, hep-th/9704145.
[12] A.A. Tseytlin, Nucl. Phys. B467 (1996) 383, hep-th/9512081.
[13] I. Antoniadis, E. Gava, K.S. Narain and T.R. Taylor, Nucl. Phys. B455 (1995) 109, hep-th/9507115.
[14] M.T. Grisaru, A.E.M. van de Ven and D. Zanon, Nucl. Phys. B277 (1986) 388, 409.

[15] M.T. Grisaru and D. Zanon, Phys. Lett. B177 (1986) 347; M.D. Freeman, C.N. Pope, M.F. Sohnius and K.S. Stelle, Phys. Lett. B178 (1986) 199; Q.-H. Park and D. Zanon, Phys. Rev. D35 (1987) 4038.

Part III

Supersymmetric Quantum Mechanics and Integrable Systems

Nilpotent Marsh and SUSY QM

V.P. Akulov[1] and Steven Duplij[2]

[1] City College of City University of New York,
138th St. Convent Ave , New York, NY 10031, USA
[2] Theory Group, Nuclear Physics Laboratory,
Kharkov State University, Kharkov 310077, Ukraine
E-mail: Steven.A.Duplij@univer.kharkov.ua
Internet: http://www-home.univer.kharkov.ua/~duplij

Abstract. We consider the nilpotent additions to classical trajectories in supersymmetric and nonsupersymmetric theories. The condition of anilpotence of action on some generalized solutions leads to the Witten supersymmetric Lagrangian. The condition of anilpotence of topological charge is the same as one of superpotential with spontaneous broken supersymmetry. We should vanish half of Grassmann constants of integration, because in this case only we obtain the same number of normalized bosonic and fermionic zero modes.

1 Introduction

This paper is dedicated to Victor Isaakovich Ogievetsky who was an outstanding scientist and brilliant man. He gave a great contribution to quantum field theory, supersymmetry and supergravity.

In recent time the interest to extended supersymmetric quantum mechanics (Witten (1981)) with N supersymmetries (Andrianov, Cannata, Dedonder, and Ioffe (1998), Diaconescu and Entin (1998), Manton (1998), Lowe (1998), Paban, Sethi, and Stern (1998), Paban, Sethi, and Stern (1998)) is greatly renewed. This is connected on one side with dimensional reduction of additional dimensions of D-brane where one can receive SQM limit, which is more simple object. Maldacena's correspondence (Maldacena (1997)) between string theory on AdS spaces and superconformal quantum theories attracts attention to the structure of extremal black holes near the horizon which is described by the supersymmetric quantum mechanics (Claus, Derix, Kallosh, Kumar, Townsend, and Van Proeyen (1998)).

On the other side the supermembrane in light cone gauge gives rise to SQM system with $SU(N)$ gauge symmetry (de Wit, Marquard, and Nicolai (1990), Lowe (1998)). D(-1) branes or D-instantons have the world volume which is just a point; they are topological objects and carry out topological charge.

Instantons as such solitons have zero modes connected with derivative by constants of integration. It is well known that these solutions (as a rule) break one half supersymmetries. For any supersymmetric theory the number of bosonic zero modes should be the same as fermionic ones.

We want to consider a full class of solutions for classical equations of motion, which consist of both bosonic and Grassmann constants of integration (Akulov (1986), Akulov and Volkov (1980), Akulov and Duplij (1988), Akulov and Pashnev (1983), Manton (1998)). Fields take their values in Grassmann algebra with an even part and an odd part. But the action have to belong only to the even part and should not have any nilpotent addition, because the action is connected with a measure defined by functional integration in the path integral approach.

We here concentrate our attention mainly on the conditions which give absence of nilpotent additions to the action on the generalized classical trajectories and receive the connection between bosonic and fermionic parts of the interaction.

The plan of the paper is the following. In Sect.2 we introduce the superfield approach to SQM with $N = 2$ supersymmetries (to confine ourselves with most simple model). In Sect.3 we consider full classical solutions and the corresponding nilpotent "marsh" — trajectories with nilpotent addition. In Sect.4 we investigate the condition of vanishing of half number of Grassmann constants, because only in this case we obtain exactly correct number of fermionic zero modes. In Sect.5 we obtain Witten's supersymmetric Lagrangian as a consequence of anilpotence of action on such trajectories. In Sect.6 we obtained SWKB rules for quantization. In Sect.7 we conclude with some comments.

2 Superfield Formulation

In (1|2) dimensional Euclidean superspace and in the chiral basis $(\tau, \theta^+, \theta^-)$ a scalar superfield $\Phi(\tau, \theta^+, \theta^-)$ has the following expansion

$$\Phi(\tau, \theta^+, \theta^-) = q(\tau) + \theta^+ \psi_-(\tau) + \theta^- \psi_+(\tau) + \theta^+\theta^- F(\tau), \tag{1}$$

where $q(\tau)$ is a bosonic coordinate, $F(\tau)$ is an auxiliary bosonic field, $\psi_\pm(\tau)$ are fermionic coordinates. Using chiral derivatives $D^\pm = \partial/\partial\theta^\mp + \theta^\pm \partial/\partial\tau$ the superfield Lagrangian of the model with a superpotential $V(\Phi)$ can be written as

$$\mathcal{L}_E = \frac{1}{2} D^+\Phi \cdot D^-\Phi + V(\Phi) \tag{2}$$

and corresponding equations of motion are[1]

$$\frac{1}{2}\left[D^+, D^-\right]\Phi - V'(\Phi) = 0. \tag{3}$$

The Euclidean action has the standard form

$$S_E = \int dt L_E, \tag{4}$$

[1] Prime is differentiation by argument.

where

$$L_E = \int d\theta^- d\theta^+ \mathcal{L}_E \tag{5}$$

is an ordinary Euclidean Lagrangian. Expanding (2) in θ-series which is finite due to nilpotence of $\theta^\pm$, using the nondynamical equation $F = -V'(q)$ and integrating by in (5) $\theta^\pm$ we obtain[2]

$$L_E = \frac{\dot{q}^2}{2} + \frac{V'^2(q)}{2} + \psi_+\dot{\psi}_- + V''(q)\,\psi_+\psi_-. \tag{6}$$

3 Classical Solutions and Even Nilpotent Directions

From (6) we find the following equations of motion

$$\ddot{q} - V'(q)\,V''(q) + V'''(q)\,\psi_+\psi_- = 0, \tag{7}$$

$$\dot{\psi}_\pm \mp V''(q)\,\psi_\pm = 0. \tag{8}$$

The classical solution of the system (7)–(8) is much simplified if one takes $\psi_\pm^{cl} = 0$ (Cooper and Freedman (1983), Solomonson and van Holten (1982)). Let us write a general instanton and antiinstanton solutions without vanishing fermionic classical coordinates (Akulov (1986), Akulov and Duplij (1988)) in the following form

$$\begin{cases} q_+ = q_{0+} - \frac{\lambda_+\lambda_-}{2V'(q_{0+})},\ \psi_+ = \lambda_+ V'(q_{0+}),\ \psi_- = \frac{\lambda_-}{V'(q_{0+})},\ inst. \\ q_- = q_{0-} + \frac{\lambda_+\lambda_-}{2V'(q_{0-})},\ \psi_+ = \frac{\lambda_+}{V'(q_{0-})},\ \psi_- = \lambda_- V'(q_{0-}),\ antiinst. \end{cases} \tag{9}$$

where $\lambda_\pm$ are Grassmann integration constants and $q_{0\pm}(\tau)$ is a standard instanton/antiinstanton solution (Solomonson and van Holten (1982)) which can be found from the equations

$$\dot{q}_{0\pm}(\tau) = \pm V'(q_{0\pm}). \tag{10}$$

The classical solutions (9) can be presented in more compact form

$$q_k = q_{0(k)} - \frac{\nu}{2W\left(q_{0(k)}\right)},\ \psi_{+(k)} = \lambda_+ W^k\left(q_{0(k)}\right),\ \psi_{-(k)} = \lambda_- W^{-k}\left(q_{0(k)}\right), \tag{11}$$

where $\nu = \lambda_+\lambda_-, \nu^2 = 0$, $W(q) = V'(q)$, $k = +1$ for instantons and $k = -1$ for antiinstantons. The corresponding scalar superfields (1) are

$$\Phi_{(k)}\left(\tau, \theta^+, \theta^-\right) = q_{0(k)}(\tau) + \theta^+\lambda_- W^{-k} + \theta^-\lambda_+ W^k + \theta^+\theta^-\left(-W + \nu\frac{W'}{2W}\right). \tag{12}$$

It is important to stress that despite the fact that the classical trajectories have a nilpotent part proportional to ν (11) the action has no such additional parts and it is equal to the action on instanton trajectories, i.e.

[2] Dote is differention by τ.

$$S_{Ek} = \int W^2\left(q_{0(k)}\right) d\tau = S_{Eq_{0(k)}}. \tag{13}$$

For the topological charge Q_k of instantons/antiinstantons we obtain

$$Q_k = \frac{1}{2q_0(\infty)} \int\limits_{-\infty}^{\infty} \dot{q}_k^2 d\tau =$$

$$Q_{0(k)} + \frac{\nu}{2q_{(k)}(\infty)}\left[\frac{1}{W\left(q_{(k)}(+\infty)\right)} - \frac{1}{W\left(q_{(k)}(-\infty)\right)}\right]. \tag{14}$$

From (14) we conclude that the nilpotent part of the topological charge vanishes in case of spontaneous symmetry breaking, i.e. when $W(q) = V'(q)$ is an even function. That gives another meaning to the conclusions of (Cooper and Freedman (1983), Solomonson and van Holten (1982)).

4 Fermionic Zero Modes

Let us consider a contribution of zero modes in quantum evolution of our system. In general we have

$$\Phi_{quant}\left(\tau,\theta^+,\theta^-\right) = \Phi\left(\tau+\tau_0,\theta^+ + \eta^+,\theta^- + \eta^-\right) + \Sigma', \tag{15}$$

where τ_0, η^+, η^- are shifts corresponding zero modes, Σ' is contribution of nonzero modes, $\Phi\left(\tau,\theta^+,\theta^-\right)$ is a classical superfield (12). Expanding the action next to the classical solution or differentiating of the classical equation of motion by corresponding parameter we find the equation for zero modes

$$\{1/2[D^+,D^-] - V''(\Phi)\}\Phi_0 = 0$$

But we can find the zero modes differentiating by the parameters of the solutions (ref9Akul). We see that the zero modes connected to differentiating of $\psi_-$by $\lambda_-$and $\psi_+$by λ_+ reduce to inverse functions. But from Cauchy-Schwarz-Bounjakovsky inequality

$$\left(\int_0^\infty \psi_{-0} dt\right)^2 \left(\int_0^\infty \psi_{+0} dt\right)^2 \leq \left(\int_0^\infty \psi_{-0} \times \psi_{+0} dt\right)^2$$

where ψ_{-0} and ψ_{+0} are fermionic zero modes, one can see that only one of two fermionic zero mode can be normalized and consequently one of two Grassmann constants have to vanish. But a vanished Grassmann constant corresponds to N=1 broken supersymmetry. Thus we see that one half of supersymmetries must be broken.

5 Supersymmetry as Anilpotence of Lagrangian

Let us consider nonsupersymmetric Lagrangian similar to (6) as

$$L_{E(nonsusy)} = \frac{\dot{q}^2}{2} + \frac{W^2(q)}{2} + \psi_+\dot{\psi}_- + U(q)\psi_+\psi_-, \tag{16}$$

where $W(q)$ and $U(q)$ are some functions. Equations of motion are

$$\ddot{q} - W(q)W'(q) + U'\psi_+\psi_- = 0, \tag{17}$$

$$\dot{\psi}_\pm \mp U(q)\psi_\pm = 0. \tag{18}$$

Fermionic classical solutions are

$$\psi_\pm = \lambda_\pm e^{\pm\int U(q)d\tau}, \tag{19}$$

and therefore $\psi_+\psi_- = \lambda_+\lambda_- = \nu = const$ on equations of motion. Let

$$q(\tau) = q_0(\tau) + \nu q_N(\tau), \tag{20}$$

where $q_0(\tau)$ is nonsupersymmetric instanton solution $\dot{q}_0 = \pm W(q_0)$ and $q_N(\tau)$ satisfies the following equation

$$\dot{q}_N = W'(q_0)q_N - \frac{U(q_0)}{W(q_0)}. \tag{21}$$

Then we obtain for on-shell Lagrangian (16)

$$L^{n-shell}_{E(nonsusy)} = W^2(q_0) + \nu W(q_0)\left[2W(q_0)W'(q_0)q_N - U(q_0)\right]. \tag{22}$$

The requirement of vanishing of the nilpotent part in (22) gives

$$q_N = \frac{U(q_0)}{2W(q_0)W'(q_0)}. \tag{23}$$

So that from (21) and (23) we finally obtain (Akulov (1986))

$$\frac{U'}{U} = \frac{W''}{W'} \tag{24}$$

which can be solved by

$$U = cW', \tag{25}$$

where c is an even constant which can be taken as 1. Substituting (25) into $L_{E(nonsusy)}$ (16) gives supersymmetric Lagrangian $L_{E(usy)}$ (cf. (6)) in the following form

$$L_{E(susy)} = \frac{\dot{q}^2}{2} + \frac{W^2(q)}{2} + \psi_+\dot{\psi}_- + W'(q)\psi_+\psi_-, \tag{26}$$

Thus, the requirement of anilpotence of the on-shell Lagrangian gives us the supersymmetry condition for the on-shell Lagrangian (Akulov (1986)).

6 Quasiclassical Quantization Rules as Anilpotence of Action

Let us return to Minkowski (1|2) superspace by the substitution $\tau = it$. Then the action instead of (4) will take the form

$$S = \int L dt, \tag{27}$$

where[3]

$$L = \frac{\dot{q}^2}{2} - \frac{V'^2(q)}{2} + i\psi_+\dot{\psi}_- + V''(q)\,\psi_+\psi_-. \tag{28}$$

The corresponding equations of motion are

$$\ddot{q} + V'(q)\,V''(q) - V'''(q)\,\psi_+\psi_- = 0, \tag{29}$$

$$\dot{\psi}_\pm \pm iV''(q)\,\psi_\pm = 0. \tag{30}$$

Their solutions in case we admit nonzero fermionic classical solutions (Akulov and Duplij (1988)) have the following form

$$q = q_0 + \nu\frac{W}{2E},\ \psi_\pm = \lambda_\pm^{\mp i \arcsin\left(\frac{W}{\sqrt{2E}}\right)}, \tag{31}$$

where E is energy of the system and $W = V'(q_0)$. Here $q_0(t)$ is a standard solution of 1-dimensional nonsupersymmetric system which can be found from the equation

$$dt = \frac{q_0}{\sqrt{2E - W^2}}. \tag{32}$$

The corresponding Maupertuis action is

$$S_M = \oint (pdq + \psi_+ d\psi_-), \tag{33}$$

where

$$p = \sqrt{2E - W^2}\left(1 + \nu\frac{W'}{2E}\right) \tag{34}$$

is a canonical momentum for q. On equations of motion the action (33) takes the form

$$S_M = \int \sqrt{2E - W^2}dq_0 + \nu\int\left(\frac{1}{E} - \frac{1}{2E - W^2}\right)\sqrt{2E - W^2}W'dq_0. \tag{35}$$

Second term with nilpotent addition vanishes after integration over a closed path and we obtain the standard super WKB rules (Comtet, Bandrauk, and Campbell (1985)).

The quasiclassical wave function is also defined by classical trajectories as follows

---

[3] In this section a dote denotes differentiation by t.

$$\Psi\left(q,\psi_{+},\psi_{-}\right)\approx A_{\psi}^{\frac{i}{\hbar}S_{\psi}}, \tag{36}$$

where

$$A_{\psi}=\operatorname{Ber}\left|-\frac{\partial^{2}S_{\psi}}{\partial q\left(0\right)\partial q\left(t\right)}\right| \tag{37}$$

is a super generalization of the van Fleck determinant. Using the equations of motion (29)–(30) we obtain

$$A_{\psi}=\frac{1}{\sqrt{2E-W^{2}}}\left[1+\frac{\nu}{2}\left(\frac{W'}{2E}+\frac{W'^{2}}{4E^{2}}e^{i\arcsin\left(\frac{W}{\sqrt{2E}}\right)}\right)\right], \tag{38}$$

$$S_{\psi}=\int dt\left(E-W^{2}\right)\left(1+\nu\frac{W'}{E}\right). \tag{39}$$

7 Conclusions

Thus we have considered the generalized solutions for the classical equations of motion for a bosonic-fermionic system which contain the nilpotent additions – "nilpotent marsh". In consequence of conservation the nilpotent constant ν the particle will "live" on such trajectory and never will have the intersections with usual classical trajectory. But on such instanton (antiinstanton) trajectories the classical action has a nilpotent addition, which vanishes only for supersymmetric Lagrangian. The nilpotent addition to the topological charge vanishes for the superpotential with spontaneous broken supersymmetry. We studied the normalized zero modes for the instanton and antiinstanton solutions. As a consequence of Cauchy-Schwarz-Bounjakovsky inequality for fermionic zero modes only half of then can be normalized. So we have to vanish half of Grassmann constants, which corresponds to $N=1$ broken supersymmetry. In contrast of paper (Manton (1998)) we have used a superfield approach to SQM (Akulov and Pashnev (1983)). We have considered a quasiclassical action on such trajectories and obtained Super WKB rules for quantization, as an application of our approach.

8 Acknowledgments

We thank A. Pashnev, A. Burinsky, A. Gumenchuk and A. Zheltukhin for discussions and interest to the work.

This work is partly (A.V.) supported by grants INTAS-93/127ext, INTAS-93/493ext, INTAS-96/0308.

References

Akulov, V. P. and Duplij, S. (1988): Quasiclassical quantization in supersymmetric quantum mechanics, *Ukrainian J. Phys.* **33**, 309–311.

Akulov, V. P. and Pashnev, A. I. (1983): Quantum superconformal model in (1, 2) space, *Theor. Math. Phys.* **56**, 862–878.

Akulov, V. P. and Volkov, D. V. (1980): Supersymmetric equations in the space (1, 2), *Theor. Math. Phys.* **42**, 10–25.

Akulov, V. P. (1986): New class of solutions for Boson-Fermion system of equations and supersymmetry, *Ukrainian J. Phys.* **31**, 1615–1618.

Andrianov, A. A., Cannata, F., Dedonder, J.-P., and Ioffe, M. V. (1998): SUSY quantum mechanics with complex superpotentials and real energy spectra, St. Petersburg preprint, *St. Petersburg State Univ.*: SPbU-IP-97-24, quant-ph/9806019.

Claus, P., Derix, M., Kallosh, R., Kumar, J., Townsend, P. K., and Van Proeyen, A. (1998): Black holes and superconformal mechanics, Stanford preprint, *Stanford Univ.*: SU-ITP-98/27, hep-th/9804177.

Comtet, A., Bandrauk, A. D., and Campbell, D. K. (1985): Exactness of semiclassical bound states, *Phys. Lett.* **B150**, 159–162.

Cooper, F. and Freedman, B. (1983): Spontaneous supersymmetry breaking in quantum mechanics, *Ann. Phys.* **146**, 262–288.

de Wit, B., Marquard, U., and Nicolai, H. (1990): Area preserving diffeomorphisms and supermembrane Lorentz invariance, *Comm. Math. Phys.* **128**, 39–56.

Diaconescu, D.-E. and Entin, R. (1998): A non-renormalization theorem for the $d = 1$, $N = 8$ vector multiplet, *Phys. Rev.* **D56**, 8045–8052.

Lowe, D. A. (1998): Eleven-dimensional Lorentz symmetry from SUSY quantum mechanics, Providence preprint, *Brown Univ.*: BROWN-HET-1133, hep-th/9807229.

Maldacena, J. (1997): The large N limit of superconformal field theories and supergravity, Cambridge preprint, *Harvard Univ.*: HUTP-97/A097, hep-th/9711200.

Manton, N. S. (1998): Deconstructing supersymmetry, Cambridge preprint, *Univ. Cambridge*: DAMTP 1998-39, hep-th/9806077.

Paban, S., Sethi, S., and Stern, M. (1998): Constraints from extended supersymmetry in quantum mechanics, Princeton preprint, *Inst. Adv. Study*: hep-th/9805018.

Paban, S., Sethi, S., and Stern, M. (1998): Supersymmetry and higher derivative terms in the effective action of Yang-Mills theories, *J. High Energy Phys.* **06**, 012–023.

Solomonson, P. and van Holten, J. W. (1982): Fermionic coordinates and supersymmetry in quantum mechanics, *Nucl. Phys.* **B196**, 509–531.

Witten, E. (1981): Dynamical breaking of supersymmetry, *Nucl. Phys.* **B188**, 513–537.

On Superconformal-Like Transformations and Their Nonlinear Realization

Steven Duplij

Theory Group, Nuclear Physics Laboratory,
Kharkov State University, Kharkov 310077, Ukraine
E-mail: Steven.A.Duplij@univer.kharkov.ua
Internet: http://www-home.univer.kharkov.ua/~duplij

Abstract. We consider nonlinear realizations of invertible and noninvertible $N = 1$ superconformal-like transformations by means of the odd curve motion technique and introduced diagrammatic method.

First papers on supersymmetry (Volkov and Akulov (1972), Volkov and Akulov (1973), Akulov and Volkov (1974)) were written in terms of nonlinear realizations (for nonsupersymmetric background of the method see (Volkov (1969), Volkov (1973), Coleman, Wess, and Zumino (1969))). Further, there were hopes that in the framework of nonlinear realizations one could solve the problems with superpartners and spontaneously supersymmetry breaking in realistic models (Samuel and Wess (1983)). From another side, nonlinearly realized two dimensional superconformal symmetry (Kunitomo (1995)) were used in the theory of superstring embeddings (Belucci, Gribanov, Ivanov, Krivonos, and Pashnev (1998), Berkovits and Vafa (1994)). Here we consider finite superconformal-like (Duplij (1997)) transformations and include noninvertibility (Duplij (1991), Duplij (1997)). We also consider the connection between "linear" and "nonlinear" realizations (Ivanov and Kapustnikov (1978), Ivanov and Kapustnikov (1982)), but from the pure kinematical viewpoint and give a transparent diagram presentation for it in our special case.

Motion of odd curve in $\mathrm{C}^{1|1}$. Let us consider $N = 1$ superanalytic transformations in $\mathrm{C}^{1|1}$

$$\begin{cases} \tilde{z} = f(z) + \theta \cdot \chi(z), \\ \tilde{\theta} = \psi(z) + \theta \cdot g(z), \end{cases} \tag{1}$$

where four component functions $f(z), g(z) : \mathrm{C}^{1|0} \to \mathrm{C}^{1|0}$ and $\psi(z), \chi(z) : \mathrm{C}^{1|0} \to \mathrm{C}^{0|1}$ satisfy supersmooth conditions generalizing C^∞ (Rogers (1980), De Witt (1992)). According to the interpretation (Wess (1984)) we can study the motion of the curve $\theta = \lambda(z)$ in $\mathrm{C}^{1|1}$. Then we obtain

$$\tilde{z} = f(z) + \lambda(z) \cdot \chi(z), \tag{2}$$

$$\tilde{\lambda}(\tilde{z}) = \psi(z) + \lambda(z) \cdot g(z), \tag{3}$$

where the second equation reflects the Einstein style of transformations.

In four dimensional case the function $\lambda(z)$ is usually called Akulov-Volkov field (Wess (1984)) and in physical applications it plays a role of *Nambu-Goldstone fermion* (Volkov and Akulov (1973), Akulov and Volkov (1974)) (and therefore it is also called a *goldstino*).

The transformation of goldstino $\lambda(z)$ is highly nonlinear as it is seen from (3). Relations of such kind always appear in nonlinear group realizations, and $\lambda(z)$ is responsible for supersymmetry breaking (Akulov and Volkov (1974)).

To find goldstino finite transformations in our case we expand $\tilde{\lambda}(\tilde{z})$ in series and iterate exploiting nilpotency

$$\tilde{\lambda}(f(z)) = \psi(z) + \lambda(z) \cdot g(z) - \tilde{\lambda}'(f(z)) \cdot \lambda(z) \cdot \chi(z)\,. \tag{4}$$

In case f^{-1} exists, we derive the finite superanalytic transformation of $\lambda(z)$ as $\tilde{\lambda} = \psi \circ f^{-1} + \lambda \circ f^{-1} \cdot g \circ f^{-1} - \tilde{\lambda}' \cdot \lambda \circ f^{-1} \cdot \chi \circ f^{-1}$,where $f \circ g = f(g(z))$.

It is not possible to find a general solution of the equation (4), and therefore we consider some particular cases.

Global SUSY. The global supersymmetry in $\mathrm{C}^{1|1}$ corresponds to the following choice $f(z) = z$, $g(z) = 1$, $\chi(z) = \varepsilon$, $\psi(z) = \varepsilon$,where ε is a constant odd parameter. Then from (2) and (3) we have

$$\tilde{\lambda}_{Glob}(z) = \varepsilon + \lambda(z) - \tilde{\lambda}'_{Glob}(z) \cdot \lambda(z) \cdot \varepsilon\,. \tag{5}$$

This equation is also difficult to solve manifestly without any additional requirements. But for infinitesimal transformations we obtain

$$\delta_\varepsilon \lambda_{Glob}(z) = \tilde{\lambda}_{Glob}(z) - \lambda(z) = \varepsilon \cdot [1 + \lambda(z) \cdot \lambda'(z)] \tag{6}$$

which satisfy the conventional supersymmetry algebra $[\delta_\varepsilon, \delta_\eta]\,\lambda_{Glob}(z) = 2\varepsilon\eta \cdot \lambda(z) \cdot \lambda'(z)$ in accordance with (Akulov and Volkov (1974)).

In finite global case we put $\tilde{\lambda}^{fin}_{Glob}(z) = \tilde{\lambda}_{Glob}(z) + \Delta(z)$,where $\tilde{\lambda}_{Glob}(z)$ is given by (6). Inserting it into (5) one derives the equation for $\Delta(z)$ as follows $\Delta'(z) \cdot \varepsilon \cdot \lambda(z) = \Delta(z)$ which can be solved by expanding on nilpotents in a given underlying superalgebra.

Let us consider *superconformal-like transformations* parametrized by two functions $g(z)$, $\psi(z)$ (see (Duplij (1996), Duplij (1997))). Then starting from the same function $\lambda(z)$ we can in general find $\tilde{\lambda}_n(z)$ from (3) as two separate solutions (corresponding to the projection n of the "reduction spin" (Duplij (1997))) of the following system of equations

$$\begin{cases} \tilde{\lambda}_n\left(f_n^{(g\psi)}(z)\right) = \psi(z) + \lambda(z) \cdot g(z) - \tilde{\lambda}'_n\left(f_n^{(g\psi)}(z)\right) \cdot \lambda(z) \cdot \chi_n^{(g\psi)}(z)\,, \\ f_n^{(g\psi)\prime}(z) = \psi'(z)\,\psi(z) + \frac{1+n}{2} g^2(z)\,, \\ \chi_n^{(g\psi)\prime}(z) = g'(z)\,\psi(z) + n g(z)\,\psi'(z)\,, \end{cases} \tag{7}$$

where prime denotes derivative by argument, $n = +1$ corresponds to SCf transformations and $n = -1$ - to TPt transformations, and TPt are so called *noninvertible transformations twisting parity of tangent space* (see (Duplij (1991), Duplij (1997))). We will call $\tilde{\lambda}_{SCf}(z) = \tilde{\lambda}_{n=+1}(z)$ a *SCf goldstino*, and $\tilde{\lambda}_{TPt}(z) = \tilde{\lambda}_{n=-1}(z)$ a *TPt goldstino*.

It is necessary to stress that equations (7) do not depend on invertibility properties of superconformal-like transformations (Duplij (1996), Duplij (1991)) and only they can be used to find TPt goldstino evolution ($n = -1$ case). As previously, it is not possible to solve the system (7) manifestly in general case.

Infinitesimal SCf. Let we parametrize infinitesimal SCf transformations by $f(z) = z + r(z)$, $g(z) = 1 + \frac{1}{2}r'(z)$, $\chi(z) = \varepsilon(z)$, $\psi(z) = \varepsilon(z)$, where $r(z), \varepsilon(z)$ are infinitesimal. Then, from (7) we obtain

$$\delta_{r,\varepsilon}\lambda_{SCf}(z) = \varepsilon(z) \cdot [1 + \lambda(z) \cdot \lambda'(z)] + \frac{1}{2}r'(z) \cdot \lambda(z) - r(z) \cdot \lambda'(z) \quad (8)$$

in agreement with (Kunitomo (1995)).

Connection between linear and nonlinear realizations from diagrammatic viewpoint. Relationship between linear and nonlinear realizations (Ivanov and Kapustnikov (1978)) plays an important role in understanding of the spontaneously supersymmetry breaking mechanisms (Ivanov and Kapustnikov (1982)). Here we investigate that in noninvertible finite case and from some another kinematical viewpoint using a clear diagrammatic approach (which is applicable to any general multidimensional case as well). Let us consider the following diagram

$$
\begin{array}{ccc}
Z_A & \xrightarrow[\text{W-Z}]{\mathcal{G}} & \tilde{Z} \\
\big\uparrow{\scriptstyle\mathcal{A}} & & \big\uparrow{\scriptstyle\mathcal{B}} \\
Z & \xrightarrow[\text{A-V}]{\mathcal{H}} & Z_H
\end{array}
$$

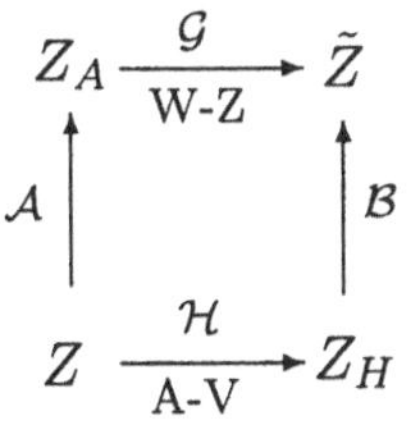

where $\mathcal{A}: Z \to Z_A$, $\mathcal{G}: Z_A \to \tilde{Z}$, $\mathcal{B}: Z_H \to \tilde{Z}$, $\mathcal{H}: Z \to Z_H$ (and $Z = (z,\theta)$) are superanalytic transformations (1). The transformation $\mathcal{G}$ plays the role of the linear transformation of Wess-Zumino type and the nonlinear transformation $\mathcal{H}$ (from a subgroup) is of Akulov-Volkov type, while $\mathcal{A}$ and $\mathcal{B}$ correspond to the transformations with Goldstone fields as parameters (Volkov (1969), Volkov (1973), Volkov and Akulov (1972)).

Global 2D supersymmetry. According to the general prescriptions (Ivanov and Kapustnikov (1978)) we can take $\mathcal{G}$ as a global linear supersymmetry transformation in two-dimensional case

$$\mathcal{G}:\begin{cases}\tilde{z}=z_A+\theta_A\cdot\varepsilon,\\ \tilde{\theta}=\varepsilon+\theta_A,\end{cases} \tag{9}$$

then we take $\mathcal{H}$ as an ordinary conformal transformation with composite parameters to be find and interpret $\mathcal{A}$ and $\mathcal{B}$ as coset transformations with the local odd parameters $\lambda(z)$ and $\tilde{\lambda}_{Glob}(z_H)$

$$\mathcal{A}:\begin{cases}z_A=z+\theta\cdot\lambda(z),\\ \theta_A=\lambda(z)+\theta,\end{cases}\quad \mathcal{B}:\begin{cases}\tilde{z}=z_H+\theta_H\cdot\tilde{\lambda}_{Glob}(z_H),\\ \tilde{\theta}=\tilde{\lambda}_{Glob}(z_H)+\theta_H.\end{cases} \tag{10}$$

Indeed, the commutativity of the diagram gives us the equation of $\lambda(z)$ evolution similar to (3) and (5) and equations for parameters of $\mathcal{H}$ in the following way. A "linear" transformation $\mathcal{G}$ is *representable* by a "nonlinear" transformation $\mathcal{H}$, iff the diagram is commutative

$$\mathcal{G}\circ\mathcal{A}=\mathcal{B}\circ\mathcal{H}. \tag{11}$$

In the group theory this construction is related to the induced representation (Kirillov (1976)). But here we, in general, do not demand invertibility of the entries in (11) and consider finite transformations. Using (11) we obtain the relations

$$\begin{aligned}\tilde{z}_{\mathcal{G}\circ\mathcal{A}}&=\tilde{z}_{\mathcal{B}\circ\mathcal{H}},\\ \tilde{\theta}_{\mathcal{G}\circ\mathcal{A}}&=\tilde{\theta}_{\mathcal{B}\circ\mathcal{H}}\end{aligned} \tag{12}$$

which are the representability condition (11) in coordinate language (as 4 component equations after expanding in θ).

In the particular case of global supersymmetry (9) the equations (12) and (10) give the parameters of the conformal transformation

$$\mathcal{H}:\begin{cases}z_H=z+\lambda(z)\cdot\varepsilon,\\ \theta_H=\theta,\end{cases} \tag{13}$$

and the evolution equation for $\tilde{\lambda}_{Glob}(z_H)=\varepsilon+\lambda(z)$. Then expanding on nilpotents

$$\varepsilon+\lambda(z)=\tilde{\lambda}_{Glob}(z)+\tilde{\lambda}'_{Glob}(z)\cdot\lambda(z)\cdot\varepsilon \tag{14}$$

which coincides with (5).

The $-\lambda$-rule in 2D. If $\mathcal{A}$ is invertible, the representability condition (11) becomes

$$\mathcal{G}=\mathcal{B}\circ\mathcal{H}\circ\mathcal{A}^{-1}. \tag{15}$$

In the global case invertibility of $\mathcal{A}$ is evident, then from (10) we derive

$$\mathcal{A}^{-1}:\begin{cases} z=z_A-\theta_A\cdot\lambda\left(z_A\right), \\ \theta=-\lambda\left(z_A\right)+\theta_A\left[1+\lambda\left(z_A\right)\cdot\lambda'\left(z_A\right)\right]. \end{cases} \tag{16}$$

This explains nature of the well-known "$-\lambda$ rule" (Ivanov and Kapustnikov (1978)) while comparing superfields of linear and nonlinear realizations (Wess (1983)). The relation (15) is a general form of the "splitting trick" (Ivanov and Kapustnikov (1978), Ivanov and Kapustnikov (1982)) according to which any linear superfield can be presented as a set of nonlinear transforming components. The analog of this trick for a noninvertible finite case is the representability condition (11), and it is not solved under $\mathcal{A}$. Thus, for a superfield $\Phi\left(z,\theta\right)$ we can write $\delta_{\mathcal{H}}\Phi\left(z,\theta\right)=\varepsilon\cdot\lambda\left(z\right)\cdot\frac{\partial\Phi(z,\theta)}{\partial z}$,where $\delta_{\mathcal{H}}$ is infinitesimal "nonlinear" transformation $\mathcal{H}$ corresponding to $\mathcal{G}$. If we use (16) and put $\Phi\left(z,\theta\right)=\Phi_A\left(z_A,\theta_A\right)$,then for infinitesimal "linear" transformation $\mathcal{G}$ we obtain the standard supersymmetry relation

$$\delta_{\mathcal{G}}\Phi_A\left(z_A,\theta_A\right)=\Phi\left(z_A+\varepsilon\cdot\theta_A,\theta_A+\varepsilon\right)-\Phi_A\left(z_A,\theta_A\right)=\varepsilon\cdot Q_A\Phi_A\left(z_A,\theta_A\right), \tag{17}$$

where Q_A is an ordinary supertranslation (Ivanov and Kapustnikov (1978)). Now we are ready to prove the "reversed" splitting trick which manifestly follows from the representability condition (11) applied to global two dimensional supersymmetry. It can be shown that a superfield $\Phi\left(z,\theta\right)$ transforming nonlinearly together with $\lambda\left(z\right)$ transforming as in (6) gives a linearly (globally) transformed superfield (17). Indeed, we see that $\Delta\Phi\left(z,\theta\right)=\delta_{\mathcal{G}}\Phi_A\left(z_A,\theta_A\right)$, where $\Delta\Phi\left(z,\theta\right)\stackrel{def}{=}\delta_{\mathcal{H}}\Phi\left(z,\theta\right)+\delta_{\mathcal{B}}\Phi\left(z,\theta\right)-\delta_{\mathcal{A}}\Phi\left(z,\theta\right)$.

Nonlinear realization of general finite $N=1$ superconformal transformations. Let us consider the representability condition (11) for a general $N=1$ superconformal-like transformations $Z_A\rightarrow\tilde{Z}$ which now play the role of "linear" ones. According to (Duplij (1996), Duplij (1997)) they can be parametrized by two functions $g\left(z_A\right)$ and $\psi\left(z_A\right)$ and have the form

$$\mathcal{G}:\begin{cases} \tilde{z}=f_n^{(g\psi)}\left(z_A\right)+\theta_A\cdot\chi_n^{(g\psi)}\left(z_A\right), \\ \tilde{\theta}=\psi\left(z_A\right)+\theta_A\cdot g\left(z_A\right), \end{cases} \tag{18}$$

where

$$\begin{aligned} f_n^{(g\psi)\prime}\left(z_A\right)&=\psi'\left(z_A\right)\psi\left(z_A\right)+\tfrac{1+n}{2}\cdot g^2\left(z_A\right), \\ \chi_n^{(g\psi)\prime}\left(z_A\right)&=g'\left(z_A\right)\psi\left(z_A\right)+n\cdot g\left(z_A\right)\psi'\left(z_A\right), \end{aligned} \tag{19}$$

where $n=\begin{cases} +1, & \text{SCf transformation,} \\ -1, & \text{TPt transformation,} \end{cases}$ is a projection of "reduction spin" switching the type of transformation (see (Duplij (1997)) for more details).

Then while trying to represent $\mathcal{G}$ in terms of nonlinear compositions we face with the following restriction which is consequence of the $N=1$ superconformal-like multiplication law (Duplij (1997)). If $\mathcal{T}$ is a superconformal-like transformation, then there are only two possibilities in the composition $z\xrightarrow{\mathcal{T}}\tilde{z}\xrightarrow{\tilde{\mathcal{T}}}\tilde{\tilde{z}}$

$$\begin{aligned} \tilde{\mathcal{T}}_{SCf} * \mathcal{T}_{SCf} &= \tilde{\tilde{\mathcal{T}}}_{SCf}, \\ \tilde{\mathcal{T}}_{TPt} * \mathcal{T}_{SCf} &= \tilde{\tilde{\mathcal{T}}}_{TPt}. \end{aligned} \tag{20}$$

Therefore, we have only two possibilities to include TPt transformations into the diagrammatic representation, viz.

$$\mathcal{G}_{SCf} \circ \mathcal{A}_{SCf} = \mathcal{B}_{SCf} \circ \mathcal{H}_{SCf}, \tag{21}$$

$$\mathcal{G}_{TPt} \circ \mathcal{A}_{SCf} = \mathcal{B}_{TPt} \circ \mathcal{H}_{SCf}. \tag{22}$$

The first one is the nonlinear representation of $N = 1$ superconformal group in analogy with the ordinary infinitesimal invertible four-dimensional case (Ivanov and Kapustnikov (1978), Ivanov and Kapustnikov (1990)) (and 11) in which $\mathcal{A}_{SCf}$ and $\mathcal{B}_{SCf}$ play the role of cosets.

Let us consider (21) in more detail. The exact shape of cosets $\mathcal{A}_{SCf}$ and $\mathcal{B}_{SCf}$ can be taken as

$$\mathcal{A}_{SCf} : \begin{cases} z_A = z + \theta \cdot \lambda(z), \\ \theta_A = \lambda(z) + \theta\sqrt{1 + \lambda(z) \cdot \lambda'(z)}, \end{cases} \tag{23}$$

$$\mathcal{B}_{SCf} : \begin{cases} \tilde{z} = z_H + \theta_H \cdot \tilde{\lambda}(z_H), \\ \tilde{\theta} = \tilde{\lambda}(z_H) + \theta_H\sqrt{1 + \tilde{\lambda}(z_H) \cdot \tilde{\lambda}'(z_H)}, \end{cases} \tag{24}$$

and for $\mathcal{H}$ we choose the following general parametrization

$$\mathcal{H}_{SCf} : \begin{cases} z_H = p(z), \\ \theta_H = \rho(z) + \theta \cdot q(z) \end{cases} \tag{25}$$

Then, expanding the coordinate form (12) into components we obtain 4 corresponding equations for 4 unknown functions $p(z), q(z), \rho(z), \tilde{\lambda}(z)$

$$p(z) + \rho(z) \cdot \tilde{\lambda}(p(z)) = f_{+1}^{(g\psi)}(z) + g(z) \cdot \lambda(z) \cdot \psi(z), \tag{26}$$

$$\tilde{\lambda}(p(z)) + \rho(z) \cdot \sqrt{1 + \tilde{\lambda}(p(z)) \cdot \tilde{\lambda}'(p(z))} = \psi(z) + g(z) \cdot \lambda(z), \tag{27}$$

$$\begin{aligned} q(z) \cdot \tilde{\lambda}(p(z)) = \lambda(z) \cdot f_{+1}^{(g\psi)\prime}(z) + \\ g(z) \cdot \psi(z) \cdot \sqrt{1 + \lambda(z) \cdot \lambda'(z)}, \end{aligned} \tag{28}$$

$$\begin{aligned} q(z) \cdot \sqrt{1 + \tilde{\lambda}(p(z)) \cdot \tilde{\lambda}'(p(z))} = \lambda(z) \cdot \psi'(z) + \\ g(z) \cdot \sqrt{1 + \lambda(z) \cdot \lambda'(z)}, \end{aligned} \tag{29}$$

where $f_{+1}^{(g\psi)}(z)$ is determined from (19).

In case $q(z)$ and $g(z)$ are invertible, these equations have the following solution for parameters of nonlinear $\mathcal{H}$ transformation in terms of parameters of "linear" $\mathcal{G}$ transformation as

$$p(z) = f_{+1}^{(g\psi)}(z) + g(z) \cdot \lambda(z) \cdot \psi(z), \tag{30}$$

$$q(z) = \sqrt{p'(z)}, \tag{31}$$

$$\rho(z) = 0, \tag{32}$$

and for goldstino transformation rule

$$\tilde{\lambda}(p(z)) = \psi(z) + g(z) \cdot \lambda(z), \tag{33}$$

that naturally coincides with the previous approach (4) with $f(z) = f_{+1}^{(g\psi)}(z)$ and $\chi(z) = g(z) \cdot \psi(z)$.

Therefore, $\mathcal{H}$ is the split $N = 1$ SCf transformation (Friedan (1986))

$$\mathcal{H}_{SCf} : \begin{cases} z_H = p(z), \\ \theta_H = \theta \cdot \sqrt{p'(z)} \end{cases} \tag{34}$$

with the composite parameter $p(z)$ from (30), which can be presented as the following commutative diagram

$$\begin{array}{ccc} Z_A & \xrightarrow[\text{full}]{\mathcal{G}_{SCf}} & \tilde{Z} \\ \mathcal{A}_{SCf} \Big\uparrow & & \Big\uparrow \mathcal{B}_{SCf} \\ Z & \xrightarrow[\text{split}]{\mathcal{H}_{SCf}} & Z_H \end{array}$$

Second relation (22) and the corresponding commutative diagram

$$\begin{array}{ccc} Z_A & \xrightarrow{\mathcal{G}_{TPt}} & \tilde{Z} \\ \mathcal{A}_{SCf} \Big\uparrow & & \Big\uparrow \mathcal{B}_{TPt} \\ Z & \xrightarrow{\mathcal{H}_{SCf}} & Z_H \end{array}$$

have no such transparent meaning, because $\mathcal{B}_{TPt}$ is noninvertible, and so it cannot be a standard coset. Nevertheless, since the final answer for the nonlinear transformation $\mathcal{H}_{SCf}$ is known from the previous approach (7), the noninvertible analog of coset $\mathcal{B}_{TPt}$ can be found in principle from the system of equations analogous to (26)–(29).

Let we write $\mathcal{B}_{TPt}$ in the form

$$\mathcal{B}_{SCf} : \begin{cases} \tilde{z} = f_{-1}^{(\widetilde{b\lambda})}(z_H) + \theta_H \cdot \chi_{-1}^{(\widetilde{b\lambda})}(z_H), \\ \tilde{\theta} = \tilde{\lambda}(z_H) + \theta_H \cdot b(z_H), \end{cases} \tag{35}$$

where

$$\begin{aligned} f_n^{(b\widetilde{\lambda})\prime}(z_H) &= \widetilde{\lambda}'(z_H)\cdot\widetilde{\lambda}(z_H)+\tfrac{1+n}{2}\cdot b^2(z_H), \\ \chi_n^{(b\widetilde{\lambda})\prime}(z_H) &= b'(z_H)\cdot\widetilde{\lambda}(z_H)+n\cdot b(z_H)\cdot\widetilde{\lambda}'(z_H), \end{aligned} \tag{36}$$

and prime denotes derivative by argument. So the corresponding system of equations now is

$$f_{-1}^{(b\widetilde{\lambda})}(p(z))+\rho(z)\cdot\chi_{-1}^{(b\widetilde{\lambda})}(p(z))=f_{+1}^{(g\psi)}(z)+\lambda(z)\cdot\chi_{+1}^{(g\psi)}(z), \tag{37}$$

$$\widetilde{\lambda}(p(z))+\rho(z)\cdot b(p(z))=\psi(z)+g(z)\cdot\lambda(z), \tag{38}$$

$$\begin{aligned} \rho(z)\cdot f_{-1}^{(b\widetilde{\lambda})\prime}(p(z))+q(z)\cdot\chi_{-1}^{(b\widetilde{\lambda})}(p(z)) = \lambda(z)\cdot f_{+1}^{(g\psi)\prime}(z)+ \\ \chi_{+1}^{(g\psi)}(z)\cdot\sqrt{1+\lambda(z)\cdot\lambda'(z)}, \end{aligned} \tag{39}$$

$$\begin{aligned} \rho(z)\cdot q(z)\cdot\widetilde{\lambda}'(p(z))+q(z)\cdot b(p(z)) = \lambda(z)\cdot\psi'(z)+ \\ g(z)\cdot\sqrt{1+\lambda(z)\cdot\lambda'(z)}. \end{aligned} \tag{40}$$

In case $\mathcal{A}_{SCf}$ is invertible we can obtain $\mathcal{G}_{TPt}=\mathcal{B}_{TPt}\circ\mathcal{H}_{SCf}\circ\mathcal{A}_{SCf}^{-1}$ which gives an analog of nonlinear realization for noninvertible TPt transformations.

The author is grateful to V. P. Akulov, A. P. Demichev, E. A. Ivanov, A. A. Kapustnikov, A. I. Pashnev and L. L. Vaksman for fruitful conversations and remarks.

References

Akulov, V. P. and Volkov, D. V. (1974): Goldstone fields with spin 1/2, *Theor. Math. Phys.* **18**, 35–54.

Belucci, S., Gribanov, V., Ivanov, E., Krivonos, S., and Pashnev, A. (1998): Nonlinear realizations of superconformal and W algebras as embeddings of strings, *Nucl. Phys.* **B510**, 477–501.

Berkovits, N. and Vafa, C. (1994): On the uniqueness of string theory, *Mod. Phys. Lett.* **A9**, 653–657.

Coleman, S., Wess, J., and Zumino, B. (1969): Structure of phenomenological lagrangians. **I**, *Phys. Rev.* **177**, 2239–2247.

De Witt, B. S. (1992): *Supermanifolds*, 2nd edition (Cambridge Univ. Press: Cambridge).

Duplij, S. (1991): On semigroup nature of superconformal symmetry, *J. Math. Phys.* **32**, 2959–2965.

Duplij, S. (1996): Ideal structure of superconformal semigroups, *Theor. Math. Phys.* **106**, 355–374.

Duplij, S. (1997): Noninvertible $N=1$ superanalog of complex structure, *J. Math. Phys.* **38**, 1035–1040.

Friedan, D. (1986): Notes on string theory and two dimensional conformal field theory, in *Unified String Theories*, Green, M. and Gross, D., eds. (World Sci.: Singapore), pp. 118–149.

Ivanov, E. A. and Kapustnikov, A. A. (1978): General relationship between linear and nonlinear realisations of supersymmetry, *J. Phys.* **A11**, 2375–2384.

Ivanov, E. A. and Kapustnikov, A. A. (1982): The non-linear realisation structure of modeals with spontaneously broken supersymmetry, *J. Phys.* **G8**, 167–191.

Ivanov, E. A. and Kapustnikov, A. A. (1990): Geometry of spontaneously broken local $N = 1$ supersymmetry in superspace, *Nucl. Phys.* **B333**, 439–470.

Kirillov, A. A. (1976): *Elements of the Theory of Representations* (Springer-Verlag: Berlin).

Kunitomo, H. (1995): On the nonlinear realization of the superconformal symmetry, *Phys. Lett.* **B343**, 144–146.

Rogers, A. (1980): A global theory of supermanifolds, *J. Math. Phys.* **21**, 1352–1365.

Samuel, S. and Wess, J. (1983): A superfield formulation of the non-linear realization of supersymmetry and its coupling to supergravity, *Nucl. Phys.* **B221**, 153–177.

Volkov, D. V. and Akulov, V. P. (1972): On the possible universal neutrino interaction, *JETP Lett.* **16**, 621–624.

Volkov, D. V. and Akulov, V. P. (1973): Is the neutrino a Goldstone particle?, *Phys. Lett.* **B46**, 109–112.

Volkov, D. V. (1969): Phenomenological interaction lagrangian of Goldstone particles, Kiev preprint, *Inst. Theor. Phys.*: ITF-69-75.

Volkov, D. V. (1973): Phenomenological lagrangians, *Sov. J. Part. Nucl.* **4**, 1–17.

Wess, J. (1983): Nonlinear realization of the $N = 1$ supersymmetry, in *Quantum Theory of Particles and Fields*, Jancewicz, B. and Lukierski, J., eds. (World Sci.: Singapore), pp. 223–234.

Wess, J. (1984): Nonlinear realization of supersymmetry, in *Mathematical Aspects of Superspace*, Seifert, H.-J., Clarke, C. J., and Rosenblum, A., eds. (D. Reidel: Dordrecht), pp. 1–14.

$N = 2$ SUSY Two-Boson KP Hierarchy, (Derivative) NLS Equation and Miura Transformations

S. Krivonos[1] and Z. Popowicz[2]

[1] JINR–Bogoliubov Laboratory of Theoretical Physics, 141980 Dubna, Moscow Region, Russia
[2] Institute of Theoretical Physics, University of Wroclaw, ul. Cybulskiego 36, 50-205 Wroclaw, Poland

Abstract. We present the $N = 2$ supersymmetric extensions of the derivative NLS equation and quadratic two-boson KP hierarchy. We propose the KP-like Lax operators in terms of the $N = 2$ superfields which reproduce all conserved currents through the non-standard Lax representations. The connection of these derivative supersymmetric NLS equations with the $N = 2$, $a = 4$ super-KdV hierarchy is established.

1 Introduction

The motivations for studying the N=2 supersymmetric hierarchies are diverse. On the one hand we have quite a good acquaintance with purely bosonic hierarchies and their connection with physical models, while our knowledge of $N = 2$ supersymmetric integrable hierarchies is still scanty. On the other hand the supersymmetrization of the solitons equations extended the class of the integrable models. In order to obtain such supersymmetric theory we have to add to a system of k bosonic equations kN fermions and $k(N-1)$ bosons fields ($k = 1, 2, .., N = 1, 2..$) in such a way that the final theory becomes supersymmetric invariant. Interestingly enough, it appeared that during the supersymmetrizations, some typical supersymmetric effects (compare with the classical theory) occurred. We mention a few of them: the non-uniqueness of the roots for the supersymmetric Lax operator (Oevel and Popowicz (1991)), the lack of bosonic reduction to the classical equations (Ivanov and Krivonos (1993)) and the occurrence of non-local conservation laws (Kersten (1988), Dagris and Mathieu (1993)). These effects rely strongly on the descriptions of the generalized classical systems of equations which we would like to supersymmetrize.

From the soliton point of view we can distinguish two important classes of the supersymmetric equations: the non-extended ($N = 1$) and extended ($N > 1$) cases. Considerations of the extended case may imply new bosonic equations whose properties need further investigation. This may be viewed as a bonus, but this extended case is in no way more fundamental than the non-extended one. Up to now many solitons equations have been supersymmetrized.

The Korteweg de Vries (KdV) hierarchy and its supersymmetric extensions were the subject of many studies for the last several years [4-10]. One of the remarkable properties of these systems is that they are related, via the second hamiltonian structure, to the classical (super) conformal algebras Labelle and Mathieu (1991). Up to now, supersymmetric KdV hierarchies have been constructed for $N = 1, 2, 3$ and 4, based on the above mentioned connection to superconformal algebras. An interesting peculiarity is that, beginning with $N = 2$, the supersymmetric KdV equations turn out to be integrable only for special choice of the parameters in the hamiltonian. There exist only three integrable $N = 2$ KdV hierarchies: the $a = 4, a = -2$ and $a = 1$ ones (Labelle and Mathieu (1991), Popowicz (1993)), with a parameter entering into the $N = 2$ KdV hamiltonian. Interestingly for any value of a the related $N = 2$ supersymmetric KdV equation possesses $N = 2$ supersymmetric conformal algebra as the second hamiltonian structure. The generalizes $N = 2$ supersymmetric systems associated with $N = 2\ W_n$ algebras have similar properties (Bonora, Krivonos and Sorin (1996), Bonora, Krivonos and Sorin (1998), Delduc and Gallot (1997), Krivonos and Sorin (1997)).

Interestingly there is a connection (Krivonos and Sorin (1995)) between the supersymmetric Nonlinear Schrödinger equation (NLS) and the special $N = 2$ supersymmetric version of KdV with the parameter $a = 4$. There exist a Backlund transformation which maps the super-NLS equation into the second flow of the hierarchy associated to the $N = 2, a = 4$, KdV equation, the super-KdV itself being associated to the third flow.

It is know that the classical NLS equation is connected with two different generalization of NLS equation. One of them is the so called Derivative NLS considered firstly by Kaup and Newell (Kaup and Newell (1978)) while second was constructed by Chen, Lee and Lin (Chen, Lee and Lin (1979)). These equations are gauge equivalent with the usual NLS. The aim of the present paper is to show that the super-NLS equation is also connected, via Miura transformations, with two different generalization of the super-NLS. We firstly postulate the supersymmetric versions of these Derivative NLS equations and then show that they are gauge equivalent to the super-NLS equation. Our Miura transformations, which give rise to the super-Derivative NLS equations, are non-polynomial in the original fields (and their (super)derivatives) . We are able to construct for the super-Derivative NLS equations their manifestly $N = 2$ supersymmetric Lax operators which generate the infinite tower of hamiltonians in involution. This construction is carried out in two steps: first we defined the supersymmetric version of the classical two-boson hierarchy (Aratyn at al (1993)) and then we constructed the corresponding supersymmetric Lax operators for the super-DNLS equations.

All results presented in this paper have been obtained using the symbolic computer languages Reduce and Mathematica (Popowicz (1997), Krivonos and Thielemans (1996)).

2 Classical Two-Boson KP Hierarchy

Let us start from the brief recalling of the main features of the quadratic two-boson KP hierarchy (Aratyn at al (1993)).

The quadratic two-boson KP hierarchy is defined via the following Lax operator

$$L_j = \partial_z + \bar{j}\,(\partial_z - j - \bar{j})^{-1}\, j\,, \tag{1}$$

and the flows

$$\frac{\partial L_j}{\partial t_n} = \left[(L_j^n)_+, L_j\right]\,, \tag{2}$$

where subscript (+) means the differential part of L_j^n.

The corresponding second Poisson structure is given by

$$P_2(j,\bar{j}) = \begin{pmatrix} 0, & \partial \\ \partial, & 0 \end{pmatrix}, \tag{3}$$

Here we used the standard notation encountered in the soliton theory but this Poisson tensor could be rewritten equivalently as the operator product expansion (OPE). In the next we will use OPE language and the last formula reads as

$$j(z_1)\bar{j}(z_2) = \frac{1}{z_{12}^2}\,,\; j(z_1)j(z_2) = \bar{j}(z_1)\bar{j}(z_2) = 0\,, \tag{4}$$

where $z_{12} = z_1 - z_2$. The first three lowest Hamiltonians have the following form

$$H_1 = \int dz j\bar{j}\,,\; H_2 = \int dz\left(-j'\bar{j} + j^2\bar{j} + j\bar{j}^2\right)\,,$$
$$H_3 = \int dz\left(j''\bar{j} - 3jj'\bar{j} + 3j\bar{j}\bar{j}' + j^3\bar{j} + 3j^2\bar{j}^2 + j\bar{j}^3\right)\,. \tag{5}$$

3 $N = 2$ Supersymmetric Quadratic Two-Boson KP Hierarchy

The key point to construct the N=2 supersymmetric extension of the quadratic two-boson KP hierarchy is its gauge equivalence to the linear two-boson KP hierarchy (Bonora and Xiong (1992)). Namely, after the following Miura transformation

$$S(z) = j + \bar{j}\,,\; R(z) = j\bar{j} + \bar{j}'\,, \tag{6}$$

the new defined currents $S(z)$ and $R(z)$ possess the following second Poisson structure (Bonora and Xiong (1992))

$$R(z_1)R(z_2) = \frac{2R(z_2)}{z_{12}} + \frac{R'(z_2)}{z_{12}^2}\,,\ S(z_1)S(z_2) = -\frac{2}{z_{12}^2}\,,$$
$$R(z_1)S(z_2) = \frac{2}{z_{12}^3} + \frac{S(z_2)}{z_{12}} + \frac{S'(z_2)}{z_{12}^2} \tag{7}$$

The Hamiltonians $H_1 - H_3$ (5) can be rewritten in terms of currents $S(z)$ and $R(z)$ as

$$H_1 = \int dz R\,,\ H_2 = \int dz RS\,,\ H_3 = \int dz\left(R^2 + RS^2 + R'S\right)\,, \tag{8}$$

while Lax operator has the form (Bonora and Xiong (1992))

$$L = \partial_z + R\frac{1}{\partial_z - S}\,. \tag{9}$$

As was shown in (Krivonos and Sorin (1995)), the Hamiltonian structure (7), which is as extension of Virasoro algebra with stress-tensor R by the spin 1 current S, can be immediately extended to the N=2 supersymmetric case by combining the currents S and R into one spin 1 N=2 bosonic supercurrent $J(Z)$ generating N=2 super conformal algebra:

$$J(Z_1)J(Z_2) = \frac{c/4}{Z_{12}^2} + \frac{\bar{\theta}_{12}\bar{D}J(Z_2)}{Z_{12}} - \frac{\theta_{12}DJ(Z_2)}{Z_{12}} + \frac{\theta_{12}\bar{\theta}_{12}J(Z_2)}{Z_{12}^2} + \frac{\theta_{12}\bar{\theta}_{12}J'(Z_2)}{Z_{12}}\,, \tag{10}$$

where

$$Z = (z,\theta,\bar{\theta}),\ \theta_{12} = \theta_1 - \theta_2,\ \bar{\theta}_{12} = \bar{\theta}_1 - \bar{\theta}_2,\ Z_{12} = z_1 - z_2 + \frac{1}{2}\left(\theta_1\bar{\theta}_2 - \theta_2\bar{\theta}_1\right)\,, \tag{11}$$

and $D, \bar{D}$ are the spinor covariant derivatives

$$D = \frac{\partial}{\partial\theta} - \frac{1}{2}\bar{\theta}\frac{\partial}{\partial z}\quad,\quad \bar{D} = \frac{\partial}{\partial\bar{\theta}} - \frac{1}{2}\theta\frac{\partial}{\partial z} \tag{12}$$
$$\{D,\bar{D}\} = -\frac{\partial}{\partial z}\quad,\quad \{D,D\} = \{\bar{D},\bar{D}\} = 0.$$

The components currents are defined as follows:

$$S = 2J|\,,\ R = \left(\left[D,\bar{D}\right]J - J'\right)|\,,\ \Xi = DJ|\,,\ \bar{\Xi} = \bar{D}J| \tag{13}$$

and the sign $|$ means putting $\theta, \bar{\theta}$ equal to zero.

The N=2 supersymmetric extension of the linear two-boson KP hierarchy (7)-(9) has been defined in (Krivonos, Sorin and Toppan (1995)) via the following Lax operator

$$\mathcal{L} = \partial_z - 2J - 2\bar{D}\partial_z^{-1}(DJ)\,, \tag{14}$$

and flows defined as

$$\frac{\partial \mathcal{L}}{\partial t_n} = \left[(\mathcal{L}^n)_{\geq 1}, \mathcal{L} \right] , \tag{15}$$

where subscript (≥ 1) means the purely derivative part of $\mathcal{L}^n$. We have explicitly the first three flows

$$\frac{\partial J}{\partial t_1} = J', \ \frac{\partial J}{\partial t_2} = \left[D, \bar{D} \right] J' + 4JJ',$$
$$\frac{\partial J}{\partial t_3} = -J''' + 6(\bar{D}JDJ)' - 6(J\left[D, \bar{D} \right] J)' - 4(J^3)'. \tag{16}$$

After fixing the unessential value of central charge in (10) equals to $c = 8$, the first three Hamiltonians read as

$$\mathcal{H}_1 = \int dz d\theta d\bar{\theta} J \ , \ \mathcal{H}_2 = \int dz d\theta d\bar{\theta} J^2 \ ,$$
$$\mathcal{H}_3 = \int dz d\theta d\bar{\theta} (J^3 + \frac{3}{4} J \left[D, \bar{D} \right] J). \tag{17}$$

It was shown recently (Krivonos and Sorin (1995)) that this N=2 supersymmetric linear two-boson KP hierarchy coincides with the N=2, $a = 4$ super-KdV hierarchy. On the other side we can embed the N=2 supersymmetric (Derivative) Nonlinear Schrödinger Equation in second flows. In order to do it let us define two different Miura transformations. The first transformation is local but not polynomial in the fermionic chiral and anti-chiral superfileds F and $\bar{F}$ and reads (Krivonos, Sorin and Toppan (1995))

$$J = \frac{1}{4} \bar{F} F - \frac{(D\bar{F})'}{2D\bar{F}}, \tag{18}$$

where $(DF) = 0$ and $(\bar{D}\bar{F}) = 0$.

Now we can quickly recognize that the N=2 Supersymmetric Nonlinear Schrödinger Equation considered in (Krivonos and Sorin (1995))

$$\frac{\partial}{\partial t} F = F'' - FD(\bar{F}\bar{D}F), \ \frac{\partial}{\partial t} \bar{F} = -\bar{F}'' + \bar{F}\bar{D}(FD\bar{F}). \tag{19}$$

could be embedded in the second flows using the previous transformation. As we will show in the next section the supersymmetric generalization of the derivative nonlinear Schrödinger equation can be also connected with the supersymmetric nonlinear Schrödinger equation.

Inserting the above Miura transformation to the Lax operator we obtain the following operator

$$L = \partial_z - \frac{1}{2} \bar{F} F + \frac{(D\bar{F})'}{DF} - \frac{1}{2} \bar{D} \partial_z^{-1} F (D\bar{F}). \tag{20}$$

which produce consistent equation of motion. We can remove the denominator, in the third term of the Lax operator utilizing the gauge transformation which maps $\mathcal{L}$ into $\bar{L}$ through

$$\bar{L} = (D\bar{F})L\frac{1}{(D\bar{F})}. \tag{21}$$

However this $\bar{L}$ does not produce consistent equation of motion but its formal adjoint $\bar{L}^*$ (Krivonos, Sorin and Toppan (1995))

$$\bar{L}^* = \partial_z + \frac{1}{2}\bar{F}F - \frac{1}{2}F\bar{D}\partial_z^{-1}(D\bar{F}), \tag{22}$$

give us correct equation (19).

The second transformation is local in the fermionic chiral and anti-chiral fermionic spin 1/2 superfields $Q(Z)$ and $\bar{Q}(Z)$

$$2J = D\bar{Q} - \bar{D}Q - Q\bar{Q}, \tag{23}$$

where $(DQ) = 0$ and $(\bar{D}\bar{Q}) = 0$ It is essentially the well known realization of N=2 superconformal algebra in terms of N=2 chiral fermionic superfields Q and $\bar{Q}$ with unique defined Poisson structure

$$\begin{aligned} Q(Z_1)\bar{Q}(Z_2) &= \frac{1}{Z_{12}} - \frac{\theta_{12}\bar{\theta}_{12}}{2Z_{12}^2}\,, \\ Q(Z_1)Q(Z_2) &= \bar{Q}(Z_1)\bar{Q}(Z_2) = 0\,. \end{aligned} \tag{24}$$

It is evident, that this transformation maps the Poisson structure (24) into (10) and the first Hamiltonians in terms of Q and $\bar{Q}$ read as

$$\begin{aligned} \mathcal{H}_1 &= \int dz d\theta d\bar{\theta} Q\bar{Q}\,,\ \mathcal{H}_2 = \int dz d\theta d\bar{\theta}\,(Q\bar{Q}' + (\bar{D}Q - D\bar{Q})Q\bar{Q}) \\ \mathcal{H}_3 &= \int dz d\theta d\bar{\theta}(-\bar{Q}'Q' + 3\bar{Q}'Q(DQ) + 3\bar{Q}Q'(D\bar{D}Q) + (D\bar{Q})(\bar{D}Q)' \\ &\quad -(D\bar{Q}'(\bar{D}Q) - \bar{Q}Q((D\bar{Q})^2 + (\bar{D}Q)^2 - 3(D\bar{Q})(\bar{D}Q))). \end{aligned} \tag{25}$$

This second Miura transformations maps the following equations

$$\begin{aligned} \frac{\partial}{\partial t}Q &= -Q'' - 2(QD\bar{Q})' + 2Q'(\bar{D}Q), \\ \frac{\partial}{\partial t}\bar{Q} &= \bar{Q}'' + 2(\bar{Q}\bar{D}Q)' - 2\bar{Q}'D\bar{Q}, \end{aligned} \tag{26}$$

into the equation (16). Applying the transformations (23) to the Lax operator (14) we obtained new Lax operator responsible for the equations (26)

$$L = \partial_z - D\bar{Q} + \bar{D}Q + Q\bar{Q} - \bar{D}\partial^{-1}(-Q' + QD\bar{Q}), \tag{27}$$

As we see in the next section these equations could be connected with supersymmetric generalizations of the two different derivative Nonlinear Schrödinger Equations.

4 SuSy Derivative Nonlinear Schrödinger Equation

As we saw in the previous section it was possible by the use of the Miura transformation to embed the supersymmetric generalizations of the Nonlinear Schrödinger Equation into the second flows of the quadratic two-boson KP hierarchy. Now we would like to proceed in the similar manner with the supersymmetric generalization of the Derivative Nonlinear Schrödinger Equation (DNLS).

The classical DNLS has important applications in such fields as plasma physics and nonlinear optics and are completely integrable Hamiltonian systems. It appeared that we have two different types of DNLS (Kaup and Newell (1978), Chen, Lee and Lin (1979)). The first one reads

$$\frac{\partial}{\partial t}q = (q' + 2q^2r)', \quad \frac{\partial}{\partial t}r = (-r' + 2r^2q)', \tag{28}$$

while second is

$$\frac{\partial}{\partial t}w = w'' + 2uww', \quad \frac{\partial}{\partial t}u = -u'' + 2wuu'. \tag{29}$$

These equations constitute the bihamiltonian systems with the following scalar Lax operators

$$\begin{aligned} L &= \partial + 2qr + q_x/q - \partial^{-1}(qr_x - q^2r^2), \\ L &= \partial + wu + w_x/w - \partial^{-1}uw_x. \end{aligned} \tag{30}$$

Moreover these equations are directly connected with the standard Nonlinear Schrödinger Equation

$$\frac{\partial}{\partial t}f = f'' - 2\bar{f}f^2, \quad \frac{\partial}{\partial t}\bar{f} = -\bar{f}'' + 2f\bar{f}^2. \tag{31}$$

via the following Miura transformations

$$f = q\exp\Big(2\int_{-\infty}^{x} qrdx\Big), \quad \bar{f} = \Big(r_x - qr^2\Big)\exp\Big(-2\int_{-\infty}^{x} qrdx\Big), \tag{32}$$

and

$$f = w\exp\Big(\int_{-\infty}^{x} wudx\Big), \quad \bar{f} = u_x\exp\Big(-\int_{-\infty}^{x} wudx\Big), \tag{33}$$

Let us now return back to the construction of the supersymmetric version of the equations (28-29). We show that these equations are connected with the equations (26). In order to end it let us define two different Miura transformations for the chiral and antichiral fields Q and $\bar{Q}$ in terms of the zero dimensional chiral and anitchiral superfermions fields $H(Z)$ and $\bar{H}(Z)$ as

$$\bar{Q} = -\frac{\bar{H}'}{D\bar{H}} - \beta\bar{D}H\bar{H}, \quad Q = D\bar{H}H, \tag{34}$$

where β takes two values 0 or 1 and $DH = 0, \bar{D}\bar{H} = 0$.

Inserting these expressions for the superfields Q and $\bar{Q}$ into the Miura transformation for J (23) we obtained

$$J = -(\beta+1)D\bar{H}\bar{D}H + \beta H'\bar{H} - \frac{D\bar{H}'}{D\bar{H}} + \beta H\bar{H}D\bar{H}\bar{D}H. \tag{35}$$

It is also possible to define this transformation in the different manner

$$\bar{Q} = -\bar{H}\bar{D}H, \; Q = -\frac{H'}{\bar{D}H} + \beta HD\bar{H}, \tag{36}$$

with the following representation for J

$$J = -(\beta+1)D\bar{H}\bar{D}H - \beta H\bar{H}' + \frac{\bar{D}H'}{\bar{D}H} + \beta H\bar{H}D\bar{H}\bar{D}H. \tag{37}$$

In the next we consider the formulas (34-35) only.

Now we can easily find the second flow equations for the superfields $H, \bar{H}$

$$\begin{aligned}\frac{\partial}{\partial t}H &= -H'' + 2\beta\bar{H}'H'H + 2\beta HD\bar{H}'\bar{D}H + 2(1+\beta)H'D\bar{H}\bar{D}H,\\ \frac{\partial}{\partial t}\bar{H} &= \bar{H}'' - 2\beta\bar{H}'H'\bar{H} + 2\beta\bar{H}D\bar{H}\bar{D}H' + 2(1+\beta)\bar{H}'D\bar{H}\bar{D}H,\end{aligned} \tag{38}$$

We quickly recognize that this systems contains the equations (28-29) as the particular case if the components of H and $\bar{H}$ are defined as follows: for $\beta = 1$ as

$$q = \bar{D}H| \,, r = D\bar{H}| \,, \tag{39}$$

while for $\beta = 0$ as

$$w = \bar{D}H| \,, u = D\bar{H}| \,, \tag{40}$$

Therefore we call the equations (38) by the Supersymmetric Derivative NLS equations. The Lax operator for these equations explicitly reads

$$\begin{aligned}L = \partial &+ (\beta+1)D\bar{H}\bar{D}H + \beta H\bar{H}' + \frac{D\bar{H}'}{D\bar{H}} + \beta H\bar{H}D\bar{H}\bar{D}H\\ &- D\partial^{-1}\left[H'D\bar{H} - \beta H(D\bar{H})^2\bar{D}H - \beta H\bar{H}H'D\bar{H}\right].\end{aligned} \tag{41}$$

To obtain the conservation currents for our super-DNLS equation one should to substitute the representation (35) into (17).

To finish this section let us notice that our Lax operator is gauge equivalent to the operator (22). It can be easily noticed that the operator

$$L := \left(G(DH)L(G(DH))^{-1}\right)^*, \tag{42}$$

coincides with the Lax operator (22) if

$$F = (H' - \beta H(D\bar{H})(\bar{D}H) - \beta H\bar{H}H')G^{-1} \,, \; \bar{F} = \bar{H}G \tag{43}$$

where

$$G = \exp\left((\beta+1)\int_{-\infty}^{x}\left((D\bar{H})(\bar{D}H) - H'\bar{H}\right)dx\right). \tag{44}$$

Thus Supersymmetric Derivative NLS equation (38) can be embedded into the supersymmetric nonlinear Schrödinger equation.

References

W. Oevel and Z. Popowicz (1991), Commun.Math.Phys. **139** 441.
E. Ivanov and S. Krivonos (1993), Phys.Lett. **B309** 63.
P. Kersten (1988), Phys.Lett. **A134** 25.
P. Dagris and P. Mathieu (1993), Phys. Lett **A176** 67.
C.A. Laberge and P. Mathieu (1988), Phys.Lett. **B215** 718.
P. Labelle and P. Mathieu (1991), J. Math. Phys. **32** 923.
Z. Popowicz (1993), Phys. Lett. **A174** 411.
F. Delduc and E. Ivanov (1993), Phys. Lett. **B309** 312.
E. Ivanov and S. Krivonos (1997), Phys. Lett. **A231** 75.
S. Bellucci, E. Ivanov and S. Krivonos (1993), J. Math. Phys. **34** 3087.
C. Yung (1993), Mod. Phys. Lett. **A8** 1161.
L. Bonora, S. Krivonos and A. Sorin (1996), Nucl. Phys. **B477** 835.
L. Bonora, S. Krivonos and A. Sorin (1998), Lett. Math. Phys. **45** 63.
F. Delduc and L. Gallot (1997), Comm. Math. Phys. **190** 395.
S. Krivonos and A. Sorin (1997), *"Extended N=2 supersymmetric matrix (1,s) KdV hierarchies"*, solv-int/9712002.
S. Krivonos and A. Sorin (1995), Phys. Lett. **B357** 94.
D. Kaup and A. Newell (1978), J. Math. Phys. **19** 798.
H. Chen, Y. Lee and C. Lin (1979), Phys. Scr. **20** 490.
H. Aratyn, L.A. Ferreira, J.F. Gomes and A.H. Zimerman (1993), Phys. Lett. **B316** 85.
Z. Popowicz (1997), Computer Physics Comm. **100** 277.
S.Krivonos and K.Thielemans (1996), Class. Quant. Grav. **13** 2899.
L. Bonora and C.S. Xiong (1992), Phys. Lett. **B285** 191; Phys. Lett. **B317** (1993) 329; Int. J. Mod. Phys. **A8** (1993) 2973.
S. Krivonos, A. Sorin and F. Toppan (1995), Phys.Lett. **A206** 146.

Third Family of $N = 2$ Supersymmetric KdV Hierarchies

S. Krivonos and A. Sorin

Bogoliubov Laboratory of Theoretical Physics, JINR, 141980 Dubna, Moscow Region, Russia

Abstract. We propose the Lax operators for $N = 2$ supersymmetric matrix generalization of the bosonic $(1, s)$-KdV hierarchies. The simplest examples – the $N = 2$ supersymmetric $a = 4$ KdV and $a = 5/2$ Boussinesq hierarchies – are discussed in detail.

1 Introduction

The existence of three different infinite families of $N = 2$ supersymmetric integrable hierarchies with the $N = 2$ super W_s algebras as their second Hamiltonian structure is a well-established fact by now (Labelle and Mathieu (1991), Yung (1993), Yung and Warner (1993)). Their bosonic limits have been analyzed in (Bonora, Krivonos and Sorin (1996)), and three different families of the corresponding bosonic hierarchies and their Lax operators have been selected. Then, a complete description in terms of super Lax operators for two out of three families has been proposed in (Delduc and Gallot (1997), Bonora, Krivonos and Sorin (1998)), and the generalization to the matrix case has been derived in (Bonora, Krivonos and Sorin (1998)).

The last remaining family of $N = 2$ hierarchies is supersymmetrization of the bosonic $(1, s)$-KdV hierarchies (Bonora, Krivonos and Sorin (1996)). We call them the $N = 2$ supersymmetric $(1, s)$-KdV hierarchies. As opposed to the bosonic counterparts of the former two hierarchies (Bonora, Krivonos and Sorin (1996)), the $(1, s)$-KdV hierarchy is irreducible (see (Bonora, Liu and Xiong (1996)) and references therein), i.e. its Lax operator cannot be decomposed into a direct sum of some more elementary components. This reduction property leads to a strong restriction of the original supersymmetric Lax operator: its bosonic limit should be irreducible. In other words, it should generate only a single operator component. This property is surely satisfied for a supersymmetric Lax operator which is a pure bosonic pseudo-differential operator with the coefficients expressed in terms of $N = 2$ superfields and their fermionic derivatives in such a way that it commutes with one of the two $N = 2$ fermionic derivatives. The Lax operator of this kind has in fact been observed in (Bonora, Krivonos and Sorin (1996)) for the $N = 2$ $a = 5/2$ Boussinesq hierarchy in the negative-power decomposition over bosonic derivative up to the ∂^{-5} order. Quite recently, its closed analytic representation has been obtained in (Gribanov, Krivonos and Sorin (1998)).

The aim of the present letter is to present a new infinite class of reductions (with a finite number of fields) of $N=2$ supersymmetric matrix KP hierarchy which includes the above-mentioned family of $N=2$ $(1,s)$-KdV hierarchies in the scalar case.

2 Extended Matrix $N=2$ Super $(1,s)$-KdV Hierarchy

The Lax operator

$$L_{KP}^{red}=I\partial+a_0+\omega_0 D+\sum_{j=-\infty}^{-1}\left(a_j\partial-[Da_j]\overline{D}+\omega_j D\partial-\frac{1}{2}[D\omega_j]\,[D,\overline{D}]\right)\partial^{j-1} \tag{1}$$

derived by reduction $[D,L_{KP}^{red}]=0$ (Sorin (1997)) of the $N=2$ supersymmetric matrix KP hierarchy has been constructed in (Bonora, Krivonos and Sorin (1998)). Here, a_j and ω_j at $j\geq 1$ (a_0 and ω_0) are generic (chiral) bosonic and fermionic square matrix $N=2$ superfields. The Lax operator (1) still contains an infinite number of fields. Its further reductions (Bonora, Krivonos and Sorin (1998)),

$$\begin{aligned}(L_{KP}^{red})^s=I\partial^s+\sum_{j=1}^{s-1}\left(J_{s-j}\partial-[DJ_{s-j}]\overline{D}\right)\partial^{j-1}-J_s-\overline{D}\partial^{-1}[DJ_s]-\\ F\overline{F}-F\overline{D}\partial^{-1}[D\overline{F}]\,,\end{aligned} \tag{2}$$

are characterized by a finite number of fields and contain two out of three families of $N=2$ supersymmetric hierarchies with $N=2$ super W_s algebras as their second Hamiltonian structure in the scalar case at $F=\overline{F}=0$ (see (Bonora, Krivonos and Sorin (1998)) for details).

It appears that besides reductions (2), there exist other reductions of the Lax operator (1) which in the scalar case correspond to the last remaining family of $N=2$ hierarchies with the $N=2$ super W_s algebras as their second Hamiltonian structure, i.e. $N=2$ $(1,s)$-KdV hierarchies. Based on the inputs given above, we are led to the following conjecture for the expression of the matrix-valued pseudo-differential operator with a finite number of superfields representing the new reductions of the Lax operator (1):

$$L_{KP}^{red}\equiv L_s=I\partial-\left[D\mathcal{L}_s^{-1}\overline{D}\mathcal{L}_s\right],\quad \mathcal{L}_s\equiv I\partial^s+\sum\nolimits_{j=0}^{s-1}J_{s-j}\partial^j+\overline{F}\partial^{-1}F \tag{3}$$

Here, $s=0,1,2,\ldots$, $F\equiv F_{aA}(Z)$ and $\overline{F}\equiv\overline{F}_{Aa}(Z)$ ($A,B=1,\ldots,k$; $a,b=1,\ldots,n+m$) are chiral and antichiral rectangular matrix-valued $N=2$ superfields,

$$DF=0,\quad \overline{D}\,\overline{F}=0, \tag{4}$$

respectively, and $J_j \equiv (J_j(Z))_{AB}$ are $k \times k$ matrix–valued bosonic $N = 2$ superfields with the scaling dimensions in length $[F] = [\overline{F}] = -(s+1)/2$, $[J_j] = -j$; I is the $k \times k$ unity matrix, $I \equiv \delta_{A,B}$, and the matrix product is understood. The matrix entries are bosonic superfields for $a = 1, \ldots, n$ and fermionic superfields for $a = n+1, \ldots, n+m$, i.e., $F_{aA}\overline{F}_{Bb} = (-1)^{d_a \overline{d}_b}\overline{F}_{Bb}F_{aA}$, where d_a and $\overline{d}_b$ are the Grassmann parities of the matrix elements F_{aA} and $\overline{F}_{Bb}$, respectively, $d_a = 1$ ($d_a = 0$) for fermionic (bosonic) entries. $Z = (z, \theta, \overline{\theta})$ is a coordinate of the $N = 2$ superspace, $dZ \equiv dz d\theta d\overline{\theta}$. In (3) the square brackets mean that the $N = 2$ supersymmetric fermionic covariant derivatives D and $\overline{D}$,

$$D = \frac{\partial}{\partial \theta} - \frac{1}{2}\overline{\theta}\frac{\partial}{\partial z}, \overline{D} = \frac{\partial}{\partial \overline{\theta}} - \frac{1}{2}\theta\frac{\partial}{\partial z}, D^2 = \overline{D}^2 = 0, \{D, \overline{D}\} = -\frac{\partial}{\partial z}, \quad (5)$$

act only on the matrix superfields inside the brackets. Let us stress the property of L_s (3) to commute with the fermionic derivative D, that is $[D, L_s] = 0$.

The flows are the standard ones,

$$\frac{\partial}{\partial t_p} L_s = [A_p, L_s], \quad A_p = (L_s^p)_+, \quad (6)$$

where $p = 1, 2, \ldots$, and the subscript $+$ means a differential part of a pseudo-differential operator. Let us remark that the Lax-pair representation (6) (with a generic matrix Lax operator of the type $[D, L] = 0$) being multiplied by the projector $-D\partial^{-1}\overline{D}$ from the right, can identically be rewritten in the same form but with the operators L and A_p replaced by the operators $\mathcal{L} \equiv -DL\partial^{-1}\overline{D}$ and $\mathcal{A}_p \equiv -DA_p\partial^{-1}\overline{D}$, respectively, obeying the chirality preserving condition $D\mathcal{L} = \mathcal{L}\overline{D} = 0$ (as opposed to the former condition $[D, L] = 0$ we started with). In the scalar (generic matrix) case, Lax operators of the last type have been considered in (Delduc and Gallot (1997), Bonora, Krivonos and Sorin (1998)). The relation of the chirality preserving scalar Lax operators of Ref. (Delduc and Gallot (1997)) with the former ones (which have been introduced in (Krivonos, Sorin and Toppan (1995), Bonora, Krivonos and Sorin (1996), Bonora, Krivonos and Sorin (1998)) was observed recently (Delduc and Gallot (1998)).

For the Lax operator L_s (3) the $N = 2$ and $N = 1$ residues[1] vanish since it does not contain the $N = 2$ fermionic derivatives acting as operators. Nevertheless, an infinite number of Hamiltonians can be obtained by using the non–standard definition of the $N = 2$ residue introduced in (Bonora, Krivonos and Sorin (1996)) for the Lax operator of the $N = 2$ $a = 5/2$ Boussinesq hierarchy which coincides with the definition of the residue for bosonic pseudo-differential operators, i.e. it is the integrated coefficient of ∂^{-1}

[1] Let us recall that the standard $N = 2$ ($N = 1$) super-residue is defined as the $N = 2$ ($N = 1$) superfield integral of the coefficient of the operator $[D, \overline{D}]\partial^{-1}$ ($D\partial^{-1}$ or $\overline{D}\partial^{-1}$).

$$H_p = \int dz\ tr(L_s^p)_{\partial^{-1}}|, \tag{7}$$

where $|$ means the $(\theta,\overline{\theta}) \to 0$ limit, the integration is over the space coordinate z, and tr means the usual matrix trace. These Hamiltonians are $N = 2$ supersymmetric. Indeed, the operators $tr(L_s^p)_{\partial^{-1}}|$ (for the L_s given by eqs. (3)) can be represented as

$$tr(L_s^p)_{\partial^{-1}}| = [D,\overline{D}]\mathcal{H}_p| + \text{full space derivative terms} \tag{8}$$

with local superfield functions $\mathcal{H}_p$. Consequently, the Hamiltonians H_p (7) can identically be rewritten in a manifestly supersymmetric form

$$H_p = \int dZ \mathcal{H}_p, \tag{9}$$

where the integration is over the $N = 2$ superspace coordinate Z.

One can easily derive the bosonic limit of the Lax operator L_s (3) at $F = \overline{F} = 0$,

$$L_s^{bos} = (I\partial^s + u_1\partial^{s-1} + \sum_{j=0}^{s-2} u_{s-j}\partial^j)^{-1}(I\partial^{s+1} + u_1\partial^s + \sum_{j=1}^{s-1}(u_{s-j+1} - v_{s-j})\partial^j - v_s), \tag{10}$$

where u_j and v_j are bosonic matrix components of the superfield matrix J_j,

$$u_j = J_j|, \quad v_j = D\overline{D}J_j|. \tag{11}$$

In the scalar case, i.e. at $k = 1$, the Lax operator L_s^{bos} (10) in fact reproduces the Lax operator $L_{[s;\alpha]}^{(1)}$ (Bonora, Krivonos and Sorin (1996)) of the $(1,s)$-KdV hierarchy. Therefore, we conclude that the supersymmetric Lax operator L_s (3) at $F = \overline{F} = 0$ corresponds to the matrix $N = 2$ supersymmetric generalization of the bosonic $(1,s)$-KdV hierarchy, while if it contains the superfield matrices F and $\overline{F}$, it generates the extended matrix $N = 2$ $(1,s)$-KdV hierarchy.

3 Examples: Scalar Case

To better understand what kind of hierarchies we have proposed, let us discuss the first four hierarchies corresponding to the values $s = 0,1,2$ and $s = 3$ in the Lax operator L_s (3) in a simpler and more studied scalar case (i.e., at $k = 1$). In this case, J_j $(j = 1,\ldots,s)$ are generic scalar $N = 2$ bosonic superfields with spins j, while the chiral (antichiral) $N = 2$ superfields F_a $(\overline{F}_a)$ contain n bosonic and m fermionic components with spin $(s+1)/2$.

1. The $s = 0$ case.

For this simplest case the Lax operator (3)

$$L_0 = \partial - \left[D \frac{1}{1 + \overline{F}\partial^{-1}F} \overline{D}(\overline{F}\partial^{-1}F)\right] \tag{12}$$

does not contain any superfields J_j, and the chiral/antichiral superfields F and $\overline{F}$ have spins 1/2. The second–flow equations (6) have the following form:

$$\frac{\partial}{\partial t_2} F = -F\,'' + 2D(F\overline{F}\,\overline{D}F), \quad \frac{\partial}{\partial t_2}\overline{F} = \overline{F}\,'' + 2\overline{D}\left((D\overline{F})F\overline{F}\right) \tag{13}$$

and coincide with the corresponding flow of the $N = 2$ GNLS hierarchy (Bonora, Krivonos and Sorin (1996).) Therefore, the Lax operator (12) provides a new description of the $N = 2$ GNLS hierarchy.

2. The $s = 1$ case.

The Lax operator (3) has the following form:

$$L_1 = \partial - \left[D \frac{1}{\partial + J_1 + \overline{F}\partial^{-1}F} \overline{D}(J_1 + \overline{F}\partial^{-1}F)\right], \tag{14}$$

and the first two non-trivial flows (6) read as

$$\begin{aligned} &\frac{\partial}{\partial t_2} J_1 = ([D,\overline{D}\,]J_1 - J_1^2 + 2\overline{F}F)\,', \\ &\frac{\partial}{\partial t_2} F = -F\,'' + 2FD\overline{D}J_1, \quad \frac{\partial}{\partial t_2}\overline{F} = \overline{F}\,'' + 2(\overline{D}DJ_1)\overline{F}, \end{aligned} \tag{15}$$

$$\begin{aligned} &\frac{\partial}{\partial t_3} J_1 = J_1\,''' - 3\left[J_1[D,\overline{D}]J_1 - (\overline{D}J_1)DJ_1 - J_1^3 + J_1\overline{F}F + \right. \\ &\left.(\overline{D}F)D\overline{F} + \overline{F}F\,' - \overline{F}\,'F\right]', \\ &\frac{\partial}{\partial t_3} F = F\,''' - 3D\left[((\overline{D}J)F)\,' + J(\overline{D}J)F - F\overline{F}\,\overline{D}F\right], \\ &\frac{\partial}{\partial t_3}\overline{F} = \overline{F}\,''' + 3\overline{D}\left[((DJ)\overline{F})\,' - J(DJ)\overline{F} + (D\overline{F})F\overline{F}\right]. \end{aligned} \tag{16}$$

From these expressions we can easily recognize that after rescaling $J_1 \to -2J_1, t_n \to -t_n$ they coincide with the corresponding flows of the $N = 2$ $a = 4$ KdV hierarchy (Labelle and Mathieu (1991)) at $F = \overline{F} = 0$. With the non-zero superfields F and $\overline{F}$ we obtain a new extension of the $N = 2$ $a = 4$ KdV hierarchy. Thus, our family of $N = 2$ hierarchies includes the well-known $N = 2$ $a = 4$ KdV hierarchy and possesses the Lax-pair representation with the new Lax operator[2] L_1 (14).

3. The $s = 2$ case.

This case is rather interesting because it corresponds to the $N = 2$ $a = 5/2$ Boussinesq hierarchy which has been a puzzle for a long time. The Lax operator (3)

$$L_2 = \partial - \left[D \frac{1}{\partial^2 + J_1\partial + J_2 + \overline{F}\partial^{-1}F} \overline{D}(J_1\partial + J_2 + \overline{F}\partial^{-1}F)\right] \tag{17}$$

[2] For alternative Lax-pair representations of the $N = 2$ $a = 4$ KdV hierarchy, see Refs. (Labelle and Mathieu (1991), Krivonos and Sorin (1995), Krivonos, Sorin and Toppan (*1995*), Delduc and Gallot (1997)).

gives rise to the following second–flow equations

$$
\begin{aligned}
&\frac{\partial}{\partial t_2} J_1 = (2J_2 - J_1\,' + 2[D,\overline{D}\,]J_1 - J_1^2)\,', \\
&\frac{\partial}{\partial t_2} J_2 = (J_2 + 2D\overline{D}J_1)\,'' + 2(\overline{F}F)' - 2J_2J_1\,' + 2J_1D\overline{D}J_1\,', \\
&\frac{\partial}{\partial t_2} F = -F\,'' + 2FD\overline{D}J_1, \quad \frac{\partial}{\partial t_2}\overline{F} = \overline{F}\,'' + 2(\overline{D}DJ_1)\overline{F}.
\end{aligned} \tag{18}
$$

In the new basis[3],

$$
t_2 \to -\frac{t_2}{3},\; J_1 \to \frac{1}{3}J_1,\; J_2 \to -J_2 + \frac{1}{2}J_1\,' - \frac{1}{6}[D,\overline{D}]J_1 + \frac{2}{9}{J_1}^2, \tag{19}
$$

at $F = \overline{F} = 0$, eqs. (18) coincide with the $N = 2$ $a = 5/2$ Boussinesq equation (Yung (1993)). Thus, we conclude that the $N = 2$ $a = 5/2$ Boussinesq hierarchy also belongs to the family of $N = 2$ super $(1, s)$-KdV hierarchies with the Lax operator (3).

4. The $s = 3$ case.

As the last example, we present the second–flow equations

$$
\begin{aligned}
&\frac{\partial}{\partial t_2} J_1 = (2J_2 - 2J_1\,' + 3[D,\overline{D}\,]J_1 - J_1^2)\,', \\
&\frac{\partial}{\partial t_2} J_2 = (2J_3 + J_2\,' + 6D\overline{D}J_1\,')\,' - 2J_2J_1\,' + 4J_1D\overline{D}J_1\,', \\
&\frac{\partial}{\partial t_2} J_3 = (J_3 + 2D\overline{D}J_1\,')\,'' + 2J_1D\overline{D}J_1\,'' + 2J_2D\overline{D}J_1\,' - 2J_1\,'J_3
\end{aligned} \tag{20}
$$

of the $N = 2$ super $(1,3)$-KdV hierarchy which possesses the $N = 2$ W_4 algebra as the second Hamiltonian structure. This hierarchy contains the $N = 2$ superfields J_1, J_2 and J_3 with the spins 1, 2 and 3, respectively, and its Lax operator looks like

$$
L_3 = \partial - \left[D\frac{1}{\partial^3 + J_1\partial^2 + J_2\partial + J_3}\overline{D}(J_1\partial^2 + J_2\partial + J_3)\right]. \tag{21}
$$

The extension of this system by the superfields F and $\overline{F}$ can be straightforwardly derived from the Lax-pair representation (3), (6), and we do not present it here.

4 Examples: Matrix Case

The construction of flows (6) in the matrix case goes without any new peculiarities. The only difference with respect to the scalar case is the appearance of some new terms in the flow equations and their ordering. To demonstrate

[3] The complexity of these transformations is the price we have to pay for the simplicity of the Lax operator (17).

the difference, we present the Hamiltonian densities[4] and the flow equations for the systems considered in the previous section without comments.

1. The $s = 0$ case.

$$\mathcal{H}_2 = tr(\overline{F}F' + \overline{F}F\overline{F}F) , \tag{22}$$

$$\frac{\partial}{\partial t_2}F = -F'' + 2D(F\overline{F}\,\overline{D}F), \quad \frac{\partial}{\partial t_2}\overline{F} = \overline{F}'' + 2\overline{D}\left((D\overline{F})F\overline{F}\right). \tag{23}$$

These equations reproduce the second–flow equations of the $N = 2$ supersymmetric matrix GNLS hierarchy (Bonora, Krivonos and Sorin (1998)).

2. The $s = 1$ case.

$$\begin{aligned}
&\mathcal{H}_1 = tr(J_1), \quad \mathcal{H}_2 = tr(\frac{1}{2}J_1^2 - \overline{F}F), \\
&\mathcal{H}_3 = tr(\frac{1}{3}J_1^3 - J_1 D\overline{D}J_1 - \overline{F}F' - \overline{F}FJ_1),
\end{aligned} \tag{24}$$

$$\begin{aligned}
&\frac{\partial}{\partial t_2}J_1 = ([D,\overline{D}\,]J_1 - J_1^2 + 2\overline{F}F)' + [J_1,[D,\overline{D}\,]J_1], \\
&\frac{\partial}{\partial t_2}F = -F'' + 2FD\overline{D}J_1, \quad \frac{\partial}{\partial t_2}\overline{F} = \overline{F}'' + 2(\overline{D}DJ_1)\overline{F},
\end{aligned} \tag{25}$$

$$\begin{aligned}
&\frac{\partial}{\partial t_3}J_1 = J_1''' + 3\left[((\overline{D}DJ_1)J_1 - D(J_1\overline{D}J_1) + \overline{F}'F - \overline{F}F' + D\overline{D}(\overline{F}F))'\right. \\
&\quad -J_1D(J_1\overline{D}J_1) - (\overline{D}(DJ_1)J_1)J_1 + \{J_1, D\overline{D}(\overline{F}F)\} + \overline{F}FD\overline{D}J_1 \\
&\quad \left.+(\overline{D}DJ_1)\overline{F}F - J_1\overline{F}F' - \overline{F}'FJ_1\right], \\
&\frac{\partial}{\partial t_3}F = F''' + 3\left[D(F\overline{F}\,\overline{D}F) - (FD\overline{D}J_1)' - FD(J_1\overline{D}J_1)\right], \\
&\frac{\partial}{\partial t_3}\overline{F} = \overline{F}''' + 3\overline{D}\left[(D\overline{F})F\overline{F} + ((DJ_1)\overline{F})' - (DJ_1)J_1\overline{F}\right],
\end{aligned} \tag{26}$$

where the brackets $(\{,\})$ $[,]$ represent the (anti)commutator.

3. The $s = 2$ case.

$$\mathcal{H}_1 = tr(J_1), \quad \mathcal{H}_2 = tr(\frac{1}{2}J_1^2 - J_2), \tag{27}$$

$$\begin{aligned}
&\frac{\partial}{\partial t_2}J_1 = (2J_2 - J_1' + 2[D,\overline{D}\,]J_1 - J_1^2)' + [J_1,[D,\overline{D}\,]J_1], \\
&\frac{\partial}{\partial t_2}J_2 = (J_2 + 2D\overline{D}J_1)'' + 2(\overline{F}F)' - \{J_2, J_1'\} + 2J_1D\overline{D}J_1' \\
&\quad +[J_2,[D,\overline{D}\,]J_1], \\
&\frac{\partial}{\partial t_2}F = -F'' + 2FD\overline{D}J_1, \quad \frac{\partial}{\partial t_2}\overline{F} = \overline{F}'' + 2(\overline{D}DJ_1)\overline{F}.
\end{aligned} \tag{28}$$

4. The $s = 3$ case.

---

[4] Let us recall that Hamiltonian densities are defined up to terms which are fermionic or bosonic total derivatives of arbitrary nonsingular, local functions of the superfield matrices.

$$\mathcal{H}_1 = tr(J_1), \quad \mathcal{H}_2 = tr(\frac{1}{2}J_1^2 - J_2), \tag{29}$$

$$\begin{aligned}
&\tfrac{\partial}{\partial t_2} J_1 = (2J_2 - 2J_1\,' + 3[D,\overline{D}\,]J_1 - J_1^2)\,' + [J_1,[D,\overline{D}\,]J_1], \\
&\tfrac{\partial}{\partial t_2} J_2 = (2J_3 + J_2\,' + 6D\overline{D}J_1\,')\,' - \{J_2, J_1\,'\} \\
&\quad +4J_1 D\overline{D}J_1\,' + [J_2,[D,\overline{D}\,]J_1], \\
&\tfrac{\partial}{\partial t_2} J_3 = (J_3 + 2D\overline{D}J_1\,')\,'' + 2J_1 D\overline{D}J_1\,'' + 2J_2 D\overline{D}J_1\,' + [J_3,[D,\overline{D}]J_1] \\
&\quad -\{J_3, J_1\,'\}.
\end{aligned} \tag{30}$$

5 Involution Properties

Equations (23) and (25)–(26) admit the involution

$$\begin{aligned}
&F^* = i^{s-1}\mathcal{I}\overline{F}^T, \quad \overline{F}^* = i^{s-1}F^T\mathcal{I}, \quad J_j^* = (-1)^j J_j^T, \\
&\theta^* = \overline{\theta}, \quad \overline{\theta}^* = \theta, \quad t_p^* = (-1)^{p+1}t_p, \quad z^* = z, \quad i^* = -i,
\end{aligned} \tag{31}$$

for $s = 0$ and $s = 1$, respectively. Here, i is the imaginary unit, the symbol T means the operation of matrix transposition, and the matrix $\mathcal{I}$ is

$$\mathcal{I} \equiv (-i)^{d_a}\delta_{ab}, \quad \mathcal{I}\mathcal{I}^* = I, \quad \mathcal{I}^3 = \mathcal{I}^*, \quad \mathcal{I}^2 = (-1)^{d_a}\delta_{ab}. \tag{32}$$

The same involution property is not straightforwardly satisfied for eqs. (28) ($s = 2$) and eqs. (30) ($s = 3$): it is satisfied in a new basis with the superfields J_2 and J_2, J_3 being replaced by

$$J_2 \Rightarrow J_2 - \frac{1}{2}J_1\,', \tag{33}$$

and

$$J_2 \Rightarrow J_2 - J_1\,', \quad J_3 \Rightarrow J_3 - \frac{1}{2}J_2\,', \tag{34}$$

respectively, while all the other superfields are unchanged. It seems reasonable to conjecture the existence of a basis in the space of superfield matrices where the involution (31) is admitted for any given value of the parameter s that parametrizes the Lax operator L_s (3).

6 Conclusion

In this letter, we constructed a new infinite variety of matrix $N = 2$ supersymmetric hierarchies by exhibiting the corresponding super Lax operators. Their involution properties are analyzed. As a byproduct, we solved the problem of a Lax-pair description for the last remaining family of $N = 2$ hierarchies with the $N = 2$ super W_s algebras as their second Hamiltonian structure and derived new extensions of such familiar hierarchies as the $N = 2$ supersymmetric $a = 4$ KdV and $a = 5/2$ Boussinesq hierarchies. New bosonic hierarchies can be obtained from the constructed supersymmetric counterparts in the bosonic limit.

Acknowledgments. We would like to thank L. Bonora for his interest in this paper and hospitality at SISSA. This work was partially supported by the Russian Foundation for Basic Research, Grant No. 96-02-17634, RFBR-DFG Grant No. 96-02-00180, INTAS Grant No. 93-127 ext..

References

P. Labelle and P. Mathieu (1991), J. Math. Phys. **32** 923

C.M. Yung (1993), Phys. Lett. **B 309** 75;
S. Bellucci, E. Ivanov, S. Krivonos and A. Pichugin (1993), Phys. Lett. **B 312** 463;
Z. Popowicz (1993), Phys. Lett. **B 319** 478

C.M. Yung and R.C. Warner (1993), J. Math. Phys. **34** 4050

L. Bonora, S. Krivonos and A. Sorin (1996), Nucl. Phys. **B 477** 835

F. Delduc and L. Gallot (1997), Commun. Math. Phys. **190** 395

L. Bonora, S. Krivonos and A. Sorin (1998), Lett. Math. Phys. **45** 63

L. Bonora, Q.P. Liu and C.S. Xiong (1996), Comm. Math. Phys. **175** 177

V. Gribanov, S. Krivonos and A. Sorin, *The Lax-pair representation for the $N = 2$ supersymmetric $a = 5/2$ Boussinesq hierarchy*, in preparation.

A. Sorin (1997), *The discrete symmetries of the $N = 2$ supersymmetric GNLS hierarchies*, JINR E2-97-37, solv-int/9701020.

S. Krivonos, A. Sorin and F. Toppan (1995), Phys. Lett. **A 206** 146

F. Delduc and L. Gallot, private communication.

S. Krivonos and A. Sorin (1995), Phys. Lett. **B 357** 94

Integrals of Motion, Supersymmetric Quantum Mechanics and Dynamical Supersymmetry

Mikhail S. Plyushchay

Departamento de Física, Universidad de Santiago de Chile, Casilla 307, Santiago 2, Chile;
Institute for High Energy Physics, Protvino, Moscow Region, Russia
E-mail: mplyushc@lauca.usach.cl

Abstract. The class of relativistic spin particle models reveals the 'quantization' of parameters already at the classical level. The special parameter values emerge if one requires the maximality of classical global continuous symmetries. The same requirement applied to a non-relativistic particle with odd degrees of freedom gives rise to supersymmetric quantum mechanics. Coupling classical non-relativistic superparticle to a 'U(1) gauge field', one can arrive at the quantum dynamical supersymmetry. This consists in supersymmetry appearing at special values of the coupling constant characterizing interaction of a system of boson and fermion but disappearing in a free case. Possible relevance of this phenomenon to high-temperature superconductivity is speculated.

1 Introduction

In ref. (Grignani, Plyushchay and Sodano (1996)) it was observed that the classical (2+1)-dimensional system given by Lagrangian (Cortés, Plyushchay and Velázquez (1993))

$$L_f = \frac{1}{2e}(\dot{x}_\mu - iv\epsilon_{\mu\nu\lambda}\xi^\nu\xi^\lambda)^2 - \frac{e}{2}m^2 + 2i\nu m v\theta_1\theta_2 - \frac{i}{2}\xi_\mu\dot{\xi}^\mu + \frac{i}{2}\theta_a\dot{\theta}_a \qquad (1.1)$$

reveals a 'classical quantization' of a real parameter ν. The nature of this phenomenon is the following. The model (1.1), described by even and odd vectors x_μ and ξ_μ, by scalar odd variables θ_a, $a = 1, 2$, and by even Lagrange multipliers e and v, admits the construction of oscillator-like odd variables $\xi^\pm$ linear in ξ_μ, whose phases at $|\nu| = 1$ evolve exactly as the phases of $\theta^\pm = \frac{1}{\sqrt{2}}(\theta_1 \pm i\theta_2)$. As a consequence, two nilpotent integrals of motion $\xi^+\theta^-$ and $\xi^-\theta^+$ can be constructed in addition to the integrals $\theta^+\theta^-$ and $\xi^+\xi^-$. So, at $\nu = \pm 1$ the system has a maximal classical global continuous symmetry. The same values of ν are separated if one requires that the system would have the maximal quantum global continuous symmetry. Moreover, only at $\nu = \pm 1$ the discrete P, T-invariance of the classical system may be preserved at the quantum level. As a result, the quantum model describes the P, T-invariant system of massive fermions realizing irreducible representation of a nonstandard superextension of the $(2+1)$-dimensional Poincaré group (Grignani, Plyushchay and Sodano (1996), Gamboa and Plyushchay (1998)).

Instead of the pair of scalar Grassmann variables θ_a, one can introduce another Grassmann vector and construct the Lagrangian similar to (1.1) (Nirov and Plyushchay (1997)),

$$L_{CS} = \frac{1}{2e}(\dot{x}_\mu - \frac{i}{2}v\epsilon_{\mu\nu\lambda}\xi^\nu_a\xi^\lambda_a)^2 - \frac{e}{2}m^2 + 2i\nu mv\xi^\mu_1\xi_{2\mu} - \frac{i}{2}\xi_{a\mu}\dot{\xi}_a{}^\mu. \quad (1.2)$$

It turns out that this system reveals the same phenomenon of 'classical quantization': there are two special values of the parameter, $\nu = \pm 2$, at which the system has a maximal global continuous symmetry. In this case the corresponding additional classical local integrals of motion are of the third order in Grassmann variables. The same values of ν are separated at the quantum level from the requirement of the maximality of global continuous symmetry and if to require preserving the classical invariance with respect to the discrete P and T transformations. Quantum mechanically, the system (1.2) describes the P,T-invariant system of Chern-Simons fields (topologically massive vector gauge fields (Jackiw and Templeton (1981))) (Nirov and Plyushchay (1998), Nirov and Plyushchay (1998a)). Analogously to the fermion system (1.1), the quantum states of the system (1.2) realize irreducible representation of a nonstandard superextension of the $(2+1)$-dimensional Poincaré group (Nirov and Plyushchay (1998), Nirov and Plyushchay (1998a)).

Here we show that the requirement of the maximality of classical global continuous symmetry applied to a non-relativistic particle with odd degrees of freedom gives rise naturally to supersymmetric quantum mechanics. Then, coupling classical non-relativistic superparticle to a 'U(1) gauge field', we arrive at the quantum dynamical supersymmetry. The latter consists in supersymmetry appearing at special values of the coupling constant characterizing interaction of a system of boson and fermion but disappearing in a free case. In conclusion, we speculate on a possible relevance of this phenomenon to high-temperature superconductivity in the context of a $U_{\sigma_3}(1)$ gauged version of the model (1.1) (Grignani, Plyushchay and Sodano (1996), Gamboa and Plyushchay (1998)).

2 Supersymmetric Quantum Mechanics

Let us consider a nonrelativistic particle of unit mass having two odd (Grassmann) degrees of freedom. Its Lagrangian may be written in general form as

$$L = \frac{1}{2}\dot{x}^2 - V(x) - R(x)N + \frac{i}{2}\theta_a\dot{\theta}_a. \quad (2.3)$$

Here $V(x)$ and $R(x)$ are two arbitrary functions and $N = -i\theta_1\theta_2 = \theta^+\theta^-$, $\theta^\pm = \frac{1}{\sqrt{2}}(\theta_1 \pm i\theta_2)$. In the next section it will be shown that a possible inclusion of additional term $\dot{x}\mathcal{A}(x)N$ does not change the system (2.3) classically but turns out to be important quantum mechanically. The nontrivial Poisson-Dirac brackets of the system (2.3) are $\{x,p\}=1$, $\{\theta_a,\theta_b\}=-i\delta_{ab}$, and the Hamiltonian $H=\frac{1}{2}p^2+V(x)+R(x)N$ generates the following equations of motion:

$$\dot{x}=p, \quad \dot{p}=-V'(x)-R'(x)N, \quad \dot{\theta}^{\pm}=\pm iR(x)\theta^{\pm}.$$

Thus, N is an obvious integral of motion additional to H. If the evolution law $x=x(t)$ is known, then the solution to equations of motion for odd variables is

$$\theta^{\pm}(t)=\theta^{\pm}(t_0)\exp\left[\pm i\int_{t_0}^{t}R(x(\tau))d\tau\right], \tag{2.4}$$

and we find that the odd quantities

$$\Theta^{\pm}=\theta^{\pm}(t)\exp[\mp i\int_{t_0}^{t}R(x(\tau))d\tau] \tag{2.5}$$

are *nonlocal in time* integrals of motion. In particular case $R=0$, odd variables have a trivial dynamics, $\dot{\theta}^{\pm}=0$, and integrals (2.5) take the *local* form: $\Theta^{\pm}=\theta^{\pm}$. One may put the question: is there any other nontrivial case characterized by *local* odd integrals of motion instead of the nonlocal integrals (2.5)? It is clear that if the system has even complex conjugate quantities $A^{\pm}$, $(A^{+})^{*}=A^{-}$, whose evolution looks up to the term proportional to N like the evolution of odd variables in (2.4), then local odd integrals of motion could be constructed in the form

$$Q^{\pm}=A^{\pm}\theta^{\mp}. \tag{2.6}$$

Let us consider the oscillator-like variables

$$A^{\pm}=\frac{1}{\sqrt{2}}(\phi(x)\mp ip) \tag{2.7}$$

with some function $\phi(x)$. We find that

$$\dot{A}^{\pm}=\frac{1}{\sqrt{2}}(\phi'p\pm i(V'+R'N)),$$

and from the relation

$$\dot{Q}^{\pm}=\pm i\left[(\phi'-R)Q^{\pm}+\frac{1}{\sqrt{2}}(V'-\phi\phi')\theta^{\mp}\right]$$

we conclude that $\dot{Q}^{\pm}=0$ when $\phi'=R$, $V'=\frac{1}{2}(\phi^2)'$. Therefore, when the functions $R(x)$ and $V(x)$ are related as $R(x)=W'(x)$, $V(x)=\frac{1}{2}W^2(x)+C$, where $W(x)=\phi(x)$ is an arbitrary function and C is a constant, then odd quantities (2.6) are integrals of motion additional to H and N. Together with Hamiltonian H, they form the classical analog of a central extension of $N=1$ supersymmetry algebra,

$$\{Q^+, Q^-\} = -i(H - C), \quad \{H, Q^\pm\} = \{Q^+, Q^+\} = \{Q^-, Q^-\} = 0,$$

with constant C playing a role of a central charge, whereas the integral N is classical analog of the grading operator,

$$\{N, Q^\pm\} = \pm iQ^\pm, \quad \{N, H\} = 0.$$

Putting $C = 0$, we arrive at the classical analog of Witten's supersymmetric quantum mechanics (Witten (1981)) given by the Lagrangian

$$L_{SUSY} = \frac{1}{2}\dot{x}^2 - \frac{1}{2}W^2(x) + i\theta_1\theta_2 W'(x) + \frac{i}{2}\theta_a\dot{\theta}_a. \tag{2.8}$$

We conclude that the system (2.8) is a special case of more general classical system (2.3), which is characterized by the presence of two additional *local* in time odd integrals of motion (2.6) being supersymmetry generators.

3 Quantum Dynamical Supersymmetry

Let us consider a system given by the Lagrangian

$$L_g = L_{SUSY} + \dot{x}\mathcal{A}(x)N. \tag{3.9}$$

Since N is integral of motion of the system (2.8), one may write

$$L_g = L_{SUSY} + \frac{d}{dt}\Delta, \quad \Delta = N\int^{x(t)} \mathcal{A}(y)dy. \tag{3.10}$$

The Hamiltonian for (3.9) is

$$H = \frac{1}{2}(p - \mathcal{A}(x)N)^2 + \frac{1}{2}W^2(x) + NW'. \tag{3.11}$$

In correspondence with lagrangian relation (3.10), the canonical transformation

$$x \to \tilde{x} = x, \quad p \to \tilde{p} = p - \mathcal{A}(x)N, \quad \theta^\pm \to \tilde{\theta}^\pm = \theta^\pm \exp(\pm i\varphi(x)),$$

$$\varphi(x) = \int^x \mathcal{A}(y)dy$$

reduces (3.11) to the Hamiltonian of the 'free' system (2.8) characterized by $\mathcal{A}(x) = 0$. Therefore, classical systems (2.8) and (3.9) are equivalent and the system (3.9) is supersymmetric for any ' U(1) gauge potential' $\mathcal{A}$. However, due to ambiguities of quantization procedure, the corresponding quantum systems are not necessarily equivalent in generic case. Let us show that using the quantization ambiguity, one can construct the quantum system with supersymmetry appearing dynamically (Das and Hott (1995)) at special values of a coupling constant characterizing interaction of a system of boson and fermion but disappearing in a free case.

We consider the simplest nontrivial system (3.11) given by $\mathcal{A} = g = const$ and $W(x) = \epsilon x$ with $\epsilon = 1$:

$$H = \frac{1}{2}[(p - gN)^2 + x^2] + \epsilon N. \tag{3.12}$$

After realizing the canonical transformation $x \to p$, $p \to -x$, the Hamiltonian takes the form

$$H = \frac{1}{2}[p^2 + (x + gN)^2] + \epsilon N. \tag{3.13}$$

This can be presented equivalently as

$$H = \frac{1}{2}(p^2 + x^2) + \epsilon N + gxN, \tag{3.14}$$

and (3.14) can be understood as the Hamiltonian of the system of bosonic and fermionic oscillators interacting through the coupling term gxN. Quantum analog of (3.14) is

$$H = a^+a^- + \epsilon c^+c^- + gxc^+c^-, \tag{3.15}$$

where $a^\pm = \frac{1}{\sqrt{2}}(x \mp ip)$, $[a^-, a^+] = 1$, $[c^-, c^+]_+ = 1$, $c^{\pm 2} = 0$, $[a^\pm, c^\pm] = 0$, and passing from (3.14) to (3.15) we have chosen the normal ordering of bosonic and fermionic creation-annihilation operators. Due to the quantum relation $N^2 = N$, $N = c^+c^-$, this quantum Hamiltonian can be written in the form $H = b^+b^- + \tilde{\epsilon}c^+c^-$ with

$$b^\pm = a^\pm + \frac{g}{\sqrt{2}}c^+c^-, \quad \tilde{\epsilon} = \epsilon - \frac{g^2}{2}.$$

Performing in correspondence with classical picture the unitary Bogoliubov transformation $a^\pm \to b^\pm$, $f^\pm = c^\pm \exp[\pm\frac{g}{\sqrt{2}}(a^- - a^+)]$, we present (3.15) equivalently as

$$H = b^+b^- + \tilde{\epsilon}f^+f^- \tag{3.16}$$

with $[b^-, b^+] = [f^-, f^+]_+ = 1$, $f^{\pm 2} = 0$, $[b^\pm, f^\pm] = 0$. The Hamiltonian (3.16) is the exact quantum analog of the classical Hamiltonian (3.13), but with ϵ changed for $\tilde{\epsilon}$. This difference happens since we constructed (3.15) proceeding from (3.14), and in contrast with classical relation $N^2 = 0$ quantum mechanically we have $N^2 = N$. The system (3.16) (and so, the system (3.15)) is supersymmetric at $\tilde{\epsilon} = 1$, i.e. only in a free case ($g = 0$). But we can interpret the system (3.15) in another way. Suppose that the quantum system (3.15) is characterized by the parameter $\epsilon = 1 + \frac{\alpha^2}{2}$ with $\alpha \neq 0$, which means that bosonic and fermionic oscillators have different frequencies. Then in a free case ($g = 0$) this quantum system has no supersymmetry, but turns into a supersymmetric system at two special values of the coupling constant: $g = \pm\alpha$. If the coupling constant may be varied effectively as a function of some external parameter(s) (temperature etc.), $g = g(\lambda)$, then the system will reveal supersymmetry at special values of parameter(s) defined by the relation $g^2(\lambda_s) = \alpha^2$.

4 Concluding Remarks

For the models (1.1) and (1.2), the values of corresponding parameters do not influence on the integrability properties: the general solution to the equations of motion can be constructed explicitly for arbitrary νs (Grignani, Plyushchay and Sodano (1996), Nirov and Plyushchay (1998a)). However, there are special values of νs for which classical global continuous symmetries of these systems happen to be maximal. The situation is similar for the system (2.3), but here the requirement of maximality of classical global continuous symmetry restricts and relates the form of two functions, $V(x)$ and $R(x)$. If instead of the generic system (2.3) we consider its much more restricted case given by the Lagrangian (2.8) with $W'(x)$ changed for $\nu W'(x)$, the requirement of maximal classical symmetry would lead us to the analogous 'quantization condition': $\nu^2 = 1$. With this observation we arrive at the interesting question whether the systems (1.1) and (1.2) may be extended in such a way that the requirement of maximality of classical global continuous symmetries would fix a functional arbitrariness but not only the values of parameters.

The 'U(1)-gauged' version of nonrelativistic supersymmetric system is given by Lagrangian (3.9). It is related classically to the initial 'free' system (2.8) by the canonical transformation, and therefore also is supersymmetric. On the other hand, the model (1.1) may be consistently coupled to a $U_{\sigma_3}(1)$ gauge field (see refs. (Grignani, Plyushchay and Sodano (1996), Gamboa and Plyushchay (1998))),

$$L_f \to L_f^g = L_f + qN\left(\dot{x}_\mu \mathcal{A}^\mu + \frac{e}{2} i\xi_\mu\xi_\nu\mathcal{F}^{\mu\nu}\right), \tag{4.17}$$

where q is a coupling constant, $\mathcal{A}_\mu = \mathcal{A}_\mu(x)$ is a U(1) gauge field, $\mathcal{F}_{\mu\nu} = \partial_\mu\mathcal{A}_\nu - \partial_\nu\mathcal{A}_\mu$, and $N = \theta^+\theta^-$. The structure of this P,T-invariant gauge interaction (see ref. (Grignani, Plyushchay and Sodano (1996))) is similar to the structure of the 'U(1) gauge interaction' considered above for non-relativistic supersymmetric particle. But now there is no trivial relationship of the U(1)-gauged system (4.17) to the initial free system (1.1) due to the presence in (4.17) of the spin-field interaction term proportional to $i\xi_\mu\xi_\nu\mathcal{F}^{\mu\nu}$. Thus, it is not obvious what happens with classical global continuous symmetries of relativistic spin particle system (1.1) under coupling it to the $U_{\sigma_3}(1)$ gauge field. However, if one conjectures that the classical global continuous symmetries of the system (1.1) survive in the case of interaction (4.17), what will happen with them at the quantum level? These questions on classical and quantum symmetries of (4.17) deserve further investigation. If the analog of the discussed phenomenon of quantum dynamical supersymmetry also exists for the system (4.17), this could find important applications in the context of hypothetical relevance of the P,T-invariant model (4.17) to the problem of describing high-temperature superconductivity (see refs. (Grignani, Plyushchay and Sodano (1996), Kovner and Rosenstein (1990))).

Acknowledgement

The work was supported in part by grant 1980619 from FONDECYT (Chile).

References

G. Grignani, M. Plyushchay and P. Sodano (1996), Nucl. Phys. **B464** 189, hep-th/9511072.
J. L. Cortés, M. S. Plyushchay and L. Velázquez (1993), Phys. Lett. **B306** 34.
G. Gamboa and M. Plyushchay (1998), Nucl. Phys. **B512** 485, hep-th/9711170.
Kh. S. Nirov and M. S. Plyushchay (1997), Phys. Lett. **B405** 114, hep-th/9707070.
Kh. S. Nirov and M. S. Plyushchay (1998), J. High Energy Phys. **02** 015, hep-th/9712097.
Kh. S. Nirov and M. S. Plyushchay (1998a), Nucl. Phys. **B512** 295, hep-th/9803221.
R. Jackiw and S. Templeton (1981), Phys. Rev. **D23** 2291;
J. Schonfeld, Nucl. Phys. **B185** (1981) 157;
W. Siegel, Nucl. Phys. **B156** (1979) 135;
S. Deser, R. Jackiw and S. Templeton, Phys. Rev. Lett. **48** (1982) 975.
E. Witten (1981), Nucl. Phys. **B188** 513; **B202** (1982) 253.
A. Das and M. Hott (1995), hep-th/9504059.
A. Kovner and R. Rosenstein (1990), Phys. Rev. **B42** 4748;
G.W. Semenoff and N. Weiss, Phys. Lett. **B250** (1990) 117;
N. Dorey and N.E. Mavromatos, Nucl. Phys. **B386** (1992) 614.

From LG to a Butterfly Resolution of Unitary $N=2$ Representations

Alexei Semikhatov

Tamm Theory Division, Lebedev Physics Institute, Russian Academy of Sciences

Abstract. I review free-field resolutions of unitary representations of the $N=2$ superconformal algebra. The one-screening resolution is related to the representation theory picture of "gravitational descendants." The realization with two fermionic screenings gives rise to a "butterfly" resolution.

1 Introduction

As is well known, free-field constructions ("bosonizations") of infinite-dimensional algebras alter the embedding structure of Verma-like modules; for irreducible representations, this changes the resolutions, as for example happens to the BGG resolution of irreducible Virasoro representations, which is replaced with the Felder resolution (Felder (1989)).

Several free-field "bosonizations" of the $N=2$ superconformal extension of the Virasoro algebra are known. One of these (Fre *et al.* (1992), Witten (1994)) has been used to analyse certain aspects of the Landau–Ginzburg (LG) models (Witten (1994), Kawai *et al.* (1994), Nemeschansky and Warner (1995)). There also exist a bosonization (Mussardo *et al.* (1989), Ohta and Suzuki (1990), Ito (1992)) with two fermionic screening operators, and the $N=2$ construction (Gato-Rivera and Semikhatov (1992), Bershadsky *et al.* (1993)) realized in the bosonic string. Despite the popularity of free-field constructions, however, the corresponding resolutions of irreducible $N=2$ representations have never been written out. In this talk, which is based on (Feigin and Semikhatov (1998)), I show how this gap can be filled.

There is a deep relation between free-field realizations of unitary $N=2$ representations and the LG formalism. The unperturbed A_{p-1} LG equations of motion (suppressing the kinetic term) have a direct $N=2$ analogue that reads as

$$\partial^{p-2}\mathcal{G}(z)\ldots\partial\mathcal{G}(z)\,\mathcal{G}(z)=0\,, \tag{1.1}$$

where $\mathcal{G}(z)$ is one of the two fermionic fields entering the $N=2$ superconformal algebra. This "operator" relation holds in any unitary $N=2$ representation. One of the $N=2$ resolutions considered in what follows can be interpreted as a means to solve (1.1) in cohomology. Once the cohomology given by the unitary $N=2$ representations, we see that these representations are *characterized* by (1.1).

The first resolution that I consider is in the space of a bosonic and a fermionic ghost systems. It turns out to be similar to the resolution constructed in (Bernard and Felder (1990), Feigin and Frenkel (1990)) for the $\widehat{s\ell}(2)$ algebra, however the actual $N=2$-details appear to be new. While the screening operator $\mathbb{Q}_0$ singles out precisely a unitary $N=2$ representation, the $\mathbb{Q}_0$-trivial "replicas" of the highest-weight vector provide the representation-theory counterpart of the gravitational descendants (Losev (1992), Eguchi *et al.* (1993), Lerche and Warner (1994)).

Another free-field realization (Mussardo *et al.* (1989), Ohta and Suzuki (1990), Ito (1992)) that we consider here involves two screening operators, which are both fermions. These give rise to the ***butterfly resolution*** (Feigin and Semikhatov (1998))

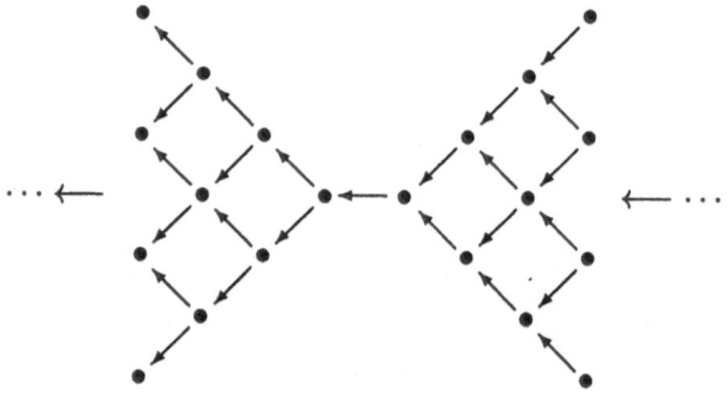

(1.2)

whose two-winged shape that has no known analogues for other algebras.

2 Generalities

The $N=2$ superconformal algebra is taken in the basis where the nonvanishing commutation relations read as

$$\begin{aligned}
&[\mathcal{L}_m, \mathcal{L}_n] = (m-n)\mathcal{L}_{m+n}\,, && [\mathcal{H}_m, \mathcal{H}_n] = \tfrac{\mathsf{c}}{3} m\delta_{m+n,0}\,,\\
&[\mathcal{L}_m, \mathcal{G}_n] = (m-n)\mathcal{G}_{m+n}\,, && [\mathcal{H}_m, \mathcal{G}_n] = \mathcal{G}_{m+n}\,,\\
&[\mathcal{L}_m, \mathcal{Q}_n] = -n\mathcal{Q}_{m+n}\,, && [\mathcal{H}_m, \mathcal{Q}_n] = -\mathcal{Q}_{m+n}\,,\\
&[\mathcal{L}_m, \mathcal{H}_n] = -n\mathcal{H}_{m+n} + \tfrac{\mathsf{c}}{6}(m^2+m)\delta_{m+n,0}\,, &&\\
&\{\mathcal{G}_m, \mathcal{Q}_n\} = 2\mathcal{L}_{m+n} - 2n\mathcal{H}_{m+n} + \tfrac{\mathsf{c}}{3}(m^2+m)\delta_{m+n,0}\,, &&
\end{aligned} \tag{2.1}$$

with $m,\ n \in \mathbb{Z}$. Here, $\mathcal{L}_n$ and $\mathcal{H}_n$ are bosons, and $\mathcal{G}_n$ and $\mathcal{Q}_n$, fermions. I do not distinguish between the central element c and its eigenvalue parametrised as $\mathsf{c} = 3(1-\frac{2}{t})$ with $t \in \mathbb{C}\setminus\{0\}$. In this basis, the spectral flow (Schwimmer and Seiberg (1987)) acts as

$$\mathsf{U}_\theta : \begin{array}{ll} \mathcal{L}_n \mapsto \mathcal{L}_n + \theta\mathcal{H}_n + \frac{\mathsf{c}}{6}(\theta^2+\theta)\delta_{n,0}\,, & \mathcal{H}_n \mapsto \mathcal{H}_n + \frac{\mathsf{c}}{3}\theta\delta_{n,0}\,,\\ \mathcal{Q}_n \mapsto \mathcal{Q}_{n-\theta}\,, & \mathcal{G}_n \mapsto \mathcal{G}_{n+\theta}\,. \end{array} \tag{2.2}$$

For $\theta \in \mathbb{Z}$, these transformations are automorphisms of the algebra. Allowing θ to be half-integral in (2.2), we obtain the isomorphism between the Ramond and Neveu–Schwarz sectors. I refer to the $N=2$ modules subjected to the action of the spectral flow as *twisted* modules.

I now briefly review several key properties of unitary $N=2$ representations following (Feigin *et al.* (1998)). These are the irreducible quotients of the twisted topological[1] $N=2$ Verma modules (see the Appendix or Semikhatov and Tipunin (1998) for more details) $\mathfrak{V}_{\mathsf{h}^+(r,1,p),p;\theta}$, where the variable parametrising the central charge is $t=p\in\mathbb{N}+1$, h^+ is defined in (A.4), r is an integer such that $1\leq r\leq p-1$, and θ is the twist (see Definition A.1). The unitary representations are periodic with period p with respect to θ: $\mathfrak{K}_{r,p;\theta+p}\approx\mathfrak{K}_{r,p;\theta}$. Moreover, there are only $p(p-1)/2$ non-isomorphic unitary representations labelled by $0\leq\theta\leq p$ and $1\leq r\leq p-1$, since there are the $N=2$ *isomorphisms*

$$\mathfrak{K}_{r,p;\theta+r}\approx\mathfrak{K}_{p-r,p;\theta}\,,\quad 1\leq r\leq p-1\,,\quad \theta\in\mathbb{Z}_p\,. \tag{2.3}$$

Thus, in order to count each unitary representation once, we can, for example, take $1\leq r\leq[p/2]$ with the full range of θ, $0\leq\theta\leq p-1$, or allow $1\leq r\leq p-1$ with $0\leq\theta\leq r-1$. The untwisted representations will also be denoted by $\mathfrak{K}_{r,p}\equiv\mathfrak{K}_{r,p;0}$.

3 The Structure of the Ghost Realization: "Gravitational Descendants" and the Resolution

Constraint (1.1) can be "solved" by setting $\mathcal{G}(z)=\gamma(z)\,C(z)$, where $\gamma(z)$ is a bosonic field subject to the constraint $\gamma(z)^{p-1}=0$ and C is a "Kazama–Suzuki" ghost field. The constraint on γ can in turn be "solved" by evaluating the cohomology of the differential

$$\mathbb{Q}_0=\oint c\gamma^{p-1}\,. \tag{3.1}$$

The known relation with the LG formulation—which generalizes to a more general case involving several fields γ_i—follows by noticing that the differential is a linear combination of the LG equations of motion $\partial W(\gamma)/\partial\gamma_i$.

The fields involved in (3.1) are viewed as the respective halves of a bc (fermionic) and a $\beta\gamma$ (bosonic) first-order systems with the operator products $b(z)\,c(w)=\frac{1}{z-w}$, $\beta(z)\,\gamma(w)=\frac{-1}{z-w}$. Then (Fre *et al.* (1992), Witten (1994)) $\mathbb{Q}_0$ commutes with the $N=2$ algebra realized in terms of the *currents* $Q(z)$ etc. as

$$\begin{aligned}&Q=-\beta c\,,\quad T=\tfrac{1}{p}\partial b\,c+(\tfrac{1}{p}-1)b\,\partial c+\tfrac{1}{p}\partial\beta\,\gamma-(1-\tfrac{1}{p})\beta\,\partial\gamma\,,\\&G=-\tfrac{1}{p}\gamma\partial b-(\tfrac{1}{p}-1)\partial\gamma b\,,\quad H=-\tfrac{1}{p}\beta\gamma+(1-\tfrac{1}{p})bc\,,\end{aligned} \tag{3.2}$$

The ghost systems in (3.2) have the conformal spin $\lambda=1-\frac{1}{p}$. I assume the point of view that twisted boundary conditions on fermions give rise to a

[1] *Chiral* modules in a different nomenclature (Lerche *et al.* (1989)).

fractional fermion number of the vacuum. I define the bc module $\Lambda_\lambda(n)$ with the cyclic vector $|n\rangle_{bc}$ and the $\beta\gamma$ module $\Xi_\lambda(\nu)$ with the cyclic vector $|\nu\rangle_{\beta\gamma}$ subjected to the annihilation conditions

$$\begin{aligned} b_{\geq 1-\lambda-n}\,|n\rangle_{bc} &= 0\,, \qquad c_{\geq \lambda+n}\,|n\rangle_{bc} = 0\,,\\ \beta_{\geq 1-\lambda-\nu}\,|\nu\rangle_{\beta\gamma} &= 0\,, \qquad \gamma_{\geq\lambda+\nu}\,|\nu\rangle_{\beta\gamma} = 0\,, \end{aligned} \tag{3.3}$$

where I use the convention that $b_{\geq 1-\lambda-n}$ means $(b_{m+1-\lambda-n})_{m\in\mathbb{N}_0}$ and $c_{\geq\lambda+n}$ means $(c_{m+\lambda+n})_{m\in\mathbb{N}_0}$ (here, $\Lambda_\lambda(n)$ and $\Lambda_\lambda(n')$ are of course isomorphic whenever $n - n' \in \mathbb{Z}$). Then, $(bc)_0\,|n\rangle_{bc} = n|n\rangle_{bc}$, $(\beta\gamma)_0\,|\nu\rangle_{\beta\gamma} = -\nu|\nu\rangle_{\beta\gamma}$. The modding of the ghost fields spanning the given picture representations is

$$\begin{aligned} b_m,\ m &\in \tfrac{1}{p} - n + \mathbb{Z}\,, \qquad c_m,\ m \in -\tfrac{1}{p} + n + \mathbb{Z}\,,\\ \beta_m,\ m &\in \tfrac{1}{p} - \nu + \mathbb{Z}\,, \qquad \gamma_m,\ m \in -\tfrac{1}{p} + \nu + \mathbb{Z}\,. \end{aligned} \tag{3.4}$$

For the $N=2$ representation generated from $|n\rangle_{bc} \otimes |\nu\rangle_{\beta\gamma}$, the $N=2$ spectral flow transform $\mathcal{U}_\vartheta$ is realized by changing the bc and $\beta\gamma$ pictures as

$$\Lambda_\lambda(n) \to \Lambda_\lambda(n - \vartheta\lambda)\,, \quad \Xi_\lambda(\nu) \to \Xi_\lambda(\nu + \vartheta(1-\lambda))\,. \tag{3.5}$$

This allows me to fix the overall twist in an arbitrary way until the very end, when the spectral flow transform with any desired θ can be applied to all the representations.

Notation for the modules. The notation $\widehat{\mathfrak{U}}_{h,\ell,p;\theta}$ for the modules used in the Theorem is chosen so as to remind of the Verma modules (see (A.7)–(A.8)), which in the free-field realization are "perverted" into the corresponding $\widehat{\mathfrak{U}}$-modules. However, this is some abuse of notation, since the properties of the $\widehat{\mathfrak{U}}_{h,\ell,p;\theta}$ modules do not directly follow from the values of h, ℓ, and θ, and it actually has to be additionally specified from which vector in the ghost realization such a module is generated.

Theorem 3.1 *Let $\mathfrak{G}_{n,\nu,p} = \Lambda_\lambda(n) \otimes \Xi_\lambda(\nu)$, $\lambda = 1 - \frac{1}{p}$, be the ghost representation space, where n, $\nu \in \frac{1}{p}\mathbb{Z}$, $\nu - n \in \mathbb{Z}$, and*

$$1 + (\nu+1)(1-p) \leq n \leq \nu(1-p)\,. \tag{3.6}$$

Then there is a complex of $N=2$ representations on $\mathfrak{G}_{n,\nu,p}$

$$\begin{aligned} \ldots \xleftarrow{\mathcal{Q}_0} \widehat{\mathfrak{U}}_{\frac{r+1}{p}+m,0,p;\nu p-1} \xleftarrow{\mathcal{Q}_0} \ldots \xleftarrow{\mathcal{Q}_0} \widehat{\mathfrak{U}}_{\frac{r+1}{p}+1,0,p;\nu p-1} \xleftarrow{\mathcal{Q}_0}\\ \xleftarrow{\mathcal{Q}_0} \widehat{\mathfrak{U}}_{\frac{r+1}{p},0,p;\nu p-1} \xleftarrow{\mathcal{Q}_0} \widehat{\mathfrak{U}}_{\frac{r+1}{p}-2,0,p;p-1+\nu p} \xleftarrow{\mathcal{Q}_0} \ldots \xleftarrow{\mathcal{Q}_0} \widehat{\mathfrak{U}}_{\frac{r+1}{p}-m-1,0,p;p-1+\nu p} \xleftarrow{\mathcal{Q}_0} \ldots\,, \end{aligned}$$

where $r = 1+\nu(1-p)-n$, the modules $\widehat{\mathfrak{U}}_{\frac{r+1}{p}+m,0,p;\nu p-1}$ with $m \geq 0$ are generated by the $N=2$ generators (3.2) from the respective states $|1+\nu(1-p)\rangle_{bc}\otimes \gamma^{mp+r}_{\lambda+\nu-1}|\nu\rangle_{\beta\gamma}$, and the modules $\widehat{\mathfrak{U}}_{\frac{r+1}{p}-m,0,p;p-1+\nu p}$, $m \geq 2$, from $|2-p+\nu(1-p)\rangle_{bc}\otimes\beta^{mp-r-1}_{-\lambda-\nu}|\nu\rangle_{\beta\gamma}$. The cohomology of (3.7) is concentrated *at the term* $\widehat{\mathfrak{U}}_{\frac{r+1}{p},0,p;\nu p-1}$ *and is given by the unitary* $N = 2$ *representation* $\mathfrak{K}_{r,p;\theta}$ *with*

$$r = 1 + \nu(1-p) - n\,, \quad \theta = \nu - n\,. \tag{3.7}$$

The data in the conditions of the Theorem are invariant under the shifts $n \mapsto n+a(1-p)$, $\nu \mapsto a$, which change the twist as $\theta \mapsto \theta + pa$. For $a \in \mathbb{Z}$, this induces an isomorphism on unitary representations. For a *fractional* a, such a shift leads to another unitary $N\!=\!2$ representation, however the difference amounts to the overall spectral flow transform, which is applied to the ghost spaces in accordance with (3.5). In particular, the theorem is *equivalent* to its $\nu = 0$-case.

Choosing, thus, $\nu = 0$ for simplicity, the modules $\widehat{\mathfrak{U}}_{\frac{r+1}{p}+m,0,p;-1}$, $m \geq 0$, are generated from the "*gravitational descendant*" states

$$|1\rangle_{bc} \otimes \gamma_{\lambda-1}^{mp+r}|0\rangle_{\beta\gamma} \equiv \sigma_{(p)}^{m}\left(|1\rangle_{bc} \otimes \gamma_{\lambda-1}^{r}|0\rangle_{\beta\gamma}\right), \quad \sigma_{(p)} = \gamma_{\lambda-1}^{p}\,, \tag{3.8}$$

which are all $\mathcal{Q}_0$-trivial except the original one with $m = 0$ (in the topological gravity, the $bc\beta\gamma$ states are tensored with primaries constructed in other sectors, after which these states become the gravitational descendants). As in the LG setting, the $m = 0$ state is singled out because $\gamma_{\lambda-1}$ (which in the invariant terms is the top mode of γ that does not annihilate the picture-zero vacuum) is raised to the power $0 \leq r-1 \leq k = p-2$. The effects due to the dressing by $\sigma_{(p)}^{m}$ are well-known and have been discussed in different languages, including that of integrable hierarchies (see Lerche (1994) and references therein, in particular, Bershadsky *et al.* (1993), Dijkgraaf *et al.* (1991), Losev (1992), Eguchi *et al.* (1993), Lerche and Warner (1994)). The "tic-tac-toe" equations relating the gravitational descendants read as

$$\mathcal{Q}_1\,|1\rangle_{bc} \otimes \gamma_{\lambda-1}^{(m+1)p+r}\,|0\rangle_{\beta\gamma} = (mp+r)\,\mathcal{Q}_0\,|1\rangle_{bc} \otimes \gamma_{\lambda-1}^{mp+r}\,|0\rangle_{\beta\gamma}\,, \tag{3.9}$$

and are an essential ingredient in the construction of the resolution.

The module $\widehat{\mathfrak{U}}_{\frac{r+1}{p},0,p;-1}$ generated from

$$|1\rangle_{bc} \otimes \gamma_{\lambda-1}^{r}\,|0\rangle_{\beta\gamma} \doteq \left|\tfrac{r+1}{p}, 0, p; -1\right\rangle \tag{3.10}$$

(where $\doteq$ means that the state on the LHS satisfies the same highest-weight conditions as the state on the RHS—in the present case, twisted massive highest-weight conditions (A.7)–(A.8)) has a submodule $\widehat{\mathfrak{V}}_{\frac{1-r}{p},p;r-1}$ generated from

$$|1-r\rangle_{bc} \otimes |0\rangle_{\beta\gamma} \doteq |\tfrac{1-r}{p}, p; r-1\rangle_{\text{top}}\,, \tag{3.11}$$

If $\widehat{\mathfrak{V}}_{\frac{1-r}{p},p;r-1}$ were a true Verma module, we would have singular vectors $|E(r,1,p)\rangle^{+,r-1}$ and $|E(p-r,1,p)\rangle^{-,r-1}$ (see (A.5) and (A.6)). In $\widehat{\mathfrak{V}}_{\frac{1-r}{p},p;r-1}$, however, we have

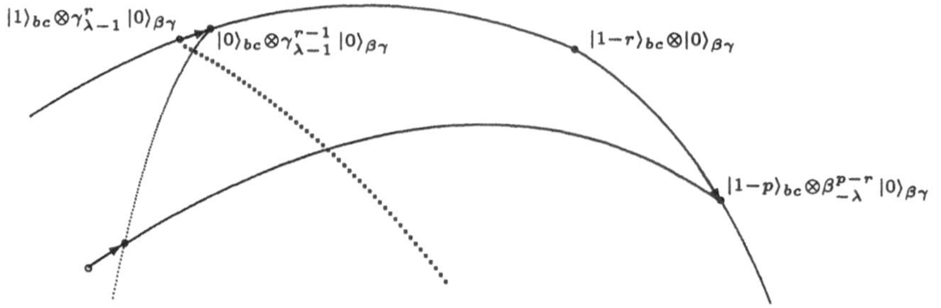

Fig. 1. Topological highest-weight states on the extremal diagram. Filled dots denote the states satisfying twisted topological highest-weight conditions. The top parabola is the extremal diagram of the $\widehat{\mathfrak{U}}_{\frac{r+1}{p},0,p;-1}$ module. The submodule $\widehat{\mathfrak{W}}_{\frac{1-r}{p},p;r-1}$ generated from $|0\rangle_{bc}\otimes\gamma^{r-1}_{\lambda-1}|0\rangle_{\beta\gamma}$ (or equivalently, from $|1-r\rangle_{bc}\otimes|0\rangle_{\beta\gamma}$) is bounded by the "vertical" parabola. The lower parabola is the extremal diagram of the submodule built on singular vector (3.13).

$$|E(r,1,p)\rangle^{+,r-1}=0\,, \tag{3.12}$$
$$|E(p-r,1,p)\rangle^{-,r-1}=|1-p\rangle_{bc}\otimes\beta^{p-r}_{-\lambda}|0\rangle_{\beta\gamma}\,. \tag{3.13}$$

Recalling the submodule $\widehat{\mathfrak{W}}_{\frac{r+1}{p}-2,p;p-1}$ generated from singular vector (3.13), we have the mappings

$$\widehat{\mathfrak{W}}_{\frac{r+1}{p}-2,p;p-1}\xrightarrow{[\mathcal{Q}_{-p+1}\ldots\mathcal{Q}_{-r}]}\widehat{\mathfrak{W}}_{\frac{1-r}{p},p;r-1}\xrightarrow{[\mathcal{Q}_{2-r}\ldots\mathcal{Q}_0\mathcal{Q}_1]}\widehat{\mathfrak{U}}_{\frac{r+1}{p},0,p;-1}\,, \tag{3.14}$$

where the square brackets in $A\xrightarrow{[\mathcal{E}]}B$ indicate that the highest-weight vector of A is mapped onto the vector obtained by applying the operator $\mathcal{E}$ to the highest-weight vector of B. It is useful to consider the extremal diagram describing relations between the modules involved, see Fig. 1. The dotted line shows the extremal diagram of the *quotient* module $\widehat{\mathfrak{U}}_{\frac{r+1}{p},0,p;-1}/\widehat{\mathfrak{W}}_{\frac{1-r}{p},p;r-1}$, in which the lower $\circ$ state becomes the topological singular vector $|E(p-r,2,p)\rangle^{+,-1}$.

The idea now is to observe that singular vector (3.13) is $\mathcal{Q}_0$-exact,

$$|1-p\rangle_{bc}\otimes\beta^{p-r}_{-\lambda}|0\rangle_{\beta\gamma}=a_{p,r}\,\mathcal{Q}_0\,|2-p\rangle_{bc}\otimes\beta^{2p-r-1}_{-\lambda}|0\rangle_{\beta\gamma}\,, \tag{3.15}$$

where $a_{p,r}^{-1}=(p-r+1)(p-r+2)\ldots(2p-r-1)$, while (3.10) is not $\mathcal{Q}_0$-closed:

$$\mathcal{Q}_0\,|1\rangle_{bc}\otimes\gamma^{r}_{\lambda-1}|0\rangle_{\beta\gamma}=|0\rangle_{bc}\otimes\gamma^{p+r-1}_{\lambda-1}|0\rangle_{\beta\gamma}\,. \tag{3.16}$$

As regards the highest-weight vector (3.11), further, we have

$$\mathcal{Q}_{\geq1-r}\,|1-r\rangle_{bc}\otimes|0\rangle_{\beta\gamma}=0\,, \tag{3.17}$$

in particular $\mathbb{Q}_0\,|1-r\rangle_{bc}\otimes|0\rangle_{\beta\gamma}=0$. Moreover, $|1-r\rangle_{bc}\otimes|0\rangle_{\beta\gamma}$ is in the cohomology of $\mathbb{Q}_0$ for $1\le r\le p-1$, because the state $|2-r\rangle_{bc}\otimes\beta_{-\lambda}^{p-1}|0\rangle_{\beta\gamma}$, which is the candidate for the $\mathbb{Q}_0$-primitive of $|1-r\rangle_{bc}\otimes|0\rangle_{\beta\gamma}$ according to the grading, would actually give the desired result only for $r=p$, which is outside the range of r for the *unitary* representations. In the cohomology of $\mathbb{Q}_0$, the bottom-right singular vector $|1-p\rangle_{bc}\otimes\beta_{-\lambda}^{p-r}\,|0\rangle_{\beta\gamma}$ in Fig. 1 vanishes, therefore $\widehat{\mathfrak{V}}_{\frac{1-r}{p},p;r-1}$ is effectively quotiened over the $\widehat{\mathfrak{V}}_{\frac{r+1}{p}-2,p;p-1}$ submodule. Since vector (3.10), further, is *not* in the cohomology of $\mathbb{Q}_0$, this suggests that the cohomology is given precisely by the *unitary representation* $\mathfrak{K}_{r,p;r-1}$. This can indeed proved to be the case (Feigin and Semikhatov (1998)).

4 The "Symmetric" Realization

I now consider the realization of the $N\!=\!2$ algebra (Mussardo *et al.* (1989), Ohta and Suzuki (1990), Ito (1992)) that has the following form in the conventions corresponding to (2.1):

$$\begin{aligned}\mathcal{G}(z)&=C(z)A(z)-\partial C(z)\,,\\ \mathcal{Q}(z)&=B(z)\overline{A}(z)-\tfrac{1}{p}\partial B(z)\,,\\ \mathcal{H}(z)&=\overline{A}(z)-\tfrac{1}{p}A(z)-B(z)C(z)\,,\\ \mathcal{T}(z)&=\partial B(z)\,C(z)+\overline{A}(z)A(z)-\partial\overline{A}(z)\,,\end{aligned}\tag{4.1}$$

where $\overline{A}(z)\,A(w)=\frac{1}{(z-w)^2}$ and $B(z)\,C(w)=\frac{1}{z-w}$. The spectral flow is realized in these terms as

$$\begin{aligned}C_n&\mapsto C_{n+\theta}\,, & A_n&\mapsto A_n+\theta\delta_{n,0}\,,\\ B_n&\mapsto B_{n-\theta}\,, & \overline{A}_n&\mapsto\overline{A}_n-\tfrac{\theta}{p}\delta_{n,0}\,.\end{aligned}\tag{4.2}$$

There are two fermionic screenings

$$S_B=\oint Be^{\int A}\,,\quad S_C=\oint Ce^{p\int\overline{A}}\,.\tag{4.3}$$

I now introduce modules over these free fields. $\Lambda=\Lambda_0(0)$ is the BC module generated from $|0\rangle_{BC}$; next, let the state $|\overline{a},a\rangle_{\overline{A}A}$ be such that

$$\overline{A}_0|\overline{a},a\rangle_{\overline{A}A}=\overline{a}|\overline{a},a\rangle_{\overline{A}A}\,,\quad A_0|\overline{a},a\rangle_{\overline{A}A}=a|\overline{a},a\rangle_{\overline{A}A}\,,\tag{4.4}$$

$$\overline{A}_{\ge 1}|\overline{a},a\rangle_{\overline{A}A}=A_{\ge 1}|\overline{a},a\rangle_{\overline{A}A}=0\tag{4.5}$$

and let $\mathfrak{H}_{\overline{a},a}$ be the corresponding Fock module. I define the "carrier" ghost representation space

$$\mathbb{G}_{r,p}=\Lambda\otimes\bigoplus_{m,n\in\mathbb{Z}}\mathfrak{H}_{n,mp+r-1}\,.\tag{4.6}$$

In $\mathbb{G}_{r,p}$, I take the state $|0\rangle_{BC}\otimes|0,r-1\rangle_{\overline{A}A}$ which satisfies the (twist-zero) topological highest-weight conditions with respect to $N\!=\!2$ generators (4.1):

$$|0\rangle_{BC} \otimes |0, r-1\rangle_{\overline{A}A} \doteq |\tfrac{1-r}{p}, p\rangle_{\mathrm{top}} \,. \tag{4.7}$$

Let $\widetilde{\mathfrak{V}}_{\frac{1-r}{p},p}$ be the $N=2$ module generated from (4.7). It is a submodule in the "central" term ◉ (and also the one carrying the cohomology) in the *butterfly resolution:*[2]

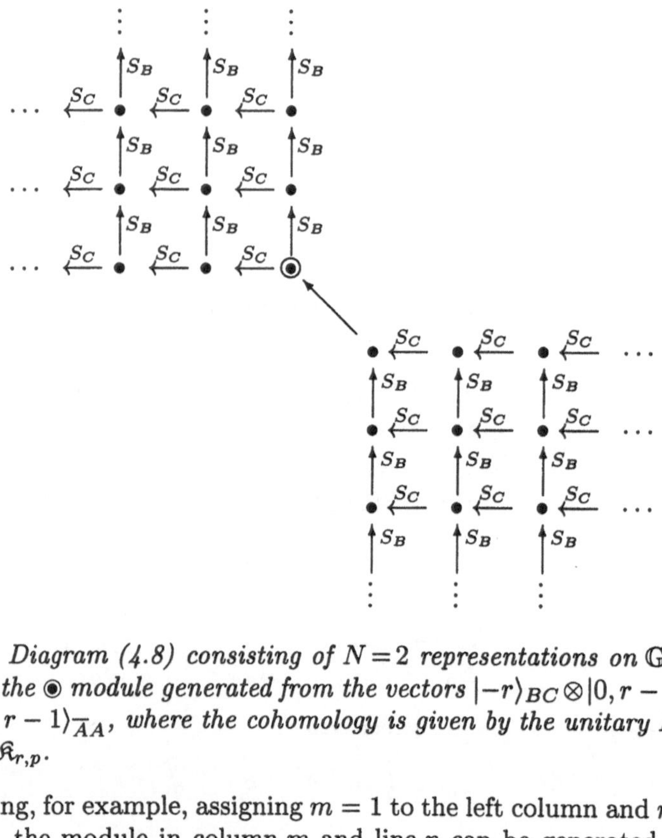

(4.8)

Theorem 4.1 *Diagram (4.8) consisting of $N=2$ representations on $\mathbb{G}_{r,p}$ is exact except at the ◉ module generated from the vectors $|-r\rangle_{BC} \otimes |0, r-1\rangle_{\overline{A}A}$ and $|1\rangle_{BC} \otimes |0, r-1\rangle_{\overline{A}A}$, where the cohomology is given by the unitary $N=2$ representation $\mathfrak{K}_{r,p}$.*

On the right wing, for example, assigning $m=1$ to the left column and $n=1$ to the top line, the module in column m and line n can be generated from $|mp-r\rangle_{BC} \otimes |-n, r-mp-1\rangle_{\overline{A}A}$.

The "central" module is obtained by gluing together the modules generated from $|-r\rangle_{BC} \otimes |0, r-1\rangle_{\overline{A}A}$ and $|1\rangle_{BC} \otimes |0, r-1\rangle_{\overline{A}A}$; these two modules have a common submodule generated from (4.7).

The $N=2$ spectral flow acts on $\mathbb{G}_{r,p}$ according to

$$|n\rangle_{BC} \otimes |\overline{a}, a\rangle_{\overline{A}A} \mapsto |n+\theta\rangle_{BC} \otimes |\overline{a} + \tfrac{\theta}{p}, a-\theta\rangle_{\overline{A}A} \,. \tag{4.9}$$

This allows me to apply the spectral flow to the data in Theorem 4.1 (with $\mathbb{G}_{r,p}$ changed appropriately) in order to obtain all the unitary representations $\mathfrak{K}_{r,p;\theta}$ from Sect. 2.

[2] Or a *sand-glass*, if rotated by $-135°$.

The butterfly construction is based on charged singular vectors (A.10) and on topological singular vectors (A.5) and (A.6). In the free-field realization, these singular vectors may vanish, in which case cosingular vectors exist in the same grade; if, on the other hand, the singular vector does not vanish, one should consider the corresponding submodule and find similar singular vectors in it. In this way, one fills the pattern on the butterfly's wings.

5 Conclusions

Two resolutions—the two-sided and the butterfly one—show a "strong dependence" on the free-field realization chosen. As regards the butterfly resolution, its shape may be interpreted as a "Felder-type" effect occurring in the "$3, 5, 7, \ldots$"-resolution of irreducible $N=2$ representations (Feigin *et al.* (1998)). The latter is constructed in terms of twisted *massive* Verma modules and goes like

$$\begin{array}{ccccccccccc} & & & & & & & & \bullet & & \\ & & & & & & \bullet & & \bullet & & \\ 0 & \leftarrow & \bullet & \leftarrow & \bullet & \leftarrow & \bullet & \leftarrow & \bullet & \leftarrow & \ldots \\ & & & & & & \bullet & & \bullet & & \\ & & & & & & & & \bullet & & \end{array} \tag{5.1}$$

Note that folding (somewhat asymmetrically) the butterfly's wings reproduces the pattern of (5.1). Thus, the moral is that the free-field realization of $N=2$ modules changes the embedding structure and, consequently, turns some of the mappings in (5.1) "inside out", thus resulting in (1.2).

This work was supported in part by the RFBR Grant 98-01-01155.

A $N=2$ Verma Modules

I first introduce the class of $N=2$ Verma modules that I call *topological* Verma modules following (Feigin *et al.* (1997), Semikhatov and Tipunin (1998)). For a fixed $\theta \in \mathbb{Z}$, I define the *twisted topological highest-weight vector* $|h, t; \theta\rangle_{\text{top}}$ to satisfy the annihilation conditions (which are referred to as the twisted topological highest-weight conditions)

$$\mathcal{Q}_{-\theta+m}|h,t;\theta\rangle_{\text{top}} = \mathcal{G}_{\theta+m}|h,t;\theta\rangle_{\text{top}} = \mathcal{L}_{m+1}|h,t;\theta\rangle_{\text{top}} = \mathcal{H}_{m+1}|h,t;\theta\rangle_{\text{top}} = 0 \tag{A.1}$$

for all $m \in \mathbb{N}_0$, with the following eigenvalues of the Cartan generators:

$$(\mathcal{H}_0 + \tfrac{\mathsf{c}}{3}\theta)\,|h,t;\theta\rangle_{\text{top}} = h\,|h,t;\theta\rangle_{\text{top}}\,, \tag{A.2}$$

$$(\mathcal{L}_0 + \theta\mathcal{H}_0 + \tfrac{\mathsf{c}}{6}(\theta^2+\theta))\,|h,t;\theta\rangle_{\text{top}} = 0\,. \tag{A.3}$$

Definition A.1 *The twisted topological Verma module* $\mathfrak{V}_{h,t;\theta}$ *is the module freely generated from the topological highest-weight vector* $|h,t;\theta\rangle_{\text{top}}$ *by* $\mathcal{Q}_{\le -1-\theta}$, $\mathcal{G}_{\le -1+\theta}$, $\mathcal{L}_{\le -1}$, *and* $\mathcal{H}_{\le -1}$.

I write $|h,t\rangle_{\mathrm{top}} \equiv |h,t;0\rangle_{\mathrm{top}}$ in the 'untwisted' case of $\theta = 0$ and also denote by $\mathfrak{V}_{h,t} \equiv \mathfrak{V}_{h,t;0}$ the untwisted module.

Submodules in a topological Verma modules are twisted topological Verma modules (or a sum of two such modules). A singular vector exists in $\mathfrak{V}_{h,t;\theta}$ if and only if $h = \mathsf{h}^+(r,s,t)$ or $h = \mathsf{h}^-(r,s,t)$, where

$$\mathsf{h}^+(r,s,t) = \tfrac{1-r}{t} + s - 1\,,\quad \mathsf{h}^-(r,s,t) = \tfrac{1+r}{t} - s\,,\qquad r,s \in \mathbb{N}\,. \tag{A.4}$$

I denote these singular vectors $|E(r,s,t)\rangle^{\pm,\theta}$, respectively (omitting the twist θ when it is equal to zero). The submodule of $\mathfrak{V}_{h,t;\theta}$ generated from $|E(r,s,t)\rangle^{\pm,\theta}$ is the twisted topological Verma module $\mathfrak{V}_{h\pm r\frac{2}{t},t;\theta\mp r}$. When $s = 1$, the topological singular vectors take a particularly simple form,

$$|E(r,1,t)\rangle^{+,\theta} = \mathcal{G}_{\theta-r}\ldots\mathcal{G}_{\theta-1}\,|\mathsf{h}^+(r,1,t),t;\theta\rangle_{\mathrm{top}}\,, \tag{A.5}$$
$$|E(r,1,t)\rangle^{-,\theta} = \mathcal{Q}_{-\theta-r}\ldots\mathcal{Q}_{-\theta-1}\,|\mathsf{h}^-(r,1,t),t;\theta\rangle_{\mathrm{top}}\,. \tag{A.6}$$

A different class of Verma-like $N=2$ modules are defined as follows (Semikhatov and Tipunin (1998)).

Definition A.2 *A twisted massive Verma module $\mathfrak{U}_{h,\ell,t;\theta}$ is freely generated from a twisted massive highest-weight vector $|h,\ell,t;\theta\rangle$ by the generators*

$$\mathcal{L}_{-m}\,,\ m\in\mathbb{N}\,,\quad \mathcal{H}_{-m}\,,\ m\in\mathbb{N}\,,\quad \mathcal{Q}_{-\theta-m}\,,\ m\in\mathbb{N}_0\,,\quad \mathcal{G}_{\theta-m}\,,\ m\in\mathbb{N}\,.$$

The twisted massive highest-weight vector $|h,\ell,t;\theta\rangle$ satisfies the conditions

$$\begin{aligned}\mathcal{Q}_{m+1-\theta}|h,\ell,t;\theta\rangle = \mathcal{G}_{m+\theta}|h,\ell,t;\theta\rangle = \mathcal{L}_{m+1}|h,\ell,t;\theta\rangle = \\ = \mathcal{H}_{m+1}|h,\ell,t;\theta\rangle = 0\,,\quad m\in\mathbb{N}_0\,,\end{aligned} \tag{A.7}$$
$$\begin{aligned}(\mathcal{H}_0 + \tfrac{\mathsf{c}}{3}\theta)\,|h,\ell,t;\theta\rangle = h\,|h,\ell,t;\theta\rangle\,,\\ (\mathcal{L}_0 + \theta\mathcal{H}_0 + \tfrac{\mathsf{c}}{6}(\theta^2+\theta))\,|h,\ell,t;\theta\rangle = \ell\,|h,\ell,t;\theta\rangle\end{aligned} \tag{A.8}$$

A *charged singular vector* occurs in $\mathfrak{U}_{h,\ell,t;\theta}$ whenever (Boucher *et al.*(1986))

$$\ell = \mathsf{l}_{\mathrm{ch}}(n,h,t) \equiv -n(h - \tfrac{n+1}{t})\,,\quad n\in\mathbb{Z}\,, \tag{A.9}$$

and reads as (Semikhatov and Tipunin (1996)), (Semikhatov and Tipunin (1998))

$$|E(n,h,t)\rangle^{\theta}_{\mathrm{ch}} = \begin{cases}\mathcal{Q}_{-\theta-n}\ \ldots\ \mathcal{Q}_{-\theta}\,|h,\mathsf{l}_{\mathrm{ch}}(n,h,t),t;\theta\rangle\,, & n\geq 0\,,\\ \mathcal{G}_{\theta+n}\ \ldots\ \mathcal{G}_{\theta-1}\,|h,\mathsf{l}_{\mathrm{ch}}(n,h,t),t;\theta\rangle\,, & n\leq -1\,.\end{cases} \tag{A.10}$$

This state on the extremal diagram satisfies the *twisted* topological highest-weight conditions with the twist $\theta + n$, the submodule generated from $|E(n,h,t)\rangle^{\theta}_{\mathrm{ch}}$ being the twisted topological Verma module $\mathfrak{V}_{h-\frac{2n}{t}-1,t;n+\theta}$ if $n\geq 0$ and $\mathfrak{V}_{h-\frac{2n}{t},t;n+\theta}$ if $n\leq -1$.

References

Bernard, D. and G. Felder (1990): Commun. Math. Phys. 127, 145–168.

Bershadsky, M., W. Lerche, D. Nemeschansky, and N.P. Warner (1993): Nucl. Phys. B401, 304.

Boucher, W., D. Friedan, and A. Kent (1986): Phys. Lett. B172, 316–322.

Dijkgraaf, R., E. Verlinde, and H. Verlinde (1991): Nucl. Phys. B352, 59.

Eguchi, T., H. Kanno, Y. Yamada and S.-K. Yang (1993): Phys. Lett. B305, 235.

Feigin, B.L. and E.V. Frenkel (1990): *Representations of Affine Kac–Moody Algebras and Bosonization*, in: Physics and Mathematics of Strings, eds. L. Brink, D. Friedan, and A.M. Polyakov, World Sci.
B.L. Feigin and E.V. Frenkel, Commun. Math. Phys. 128 (1990) 161–189.

Feigin, B.L., A.M. Semikhatov, and I.Yu. Tipunin (1997): *Equivalence between Chain Categories of Representations of Affine $s\ell(2)$ and $N=2$ Superconformal Algebras*, hep-th/9701043.

Feigin, B.L., A.M. Semikhatov, V.A. Sirota, and I.Yu. Tipunin (1998): *Resolutions and Characters of Irreducible Representations of the $N=2$ Superconformal Algebra*, hep-th/9805179.

Feigin, B.L. and A.M. Semikhatov (1998): *Free-Field Resolutions of the Unitary $N=2$ Representations*, to appear.

Felder, G., (1989): Nucl. Phys. B317, 215, E: B324, 548 (1989).

Fré, P., L. Girardello, A. Lerda, and P. Soriani (1992): Nucl. Phys. B387, 333.

Gato-Rivera, B., and A.M. Semikhatov (1992): Phys. Lett. B293, 72.

Ito, K., (1992): Nucl. Phys. B370, 123.

Kawai, T., Y. Yamada, and S.-K. Yang (1994): Nucl. Phys. B414, 191.

Lerche, W., *Generalized Drinfeld-Sokolov Hierarchies, Quantum Rings, and W-Gravity*, Nucl. Phys. B434 (1995) 445–474.

Lerche, W., C. Vafa, and N.P. Warner (1989): Nucl. Phys. B324 427.

Lossev (Losev), A.S. (1992): *Descendants Constructed from Matter Field and K. Saito Higher Residue Pairing in Landau–Ginzburg Theories Coupled to Topological Gravity*, hep-th/9211090.

Lerche, W. and N.P. Warner (1994): *On the Algebraic Structure of Gravitational Descendants in $CP(n-1)$ Coset Models*, hep-th/9409069.

Mussardo, G., G. Sotkov, and M. Stanishkov (1989): Int. J. Mod. Phys. A4, 1135.

Nemeschansky, D. and N.P. Warner (1995): Nucl. Phys. B442, 623.

Ohta, N. and H. Suzuki (1990): Nucl. Phys. B332, 146.

Schwimmer, A. and N. Seiberg (1987): Phys. Lett. B184, 191.

Semikhatov, A.M. and I.Yu. Tipunin (1998): Commun. Math. Phys., 195, 129–173.

Semikhatov, A.M. and I.Yu. Tipunin (1996): *All Singular Vectors of the $N=2$ Superconformal Algebra via the Algebraic Continuation Approach*, hep-th/9604176.

Witten, E. (1994): Int. J. Mod. Phys. A9, 4783.

Super-Affine Hierarchies and Their Poisson Embeddings

Francesco Toppan

UFES, CCE Depto de Física, Goiabeiras cep 99060-900, Vitória (ES), Brasil

Abstract. The link between (super)-affine Lie algebras as Poisson brackets structures and integrable hierarchies provides both a classification and a tool for obtaining superintegrable hierarchies. The lack of a fully systematic procedure for constructing matrix-type Lax operators, which makes the supersymmetric case essentially different from the bosonic counterpart, is overcome via the notion of Poisson embeddings (P.E.), i.e. Poisson mappings relating affine structures to conformal structures (in their simplest version P.E. coincide with the Sugawara construction). A full class of hierarchies can be recovered by using uniquely Lie-algebraic notions. The group-algebraic properties implicit in the super-affine picture allow a systematic derivation of reduced hierarchies by imposing either coset conditions or hamiltonian constraints (or possibly both).

1 Introduction

Affine Lie algebras and conformal algebras have received a great attention in the physicists community in the last several years, mainly due to their relevance to phenomenological models ($2d$ σ-models in the WZNW description), as well as the more fundamental string approach to the unification of interactions. It is well understood by now that conformal algebras (even the non-linear W-type ones) are the output of affine algebras after some construction, hamiltonian reductions or cosets, are performed on them.

While affine-Lie and conformal algebras are universally appreciated, not so much attention has received a truly remarkable property they share, i.e. that they support hierarchies of integrable equations in $1+1$ dimension in the sense that they provide (one of the) Poisson Brackets (PB for short in the following) for the associated hierarchy.

To my knowledge such a property has not yet found a direct implementation in the string-theory program, however has already found application to physically motivated theories like discretized $2d$ gravity in the matrix-model approach (see (P. Di Francesco, P. Ginsparg and J. Zinn-Justin (1995)) and references therein). There, essentially, W-algebras arise as Ward identities known as W-constraints and the partition function is a tau-function of an associated integrable hierarchy.

Moreover integrable hierarchies underline such exactly solvable models as $4d$ $N=2$ Seiberg-Witten SYM theories (see e.g. (N. Seiberg and E. Witten (1994))).

Due to the above-mentioned results it is clear why a lot of attention continues to be focused on the supersymmetric extensions. It is hoped that their understanding will provide the basis for discretized $2d$ supergravity (see (L. Alvarez-Gaumé et al (1992))). However, unlike the purely bosonic theories, the supersymmetric extensions, for reasons we will discuss later, have so far failed being accomodated into a single unifying picture. Due to this basic problem super-hierarchies have been produced by using all sort of tools available, i.e. by direct construction, via Lax operators, bosonic as well as fermionic and both in scalar or matrix form, by coset procedure (Yu. I. Manin and A.O. Radul (1985), P. Mathieu (1988), T. Inami and H. Kanno (1991), J. C. Brunelli and A. Das (1995), F. Toppan (1995), F. Delduc, E. Ivanov and S. Krivonos (1996), L. Bonora, S. Krivonos and A. Sorin (1996)) and so on. We have by now an impressive list of "zoological" data concerning superhierarchies. There is an overwhelming evidence that some order should be made and a single unifying picture should be provided. In this paper we wish to point out a possible tool for both classifying and explicitly constructing a class of super-hierarchies.

Such a tool is based on the super-affine framework and Poisson Embedding (or shortly PE). This means the derivation of super-hierarchies by regarding as fundamental ingredient the Poisson brackets structure furnished by the supersymmetric affinization of a (super)-Lie algebra (more on this notion later). Poisson Embeddings are a special class of Poisson mappings, (i.e. maps between two sets of (super)-fields $f_P : \{\Phi_i\}_I \mapsto \{\widetilde{\Phi}_j\}_{II}$ which preserves the PB structures between sets $\{I\}$ and $\{II\}$) having the further property that the PB structure in $\{II\}$ is the one of a given integrable hierarchy.

As a consequence it is possible to define on the super-fields in $\{I\}$ a hierarchy of equations which inherit the integrability property from the one defined on the second set $\{II\}$. This seemingly innocent remark has indeed far-reaching consequences and allows us to produce and identify new hierarchies of equations, even generalizing the set of hierarchies produced in the literature with more dispendious and time consuming methods.

Well-known examples of Poisson maps are the Wakimoto free (super)-fields realization of affine algebras, and the Sugawara-type construction relating affine(super-)fields to the (super)-stress energy tensor. The latter is also a Poisson Embedding, and therefore integrable hierarchies are induced both at the level of affine and of Wakimoto free superfields.

A real breakthrough in this context appears to be the realization in (E. Ivanov, S. Krivonos and F. Toppan (1997)) that among such mappings there is a (differential) polynomial Poisson map which is an $N = 4$ extension of the Sugawara construction based on super-affine $sl(2) \oplus u(1)$. Such a mapping, besides being a PB one, is a Poisson Embedding since the Sugawara-produced hierarchy turns out to be the small $N = 4$ SCA carrying the $N = 4$ KdV hierarchy.

The key observation is that superconformal algebras can be more directly identified with the PB structure of a given hierarchy since they are easily accomodated into scalar-type Lax operators (see (F. Delduc, E. Ivanov and S. Krivonos (1996))), which can be constructed with a systematic procedure.

Focus can therefore be put into the construction of generalized Sugawara mappings. Here however we advocate the point of view that we can investigate the properties of *already existing* Sugawara constructions to identify and classify series of new hierarchies. In particular one of such mappings sends any $N=2$ (super)-affine algebra into $N=2$ Virasoro. This is the PB structure for three distinct $N=2$ KdV hierarchies (P. Mathieu (1988)) (associated to a value of parameter $a=4,-2,1$). Accordingly, three induced hierarchies are associated to the $N=2$ affine superfields realizing the PE. In some cases, the three hierarchies collapse into a single one.

The full power of affine algebras gets really appreciated when one realizes that due to stringent group-theoretical reasons one can further reduce such algebras with coset procedures and/or hamiltonian reductions. On conformal algebras themselves these procedures cannot be carried out. For instance the Virasoro algebra itself is already a hamiltonian reduction of affine $sl(2)$ (A.M. Polyakov (1990)).

A full bunch of "popping out" hierarchies can therefore find a natural group-theoretical explanation and interpretation.

2 Notations and Preliminary Remarks

The class of Poisson brackets structure we will consider is given by the superaffinization of any given semisimple (super)-Lie algebra, defined as follows: Let $\mathcal{G}$ be any finite semisimple Lie algebra, either purely bosonic or supersymmetric, with n_b bosonic and n_f fermionic generators ($n_f=0$ for standard Lie algebras) collectively denoted as τ_α, for $\alpha=1,...,n_b+n_f$, and let ${f^\gamma}_{\alpha\beta}$ denote the structure constants.

We can introduce the $N=1$ superfields (in the superspace coordinate $X=x,\theta$, with θ Grassmann) $\Psi_\alpha(X)$, associated to each generator τ_α and with opposite statistics w.r.t. τ_α.

The superaffine algebra is defined by assuming the following Poisson brackets

$$\{\Psi(X)_\alpha,\Psi(Y)_\beta\} =_{def} {f^\gamma}_{\alpha\beta}\Psi(Y)_\gamma\delta(X,Y)+cK_{\alpha\beta}D_Y\delta(X,Y) \qquad (1)$$

where we introduced the supersymmetric Dirac's δ-function

$$\delta(X,Y)=\delta(x-y)(\theta-\eta)$$

and the $N=1$ superderivative

$$D_Y=\frac{\partial}{\partial\eta}+\eta\frac{\partial}{\partial y}$$

for $Y \equiv y, \eta$. c is the central extension and $K_{\alpha\beta}$ is defined as a supertrace $K_{\alpha\beta} = Str(\tau_\alpha \tau_\beta)$ in a given representation for $\mathcal{G}$, let's say the adjoint.

In the above relation the Jacobi identities are satisfied and therefore (1) indeed defines a PB structure.

We mention that if the (super)-algebra $\mathcal{G}$ of departure admits a complex structure, then (1) is indeed $N = 2$ supersymmetric and can be recasted into a manifestly $N = 2$ formalism (see (C. Ahn, E. Ivanov and A. Sorin (1997)), here however we do not need such technical improvement.

The bosonic limit of the above (1) affine algebra (realized on a bosonic $\mathcal{G}$ and via purely bosonic fields) is the building block for constructing generalized Drinfeld-Sokolov hierarchies, via the association to a matrix type Lax operator L of the kind

$$L = \partial + \sum_\alpha J_\alpha(x)\tau_\alpha + \Lambda \tag{2}$$

where Λ is a constant element in the $\widetilde{\mathcal{G}} = \mathcal{G} \otimes \mathbf{C}(\lambda, \lambda^{-1})$ loop algebra of $\mathcal{G}$ realized on an auxiliary variable λ which plays the role of a spectral parameter. By DS construction, if Λ is chosen in such a way to realize the decomposition

$$\widetilde{\mathcal{G}} = \mathcal{K} \oplus \mathcal{M} \tag{3}$$

with $\mathcal{K} = Ker_{ad-\Lambda}$ and $\mathcal{M} = Im_{ad-\Lambda}$ over the adjoint action of Λ, and if furthermore $\mathcal{K}$ is abelian

$$[\mathcal{K}, \mathcal{K}] = 0 \tag{4}$$

then we are guaranteed about the existence of infinite integrals of motion in involution.

The same kind of construction has been generalized in the supersymmetric case by Inami and Kanno in a series of papers (see e.g. (T. Inami and H. Kanno (1991)) and references therein). Now L assumes the form

$$L = D + \sum_\alpha \Psi_\alpha(X)\tau_\alpha + \Lambda \tag{5}$$

the first and second terms in the r.h.s. are fermionic and so Λ must be fermionic as well. This constraint puts a very strong restriction on the superhierarchies which can be obtained through DS procedure. In the bosonic case for instance generalized-KdV hierarchies which include among others Boussinesq are defined by taking Λ to be the sum over all simple roots of the $\widetilde{\mathcal{G}}$ algebra[1]. Consequently generalized super-KdV hierarchies can be obtained solely from those superalgebras which admit a Dynkin diagram presentation

[1] This is a slightly imprecise way of saying, in order to talk about simple roots we should introduce the extended affine algebra over the auxiliary loop space parameter λ, but let's avoid these technical complications here.

involving only fermionic simple roots. Admittedly, this is a rather restrict class of superalgebras.

For them however we have a viable and systematic procedure to construct superhierarchies.

A problem arises because very natural hierarchies like the supersymmetric extensions of NLS fail to be accomodated into this scheme due to the fact that their bosonic equivalents are recovered from a Λ in (5) belonging to the Cartan generators of $\mathcal{G}$. Since the Cartan sector of whatever simple bosonic Lie or super-Lie algebra is in any case bosonic, we have no possibility at all to construct a fermionic Λ with the desired properties.

In effect there exists a class of superalgebras, the so-called strange superalgebras of the $Q(n)$ series, which admit a fermionic Cartan sector (see (L. Frappat, P. Sorba and A. Sciarrino (1996))). But these superalgebras are of no use here for another reason. The fermionic Λ taking value in the fermionic Cartan always fail to satisfy either condition (3) or condition (4).

There are some *ad hoc* procedures to overcome this difficulty (see (F. Toppan (1996))), but it must be said that even if viable for the practical purpose of computing higher order hamiltonians, they lack a clear and compelling motivation which makes them not attractive for the purpose of classifying hierarchies. Similarly, matrix-type Lax operators have been produced for hierarchies of super-NLS type (or obtainable from super-NLS) reduction. Such Lax operators, unlike the bosonic ones, have entries which are composite superfields. Here again it is hard to justify their appearance in terms of fundamental principles. Rather, they are the signal that a simpler structure should be found behind them. In the next section we will see how an answer (at least a partial one) can be provided.

3 Poisson Maps and Poisson Embeddings

In the previous section we have discussed some problems arising in the construction of superintegrable hierarchies. Now we will show how to overcome such problems via the introduction of the notion of Poisson maps and Poisson embeddings, already outlined in the introduction. Evidence will be furnished that superaffine algebras are the right setting to deal and classify superhierarchies.

A point should be clear, the maps we are investigating are *polynomial differential* maps. In literature non-polynomial maps relating different hierarchies have been considered (in some cases they are not even Poisson maps), but in all known examples they can be recasted into (or derived from) polynomial differential maps. So it seems there is no compelling reason to look beyond the realm of polynomial differential Poisson maps.

Under an f_P Poisson Embedding the infinite series of H_k hamiltonians in involution of the integrable hierarchy ($\{H_k, H_{k'}\}_{(II)} = 0$) can be regarded as hamiltonians in involution w.r.t. the first PB structure (for any couple is indeed $\{H_k, H_{k'}\}_{(I)} = 0$) and it makes sense to define an infinite series of compatible flows for the superfields Φ_i as:

$$\frac{\partial}{\partial t_k}\Phi_i = \{H_k(f_P(\Phi_j)), \Phi_i\}_{(I)} \tag{6}$$

We will refer to hierarchies of this kind either as the induced or as the embedded hierarchy.

Examples of PE are given by Sugawara-type construction. We recall that for any bosonic and $N = 1$ supersymmetric affine algebra there exists a well-defined procedure which allows us to produce conformal fields satisfying a closed $\mathcal{W}$-algebra structure. They are expressed in terms of the enveloping algebra of the (super)-affine algebra $\mathcal{G}$ and are in 1-to-1 correspondence with each Casimir of $\mathcal{G}$. The most relevant or leading term in the enveloping algebra being given by

$$d^{i_1 \dots i_n} J_{i_1}(x) \dots J_{i_n}(x)$$

in the bosonic case and

$$d^{i_1 \dots i_n}(D_X \Psi_{i_1}) \cdot \dots \cdot (D_X \Psi_{i_{n-1}}) \cdot \Psi_{i_n}(X) \tag{7}$$

in the $N = 1$ super-case.

Here $d^{i_1 \dots i_n}$ is the symmetric tensor denoting an n-th order Casimir and J_{i_k} (Ψ_{i_k}) are spin 1 fields (spin $\frac{1}{2}$ superfields respectively).

A full bunch of improvements or covariantization terms must be added to the above "leading order terms" to render the (super)-field a primary or conformal field. For instance, in the case of the order 2 Casimir (which exists for any (super)-Lie algebra) the terms to be added are just the Feigin-Fuchs terms which provide a non-vanishing central charge so that the Sugawara-constructed field satisfies the full Virasoro ($N = 1$ superVirasoro) algebra.

As an example, in the case of the $sl(n)$ series the W-algebra so produced is the W_n algebra (or its $N = 1$ supersymmetrization). The "miracle" here, which has a group-theoretical explanation, lies in the fact that this is the same algebra arising from Dirac brackets after the Drinfeld-Sokolov hamiltonian reduction sketched in the previous section has been taken into account.

For the moment let me just point out that the (super)-affine algebra, which at the beginning was not associated to any evident hierarchy, has now acquired the status of a PB structure for the induced hierarchy. Any Poisson map onto the super-affine fields is therefore also a Poisson Embedding.

There exists a well-defined prescription on how to realize any super-affine algebra in terms of free (super)-fields. The result is given by the (generalized) (super)-Wakimoto realizations whose origin is traced on the theory of non-linear realizations of algebraic structures. The simplest case of a super-Wakimoto construction is the realization of the $N = 1$ affine $sl(2)$ algebra (F. Toppan (1995)). The generalized (super)-Wakimoto realizations are all examples of Poisson maps. In full generality the Wakimoto free (super)-fields satisfy induced hierarchies whose origin arises from their "double" embedding into the generalized KdV hierarchies.

4 Induced $N = 2$ and $N = 4$ Hierarchies

In the previous section we have discussed the general theory of Poisson Embeddings and have shown that super-affine Lie algebras are associated with (at least $N = 1$) integrable hierarchies.

Here we further discuss properties of PE and study them in the context of $N = 2$ hierarchies. This is indeed a very interesting case since for the first time appears (as shown by explicit construction in (P. Mathieu (1988))) that one and the same PB structure is associated with different series of integrable hierarchies.

Indeed if we denote as $J(x, \theta, \overline{\theta})$ the real $N = 2$ Virasoro field consisting of two (spin $= 1, 2$) boson components and a couple of spin $\frac{3}{2}$ fermions, three inequivalent $N = 2$ KdV hierarchies can be recovered (for a parameter a taking values $a = 4, -2, 1$ respectively).

Therefore any Poisson map onto $N = 2$ Virasoro induces three different and in principle inequivalent series of hierarchies.

The case $N = 2$ is interesting also for another reason. While generalized $N = 1$ Sugawara poses no problem in its construction (and in its association with super-KdV hierarchies), much less is known concerning the $N = 2$ case. Apart from the Sugawara construction of the $N = 2$ Virasoro discussed below, to my knowledge no general theorem has been given so far and only explicit examples have been worked out on how to perform Sugawara construction of $N = 2$ $\mathcal{W}$-algebras (the main question concerns the feasibility of adding Feigin-Fuchs terms, while mantaining a full $N = 2$ $\mathcal{W}$ algebra structure).

Neverthless what we already have can be exploited to produce and identify interesting classes of induced $N = 2$ hierarchies.

Let us for the moment discuss just the simplest examples of superaffine algebras producing PE onto $N = 2$ Virasoro. They are given respectively by the $N = 1$ affinization (the following examples are $N = 2$ supersymmetric and can be reformulated with an $N = 2$ formalism as stated before) of

i) the $u(1) \oplus u(1)$ algebra;

ii) the $sl(2) \oplus u(1)$ algebra and

iii) the $sl(2|1)$ superalgebra.

In the first two examples the original algebra is not a simple one, this is just because we need a complex structure (provided by the extra $u(1)$) which enables us to have a second supersymmetry.

$sl(2|1)$ is the simplest example of an $N = 2$ Lie superalgebra; it contains $sl(2)$ as subalgebra and four extra fermionic simple roots.

The first case coincides with the well-known $N = 2$ version of (the three induced hierarchies of) m-KdV.

Much more interesting and rich of structure is case *ii)*. The algebra here admits a quaternionic structure which allows a full $N = 4$ Sugawara construction (E. Ivanov, S. Krivonos and F. Toppan (1997)). This has the consequence that, besides the induced $N = 2$ hierarchies (for $a = 4, 2, -1$) recovered from the $N = 2$ Sugawara through, respectively, *a)* the $N = 2$ Virasoro $J_{u(1)\oplus u(1)}$ associated to the $u(1) \oplus u(1)$ subalgebra (it coincides with the previous case), *b)* the full $N = 2$ Virasoro J_{full} given by $J_{full} = H\overline{H} + F\overline{F} + cD\overline{H} + \overline{c}\overline{D}H$ (where the last two are the Feigin Fuchs terms), *c)* the coset $N = 2$ Virasoro given by $J_{coset} = J_{full} - J_{u(1)\oplus u(1)} = F\overline{F} + c'D\overline{H} + \overline{c}'\overline{D}H$,
we can defined induced an induced hierarchy based on $N = 4$ KdV. This hierarchy has a very interesting property, namely that it is consistent with the equations of motion to set

$$H = \overline{H} = 0. \tag{8}$$

The reduced hierarchy on the superfields $F, \overline{F}$ coincides with the $N = 2$ NLS.

In this framework the NLS hierarchy can therefore be directly obtained from its Poisson Embedding properties on $N = 4$ KdV, in contrast for instance with the original coset construction of $N = 2$ NLS (F. Toppan (1995)), whose superfields were mapped onto an $N = 2$ Virasoro *without* central charge and for that reason integrability had to be proven separately and with different methods.

The *iii)* case can be treated similarly. This superalgebra contains $sl(2) \oplus u(1)$ as a subalgebra. All the induced hierarchies defined in the previous case can therefore be consistently extended. They involve extra equations of motions associated to the $N = 2$ non-linearly (anti)-chiral bosonic superfields of spin $\frac{1}{2}$ (of "wrong" statistics) associated to the fermionic roots.

It is clear at this point that a full class of $N = 4$ induced hierarchies can be produced from the superaffinization of any quaternionic super-Lie algebra $\mathcal{G}_Q$ (i.e. with $N = 4$ supersymmetric PB), with the following recipe:

i) individuate any $sl(2) \oplus u(1)$ subalgebra,

ii) construct from the given subalgebra the $N = 4$ Sugawara leading to $N = 4$ SCA,

iii) use this concrete realization of the $N = 4$ SCA as PB structure for the $N = 4$ KdV.

An $N = 4$ hierarchy is automatically induced on the affine superfields generating the full $\mathcal{G}_Q$ algebra. The induced hierarchy is automatically $N = 4$ invariant because by construction *both* the hamiltonians and the superaffine-$\mathcal{G}_Q$ PB structure are $N = 4$ supersymmetric.

Furthermore, on these induced hierarchies it is possible to investigate whether consistent reductions can be imposed both as hamiltonian constraints or coset construction.

5 Conclusions

In this talk I have investigated the role of Lie-algebraic methods in classifying integrable hierarchies. One of the nice features of the approach based on super-affine Lie algebras is that it allows investigating systematically hamiltonian constraints and coset reductions. The structure of $N = 4$ integrable hierarchies is currently under investigation, with the methods here outlined, in a collaboration with E. Ivanov and S. Krivonos.

Acknowledgments I express my gratitude to the organizers of the workshop in memory of V.I. Ogievetsky for their invitation.

This work has been written while the author was under a JSPS (Japan Society for the Promotion of Science) contract. I am very grateful to the members of the Physics Dept. at Shizuoka University for their warm and kind hospitality. I am especially pleased to thank Shogo Aoyama for his much needed constant help and support. Finally, I have profited of clarifying discussion with E. Ivanov and S. Krivonos.

6 Appendix: the Classification of Cosets

One of the nice features of the approach based on affine superfields is the fact discussed in the text that it allows to construct reduced hierarchies by exploiting symmetries and group-theoretical properties of the affine algebras. Here we will discuss the class of reductions known as "cosets", which arises when superfields associated to semisimple (super)-Lie subalgebras are set equal to zero. This class of reductions are classified by all inequivalent (super)-Lie subalgebras which can be embedded into $\mathcal{G}$, the (super)-Lie algebra whose superaffinization furnishes the original hierarchy.

The classification scheme to find all inequivalent Lie subalgebra of a given Lie algebra has been given by Dynkin. This scheme provides all possible cosets we can construct out of a given hierarchy. To be specific we specialize our discussion here to the $sl(3)$ case which already contains all the features we are interested in.

The full list of $sl(3)$ subalgebras is given by

$$u(1), \quad u(1) \oplus u(1), \quad sl(2), \quad sl(2) \oplus u(1) \tag{9}$$

The corresponding coset hierarchies involve $8 - 1 = 7$, $8 - 2 = 6$, $8 - 3 = 5$ and $8 - 4 = 4$ (super)-fields respectively (8 is the order of $sl(3)$). This is not the full story however because an abelian $u(1)$ generator ($= v$) can always be chosen to lie in the Cartan sector of $sl(3)$ and be specified by an angle ϕ:

$$v = cos\phi H_1 + sin\phi H_2, \tag{10}$$

where H_1, H_2 are the two Cartan generators of $sl(3)$.

Coset hierarchies w.r.t. an abelian $u(1)$ are therefore labelled by such an angle ϕ. The angle however is not completely arbitrary and is further required to lie on the interval

$$0 \leq \phi < \frac{\pi}{6} \tag{11}$$

The reason is the presence of the extra discrete symmetries which involve the Cartan decomposition of a Lie algebra. In case of $sl(3)$ there are two such sources of symmetries:

i) the $Out = \frac{Aut}{Int}$ automorphism group, which coincides here with $\mathbf{Z}_2$ and is related to the symmetry of the Dynkin diagram (i.e. the exchange of the two simple roots);

ii) the Weyl group of $sl(3)$, which is the 6-elements S_3 permutation group.

The full group of discrete automorphisms coincides here with the 12-elements group, direct product of S_3 and $\mathbf{Z}_2$. It can be easily realized that its action on the Cartan subspace spanned by the H_1, H_2 generator is that of the finite rotation group generated by 30-degrees rotations, which finally leads to the above restrictions on the angle ϕ.

For what concerns $sl(2)$ subalgebras there exists only two inequivalent ways of embedding $sl(2)$ on $sl(3)$, up to the *Adj*-action of $sl(3)$ internal automorphism group. They correspond to the two decompositions of the 8 $sl(3)$ generators in terms of the $sl(2)$ representations.

Only the latter $sl(2)$ allows to accomodate a further abelian $u(1)$ subalgebra, and therefore the $sl(3)$ coset over $sl(2) \oplus u(1)$ is unique.

In this way we have listed the full class of coset-hierarchies arising from $sl(3)$. The case of the coset over $sl(2) \oplus u(1)$ is of particular interest because both $sl(3)$ and $sl(2) \oplus u(1)$ are quaternionic algebras. It is likely that its supersymmetric extension (under the appropriate $N = 4$-invariant hamiltonians) would correspond to an $N = 4$ coset hierarchy. This case is currently under investigation in a collaboration with Ivanov and Krivonos .

The above procedure can be carried in full generality and cosets can be classified according to the Dynkin scheme.

It is worth to notice that no restriction involving e.g. symmetric space is required to define consistent coset hierarchies, they are just (a more specialized) example of coset-construction.

References

P. Di Francesco, P. Ginsparg and J. Zinn-Justin (1995), Phys. Rep. **254** 1

N. Seiberg and E. Witten (1994), Nucl. Phys. **B 426** 19;
A. Gorskii, I. Krichever, A. Marshakov, A. Mironov and A. Morozov (1995) , Phys. Lett. **B 355** 466

L. Alvarez-Gaumé, H. Itoyama, J.L. Manes and A. Zadra (1992), Int. Jou. Mod. Phys. **A 7** 5337

Yu. I. Manin and A.O. Radul (1985), Commun. Math. Phys. **98** 65;
P.P. Kulish (*1985*), Lett. Math. Phys **10** 87

P. Mathieu (1988), Phys. Lett. **203** 287
T. Inami and H. Kanno (1991), Comm. Math. Phys. **136** 519;
Int. Jou. Mod. Phys. **A 7** Suppl. 1A (1992) 419
J. C. Brunelli and A. Das (1995), Jou. Math. Phys. **36** 268
F. Toppan (1995), Int. Jou. Mod. Phys. **A 10** 895
F. Delduc, E. Ivanov and S. Krivonos (1996), Jou. Math. Phys. **37** 1356
L. Bonora, S. Krivonos and A. Sorin (1996), Nucl. Phys. **B 477** 835
E. Ivanov, S. Krivonos and F. Toppan (1997), Phys. Lett. **405** 85
A.M. Polyakov (1990), Int. Jou. Mod. Phys. **A 5** 833
C. Ahn, E. Ivanov and A. Sorin (1997), Commun. Math. Phys. **183** 205
L. Frappat, P. Sorba and A. Sciarrino (1996), Dictionary on Lie Superalgebra, hep-th/9607161.
F. Toppan (1996), Int. Jou. Mod. Phys. **A 11** 3257

Part IV

Quantum Field Theory and Quantum Groups

1+1 Dimensional Models of Dilaton Gravity Coupled to Bosons and Fermions

A.T. Filippov[1]

Joint Institute for Nuclear Research, Dubna, Russia
E-mail: filippov@thsun1.jinr.dubn.su

Abstract. Main properties of 1+1 dimensional dilaton gravity coupled to different sorts of matter are summarized, with a special emphasis on scalar bosons and spinor fermions. General dilaton gravity coupled to gauge fields is an essentially finite (0+1) dimensional theory and describes space - time manifolds with horizons. Coupling to scalar fields restores the field degrees of freedom but removes the static horizons. We propose that a combination of the scalar and spinor couplings may restore the static horizons, if the spinor field is properly adjusted to the scalar one.

The general 1+1 dimensional dilaton gravity (DG) treated here[1] is given by the Lagrangian

$$\mathcal{L} = \mathcal{L}_{DG} + \mathcal{L}_M, \tag{1}$$

where the dilaton gravity Lagrangian describes the gravitational degrees of freedom, which are the metric tensor field g_{ij} and the scalar dilaton field ϕ,

$$\mathcal{L}_{DG} = \sqrt{-g}\,[UR + V + Wg^{ij}\phi_i\phi_j], \tag{1a}$$

and the matter Lagrangian describes the coupling of the dilaton gravity to gauge fields $F_{ij}^{(a)}$, scalar matter fields $\psi^{(m)}$, and spinor matter fields $\Theta^{(n)}$ (below, we omit the superscripts),

$$\mathcal{L}_M = \sqrt{-g}\,[XF_{ij}F^{ij} + Y + Z_B g^{ij}\psi_i\psi_j + iZ_F(\nabla_j\overline{\Theta}\gamma^j\Theta - \overline{\Theta}\nabla_j\gamma^j\Theta)]. \tag{1b}$$

Here U, V, W, X, Y and Z are arbitrary real functions of the dilaton field ϕ while Y may in addition depend on ψ and Θ; g^{ij} is the contravariant metric tensor, and R is the scalar curvature; the lower letter indices denote partial derivatives ($\phi_i = \partial_i\phi$, etc.), except when used in g_{ij} or F_{ij}.

Theories of this sort may be generated by Kaluza - Klein reductions of gravity and supergravity from higher dimensions, by superstrings and membranes, etc. In particular, they may describe spherically symmetric black holes (in any dimension) coupled to matter fields. From the higher - dimensional point of view, the 1+1 dimensional metric g_{ij} and the dilaton field

[1] This text contains only that part of the report presented at the conference which is related to the apparently new idea of a possible role of fermions in gravity models. The main concepts and results necessary to formulate this idea are only briefly reviewed. For notation, further details and references see (Filippov (1996,1997))

ϕ are geometric (gravitational) variables. The three different matter fields interact with the gravitational fields very differently, as will be shown below.

In the absence of matter fields, we have the general dilaton gravity model (1a), which is explicitly integrable (for arbitrary potentials U, V, W). Moreover, its solution may be expressed in terms of one free, massless (D'Alembert) field[2]. To write the solution we first define a new function $w(\phi)$ by the equation $w'/w \equiv W/U'$ (the prime always denotes the derivative of a function depending on one variable, thus $U' \equiv dU/d\phi$, etc.). Expressing w as a function of U (i.e. $w(\phi(U))$), we then define the prepotential $N(U)$ by the equation $N'(U) = V/w \equiv \widetilde{V}$. Now consider the equations derived by varying (1) w.r.t. g_{jj}. It is most instructive to write these equations using the light - like metric $ds^2 = -4f(u,v)dudv$

$$\widetilde{f}(U_j/\widetilde{f}w)_j = Z_B\psi_i^2 + iZ_F(\nabla_j\overline{\Theta}\gamma_j\Theta - \overline{\Theta}\nabla_i\gamma_j\Theta)], \tag{2}$$

where $\widetilde{f} \equiv fw$, $j = u, v$. For the pure dilaton gravity (1a), the right - hand side (r.h.s.) of this equation identically vanishes and it can be solved in terms of a free field $\Phi = a(u) + b(v)$:

$$U = F(\Phi), \quad \widetilde{f} = F'(\Phi)\Phi_u\Phi_v \, , \tag{3}$$

where F, a and b are arbitrary functions of their arguments. Thus, the solution essentially depends on one variable. This fact constitutes a generalization of the Birkhoff theorem concerning uniqueness of the Schwarzschild black hole (SBH) solution in the spherically symmetric sector of the Einstein theory of gravity.

Now, using the equation derived by varying w.r.t g_{uv} (it is independent of Z_B and Z_F), we find that $M \equiv F'(\Phi) + N[F(\Phi)]$ is a local integral of motion, i.e. $M_u = M_v = 0$ on the equations of motion. This allows us to determine $\Phi(F)$ and $F(\Phi)$ from the equation $dF/d\Phi = M - N$. The important thing is that the r.h.s. of this equation has zeroes. They are called horizons because they define zeroes of the metric,

$$\widetilde{f} = (M - N)a'(u)b'(v). \tag{4}$$

In particular, with just one horizon, we may reproduce the SBH solution in any dimension; in this case, M is the mass of the black hole. Note that the SBH is static (in a special coordinate system). For this reason, we call the obtained general solution "static". Note also that M is invariant w.r.t. general coordinate transformations because

$$M \equiv N(U) + U_uU_v/\widetilde{f} = N(U) - g^{ij}U_iU_j/w \, . \tag{5}$$

[2] Moreover, the exact solution of the field theory (1a) may be exactly reconstructed from the solutions of the one-dimensional dynamical system obtained by a special gauge fixing (or further dimensional reduction to 0+1 dimension) of the field theory; see (Filippov (1996,1997)), (Cavagliá, de Alfaro and Filippov (1995,1996)).

The above solution may be viewed as a Bäcklund transformation from the variables f and ϕ to the variables M and Φ, satisfying simple decoupled equations of motion: $M_i = 0$, $\Phi_{uv} = 0$, $i = u, v$.

The solution of the general DG was possible because there exists an additional Killing symmetry which effectively reduces the 1+1 dimensional field theory to a finite dimensional system. In fact, M and the canonically conjugate momentum P_M are the only physical variables of the theory. Another expression of this property of DG is that it can be rewritten as a topological gauge theory. The additional coupling to the gauge fields (X term) does not destroy the Killing symmetry and the theory remains topological and finite dimensional, although now it has additional physical variables (charges) describing gauge degrees of freedom. The above construction of the solutions is trivially generalized to the case of coupling to Abelian gauge fields (Filippov (1996,1997)). The topological nature of DG coupled to nonabelian gauge fields is discussed in (Klösch and Strobl (1996)). Let me stress that we need not know anything about Killing vectors or topological gauge theories to find the exact solution written above. The simple fact that the r.h.s. of equations (2) vanish identically gives all the necessary information.

Now, if there is a coupling to scalar fields (but no spinor fields), everything changes drastically. Only if all scalar fields are constant (independent of u and v) we still have the Killing symmetry. However, in general, all the solutions will essentially depend on two variables and thus are not static. In other words, for a generic solution there exists no coordinate system in which

$$\widetilde{f} = h(a+b)a'(u)b'(v), \quad \phi = \phi(a+b). \tag{6}$$

Thus, the static solutions (having this property) form only a small subclass among inequivalent solutions of DG coupled to scalars (DGB). Moreover, static solutions, for which not all scalar fields are constant, cannot have horizons. A very simple proof of this "No horizon theorem" was given in (Filippov (1996,1997)).

The proof of the theorem does not require knowledge of the solutions or even integrability. In fact, the DGB models are generally not integrable. Even their static solutions are generally described by not integrable finite - dimensional Hamiltonian systems. Nevertheless, there exists a wide enough class of integrable DGB models. For example, if $X = Y = 0$ and Z_B is constant (independent of ϕ) the theories with the potentials

$$\widetilde{V} = g_+ e^{g\phi} + g_- e^{-g\phi}. \tag{7}$$

are integrable and their solutions can be expressed in terms of two D'Alembert fields (Filippov (1996,1997)). Note that there exist integrable models with nonminimal coupling of gravity to scalar fields (Filippov and Ivanov (1998)). Note also that the static sector is integrable for a wider choice of models (see (Filippov (1996,1997)), (Cavagliá and de Alfaro (1997))).

Let us summarize main properties of the dilaton gravity:

1. Dilaton gravity (also coupled to gauge fields) is integrable and, being topological, it can be reduced to a finite - dimensional system. It has horizons and can describe black holes and other topologically nontrivial objects.
2. Dilaton gravity coupled to scalar fields is not integrable in general, not a topological theory, has no static horizons (no static black holes). Note that the transition to the case of vanishing scalar fields is not analytic and thus the Z_B term cannot be treated as perturbation.

Now introducing spinors may dramatically change the behaviour of the DGM models. Consider the simplest case when Z_B and Z_F do not vanish but do not depend on ϕ (a "minimal" coupling). The scalar contribution to the r.h.s. of (2) is positive (scalar fields with negative Z_B are unphysical, having negative kinetic energy) but the spinor contribution may change its sign. One can explicitly check this by using the results of (Cavagliá, Fatibene and Francaviglia (1998)). In the gauge used in (Cavagliá, Fatibene and Francaviglia (1998)) the fermion solution is given by the spinor with the components

$$\tilde{f}^{-1/2}\Theta_1(u), \quad \tilde{f}^{-1/2}\Theta_2(v).$$

This gives the r.h.s. of (2) in the form

$$i(\overline{\Theta}_n'\Theta_n - \overline{\Theta}_n\Theta_n'), \quad n = 1, 2,$$

which is proportional to the derivative of the phase of the complex function Θ_n. It is easy to see that the scalar and spinor contributions may cancel each other. The scalar field is simply D'Alembert field, $\psi = A(u) + B(v)$ and one may choose, for instance,

$$\Theta_1 \sim A'(u)e^{ik_1u}, \quad \Theta_2 \sim B'(v)e^{ik_2v},$$

where k_1 and k_2 are real constants.

When the r.h.s. of equations (2) vanish, we may apply the above construction to finding ϕ and $\tilde{f}$ and obtain the explicit, exact solution of DG coupled to scalar and spinor fields. This can be done for arbitrary potentials U, V, W (in the Abelian case, X may also by an arbitrary function). For these solutions, the geometric variables $\tilde{f}$ and ϕ completely decouple from the scalar fields. However, although the equations for the geometric variables do not depend on the matter fields, the spinor fields feel the metric, they even become singular on the horizons. In this sense, there is no complete decoupling, especially, for global solutions (extended through horizons).

With the nonvanishing r.h.s. of eq.(2), the models with the potential $\tilde{V}$ given by (7) are explicitly integrable for any number of minimally coupled scalar and spinor fields[3]. Indeed, the equations for ϕ and $\tilde{f}$ (without loss of generality we may choose the parametrization $U = \phi$), can be written as

[3] Note that the models with $\tilde{V} = g_0 + g_1\phi$ and $\tilde{V} = g_0 e^{g\phi}$, special solutions of which have been found in (Cavagliá, Fatibene and Francaviglia (1998)), are particular cases of the models (7), and thus may be exactly solved.

$$\phi_{uv} + \widetilde{f}\widetilde{V} = 0, \tag{8}$$

$$(\log \widetilde{f})_{uv} + \widetilde{f}\widetilde{V}' = Z'_B \psi_u \psi_v + iZ'_F(\nabla_u \overline{\Theta}\gamma_v \Theta - \overline{\Theta}\nabla_u \gamma_v \Theta)]. \tag{9}$$

If $Z' \equiv 0$, they can be reduced to two Liouville equations and solved in terms of two D'Alembert fields (Filippov (1996,1997)). The equations for the scalar and spinor fields are trivial and one of them should be related to the geometric variables ϕ and $\widetilde{f}$ by using equations (2). So, in this case, there is a stronger correlation between the matter fields and geometry than in the previous one.

In more interesting models, Z_B and Z_F depend on the dilaton field (for instance, in the models obtained by spherical reduction they are proportional to U). With ϕ-dependent Z-functions, the equations for the spinor and scalar fields are no longer free but our idea of mutual cancellations between bosons and fermions may still be applicable. Recently, some explicitly integrable models with nonminimal coupling of DG to scalar fields have been found (Filippov and Ivanov (1998)). I hope that the results of (Filippov and Ivanov (1998)) can be extended to some nonminimal couplings of spinors.

However, the physical meaning and applications of the spinor - scalar cancellation may be seriously discussed in the quantum framework only. Thus, the main problem is to construct a quantum version of the above constructions. Many papers have been devoted to quantizing the simplest dilaton gravity minimally coupled to scalar matter and this (very special) model is more or less understood (for brief reviews and references see e.g., (Jackiw (1997)), (Cavagliá, de Alfaro and Filippov (1998))). Much less is known about quantizing more general models while the models with spinors were not systematically studied (except very special simple models emerging in describing ground states in string models). I think that the above discussion of the dilaton gravity coupled to bosons and fermions strongly suggests that quantum supersymmetric models of the 1+1 dimensional dilaton gravity, which are yet to be constructed and/or systematized, may be of significant interest for physics of macroscopic and microscopic black holes. They may also be interesting as simplified models of supergravity and superstrings.

While preparing this report and its written version I often imagined that I was discussing it with Victor Ogievetsky. We all, his colleagues and friends, miss very much his deep and genuine interest in our scientific work, his sharp but friendly criticism. I only hope that he would not find a major loophole in my argument.

I am grateful to Vittorio de Alfaro and Marco Cavaglia for useful discussions and to Evgenii Ivanov for his patient insistence on publishing this report. The support from the Turin section of INFN and Turin University is greatly appreciated. This investigation was partially supported by RFBR grant 97-01-01041 and by INTAS grant 93-127-ext.

References

A.T. Filippov (1996,1997): MPL **A11** (1996) 1691; IJMP **A12** (1997) 13.
M. Cavagliá, V. de Alfaro and A.T. Filippov (1995,1996): IJMP **D4** (1995) 661; **A10** (1995) 611; **D5** (1996) 227.
T.Klösch and T.Strobl (1996): Class. Quant. Grav. **13** (1996) 965, 2395.
M. Cavagliá and V. de Alfaro (1997): IJMP **D6** (1997) 39.
M. Cavagliá, L. Fatibene and M. Francaviglia (1998): hep-th/9801155.
A.T. Filippov and V.G.Ivanov (1998): hep-th/9803059.
R. Jackiw (1997): Nucl. Phys. (Proc. Suppl.) **57** (1997) 162.
M. Cavagliá, V. de Alfaro and A.T. Filippov (1998): Phys. Lett. **B424** (1998) 265.

Finite-Size Energy Levels of the Superintegrable Chiral Potts Model

Günter v. Gehlen

Physikalisches Institut der Universität Bonn, Nussallee 12, 53115 Bonn, Germany

Abstract. In the solution of the superintegrable chiral Potts model special polynomials related to the representation theory of the Onsager algebra play a central role. We derive approximate analytic formulae for the zeros of particular polynomials which determine sets of low-lying energy eigenvalues of the chiral Potts quantum chain. These formulae allow an analytic determination of finite-size energy eigenvalues without resorting to a numerical determination of the zeros.

1 Introduction

The chiral Z_N-symmetrical Potts model, apart from showing a rich phase structure, has attracted considerable interest because for appropriate parameter choices it has very special integrability properties. In the "superintegrable" case it is integrable *both* because it provides a representation of Onsager's algebra (in this respect being a natural generalization of the Ising model), and also because its Boltzmann weights satisfy a new type of Yang-Baxter equation. Both properties guarantee the existence of an infinite set of commuting charges. In order to calculate the spectrum, functional relations are needed. Such relations had first been conjectured by Albertini *et al.* (1989), and later derived by exploring relations to the six-vertex model at $q = e^{2\pi i/N}$ (Bazhanov and Stroganov 1990, Baxter *et al.* 1990). The formulae which give the spectrum in the superintegrable case have been given by Albertini *et al.* (1989) for Z_3 and for general Z_N by Baxter (1994). Little is known about correlation functions. For the general integrable case (not satisfying the Onsager algebra), the free energy and interface tensions have been obtained in the thermodynamic limit (O'Rourke and Baxter 1996). The analytic derivation of the order parameter still evades an analytical derivation (Baxter 1998). General integrable boundary conditions have been derived recently by Zhou (1998).

In the following we concentrate on the solution of the superintegrable case, in which special polynomials related to the representation theory of the Onsager algebra play a central role. The energy eigenvalues are determined in terms of the zeros of these polynomials. For the lowest states the polynomials are of order $m_E = [((N-1)L - Q)/N]$. L denotes the number of sites of the chain and Q the Z_N-charge sector. $[x]$ stands for the integer part of a rational number x. In the limit $L \to \infty$ the sum over the terms depending on the zeros can be expressed as a contour integral without explicitly calculating the zeros.

This method has been used to obtain the ground state energy and excited levels in the thermodynamic limit. However, for finite L one has to resort to numerical methods. In this note we obtain finite-L analytic approximate expressions for the zeros.

2 Integrability of the Chiral Potts Model

2.1 The Integrable Model

For many discussions of the chiral Potts quantum chain, it turns out to be sufficient to consider the following Z_N-symmetric hamiltonians which depend on three real parameters λ, ϕ, φ:

$$H = -\sum_{j=1}^{L}\sum_{l=1}^{N-1} \frac{1}{\sin(\pi l/N)} \left(e^{i\varphi(2l-N)/N}\, Z_j^l + \lambda\, e^{i\phi(2l-N)/N}\, X_j^l\, X_{j+1}^{N-l} \right).$$

The first sum is over the sites j. At each site j there are operators X_j and Z_j acting in a vector space C^N and satisfying

$$Z_i X_j = X_j\, Z_i\, \omega^{\delta_{i,j}}; \qquad Z_j^N = X_j^N = 1; \qquad \omega = e^{2\pi i/N}.$$

Often it is useful to represent the operators X_j and Z_j by $(X_j)_{l,m} = \delta_{l,m+1}$ and $Z_j = diag(1,\omega,\omega^2,\ldots,\omega^{N-1})_j$. H commutes with the Z_N-charge operator $\hat{Q} = \prod_{j=1}^{L} Z_j$. We write the eigenvalues of $\hat{Q}$ as ω^Q where $Q = 0,1,\ldots,N-1$. λ is the inverse temperature. We shall consider the periodic case $X_{N+1} = X_1$, although, as shown by Baxter (1994), twisted toroidal boundary conditions require little additional effort.

2.2 Particular Cases: Ising, Parafermionic

For $N = 2$ the sum over l has only a single term. The angles ϕ, φ drop out, Z_j and X_j become Pauli matrices and we obtain the Ising quantum chain. If we put $\phi = \varphi = 0$ then H describes the Fateev-Zamolodchikov parity invariant parafermionic Z_N-quantum chain. For $N\mathfrak{e}3$ and $\phi \neq 0$ or $\varphi \neq 0$ parity is broken (therefore ϕ and φ are called chiral angles) and this gives rise to several interesting features of the model, e.g. to the appearance of incommensurate phases and oscillating correlation functions.

2.3 Yang-Baxter-Integrable Case

If $\cos\varphi = \lambda\cos\phi$ then H can be derived from a two-dimensional lattice model defined by the following transfer matrix which depends on rapidities p and q:

$$T_{p,q}(\{l\},\{l'\}) = \prod_{j=1}^{L} W_{p,q}(l_j - l'_j)\, \overline{W}_{p,q}(l_j - l'_{j+1}) \tag{1}$$

$\{l\}$ and $\{l'\}$ are the sets of N-valued spin variables at alternating vertex rows of the diagonally drawn lattice. The rapidity lines p run horizontally on the dual lattice, the q vertically. The Boltzmann weights W and $\overline{W}$ are defined in terms of the functions x_p, y_p, μ_p; x_q, y_q, μ_q of the rapidities p and q by (Au-Yang *et al.* 1987)

$$W_{p,q}(n) = \left(\frac{\mu_p}{\mu_q}\right)^n \prod_{j=1}^{n} \frac{y_q - \omega^j x_p}{y_p - \omega^j x_q}; \qquad \overline{W}_{p,q}(n) = (\mu_p\mu_q)^n \prod_{j=1}^{n} \frac{\omega x_p - \omega^j x_q}{y_q - \omega^j y_p}$$

where $n = 0, 1, \ldots, N-1$. The requirement of Z_N-symmetry imposes several restrictions on the six functions, e.g.

$$\lambda(x_q^N y_q^N + 1) = x_q^N + y_q^N.$$

where λ is a parameter describing the inverse temperature. At fixed λ only one of the three functions x_q, y_q, μ_q is independent, the other two are determined by nonlinear relations.

In the Ising case $N = 2$, the functions x_q etc. can be simply expressed in terms of the meromorphic elliptic functions of modulus λ:

$$x_q = -\sqrt{\lambda}\,\mathrm{sn}\, q; \qquad y_q = \sqrt{\lambda}\,\mathrm{cn}\, q/\mathrm{dn}\, q, \qquad \mu_q = \sqrt{1-\lambda^2}/\mathrm{dn}\, q$$

and one obtains the parametrization of the Ising Boltzmann weights known e.g. from Baxter (1982). Integrability is guaranteed by the following Yang-Baxter-equation, which contains the p- and q-variables *separately*

$$\sum_{l=0}^{N-1} W_{p,q}(l'-l)\overline{W}_{r,q}(l''-l)W_{r,p}(l-l''')$$
$$= R_{pqr}\overline{W}_{r,p}(l''-l')W_{r,q}(l'-l''')\overline{W}_{p,q}(l''-l''').$$

The explicit expression for R_{pqr} can be found e.g. in Albertini *et al.* (1989). The superintegrable case corresponds to fixing the p-dependent functions to $x_p = y_p$ and $\mu_p = 1$ so that we are left with only one independent function, e.g. x_q.

H is obtained if at fixed p we expand the transfer matrix for small $q - p$.

2.4 The Superintegrable Case, Onsager's Algebra

The choice $\phi = \varphi = \pi/2$ gives the "superintegrable" case of the hamiltonian. Writing $H = A_0 + \lambda A_1$ we have

$$A_0 = -\frac{4}{N}\sum_{j=1}^{L}\sum_{l=1}^{N-1} \frac{X_j^l}{1-\omega^{-l}}; \qquad A_1 = -\frac{4}{N}\sum_{j=1}^{L}\sum_{l=1}^{N-1} \frac{Z_j^l Z_{j+1}^{N-l}}{1-\omega^{-l}}. \tag{2}$$

It has been shown by v.Gehlen and Rittenberg (1985) that A_0 and A_1 satisfy the Dolan and Grady (1982) conditions

$$[A_0,[A_0,[A_0,A_1]]] = 16\,[A_0,A_1]; \quad [A_1,[A_1,[A_1,A_0]]] = 16\,[A_1,A_0]. \quad (3)$$

Due to these conditions, A_0 and A_1 generate Onsager's algebra $\mathcal{A}$:

$$[A_l, A_m] = 4\,G_{l-m}$$

$$[G_l, A_m] = 2\,A_{m+l} - 2\,A_{m-l}; \qquad [G_l, G_m] = 0; \qquad l,\, m \in \mathbb{Z}.$$

which has been an essential tool in the Onsager (1944) original solution of the Ising model. $\mathcal{A}$ implies an infinite set of constraints:

$$[A_2, A_1] = [A_1, A_0] = [A_0, A_{-1}] = \dots$$

and the existence of the infinite set of commuting charges Q_m:

$$Q_m = \tfrac{1}{2}\left(A_m + A_{-m} + \lambda(A_{m+1} + A_{-m+1})\right) \qquad \text{with} \qquad Q_0 = H.$$

Finite dimensional representations of $\mathcal{A}$ are obtained if there is a recurrence relation of finite length among the A_l (Davies 1990, Roan 1991):

$$\sum_{k=-n}^{n} \alpha_k\, A_{k-n} = 0.$$

If z_j $(j = 1, \dots, m_E)$ are the zeros of the polynomial $f(z) = \sum_{k=-n}^{n} \alpha_k z^{k+n}$, then we can express the A_m and G_m in terms of a set of $sl(2,C)$-generators

$$A_m = 2\sum_{j=1}^{n} \left(z_j^m\, E_j^+ + z_j^{-m} E_j^-\right); \qquad G_m = \sum_{j=1}^{n} \left(z_j^m - z_j^{-1}\right) H_j$$

where

$$[E_j^+, E_k^-] = \delta_{jk}\, H_k; \qquad [H_j,\, E_k^\pm] = \pm 2\, \delta_{jk}\, E_k^\pm.$$

So, $\mathcal{A}$ is isomorphic to a subalgebra of the loop algebra of a direct sum of $sl(2,C)$ algebras.

Each sector of the model is characterized by a set of points z_j, which, of course, are not fixed by the algebra $\mathcal{A}$ alone. For the hamiltonian (2) the polynomials $f(z)$, or rather related polynomials $\mathcal{P}_Q(s)$, have been obtained by Albertini *et al.* (1989) and Baxter (1994) from functional equations for the transfer matrix (1) of the two-dimensional model, as will be reviewed in the next section.

As B. Davies (1990) has shown, from the property (3) of H it follows that *all* eigenvalues E of H depend on λ in the form

$$E = a + b\lambda + N\sum_{j=1}^{m_E} \pm\sqrt{1 + 2\lambda\cos\theta_j + \lambda^2} \qquad (4)$$

where a and b are integers and $\cos\theta_j = \frac{1}{2}\left(z_j + z_j^{-1}\right)$. So the main task of the diagonalization of the superintegrable H reduces to the calculation of the $\cos\theta_j$ or equivalently, of the zeros of the polynomials $f(z)$. For hermitian hamiltonians, the z_j must be on the unit circle.

3 Solution of the Z_3-Superintegrable Model

3.1 Functional Equation and Ansatz of Albertini et al.

Albertini *et al.* (1989) discovered the following functional equation for the Z_3 superintegrable case with the p-variables fixed to $x_p = y_p = \left(\frac{1-\lambda}{1+\lambda}\right)^{1/6}$: Splitting off some poles from $T_{p,q}$ they get a meromorphic function $\overline{T}_q$ which satisfies

$$\begin{aligned}\overline{T}_q\,\overline{T}_{Rq}\,\overline{T}_{R^2q} &= 3^L e^{-iP}\ \{\ (t-1)^L(\omega^2 t-1)^L\overline{T}_q \\ &+(\omega t-1)^L(\omega^2 t-1)^L\overline{T}_{R^2q}+(t-1)^L\ (\omega t-1)^L\ \overline{T}_{R^4q}\ \}\end{aligned} \quad (5)$$

where P is the momentum operator, $t = x_q y_q/x_p^2$ and R is the mapping $R(x_q, y_q, \mu_q) = (y_q, \omega x_q, \mu_q^{-1})$. Solutions to this functional equation are then obtained from the ansatz

$$\overline{T}_q = \ \ldots \prod_{l=1}^{m_P}\left(\frac{1+\omega v_l t}{1+\omega v_l}\right)\prod_{j=1}^{m_E}\{(1-\lambda)(\mu_q^3+1)\pm 2w_j(\mu_q^3-1)\}\,. \quad (6)$$

We skip here writing some factors relevant for the determination of the linear terms in (4): for details see Albertini and McCoy (1990) or Baxter (1993). The ansatz (6) contains m_P excitations with Bethe-type rapidities v_l, and in the last factor, m_E functions w_j. It requires some calculation to see that if we put

$$w_j = \tfrac{1}{2}\sqrt{1+2\lambda\frac{1+t_j^3}{1-t_j^3}+\lambda^2}$$

then the left hand side of (5) vanishes and in order to get the right hand side vanishing too, t_j must be zero of the polynomial

$$\begin{aligned}\mathcal{P}_Q(t^3) \;=\; t^{-c}\Bigg\{&(t^2+t+1)^L\prod_{l=1}^{m_P}\frac{1+v_l t}{1+v_l^3 t^3}+\omega^Q(t^2+\omega^2 t+\omega)^L\prod_{l=1}^{m_P}\frac{1+\omega v_l t}{1+v_l^3t^3} \\ &+\ \omega^{2Q}(t^2+\omega t+\omega^2)^L\prod_{l=1}^{m_P}\frac{1+\omega^2 v_l t}{1+v_l^3 t^3}\Bigg\}\end{aligned} \quad (7)$$

The integer c is chosen such that $\mathcal{P}_Q$ becomes invariant against $t \to \omega t$ and $\mathcal{P}_Q$ depends only on $s \equiv t^3$. Despite its denominators, $\mathcal{P}_Q(s)$ will be a polynomial if the v_l satisfy the Bethe-type equations:

$$\left(\frac{1+\omega\, v_j}{1+\omega^2 v_j}\right)^L = (-1)^{m_P+1}\omega^{Q-m_P}\prod_{l=1}^{m_P}\left(\frac{v_l-\omega\, v_j}{v_j-\omega\, v_l}\right).$$

These Bethe-type equations resemble those of the spin-1 XXZ-chain.
As mentioned already above, the hamiltonian is obtained by expanding the

transfer matrix for small $p-q$ around a fixed value of p. If $\mathcal{P}$ has degree m_E, then from the m_E zeros of (7) we get *a set* of 2^{m_E} energy eigenvalues

$$E_i = a(1+\lambda) + b(1-\lambda) - 6\,(\pm w_1 \pm w_2 \pm \ldots \pm w_{m_E})\,. \tag{8}$$

where e.g. for the lowest sector $a = (L+Q-1 \bmod 3) - 2$. Comparing to the formula eq.(4) obtained from the Onsager algebra, we see that $\cos\theta_j = (1+t_j^3)/(1-t_j^3)$. Our hamiltonian is hermitian and so we should have $|\cos\theta_j| \le 1$ and all zeros $s_j = t_j^3$ must be on the negative real s-axis.

A difficult question is whether all solutions to these equations really give eigenvalues of our H, or whether there are spurious solutions. This problem has been investigated in detail by Dasmahapatra *et al.* (1993).

3.2 Example: Sector $m_P = 0$

In general, in order to obtain energy eigenvalues of H for chain length L in the charge sector Q, one first has to choose a sector of given m_P and to solve the Bethe-type equations obtaining a set of Bethe-rapidities v_l. Then one has to use these v_l to build up $\mathcal{P}_Q$ and then to calculate its zeros s_j.

Obviously, the simplest case is the sector $m_P = 0$ where there are no v_l and one can directly start with building the $\mathcal{P}_Q$ and looking for the zeros s_j. E.g. for Z_3 with $L = 20$ and $Q = 1$ we calculate first

$$\begin{aligned}\mathcal{P}_1(s) = {} & 20\,s^{13} + 8455\,s^{12} + 484500\,s^{11} + 8533660\,s^{10} + 61757600\,s^{9} \\ & + 210859245\,s^{8} + 363985680\,s^{7} + 326527350\,s^{6} + 151419816\,s^{5} \\ & + 34880770\,s^{4} + 3656360\,s^{3} + 146490\,s^{2} + 1520\,s + 1,\end{aligned}$$

then solve numerically for its 13 zeros. Since the largest appearing coefficients grow exponentially with L, this procedure becomes impractical for $L \longrightarrow 50$. and e.g. energy eigenvalues for $L = 1000$ cannot be obtained this way. Can we get an analytic expression for the zeros of the $\mathcal{P}(s)$ for large *finite* L?

Baxter (1988) and Albertini *et al.* (1989) proceeded as follows to obtain energy levels *in the limit* $L \to \infty$: One rewrites the sum over the terms determined by the zeros (chosing all $\pm$-signs to be $+$ and so obtaining the lowest eigenvalue E) as an integral over a contour C enclosing the negative real t-axis:

$$\sum_{j=1}^{m_E} \sqrt{1+2\lambda\cos\theta_j+\lambda^2} = -\frac{1}{2\pi i}\int_C dt\,\sqrt{(1-\lambda)^2+\frac{4\lambda}{1-t^3}}\,\frac{d}{dt}\ln\mathcal{P}_Q(t)$$

Then the contour is opened and gets caught by three 120° symmetrically oriented cuts, which arise from the square root of the integrand. The rightmost of these cuts is on the positive real axis at $1 \le t \le |(1+\lambda)/(1-\lambda)|^{1/3}$. All three cuts give the same contribution, and the rightmost one can be evaluated easily, since for $L \to \infty$ only the term proportional $(t^2+t+1)^L$ of $\mathcal{P}$, eq.(7), is non-oscillating and needs to be kept. In this way Baxter (1988) calculated the ground-state energy for general Z_N and finds (see also Albertini *et al.* 1989):

$$\lim_{L\to\infty} \frac{E_0}{L} = -\frac{N}{\pi}\int_0^{\pi(1-N^{-1})} dx\,\sqrt{1-2\lambda\frac{\sin^N(x+\frac{\pi}{N})-\sin^N x}{\sin^N(x+\frac{\pi}{N})+\sin^N x}+\lambda^2}. \tag{9}$$

4 Determination of the Zeros of the $\mathcal{P}_Q(s)$

We attempt to find the zeros of the $\mathcal{P}_Q(s)$ for *finite* L on the negative real s-axis analytically. From the explicit expressions and numerical studies we observe:

- For $L = Q \bmod 3$ the zeros come in pairs which are reflection symmetric with respect to $s = -1$: with s_j also $1/s_j$ is a zero. This remains approximately (within a shift) true for the other charge sectors.
- The distances between adjacent zeros alternate slightly in size so that a smooth curve $s_j(k)$ is obtained only if we connect every second zero.

4.1 Case Z_3, $m_P = 0$

A useful variable transformation to be applied to $\mathcal{P}_Q(s)$ turns out to be $t = (u-\sqrt{3})/(u+\sqrt{3})$ or, writing $u = \tan\beta$:

$$t = \frac{\sin(\beta-\frac{\pi}{3})}{\sin(\beta+\frac{\pi}{3})}.$$

This maps $-\infty < s < 0$ to e.g. $-\frac{\pi}{3} < \beta < \frac{\pi}{3}$ or $\frac{2\pi}{3} < \beta < \frac{4\pi}{3}$ (for definiteness let us choose the latter interval), giving

$$t^2+t+1 = \frac{3}{\left(2\sin(\beta+\frac{\pi}{3})\right)^2} \quad\text{and}\quad t^2+\omega t+\omega^2 = \frac{-3\,\omega\,\cos\beta\, e^{-i\beta}}{2\left(\sin(\beta+\frac{\pi}{3})\right)^2}.$$

so that

$$\mathcal{P} \sim (t^2+t+1)^L + \omega^Q(t^2+\omega^2 t+\omega)^L + \omega^{-Q}(t^2+\omega t+\omega^2)^L = 0,$$

multiplying by the non-vanishing factors $\left(\sin(\beta+\frac{\pi}{3})\right)^{2L}$ becomes

$$-(-2\cos\beta)^{-L} = 2\cos\left(L\left\{\beta-\frac{2\pi}{3}\right\}+\frac{2\pi Q}{3}\right). \tag{10}$$

Now $\frac{2\pi}{3} < \beta < \frac{4\pi}{3}$ means $|2\cos\beta| > 1$ so that for $L \gg 1$ the left hand side of (10) will be small, except near the corners of the interval in β, as is shown in Fig. 1. In order to get a solution, the fast oscillating right hand term has to coincide with the non oscillating left hand term. We see that all but the solutions near $\beta_k \approx \frac{2\pi}{3}$ and $\beta_k \approx \frac{4\pi}{3}$ are practically equally

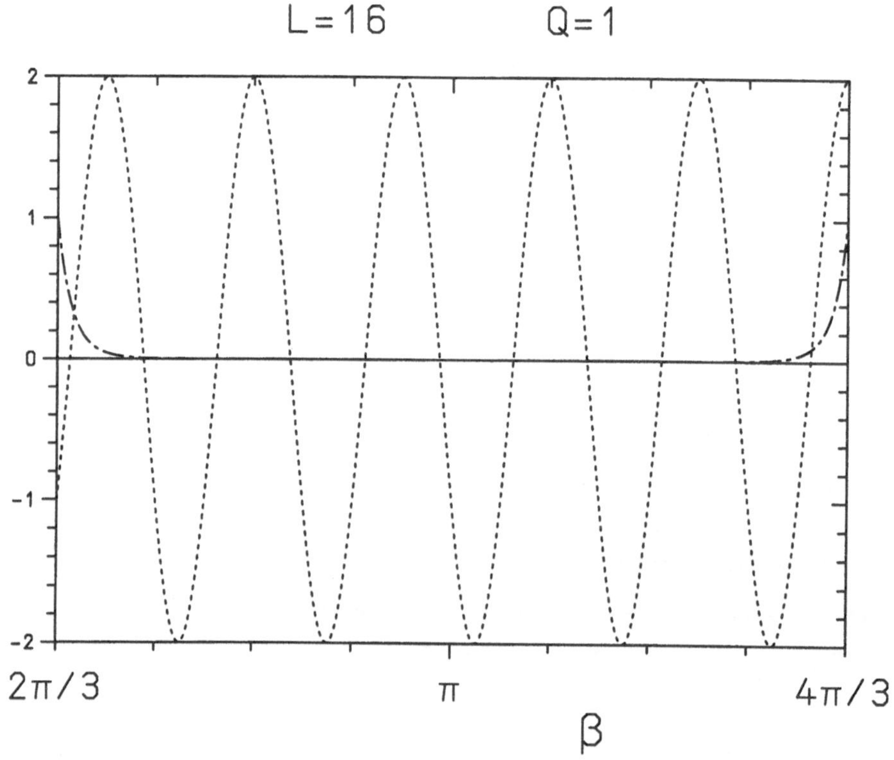

Fig. 1. Left-hand (dash-dotted line) and right hand (dotted) parts of (10) for $L = 16$ sites and the charge sector $Q = 1$. The intersection points give the solutions β_k.

spaced in the variable β. Neglecting the left-hand side altogether, we get the approximate solutions

$$\beta_k \approx K - \frac{\pi}{3}; \qquad K = \frac{6k + 2Q - 3}{6L}\pi; \qquad k = 1, 2, \dots, m_E. \tag{11}$$

or, expressed in terms of the $\cos\theta_k$ which we need for the energy eigenvalues:

$$\cos\theta_k \equiv \frac{1 + t_k^3}{1 - t_k^3} = -\frac{\sin^3(K + \frac{\pi}{3}) - \sin^3 K}{\sin^3(K + \frac{\pi}{3}) + \sin^3 K}. \tag{12}$$

It is easy to see that using (11, 12) and replacing the summation in (8) by an integral leads to Baxter's formula (9) for the thermodynamic limit. So the finite-size effects are due to the deviations from the approximation (11), which means that the leading finite-size effects come from the deviation of the largest and smallest solutions β_k from the value given by (11).

In order get an idea of the precision of (11), consider e.g. $L = 30,\ Q = 0$. With $d_k = 1 - \beta_k^{exact}/\beta_k^{appr}$ where β_k^{appr} is the approximation (11), we find

$$d_1 = 0.0012;\ d_2 = -1.89 \cdot 10^{-5};\ d_3 = 6.48 \cdot 10^{-7};\ d_4 = -4.39 \cdot 10^{-8};\ \ldots$$

Taking the arcsin of (10) we get a pair of useful equations:

$$\mp \arcsin \frac{1}{2\,(-2\cos\beta_k)^L} + L\left(\beta_k - \frac{2\pi}{3}\right) + \frac{2\pi Q}{3} - \pi(2I_k^{\pm}+1) \pm \frac{\pi}{2} = 0. \quad (13)$$

The two sets of counting integers $I_k^{\pm}$ vary within the limits

$$I_k^{\pm} = I_a^{\pm},\ I_a^{\pm}+1, \ldots,\ I_b^{\pm}$$

with

$$I_a^{\pm} = \left[\frac{Q+1+a^{\pm}}{3}\right];\quad I_b^{\pm} = \left[\frac{L+Q-3+a^{\pm}}{3}\right];\quad a^{\pm} = \begin{Bmatrix}1\\0\end{Bmatrix}.$$

The arcsin-equation has the great advantage that from it each zero can be calculated separately by fixing $I_k^{\pm}$.
Numerically on a small PC we can easily get zeros with 40 digits precision for up to $L = 300000$ sites. Recall that directly from $\mathcal{P}(t)$ we were limited to $L \longrightarrow 50$. We shall see that in addition, the arcsin-equation lends itself to excellent analytical approximations for the "corner"-solutions: Let us call ξ the term which is important only close to the corners:

$$\xi \equiv \arcsin \frac{1}{2(-2\cos\beta)^L}$$

We find numerically that $\xi_k^{\pm} \ll 1$ even for the extreme cases $I_k^{\pm} = I_{a,b}^{\pm}$. We use this feature to approximate ξ_k. First, using $\arcsin x \approx x$ we get

$$\xi \approx \tfrac{1}{2}\,(-2\cos\beta)^{-L} \quad (14)$$

On the other hand, solving (13) for the β_k which is not in ξ we get (omitting the superscripts $\pm$):

$$\beta_k = \frac{2\pi}{3} + \frac{\pi(\gamma + 2\Delta I_k)}{L} \pm \frac{\xi_k}{L}.$$

where

$$\gamma = 2I_a - \frac{2\pi Q}{3};\qquad \Delta I_k = I_k - I_a.$$

Let us consider the lower end of the interval in β, i.e. $I_k = I_a$. Then we get a behaviour smooth in L when keeping Q fixed. At the upper end $I_k = I_b$ the behaviour is smooth only if we keep $L - Q$ fixed mod 3.
Expanding around $\beta = 2\pi/3$ gives:

$$-2\cos\beta_k \approx 1 + \frac{\sqrt{3}(\xi_k + \gamma\pi)}{L} \quad (15)$$

Using this in (14) and assuming $L \gg 1$ we get

$$\xi_k \approx \frac{1}{2}\left(1 - \frac{\sqrt{3}(\xi_k + \gamma\pi)}{L}\right)^L \approx \frac{1}{2}\exp\left\{-\sqrt{3}(\gamma\pi + \xi_k)\right\}. \qquad (16)$$

This is a nonlinear equation for ξ_k from which we can determine the values β_k close to $\frac{2\pi}{3}$ through an iteration procedure. A first iteration gives for the lowest β-value β_1:

$$\xi_1^{\pm} = \pm\frac{1}{2}\exp\left\{-\sqrt{3}\left(\gamma^{\pm}\pi \pm \frac{\sqrt{3}}{2}e^{-\sqrt{3}\gamma^{\pm}\pi}\right)\right\}.$$

Collecting results and considering also $\Delta I_k \neq 0$, i.e. not only the smallest β_k, but also the next, we get, writing the 2nd iteration (again leaving out the superscripts $\pm$):

$$\beta_k = \frac{2\pi}{3} + \frac{\pi(\gamma + 2\Delta I_k)}{L}$$
$$+ \frac{1}{2L}\exp\left\{-\sqrt{3}\left((\gamma\pi + 2\Delta I_k) \pm \frac{\sqrt{3}}{2}e^{-\sqrt{3}(\gamma\pi + 2\Delta I_k)}\right)\right\}. \qquad (17)$$

Notice that the approximation becomes exponentially better for increasing ΔI_k, i.e. if we consider $\beta_k's$ away from the corners. This is true up to the middle of the interval, i.e. $\beta_k \approx \pi$ where already the approximation $\xi_k = 0$ is very accurate. Beyond this we better use the corresponding approximation down from the upper corner.

One may be worried how much the results for the β_k are affected by approximations made in (14) and (15), and by the use of only few iterations. In Table 1 we give numerical values from our iteration formula in order to compare them to the following numerical results from the exact formula (13). A fit of the L-dependence of the numerical solution, obtained from $L = 25000, \ldots, 125000$ sites is, expressed in terms $\xi^{\pm}/\pi$:

$$Q = 0+: \; +0.009927623657298 + 0.04835895/L + 0.0315/L^2 + \ldots$$
$$Q = 0-: \; -0.000045412321296 - 0.002017284/L - 0.0338/L^2 - \ldots$$
$$Q = 1+: \; +0.00000740160385053 + 0.0004910486/L + 0.0130/L^2 + \ldots$$
$$Q = 1-: \; -0.0017242062241336 - 0.023760499/L - 0.0922/L^2 - \ldots$$
$$Q = 2+: \; +0.0002780560828110 + 0.007462876/L + 0.0685/L^2 + \ldots$$
$$Q = 2-: \; -0.00000120676616449 - 0.000111825/L - 0.00430/L^2 - \ldots$$

We see that for large L our nonlinear formula (16) describes the zeros with excellent precision, but the second iteration should be used.

If we start from the upper corner $\beta = 4\pi/3$, we get exactly the same numbers and formulae, just signs are changed and sectors are permuted.

Table 1. Values of ξ_k for the lowest value of β_k, as obtained from (16) at different iteration levels. We have multiplied the values by $10^6/\pi$ in order to avoid too small numbers.

Iteration	$Q=0+$	$Q=0-$	$Q=1+$	$Q=1-$	$Q=2+$	$Q=2-$
0st	10477.0	-45.401101	7.401901956	-1708.097	278.47707	-1.20675824029
1st	9896.39	-45.412318	7.401603838	-1724.047	278.05541	-1.20676616444
2nd	9927.70	-45.412321	7.401603850	-1724.196	278.05605	-1.20676616449

4.2 Finite-Size Correction Formulae

The formulae of the last section are in principle sufficient to calculate finite-size eigenvalues. But often it is convenient to transform the sums in (8) into integrals in order to have a compact expression also for finite L eigenvalues or to get the leading finite-size corrections to the thermodynamic limit expressions. This can be done e.g. applying Euler-McLaurin-techniques as is common in calculations for the XXZ-quantum chain. Following Woynarovich and Eckle (1987) and Hamer *et al.* (1987), but taking care of the alternating spacing of the zeros, we define two functions $Z_L^{\pm}$

$$
\begin{aligned}
Z_L^{\pm}(\beta) &= I_k^{\pm}/L \\
&= \frac{1}{2\pi L}\left\{\mp\arcsin\frac{1}{2\,(-2\cos\beta)^L} + L\left(\beta-\frac{2\pi}{3}\right)+\frac{2\pi Q}{3} - \pi \pm \frac{\pi}{2}\right\}.
\end{aligned}
$$

In $Z_L^{\pm}$ the roots β_k are exactly equidistant. We define the root density

$$
\sigma_L^{\pm}(\beta) = dZ_L^{\pm}/d\beta = \frac{1}{2\pi}\left\{1 \mp (-1)^L \frac{\tan\beta}{\sqrt{4\,(2\cos\beta)^{2L}-1}}\right\}.
$$

Obviously, $\sigma_\infty^+(\beta)+\sigma_\infty^-(\beta)=\frac{1}{2\pi}\equiv\sigma_\infty$. Then finite-size correction to a quantity $\int_{2\pi/3}^{4\pi/3} d\beta\,\sigma_\infty\,w(\beta)$ (e.g. the w of (8)) is

$$
\begin{aligned}
&\frac{1}{L}\left(\sum_{k^+=0}^{m_E^+} w(\beta_{k^+}) + \sum_{k^-=0}^{m_E^-} w(\beta_{k^-})\right) - \int_{2\pi/3}^{4\pi/3} d\beta\,\sigma_\infty\,w(\beta) \\
&\approx \int_{2\pi/3}^{4\pi/3} d\beta\,(\sigma_L^+ + \sigma_L^- - \sigma_\infty)\,w(\beta) - \sum_{\pm}\left(\int_{2\pi/3}^{\beta_a^{\pm}} + \int_{\beta_b^{\pm}}^{4\pi/3}\right) d\beta\,\sigma_L^{\pm}\,w(\beta) \\
&+\frac{1}{2L}\sum_{\pm}\{w(\beta_b^{\pm})+w(\beta_a^{\pm})\} + \frac{1}{12L^2}\sum_{\pm}\left(\frac{w'(\beta_b^{\pm})}{\sigma_L(\beta_b^{\pm})} - \frac{w'(\beta_a^{\pm})}{\sigma_L(\beta_a^{\pm})}\right).
\end{aligned}
\tag{18}
$$

Here

$$
m_E^+ + m_E^- = m_E; \qquad \beta_a^{\pm} = \beta^{\pm}(I_a^{\pm}); \qquad \beta_b^{\pm} = \beta^{\pm}(I_b^{\pm}).
$$

The first right hand term of (18) vanishes because $\sigma_L^+ + \sigma_L^- = \sigma_\infty$.

5 The Zeros for Higher Z_N-Superintegrable Cases

Analogous trigonometic formulae can be derived for higher Z_N-cases. With increasing N, the equations become more involved, but have a similar structure.

5.1 The Case Z_4

Since we transform the variable t and not $s = t^4$, in order to map the negative real s-axis to a range of real β, we have to include a phase and use

$$t = e^{i\pi/4}\,\frac{\sin\left(\beta - \frac{\pi}{4}\right)}{\sin\left(\beta + \frac{\pi}{4}\right)} \tag{19}$$

so that $-\infty < s < 0$ is mapped to e.g. $-\frac{\pi}{4} < \beta < \frac{\pi}{4}$. After some algebra, we get that the condition for the zeros of the $m_P = 0$-functions $\mathcal{P}_Q(s)$

$$\begin{aligned}\mathcal{P}_Q(s) = t^{-c}\,\{&(t^3+t^2+t+1)^L + (-1)^Q(t^3-t^2+t-1)^L \\ &+ i^{-Q}(t^3+it^2-t-i)^L + i^Q(t^3-it^2-t+i)^L\} = 0\end{aligned} \tag{20}$$

translates to

$$\mathcal{A}_-^L\,\cos\left\{L(\rho_+ + \tfrac{\pi}{8}) + \frac{\pi Q}{4}\right\} + (-1)^{L+Q}\,\mathcal{A}_+^L\,\cos\left\{L(\rho_- + \tfrac{\pi}{8}) + \frac{\pi Q}{4}\right\} = 0 \tag{21}$$

where

$$\mathcal{A}_\pm = \sqrt{\sqrt{2} \pm \cos(2\beta)}; \qquad \rho_+ = \arctan\left\{\left(\sqrt{2}-1\right)\tan\beta\right\};$$
$$\rho_- = \arctan\left\{\left(\sqrt{2}-1\right)\cot\beta\right\} + \pi\,\theta(-\beta)$$

$\theta(x)$ denotes the step function, which is needed because otherwise with the standard convention the arctan would jump when β passes through zero. For $L \gg 1$ and $|\beta|$ not too close to $\frac{\pi}{4}$ we get $\mathcal{A}_-^L \ll \mathcal{A}_+^L$: the second term of (21) dominates and the solutions come where

$$\cos\left\{L(\rho_- + \tfrac{\pi}{8}) + \tfrac{\pi Q}{4}\right\} \approx 0$$

After some rearrangement this leads to

$$\tan\beta_k \approx -\tan\frac{\pi}{4}\tan\frac{\pi}{8}\cot\left(K + \frac{\pi}{8}\right) \quad \text{with} \quad K = \frac{(4k+Q-2)\pi}{4L}, \tag{22}$$

$k = 1, 2, \ldots, [(3L-Q)/4]$ and e.g. $-\frac{\pi}{4} < \beta_k < \frac{\pi}{4}$.
(19) and (22) are equivalent to

$$\cos\theta_k \equiv \frac{1+t_k^4}{1-t_k^4} \approx -\frac{\sin^4(K+\frac{\pi}{4}) - \sin^4(K)}{\sin^4(K+\frac{\pi}{4}) + \sin^4(K)}.$$

As in the Z_3-case, in the thermodynamic limit, this approximation leads back to Baxter's ground-state energy formula.

5.2 The Case Z_5

Transforming $t = \sin(\beta - \frac{\pi}{5})/\sin(\beta + \frac{\pi}{5})$, the condition

$$\mathcal{P}_Q \sim \mathcal{J}_0^L + \omega^{-Q}\mathcal{J}_1^L + \omega^{Q}\mathcal{J}_4^L + \omega^{-2Q}\mathcal{J}_2^L + \omega^{2Q}\mathcal{J}_3^L = 0,$$

with

$$\mathcal{J}_k = t^4 + \omega^k t^3 + \omega^{2k} t^2 + \omega^{3k} t + \omega^{4k}$$

takes the form

$$\tfrac{1}{2}\left(-\frac{\cos\beta}{\cos\frac{\pi}{5}}\right)^{-L} + \cos\left\{L(\beta - \frac{4\pi}{5}) + \frac{2\pi Q}{5}\right\}$$
$$+ (-B)^{-L}\cos\left\{L(\chi + \frac{2\pi}{5}) + \frac{4\pi Q}{5}\right\} = 0.$$

where

$$\tan\chi = \cot\frac{\pi}{5}\cot\frac{\pi}{10}\tan\beta \qquad \text{and} \qquad B = \sqrt{\frac{4\sin^2\beta + \sqrt{5} - 2}{2\cos\frac{\pi}{5}}}.$$

The solutions come for $-\infty < s < 0$ corresponding to e.g. $-\frac{\pi}{5} < \beta < \frac{\pi}{5}$. Since for $L \gg 1$ and $|\beta|$ not too close to $\frac{\pi}{5}$ we have $1 \ll (\cos\beta/\cos\frac{\pi}{5})^{-L} \ll B^{-L}$, the solutions come close to where the cosine containing χ vanishes, i.e. at

$$\tan\beta_k \approx -\tan\frac{\pi}{5}\tan\frac{\pi}{10}\cot\left(K + \frac{\pi}{10}\right); \qquad K = \frac{(10k + 2Q - 5)\pi}{10L}$$

and integers $k = 1, 2, \ldots$ such that e.g. $|\beta_k| < \frac{\pi}{5}$. Equivalently, we can write

$$\cos\theta_k \equiv \frac{1 + t_k^5}{1 - t_k^5} \approx -\frac{\sin^5(K + \frac{\pi}{5}) - \sin^5 K}{\sin^5(K + \frac{\pi}{5}) + \sin^5 K}. \qquad (23)$$

5.3 The Case Z_6

We conclude with the case Z_6: As in the case of Z_4, because $N = 6$ is even, we include a phase and transform $-\infty < s < 0$ to $-\frac{\pi}{6} < \beta < \frac{\pi}{6}$ using

$$s = t^6; \qquad t = e^{i\pi/6}\frac{\sin(\beta - \frac{\pi}{6})}{\sin(\beta + \frac{\pi}{6})}.$$

Now $\mathcal{P}_Q(s) = 0$ (we don't give the explicit Z_6 formula for $\mathcal{P}_Q$, since this is easily generalized from the Z_5-case) after some algebra leads to

$$\left(\frac{1-\alpha}{1+\alpha}\right)^{L/2}\cos\left(L\rho - \frac{\pi Q}{6}\right) + \cos\left(L\chi + \frac{5\pi Q}{6}\right)$$
$$+ (1-\alpha)^{L/2}\cos\left(L\xi - \frac{\pi Q}{2}\right) = 0 \qquad (24)$$

with

$$\rho = \arctan\left((1+\frac{2}{\sqrt{3}})\cot\beta\right) + \pi\,\theta(-\beta) - \frac{7\pi}{12}; \qquad \alpha = \frac{\sqrt{3}}{2}\,\frac{\tan^2\frac{\pi}{6} - \tan^2\beta}{\tan^2\frac{\pi}{6} + \tan^2\beta};$$

$$\xi = \arctan\left(\frac{\cot\beta}{\sqrt{3}}\right) + \pi\,\theta(-\beta) + \frac{5\pi}{4}; \; \chi = \arctan\left(\left(2\sqrt{3}+3\right)\tan\beta\right) + \frac{5\pi}{12};$$

For $L \gg 1$ and $|\beta|$ not too close to $\frac{\pi}{6}$ the middle term containing $L\chi$ dominates and the solutions β_k approach

$$\tan\beta_k \;\approx\; -\tan\frac{\pi}{6}\tan\frac{\pi}{12}\cot\left(K+\frac{\pi}{12}\right); \qquad K = \frac{(6k+Q-3)\pi}{6\,L}$$

and $k = 1, 2, \ldots, m_E$ with $m_E = [(5L-Q)/6]$. We can rewrite this like (23). A general pattern is already emerging from our Z_3 to Z_6 examples, but we still hesitate to guess e.g. the Z_7 analogue of (24).

6 Conclusions

In the analytic solution of the superintegrable chiral Z_N-Potts quantum chain the energy eigenvalues are determined by the zeros of special polynomials. These polynomials also play a central role in the representation theory of the Onsager algebra. In the thermodynamic limit, sums over functions of the zeros can be evaluated by contour integrations. For finite systems we can try to calculate the zeros numerically. However, applying standard computer routines directly to the explicit form of the polynomials will run into trouble beyond some chain length L, because the largest coefficients of the polynomials are strongly growing with increasing L. So we were looking for an alternative and possibly analytic approach.

By a suitable change of variable we transform the equations for the zeros into trigonometric equations, for which approximate analytic solutions can be obtained. We give explicit formulae for the simplest polynomials for the cases Z_3 through Z_6.

In the new variable β the solutions corresponding to the zeros of the original polynomials appear almost equidistant and are confined to a finite interval. A linear formula describes the solutions, except those near at the boundaries, with exponential precision in L. The leading finite-size effects are due to the solutions at the extreme ends of the interval in β. For the case Z_3 we derive a special formula which describes also the extreme solutions, thus allowing to calculate finite-size energy eigenvalues for the low-lying sectors.

Acknowledgements

I am grateful to Roger E. Behrend for many fruitful discussions.
Part of this work was performed during a visit of the author to SISSA, Trieste where it was supported by the TIMR-Network Contract FMRX-CT96-0012 of the European Commission. I thank G. Mussardo for his kind hospitality at SISSA.

References

Albertini G., McCoy B.M. and Perk J.H.H. (1989): Phys.Lett. **A135** 159; Advanced Studies in Pure Math., Vol.**19** (Kinokuniya-Academic) 1
Albertini G. and McCoy B.M. (1990): Nucl.Phys. **B350** 745
Au-Yang H., McCoy B.M., Perk J.H.H., Tang S. and Yan M.L. (1987): Phys.Lett. **A123** 219
Baxter R.J. (1982): Exactly Solved Models in Statistical Mechanics, Academic Press
Baxter R.J. (1988): Phys.Lett. **A133** 185
Baxter R.J., Bazhanov V.V. and Perk J.H.H. (1990): Int.J.Mod.Phys. **B4** 803
Baxter R.J. (1993): J.Stat.Phys. **73** 461
Baxter R.J. (1994): J.Phys.A: Math.Gen. **27** 1837
Baxter R.J. (1998): cond-mat/9808120
Bazhanov V.V. and Stroganov Yu.G. (1990): J.Stat.Phys. **59** 799
Dasmahapatra S., Kedem R. and McCoy B.M. (1993), Nucl.Phys. **B396** 506
Davies B. (1990): J.Phys.A: Math.Gen. **23** 2245
Dolan L. and Grady M. (1982): Phys. Rev. **D 25** 1587
Gehlen G.v. and Rittenberg V. (1985): Nucl.Phys. **B257** 351
Hamer C.J., Quispel G.R.W. and Batchelor M.T. (1987): J.Phys.A, Math.Gen. **20** 5677
Onsager L. (1944): Phys. Rev. **65** 117
O'Rourke M.J. and Baxter R.J. (1996): J.Stat.Phys. **82** 1
Roan S.-S. (1991): preprint Max-Plank-Inst. für Math. Bonn MPI/91-70
Woynarovich F. and Eckle H.-P. (1987): J.Phys.A, Math.Gen. **20** L97
Zhou Y.-K. (1998): Nucl.Phys. **B522** 550

Characteristic Polynomials for Quantum Matrices

A. Isaev[1], O. Ogievetsky[2,3], P. Pyatov[1], and P. Saponov[4]

[1] Bogoliubov Laboratory of Theoretical Physics, JINR, 141980 Dubna, Moscow Reg., Russia
[2] Center of Theoretical Physics, Luminy, 13288 Marseille, France
[3] P. N. Lebedev Physical Institute, Theoretical Department, Leninsky pr. 53, 117924 Moscow, Russia
[4] Theory Department, IHEP, 142284 Protvino, Moscow region, Russia

Abstract. A quantum version of the Cayley–Hamilton theorem is found for the matrix T of the generators of the RTT-algebra. In the quasitriangular case, a connection between the characteristic identities in the RTT and RE-algebras is established.

1 Introduction

To each Hecke type $\hat{R}$-matrix one may associate two different quantum algebras of the "matrix" type. The first is the famous "RTT" algebra which realizes the quantization of the functions on the corresponding Lie groups (Reshetikhin, Takhtadjan and Faddeev (1989)). The second is the so-called Reflection Equation (RE) algebra, which appeared first in the context of the factorized scattering on a half-line. Both these algebras possess a very compact and technically convenient formulation in terms of the matrix conventions of (Reshetikhin, Takhtadjan and Faddeev (1989)) which we use in the present paper.

However an analogy between generating elements of these algebras and the matrix elements of the usual matrices is much deeper.

In recent papers (Nazarov and Tarasov (1994), Pyatov and Saponov (1995), Gurevich, Pyatov and Saponov (1996)) a quantum analog of the Cayley–Hamilton theorem for the RE algebra was established.

The aim of the present paper is to establish a quantum analog of the Cayley–Hamilton theorem for the RTT algebra. A different version of the Cayley–Hamilton theorem in the 2-dimensional case was found in (Ewen, Ogievetsky and Wess (1991)).

Also, we discuss a relation between the Cayley–Hamilton theorems for the RE and RTT algebras in the quasitriangular case.

2 Notation

Let $\hat{R}^{ab}_{cd}$ be an $\hat{R}$-matrix, that is, a solution of the Yang-Baxter equation

$$\hat{R}_1\hat{R}_2\hat{R}_1 = \hat{R}_2\hat{R}_1\hat{R}_2 . \tag{2.1}$$

Here and below $\hat{R}_k$ denotes the $\hat{R}$-matrix acting nontrivially in the spaces $k, k+1$.

We shall employ a number of restrictions on $\hat{R}$-matrix, like closure, finiteness of height and Hecke conditions.

1. *The Hecke condition.* An $\hat{R}$-matrix is called the *Hecke $\hat{R}$-matrix* if it satisfies

$$\hat{R}^2 = \mathbf{I} + (q - q^{-1})\hat{R} . \tag{2.2}$$

Here q is a nonzero number and $\mathbf{I}$ is the identity operator.

2. *q-antisymmetrizers.* For a Hecke $\hat{R}$-matrix antisymmetrizers are defined by (Gurevich (1991))

$$A^{(1)} = \mathbf{I} , \; A^{(k)} = \frac{1}{k_q} A^{(k-1)} \left(q^{k-1} - (k-1)_q \hat{R}_{k-1}\right) A^{(k-1)} , \tag{2.3}$$

Here $k_q = (q^k - q^{-k})/(q - q^{-1})$ are the usual q-numbers, and we assume $k_q! \neq 0$ (q is not a root of unity).

The Hecke $\hat{R}$-matrix is called *even* if there exists such n that $A^{(n+1)} = 0$ and rank $A^{(n)} = 1$. The number n is called then the *height* of the $\hat{R}$-matrix.

For an $\hat{R}$-matrix of finite height n one introduces the following two matrices

$$\mathcal{D}_r = \frac{n_q}{q^n} \mathrm{Tr}_{(2\ldots n)} A^{(n)} , \qquad \mathcal{D}_\ell = \frac{n_q}{q^n} \mathrm{Tr}_{(1\ldots n-1)} A^{(n)} , \tag{2.4}$$

Here and below we use notation $\mathrm{Tr}_{(i_1\ldots i_k)}$ to denote the operation of taking traces in the spaces on places $(i_1 \ldots i_k)$.

3. *Closed $\hat{R}$-operators and quantum traces.* An $\hat{R}$-operator is called closed if its matrix $\hat{R}^{ab}_{cd}$ is invertible in indices (a, c), that is if a matrix Ψ^{ab}_{cd} satisfying $\Psi^{af}_{cg}\hat{R}^{gb}_{fd} = \delta^a_d\delta^b_c$ [1] exists.

For a closed $\hat{R}$-matrix one can give a different expression for the matrices (2.4):

$$(\mathcal{D}_r)^a_b = \Psi^{ac}_{bc} , \quad (\mathcal{D}_\ell)^a_b = \Psi^{ca}_{cb} , \tag{2.5}$$

and prove the following relations:

$$\mathrm{Tr}_{(2)}\mathcal{D}_{r2}\hat{R} = \mathbf{I}_1 , \qquad \mathrm{Tr}_{(1)}\mathcal{D}_{\ell 1}\hat{R} = \mathbf{I}_2 , \tag{2.6}$$

$$\hat{R}\mathcal{D}_{r1}\mathcal{D}_{r2} = \mathcal{D}_{r1}\mathcal{D}_{r2}\hat{R} , \qquad \hat{R}\mathcal{D}_{\ell 1}\mathcal{D}_{\ell 2} = \mathcal{D}_{\ell 1}\mathcal{D}_{\ell 2}\hat{R} . \tag{2.7}$$

One associates a version of a quantum trace with each of the matrices $\mathcal{D}_\ell$ and $\mathcal{D}_r$. In the present note we shall use only one version,

$$\mathrm{Tr}_q X = \mathrm{Tr}\, \mathcal{D}_r X . \tag{2.8}$$

---

[1] Summation over repeated indices is always assumed.

3 Characteristic Relations for RTT-Algebra

In this section we give a q-generalization of the Cayley–Hamilton theorem.

1. We start by recalling the classical picture.

With an operator X acting in a vector space V of dimension n one can associate the set of elementary symmetric functions on its spectrum, $\sigma_k(X)$,

$$\sigma_k(X) = \mathrm{Tr}_{\Lambda^k V}(\Lambda^k X) \ .$$

Here $\Lambda^k V$ is a k-th wedge power of V and $\Lambda^k X$ is an induced operator in $\Lambda^k V$ defined on polyvectors by $\Lambda^k X(v_1 \wedge v_2 \wedge \ldots \wedge v_k) = Xv_1 \wedge Xv_2 \wedge \ldots \wedge Xv_k$ If $\{x_i\}$ is the spectrum of the semisimple part of X then

$$\sigma_k = \sum_{i_1<\ldots<i_k} x_{i_1} \ldots x_{i_k} \ .$$

As well one can say that σ_k is the sum of all principal k-minors of X. This can be expressed shortly by a formula

$$\begin{aligned}\sigma_k &= \frac{1}{k!(n-k)!}\epsilon_{a_1\ldots a_k c_{k+1}\ldots c_n} X^{a_1}_{\ b_1} \ldots X^{a_k}_{\ b_k} \epsilon^{b_1\ldots b_k c_{k+1}\ldots c_n} \\ &= \mathrm{Tr}_{(1\ldots k)}(A^{(k)} X_1 \ldots X_k) \ .\end{aligned} \tag{3.9}$$

Here ϵ's are the usual totally antisymmetric ϵ-symbols.

Cayley-Hamilton theorem says that X satisfies the equation $\chi(X) = 0$ where $\chi(t) = det\ (t\mathbf{I} - X)$ is the characteristic polynomial of the operator X.

The explicit form of the polynomial $\chi(t)$ includes the elementary symmetric functions σ_k,

$$\chi(t) = (-1)^n \sum (-t)^{n-j} \sigma_j(X) \ . \tag{3.10}$$

2. *Definitions.* Let us recall that the RTT-algebra is a unital associative algebra which is generated by elements of a "q-matrix" $T = ||T^a_b||$ subject to the relations

$$\hat{R}_1 T_1 T_2 = T_1 T_2 \hat{R}_1 \ . \tag{3.11}$$

To define q-analogues of the invariants σ_k for the RTT-algebra one just replaces the antisymmetrizers and the matrices X in (3.9) by their quantum analogs,

$$\sigma_0(T) = 1 \ , \ \sigma_k(T) = q^k \, \mathrm{Tr}_{(1\ldots k)}(A^{(k)} T_1 \ldots T_k) \ , \ \text{for } k \geq 1 \ . \tag{3.12}$$

The coefficient q^k in front of the trace is introduced for the later convenience.

If $\hat{R}$ is even of height n, then for $k > n$ we have $\sigma_k(T) = 0$ and the last nontrivial element $\sigma_n(T)$ is proprotional to a q-determinant of matrix T:

$$\sigma_n(T) = q^n det_q T \ . \tag{3.13}$$

One can check that the elements $\sigma_k(T)$ enjoy an important non-classical property: they are mutually commutative[2].

We need an appropriate generalization of the matrix powers in the RTT-algebra.

Definition.

$$T^{\underline{k}} = \mathrm{Tr}_{(1\dots k-1)}\left(\hat{R}_1\hat{R}_2\dots\hat{R}_{k-1}T_1T_2\dots T_k\right) , \tag{3.14}$$

$$T^{\overline{k}} = \mathrm{Tr}_{(2\dots k)}\left(\hat{R}_1\hat{R}_2\dots\hat{R}_{k-1}T_1T_2\dots T_k\right) . \tag{3.15}$$

We use the superscripts $\underline{k}$ and $\overline{k}$ here for denoting different types of the k-th power of matrix T. This should not make a confusion with the usual matrix power (they coincide in the classical limit only).

Taking the last trace in 3.14 or the first trace in 3.15 one obtains the quantum analog of the power sums which was introduced by Maillet (Maillet (1990)).

3. With these definitions we can formulate the main result.

Theorem (Cayley-Hamilton theorem for the RTT-algebra).
Let $\hat{R}$ be even $\hat{R}$-matrix of height n. The quantum matrix T satisfies identities

$$\textstyle\sum_{k=0}^{n-1}\sigma_k(T)(-T)^{\underline{n-k}} + \sigma_n(T)\mathcal{D}_\ell = 0 , \tag{3.16}$$

$$\textstyle\sum_{k=0}^{n-1}(-T)^{\overline{n-k}}\sigma_k(T) + \sigma_n(T)\mathcal{D}_r = 0 . \tag{3.17}$$

Proof. We shall give the details of the proof of eq.(3.16). The relation (3.17) can be proved analogously.

For $k = 1, \dots, n-1$ we have

$$\begin{aligned}
&\sigma_k(T)T^{\underline{n-k}} = \\
&q^k\mathrm{Tr}_{(1\dots k)}(A^{(k)}T_1\dots T_k)\mathrm{Tr}_{(k+1\dots n-1)}(R_{k+1}\dots R_{n-1}T_{k+1}\dots T_n) = \\
&q^k\mathrm{Tr}_{(1\dots n-1)}(A^{(k)}T_1\dots T_k\hat{R}_{k+1}\dots\hat{R}_{n-1}T_{k+1}\dots T_n) = \\
&q^k\mathrm{Tr}_{(1\dots n-1)}(A^{(k)}\hat{R}_{k+1}\dots\hat{R}_{n-1}T_1\dots T_n)
\end{aligned} \tag{3.18}$$

We use (2.3) in the form $q^kA^{(k)} = (k+1)_qA^{(k+1)} + k_qA^{(k)}\hat{R}_kA^{(k)}$ to rewrite (3.18) as

$$\begin{aligned}
&= (k+1)_q\mathrm{Tr}_{(1\dots n-1)}(A^{(k+1)}\hat{R}_{k+1}\dots\hat{R}_{n-1}T_1\dots T_n) \\
&+ k_q\mathrm{Tr}_{(1\dots n-1)}(A^{(k)}\hat{R}_kA^{(k)}\hat{R}_{k+1}\dots\hat{R}_{n-1}T_1\dots T_n) .
\end{aligned}$$

[2] For a set of invariants which classically reduces to the power sums, the commutativity was proved in (Maillet (1990)). The commutativity of $\sigma_k(T)$ is then a byproduct of Newton relations established in (Pyatov and Saponov (1996))

In the last term the right antisymmetrizer $A^{(k)}$ commutes with the expression $R_{k+1}\dots R_{n-1}T_1\dots T_n$. Next, we can move $A^{(k)}$ to the left using the cyclic property of the trace. Finally, $(A^{(k)})^2 = A^{(k)}$ and we obtain

$$\sigma_k(T)T^{\underline{n-k}} = (k+1)_q \mathrm{Tr}_{(1\dots n-1)}(A^{(k+1)}\hat{R}_{k+1}\dots\hat{R}_{n-1}T_1\dots T_n) + k_q \mathrm{Tr}_{(1\dots n-1)}(A^{(k)}\hat{R}_k\dots\hat{R}_{n-1}T_1\dots T_n)\ . \qquad (3.19)$$

We have also $\sigma_0(T)T^{\underline{n}} = T^{\underline{n}}$.

For $k = n-1$ the first term in (3.19) can be rewritten using $A^{(n)}T_1\dots T_n = A^{(n)}det_qT$ and applying (2.4) and (3.13).

Taking the alternative sum for k from 0 to $n-1$, we obtain the relation (3.16). ■

4. In the formulation of the Cayley–Hamilton theorem for the RTT-algebra one can use quantum traces instead of the usual ones. Indeed, in view of (2.7), the map

$$T \longrightarrow T_r = \mathcal{D}_r T \qquad (3.20)$$

generates an automorphism of the RTT-algebra. Substituting T_r for T changes the definitions of the quantum power and the formulation of the Cayley–Hamilton theorem. For the quantum power we obtain

$$(T^{\overline{k}})_r = \mathrm{Tr}_{q\,(2\dots k)}\left(\hat{R}_1\hat{R}_2\dots\hat{R}_{k-1}T_1T_2\dots T_k\right)\ , \qquad (3.21)$$

The characteristic identity (3.17) becomes

$$\sum_{k=0}^{n-1}(-)^{n-k}\,(T^{\overline{n-k}})_r\,\sigma_k(T_r) + \sigma_n(T_r)\,\mathrm{I} \quad = 0\ . \qquad (3.22)$$

The elementary functions σ_k can be reexpressed in terms of quantum traces as well:

$$\sigma_k(T_r) = q^k\,\mathrm{Tr}_{q\,(1\dots k)}(A^{(k)}T_1\dots T_k)\ . \qquad (3.23)$$

The version (3.22) of the Cayley–Hamilton theorem will be useful in the next section for establishing a correspondence with the Cayley–Hamilton theorem for the RE-algebra.

4 Quasitriangular Case

In the previous section we were working with the RTT-algebra defined by a numerical R-matrix. In the case when the R-matrix is obtained from a universal $\mathcal{R}$-matrix of a quasitriangular Hopf algebra $\mathcal{U}$, one can construct elements $L^i_j \in \mathcal{U}$ satisfying the RE-algebra. In (Nazarov and Tarasov (1994), Pyatov and Saponov (1995), Gurevich, Pyatov and Saponov (1996)) the Cayley-Hamilton theorem for the RE-algebra was demonstrated. In the present section we show that in the quasitriangular case the Cayley-Hamilton theorem for T implies the Cayley-Hamilton theorem for L. This gives an independent proof for the Cayley-Hamilton theorem for the realization of the RE-algebra by the elements $L^i_j \in \mathcal{U}$.

We remind that the elements L^i_j are defined by

$$L^i_j = S((L^-)^i_k)\,(L^+)^k_j\,, \tag{4.24}$$

where S is the antipode and $L^\pm \in \mathcal{U}$ are constructed as (Reshetikhin (1989))

$$<\mathcal{R}^{12}\,, id\otimes T^i_j>=(L^+)^i_j \quad , \quad <\mathcal{R}^{21}\,, id\otimes T^i_j>=S\left((L^-)^i_j\right)\,. \tag{4.25}$$

The elements $L^\pm$ satisfy

$$\hat{R}\,L^\pm_2\,L^\pm_1 = L^\pm_2\,L^\pm_1\,\hat{R}\,,\;\; \hat{R}\,L^+_2\,L^-_1 = L^-_2\,L^+_1\,\hat{R}\,. \tag{4.26}$$

As a consequence, the elements L in (4.24) satisfy the reflection equation,

$$\hat{R}L_1\hat{R}L_1 = L_1\hat{R}L_1\hat{R}\,. \tag{4.27}$$

In (Reshetikhin and Semenov-Tian-Shansky (1988), Drinfeld (1989)) (see also (Schupp, Watts and Zumino (1993))) a map $\phi : \mathcal{U}^* \to \mathcal{U}$ was introduced,

$$\phi(\alpha) =<\mathcal{R}^{21}\,\mathcal{R}^{12}\,, id\otimes\alpha>=a\,. \tag{4.28}$$

One sees immediately that

$$\phi(T) = L\,. \tag{4.29}$$

In general the map (4.28) is not homomorphic. In (Drinfeld (1989)) it was shown that the restriction of (4.28) is homomorphic on the inverse image of the center $Z(\mathcal{U})$. We shall use a slight generalization of this statement

Lemma. *If $\phi(\alpha) \in Z(\mathcal{U})$, then for any $\beta \in \mathcal{U}^*$ one has*

$$\phi(\beta\alpha) = \phi(\beta)\,\phi(\alpha)\,. \tag{4.30}$$

Proof. By quasitriangularity, one has $(id\otimes\Delta)\,\mathcal{R}^{21}\,\mathcal{R}^{12} = \mathcal{R}^{21}\,\mathcal{R}^{31}\,\mathcal{R}^{13}\,\mathcal{R}^{12}$. Using this equality and the condition that $\phi(\alpha)$ is central one obtains

$$\phi(\beta\alpha) =<\mathcal{R}^{21}\,\mathcal{R}^{12}\,, id\otimes\beta\alpha>=<\mathcal{R}^{21}\,\mathcal{R}^{31}\,\mathcal{R}^{13}\,\mathcal{R}^{12}\,, id\otimes\beta\otimes\alpha>=$$

$$= b_{(1)}\phi(\alpha)b_{(2)} = \phi(\beta)\phi(\alpha)$$

where $b_{(1)} =<\mathcal{R}^{21}\,, id\otimes\beta_{(1)}>$, $b_{(2)} =<\mathcal{R}^{12}\,, id\otimes\beta_{(2)}>$ and $\Delta(\beta) = \beta_{(1)}\otimes\beta_{(2)}$. ■

Now we are ready to describe the correspondence between the Cayley–Hamilton theorems for the RTT and RE-algebras.

Proposition. *The map ϕ transforms the Cayley–Hamilton identity (3.22) for the RTT-algebra into the Cayley–Hamilton identity for the RE-algebra,*

$$\sum_{k=0}^{n-1}(-L)^{n-k}\,\sigma_k(L) + \sigma_n(L)\,\mathbf{I} \;= 0\,, \tag{4.31}$$

where

$$\sigma_k(L) = q^k \, \mathrm{Tr}_{q\,(1\dots k)} \left(A^{(k)} \, L_{\overline{1}} \cdots L_{\overline{k}} \right) \tag{4.32}$$

with the $L_{\overline{k}}$ defined inductively by

$$L_{\overline{1}} = L_1 \ , \ L_{\overline{k}} = \hat{R}_{k-1} L_{\overline{k-1}} \hat{R}_{k-1}^{-1} \ . \tag{4.33}$$

Proof. *Step 1.* The following relation holds

$$\phi(T_1 \cdots T_k) = L_{\overline{1}} \cdots L_{\overline{k}} \ . \tag{4.34}$$

We prove this relation by induction. Indeed, for $k = 1$ this is eqn.(4.29). Let (4.34) be valid for $k = r - 1$. Then, for $k = r$ one has

$$\phi(T_1 \cdots T_r) = < \mathcal{R}^{21} \mathcal{R}^{12}, id \otimes T_1 \cdots T_r > \ . \tag{4.35}$$

Here the subscripts and superscripts signify different things. The superscripts refer to a copy of $\mathcal{U}$ while each subscript encodes a pair of matrix indices, as in the previous section.

The rhs of (4.35) equals to

$$< (id \otimes \Delta) \mathcal{R}^{21} \mathcal{R}^{12}, id \otimes T_1 \otimes T_2 \cdots T_r > \tag{4.36}$$

which by quasitriangularity is

$$< \mathcal{R}^{21} \mathcal{R}^{31} \mathcal{R}^{13} \mathcal{R}^{12}, id \otimes T_1 \otimes T_2 \cdots T_r > \ . \tag{4.37}$$

Now we apply the coproduct to T_1 in (4.37) to obtain

$$< \mathcal{R}^{21} \mathcal{R}^{41} \mathcal{R}^{14} \mathcal{R}^{13}, id \otimes T_1 \otimes T_1 \otimes T_2 \cdots T_r > \ . \tag{4.38}$$

This expression can be splitted into

$$\begin{aligned} &< \mathcal{R}^{21}, id \otimes T_1 >< \mathcal{R}^{21} \mathcal{R}^{12}, id \otimes T_2 \cdots T_r >< \mathcal{R}^{12}, id \otimes T_1 >= \\ &= S(L_{\overline{1}}^-) P_{(1\to r)} L_{\overline{1}} \cdots L_{\overline{r-1}} P_{(1\to r)}^{-1} L_{\overline{1}}^+ \ . \end{aligned} \tag{4.39}$$

Here we have applied relation (4.34) for $k = r - 1$ and introduced a notation $P_{(1\to r)} \equiv P_{12} P_{23} \dots P_{r-1\,r}$, where P_{12}, P_{23}, ... are permutation matrices. Using the commutation relation

$$L_2 L_{\overline{1}}^+ = L_{\overline{1}}^+ \hat{R} L_1 \hat{R}^{-1} \ , \tag{4.40}$$

which is implied by (4.26) we obtain that the rhs of (4.39) is $L_{\overline{1}} \dots L_{\overline{r}}$. Thus, (4.34) holds for $k = r$ if it is valid for $k = r - 1$. This completes the induction.

The eqn.(4.34) immediately implies that

$$\phi\left(\sigma_k(T_r)\right) = q^k \, \mathrm{Tr}_{q\,(1\dots k)} \left(A^{(k)} \, L_{\overline{1}} \cdots L_{\overline{k}} \right) \equiv \sigma_k(L) \ . \tag{4.41}$$

Step 2. Now we shall find an expression for $\phi((T^{\overline{k}})_r)$ in terms of L's. The calculation uses a formula

$$\hat{R}_k L_{\overline{k}} L^p_{\overline{k+1}} = L^p_{\overline{k}} L_{\overline{k+1}} \hat{R}_k \equiv L^p_{\overline{k}} \hat{R}_k L_{\overline{k}} \tag{4.42}$$

which can be checked by induction in p.

We have

$$\begin{aligned}\phi((T^{\overline{k}})_r) &= \mathrm{Tr}_{q\,(2\ldots k)} \left(\hat{R}_1 \ldots \hat{R}_{k-1}\, L_{\overline{1}} \cdots L_{\overline{k}}\right) \\ &= \mathrm{Tr}_{q\,(2\ldots k)} \left(\hat{R}_1 \ldots \hat{R}_{k-2}\, L_{\overline{1}} \cdots L_{\overline{k-1}} \hat{R}_{k-1} L_{\overline{k-1}}\right) .\end{aligned} \tag{4.43}$$

Here we used (4.42) with $p = 1$. Now one can take the trace in the k-th space explicitely, using (2.6) to obtain

$$\mathrm{Tr}_{q\,(2\ldots k-1)} \left(\hat{R}_1 \ldots \hat{R}_{k-2}\, L_{\overline{1}} \cdots L_{\overline{k-2}} L^2_{\overline{k-1}}\right) . \tag{4.44}$$

Repeating this calculation to take traces consequently in spaces number $k - 1, \ldots, 1$, one ends up with the formula

$$\phi((T^{\overline{k}})_r) = L^k . \tag{4.45}$$

Step 3. The elements $\sigma_i(L)$ are central (Reshetikhin, Takhtadjan and Faddeev (1989), Pyatov and Saponov (1995)) and we can apply Lemma to finish the proof. ■

Acknowledgements: We are indebted to G. Arutyunov, D. Gurevich, P. Kulish and S.Pakuliak for valuable discussions. The work of P.P. and A.I. is supported in part by the RFBR grant No. 97-01-01041 and by INTAS grant 93-127-ext. The work of P.S. is supported by the grant INTAS-RFBR 95-829.

References

Reshetikhin N., Takhtadjan L. and Faddeev L., *Quantization of Lie groups and Lie algebras*, Algebra i Analiz, **1** No. 1 (1989) 178- 206. English translation in: Leningrad Math. J. **1** (1990) 193-225.

Nazarov M. and Tarasov V., *Yangians and Gelfand-Zetlin bases*, Publications RIMS **30** (1994) 459-478.

Pyatov P.N. and Saponov P.A., *Characteristic Relations for Quantum Matrices*, J. Phys. A: Math. Gen., **28** (1995) 4415-4421.

Gurevich D.I., Pyatov P.N. and Saponov P.A., *Hecke Symmetries and Characteristic Relations on Reflection Equation Algebras*, Preprint JINR E-95-202, q-alg/9606.

Ewen H., Ogievetsky O. and Wess J., *Quantum matrices in two dimensions*, Lett. Math. Phys., **22** (1991) 297-305.

Gurevich D.I., *Algebraic aspects of the quantum Yang-Baxter equation*, Algebra i Analiz, **2** (1990) 119 – 148. English translation in: Leningrad Math. J. **2** (1991) 801 – 828.

Maillet J.M., *Lax equations and quantum groups*, Phys. Lett. **B245** (1990) 480-486.
Pyatov P.N. and Saponov P.A., *Newton relations for quantum matrix algebras of RTT-type*, Preprint IHEP 96-76.
Reshetikhin N., *Quasitriangular Hopf algebras and invariants of tangles*, Algebra i Analiz, **1** No. 2 (1989) 169-188. English translation in: Leningrad Math. J. **1** (1990) 491-513.
Reshetikhin N.Yu. and Semenov-Tian-Shansky M.A., *Quantum R- matrices and factorization problem*, Geometry and Physics **5** No.4 (1988) 533-550.
Drinfeld V.G., *On almost cocommutative Hopf algebras*, Algebra i Analiz, **1** No. 2 (1989) 30-46. English translation in: Leningrad Math. J. **1** (1990) 321-342.
Schupp P., Watts P. and Zumino B., *Bicovariant quantum algebras and quantum Lie algebras*, Comm.Math.Phys. **157** (1993) 305-329.

Arbitrary Spin Massless Bosonic Fields in d-Dimensional Anti-de Sitter Space

R. R. Metsaev

Department of Theoretical Physics, P.N. Lebedev Physical Institute, Leninsky prospect 53, 117924, Moscow, Russia

Abstract. Arbitrary spin free massless bosonic fields propagating in even d - dimensional anti-de Sitter spacetime are investigated. Free wave equations of motion, subsidiary conditions and the corresponding gauge transformations for such fields are proposed. The lowest eigenvalues of the energy operator for the massless fields and the gauge parameter fields are derived. The results are formulated in $SO(d-1,2)$ covariant form as well as in terms of intrinsic coordinates. An interrelation of two definitions of masslessness based on gauge invariance and conformal invariance is discussed.

Motivation. Some time ago a completely self-consistent interacting equations of motion for higher massless fields of all spins have been discovered (Vasiliev (1990)). First these equations have been formulated for the case of four dimensional $d = 4$ AdS spacetime. Then because the equations allow very natural generalization to higher spacetime dimensions $d > 4$ they have immediately been extended to such the dimensions (Vasiliev (1991)). These equations are formulated in terms of wavefunctions $\Psi(x, Z)$ which depend on usual spacetime coordinates x^μ and certain twistor like variables Z^α. Usual physical fields as well as certain auxiliary fields are obtainable by expanding $\Psi(x, Z)$ in powers of Z: $\Psi(x, Z) = \sum_0^\infty Z^{\alpha_1} \dots Z^{\alpha_n} \Phi(x)_{\alpha_1 \dots \alpha_n}$, where $\Phi = \{\Phi_{phys}, \Phi_{aux}\}$. For the case of $4d$ theories it is well-established (Vasiliev (1991)) that Φ_{phys} satisfy the equations of motion which at free level are equivalent to those investigated in (Fronsdal (1978)). As to $d > 4$ theories, although such a statement is not proved, it is believed that equations of motion suggested also describe massless higher spin fields in a self-consistent way. Unfortunately in contrast to the completeness of description for $d = 4$ little was known about the higher spin massless spin fields in arbitrary $d > 4$ even at the level of free fields, unless considerations are restricted to totally symmetric tensor (or tensor-spinor) fields (Lopatin and Vasiliev (1988),Vasiliev (1988)). Filling this gap was a motivation of our investigation (Metsaev (1994)-Metsaev (1997)). Here we report summary of results.

Setting up the problem. First, let us remind the main fact about representations of anti-de Sitter algebra $so(d-1,2)$

$$[J^{AB}, J^{CD}] = \eta^{BC} J^{AD} \pm 3 \text{ terms}, \qquad \eta^{AB} = (-,-,+,\dots,+),$$

$A, B = 0', 0, 1, \dots, d-1$, which are relevant for elementary particles. A positive-energy lowest weight irreps of $so(d-1,2)$ algebra denoted as $D(E_0, \mathbf{h})$, are defined by E_0 which is lowest eigenvalue of energy operator $\mathrm{i}J^{00'}$ and by

$$\mathbf{h} = (h_1, \ldots, h_\nu), \qquad h_1 \geq \ldots \geq h_\nu \geq 0, \qquad \nu \equiv (d-2)/2 \tag{1}$$

which is the highest weight of representation of the $so(d-1)$ algebra. Since in the representations under considerations the energy is by definition bounded from below the $D(E_0, \mathbf{h})$ contains the vacuum $|\phi^{E_0}(\mathbf{h})\rangle$ annihilated by all those elements of $so(d-1,2)$ which decrease the energy. This vacuum forms a linear space which is invariant under the action of the energy operator $\mathrm{i}J^{00'}$ and elements of the $so(d-1)$ algebra. In other words $|\phi^{E_0}(\mathbf{h})\rangle$ is (a) the eigenvalue vector of $\mathrm{i}J^{00'}$; (b) a weight $\mathbf{h}$ representation of the $so(d-1)$ algebra. To expose these properties of $|\phi^{E_0}(\mathbf{h})\rangle$ it is convenient a usage of the coordinates $(z,\bar{z},y^I)$, $I = 1, \ldots d-1$, where $z = (y^{0'} + \mathrm{i}y^0)/\sqrt{2}$, $\bar{z} = z^*$ and the η^{AB} takes the nonvanishing elements $\eta^{z\bar{z}} = -1$, $\eta^{IK} = \delta^{IK}$. In these coordinates the generators J^{AB} split into $J^{z\bar{z}}$-energy operator, J^{zI}–spin deboost operator, $J^{\bar{z}I}$–spin boost operator and J^{IK}- generators of the $SO(d-1)$ group. Now the $|\phi^{E_0}(\mathbf{h})\rangle$ is defined by the relations

$$J^{z\bar{z}}|\phi^{E_0}(\mathbf{h})\rangle = E_0|\phi^{E_0}(\mathbf{h})\rangle, \qquad J^{zI}|\phi^{E_0}(\mathbf{h})\rangle = 0.$$

Then the representation space $D(E_0, \mathbf{h})$ can be built by acting with boost operator $J^{\bar{z}I}$ on the vacuum $|\phi^{E_0}(\mathbf{h})\rangle$:

$$D(E_0, \mathbf{h}) = \sum_{n=0}^{\infty} \oplus J^{\bar{z}I_1} \ldots J^{\bar{z}I_n}|\phi^{E_0}(\mathbf{h})\rangle.$$

It turns out that for certain value of E_0 there is singular vector on the first energy level. If we factorise whole space by space built on this singular vector then we get irreducible representations which is, by definition, massless representation. Now the problem solution to which we are going to provide can be formulated as follows. Find (i) E_0 corresponding to massless representation; (ii) second order relativistic equations of motion whose space of solution is a carrier for massless representations; (iii) corresponding gauge transformations. The relevant E_0 for $d = 4$ and corresponding field theoretical realization has been found in (Evans (1967)) and (Fronsdal (1978)) respectively (for review see Nicolai (1984)).

Summary of results. We describe the AdS spacetime as a hyperboloid

$$\eta_{AB}y^A y^B = -1, \tag{2}$$

in $d+1$- dimensional pseudo-Euclidean space with metric tensor η_{AB}. The indices A, B are raised and lowered by η^{AB} and η_{AB} respectively. In what follows to simplify our expressions we will drop the metric tensor η_{AB} in scalar products. As is usual, we split the generators J^{AB} into an orbital part L^{AB} and a spin part M^{AB}: $J^{AB} = L^{AB} + M^{AB}$. The realization of L^{AB} in terms of differential operators defined on the hyperboloid (2) is:

$$L^{AB} = y^A\nabla^B - y^B\nabla^A\,, \qquad \nabla^A \equiv \theta^{AB}\frac{\partial}{\partial y^B}\,, \qquad \theta^{AB} \equiv \eta^{AB} + y^A y^B\,.$$

The tangent derivative ∇^A satisfies relations

$$[\nabla^A, y^B] = \theta^{AB}\,, \qquad [\nabla^A, \nabla^B] = -L^{AB}\,, \qquad y^A\nabla^A = 0\,, \qquad \nabla^A y^A = d\,.$$

A form for M^{AB} depends on the realization of the representation. We will use the tensor realization of representation. As the carrier for $D(E_0, \mathbf{h})$ we use of tensor field of the $SO(d-1,2)$ group

$$A^{C(\mathbf{h})} = A^{C_1^1,\ldots,C_1^{h_1},\ldots,C_\nu^1,\ldots,C_\nu^{h_\nu}} \tag{3}$$

defined on the hyperboloid (2). By definition, $A^{C(\mathbf{h})}$ is a tensor field whose $SO(d-1,2)$ indices $C(\mathbf{h})$ have the structure of the Young tableaux (YT) corresponding to the irreps of the $SO(d-1,2)$ group labeled by $\mathbf{h}$. In what follows we use the notation $\mathsf{YT}(\mathbf{h})$ to indicate such YT. The h_i indicates the number of boxes in the i-th row of $\mathsf{YT}(\mathbf{h})$. To simplify our expressions we introduce ν creation and annihilation operators a_l^A and $\bar{a}_l^A$, $l = 1,\ldots,\nu$, and construct a Fock space vector

$$|A\rangle \equiv \prod_{l=1}^{\nu}\prod_{i_l=1}^{h_l} a_l^{C_l^{i_l}} A^{C_1^1,\ldots,C_1^{h_1},\ldots,C_\nu^1,\ldots,C_\nu^{h_\nu}}|0\rangle\,, \quad [\bar{a}_i^A, a_j^B] = \eta^{AB}\delta_{ij}\,, \quad \bar{a}_n^A|0\rangle = 0\,.$$

For a realization of this kind, M^{AB} has the form

$$M^{AB} = \sum_{l=1}^{\nu}(a_l^A\bar{a}_l^B - a_l^B\bar{a}_l^A)\,.$$

Throughout of the paper, unless otherwise specified, the indices i, j, l, n run over $1,\ldots,\nu$. For these indices we drop the summation over repeated indices. Because the $A^{C(\mathbf{h})}$ is associated with $\mathsf{YT}(\mathbf{h})$ then the $|A\rangle$ should satisfy the constraints

$$(a_{ii} - h_i)|A\rangle = 0\,, \quad a_{ij}^-|A\rangle = 0\,, \quad \varepsilon^{ij}a_{ij}|A\rangle = 0\,, \tag{4}$$

where in (4) and below we use the notation

$$a_{ij} \equiv a_i^A\bar{a}_j^A\,, \qquad a_{ij}^- \equiv \bar{a}_i^A\bar{a}_j^A\,, \qquad a_{ij}^+ \equiv a_i^A a_j^A\,, \tag{5}$$

and $\varepsilon^{ij} = 1(0)$ for $i < j (i \geq j)$. The 1st equation in (4) tells us that a_i occurs h_i times in $|A\rangle$. Tracelessness of $A^{C(\mathbf{h})}$ is reflected in the 2nd equation in (4). The 3rd equation in (4) implies that the generic tensor field (3) is antisymmetric with respect to indices in columns. As a result the $|A\rangle$ is obtainable from YT by making use of the following symmetrization rule: (i) first we perform alternating with respect to indices in all columns, (ii) then we perform symmetrization with respect to indices in all rows. Note that usual one uses the symmetrization rule when first one performs (ii) and then (i). Such kind of $|A\rangle$ could be described by using generic tensor field (3) which is symmetric with respect to indices in columns and by using anticommuting oscillators in place of commuting ones.

Because, by assumption, the $|A\rangle$ is a carrier for $D(E_0, \mathbf{h})$ it should satisfy the equation

$$(Q - \langle Q\rangle)|A\rangle = 0\,,$$

where Q is the second order Casimir operator of the $so(d-1,2)$ algebra while $\langle Q\rangle$ is its eigenvalue for $D(E_0, \mathbf{h})$

$$Q \equiv \frac{1}{2} J^{AB} J^{AB}\,, \qquad \langle Q\rangle = -E_0(E_0 + 1 - d) - \sum_{l=1}^{\nu} h_l(h_l - 2l + d - 1)\,.$$

In addition we impose on $|A\rangle$ the following subsidiary constraints

$$\overline{\nabla}_n|A\rangle = 0 \qquad \text{(divergenelessness)}\,, \tag{6}$$

$$\bar{y}_n|A\rangle = 0 \qquad \text{(transversality)}\,. \tag{7}$$

Here and below we use the notation

$$\overline{\nabla}_n \equiv \nabla^A \bar{\mathbf{a}}_n^A\,, \quad \nabla_n \equiv \mathbf{a}_n^A \nabla^A\,, \quad \bar{y}_n \equiv \bar{a}_n^A y^A\,, \quad y_n \equiv a_n^A y^A\,,$$

$$\mathbf{a}_n^A \equiv \theta^{AB} a_n^B\,, \quad \bar{\mathbf{a}}_n^A \equiv \theta^{AB} \bar{a}_n^B\,.$$

The constraint (6) is a $SO(d-1,2)$ covariant analog of usual divergenelessness condition $\partial_\mu A^{\mu\cdots} = 0$. The $SO(d-2,1)$ tensor decomposes into the same rank tensor of Lorentz subgroup $SO(d-1,1)$ and a lower rank tensor of $SO(d-1,1)$. The constraint (7) implies that the lower rank tensor is set to zero. In other words we use a $SO(d-1,2)$ tensor which is irreducible when reducing to Lorentz subgroup. Taking into account the transversality (7) and the relations

$$a_{ij} = \mathbf{a}_{ij} - y_i \bar{y}_j\,, \qquad a_{\bar{i}\bar{j}} = \mathbf{a}_{\bar{i}\bar{j}} - \bar{y}_i \bar{y}_j, \tag{8}$$

we transform the constraints (4) to form which is more convenient in practical calculations

$$(\mathbf{a}_{ii} - h_i)|A\rangle = 0\,, \quad \mathbf{a}_{\bar{i}\bar{j}}|A\rangle = 0\,, \quad \varepsilon^{ij}\mathbf{a}_{ij}|A\rangle = 0\,. \tag{9}$$

Making use of constraints above the equations of motion may be simplified. To this end we rewrite the Q as follows

$$Q = -\nabla^2 + M^{AB} L^{AB} + \frac{1}{2} M^{AB} M^{AB}\,, \qquad \nabla^2 \equiv \nabla^A \nabla^A\,,$$

use then the relations

$$M^{AB} L^{AB}|A\rangle = 2\sum_{l=1}^{\nu} h_l|A\rangle\,, \qquad M^{AB} M^{AB}|A\rangle = -2\sum_{l=1}^{\nu} h_l(h_l - 2l + d + 1)|A\rangle$$

and get the desired form of equations of motion

$$(\nabla^2 - m^2)|A\rangle = 0\,, \qquad m^2 \equiv E_0(E_0 + 1 - d)\,. \tag{10}$$

To define E_0 corresponding to massless representations we should construct gauge transformations and choose such the E_0 that the equations (10) to be invariant with respect to gauge transformations. In order to formulate gauge transformations we use the gauge parameters fields whose spacetime indices correspond to the YT which one can make by removing one box from the $\mathsf{YT}(\mathbf{h})$. The most general gauge transformations we start with are

$$\delta_{(n)}|A\rangle \sim \nabla_n|\Lambda_n\rangle + y_n|R_n\rangle\,, \tag{11}$$

where the gauge parameters fields $|\Lambda_n\rangle$ and $|R_n\rangle$ are associated with $\mathsf{YT}(\mathbf{h}_{(n)})$, and i-th component of the $\mathbf{h}_{(n)}$ is equal to $h_{i(n)} = h_i - \delta_{in}$. The $\mathsf{YT}(\mathbf{h}_{(n)})$ is obtained by removing one box from n-th row of the $\mathsf{YT}(\mathbf{h})$. We assume that only those $|\Lambda_n\rangle$ and $|R_n\rangle$ are non-zero whose $\mathbf{h}_{(n)}$ satisfy the inequalities

$$h_{1(n)} \geq \ldots \geq h_{\nu(n)} \geq 0\,. \tag{12}$$

Given the $\mathbf{h}$, the set of those n whose $\mathbf{h}_{(n)}$ satisfy (12) will be referred to as $\mathsf{S}(\mathbf{h})$. We impose on the $|\Lambda_n\rangle$, $|R_n\rangle$ and the constraints similar to those for $|A\rangle$

$$\overline{\nabla}_i|\Lambda_n\rangle = 0\,, \qquad \overline{y}_i|\Lambda_n\rangle = 0 \tag{13}$$

and constraints obtained from (13) by replacing $\Lambda \to R$. Since $|\Lambda_n\rangle$, $|R_n\rangle$ correspond to $\mathsf{YT}(\mathbf{h}_{(n)})$, they satisfy the constraints

$$(a_{ii} - h_{i(n)})|\Lambda_n\rangle = 0\,, \quad a_{ij}^{-}|\Lambda_n\rangle = 0\,, \quad \varepsilon^{ij}a_{ij}|\Lambda_n\rangle = 0\,, \tag{14}$$

and those which are obtainable from (14) by replacing $\Lambda \to R$. In practical calculation it is convenient to rewrite (14) in the form

$$(\mathbf{a}_{ii} - h_{i(n)})|\Lambda_n\rangle = 0\,, \quad \mathbf{a}_{ij}^{-}|\Lambda_n\rangle = 0\,, \quad \varepsilon^{ij}\mathbf{a}_{ij}|\Lambda_n\rangle = 0\,. \tag{15}$$

which can be obtained by using the constraints (13).

It turns out that the invariance requirement of constraints (7),(9) with respect to gauge transformations fixes the form of gauge transformations

$$\delta_{(n)}|A\rangle = \mathcal{D}_n|\Lambda_n\rangle\,, \qquad \mathcal{D}_n \equiv \sum_{j=0}^{n-1}(-)^j \sum_{l_1,\ldots,l_{j+1}=1}^{n} \delta_{nl_{j+1}} \prod_{i=1}^{j} \frac{\varepsilon^{l_i l_{i+1}}}{\lambda_{l_i n}} \mathbf{a}_{l_{i+1}l_i} D_{l_1}\,, \tag{16}$$

$$D_n \equiv \nabla_n + \sum_{l=1}^{\nu}(-y_l\mathbf{a}_{nl} + \mathbf{a}_{nl}^{+}\overline{y}_l)\,, \qquad \lambda_{ln} \equiv h_l - h_n + n - l + 1\,.$$

In (16) the $\mathbf{a}_{l_{i+1}l_i}$ are ordered as follows: $\mathbf{a}_{l_{j+1}l_j}\ldots\mathbf{a}_{l_2l_1}$. Then from the invariance requirement of (6) with respect to (16), i.e. $\overline{\nabla}_n\delta_{(n)}|A\rangle = 0$, we find the equation of motion for gauge parameter field

$$(\nabla^2 - (h_n - n)(h_n - n - 1 + d))|A_n\rangle = 0\,. \tag{17}$$

Finally from the invariance requirement of equation of motion (10) with respect to gauge transformations (16), i.e. $(\nabla^2 - m^2)\delta_{(n)}|A\rangle = 0$, we get the equation for E_0

$$E_0(E_0 + 1 - d) = (h_n - n - 1)(h_n - n - 2 + d)\,, \quad n \in \mathsf{S}(\mathbf{h})\,. \tag{18}$$

Note that in deriving (18) the equations of motion for gauge parameter has been used. Solutions to the quadratic equation for E_0 (18) read:

$$E^{(1)}_{0(n)} = h_n - n - 2 + d\,, \qquad E^{(2)}_{0(n)} = n + 1 - h_n\,. \tag{19}$$

As seen from (19) there exists an arbitrariness of E_0 parametrized by subscript n which labels gauge transformations and by superscripts (1), (2) which label two solutions of equation (18). Because the values of E_0 have been derived by exploiting gauge invariance we can conclude that the gauge invariance by itself does not uniquely determine the physical relevant value of E_0. To choose physical relevant value of E_0 we exploit the unitarity condition, that is: 1) hermiticity $(\mathrm{i}J^{AB})^\dagger = \mathrm{i}J^{AB}$; 2) the positive norm requirement. For details of resulting procedure we refer to (Metsaev (1995)) and now let us formulate the result.

Given $\mathsf{YT}(\mathbf{h})$ let k, $k = 1 \ldots \nu$, indicates maximal number of upper rows which have the same number of boxes. We call such Young tableaux the level-k $\mathsf{YT}(\mathbf{h})$. For the case of level-k Young tableaux the inequalities (1) can be rewritten as

$$h_1 = \ldots = h_k > h_{k+1} \geq h_{k+2} \geq \ldots \geq h_\nu \geq 0. \tag{20}$$

Then making use of unitarity condition one proves (Metsaev (1995)) that for the level-k Young tableaux the E_0 should satisfy the inequality [1]

$$E_0 \geq h_k - k - 2 + d\,. \tag{21}$$

Comparing (19) with (21) we conclude that only $E^{(1)}_{0(n=k)}$ satisfies the unitarity condition. Thus anti-de Sitter bosonic massless particles described by level-k $\mathsf{YT}(\mathbf{h})$ takes lowest value of energy equal to

$$E_0 = h_k - k - 2 + d\,. \tag{22}$$

Note that it is gauge transformation with $n = k$ (16) that leads to relevant E_0, i.e. given level-k $\mathsf{YT}(\mathbf{h})$ only the gauge transformation $\delta_{(k)}$ respects the unitarity. Therefore only the $\delta_{(k)}$ will be used in what follows. From now

[1] This bound for $d = 4$ has been found by Evans (1967). For the case $d = 5$ see Mack (1977) and references therein. Note that to extend (21) to odd d we should simply replace $h_k \to |h_k|$. In view that for odd d all $h_i \geq 0$ with exception of $h_{(d-1)/2} \neq 0$ our result are valid also for odd d when $h_{(d-1)/2} = 0$. At present time field theoretical description of representations for arbitrary $h_{(d-1)/2}$ is absent.

on we use letter k to indicate level of $\mathsf{YT}(\mathbf{h})$. Thus the final form of gauge transformation is

$$\delta_{(k)}|A\rangle = \sum_{j=0}^{k-1}(-)^j \sum_{l_1,\ldots,l_{j+1}=1}^{k} \delta_{kl_{j+1}} \prod_{i=1}^{j} \frac{\varepsilon^{l_i l_{i+1}} \mathbf{a}_{l_{i+1} l_i}}{k+1-l_i} D_{l_1}|\Lambda_k\rangle \,. \tag{23}$$

As an illustration of (23) we write down $\delta_{(k)}$ for $k = 1, 2, 3$:

$$\delta_{(1)}|A\rangle = D_1|\Lambda_1\rangle \,, \qquad \delta_{(2)}|A\rangle = (D_2 - \frac{1}{2}\mathbf{a}_{21}D_1)|\Lambda_2\rangle \,,$$

$$\delta_{(3)}|A\rangle = (D_3 - \frac{1}{2}\mathbf{a}_{32}D_2 - \frac{1}{3}\mathbf{a}_{31}D_1 + \frac{1}{6}\mathbf{a}_{32}\mathbf{a}_{21}D_1)|\Lambda_3\rangle \,.$$

Note that from the equation for gauge parameter field (17) we get the following lowest energy value for Λ_k: $E_0^\Lambda = E_0 + 1$, i.e.

$$E_0^\Lambda = h_k - k - 1 + d \,. \tag{24}$$

With the values for E_0^Λ at hand we are ready to provide an answer to the question: do the gauge parameter fields meet the masslessness criteria? Because the inter-relation between of spin $\mathbf{h}$ and energy value E_0 for massless field is given by (22) we should express the E_0^Λ in terms of $\mathbf{h}^\Lambda$ and k^Λ, where k^Λ is a level of $\mathsf{YT}(\mathbf{h}^\Lambda)$. Due to relations $k^\Lambda = k-1$, $h^\Lambda_{k^\Lambda} = h_k - \delta_{k1}$ we cast (24) to

$$E_0^\Lambda = h^\Lambda_{k^\Lambda} - k^\Lambda - 2 + d + \delta_{k1} \,.$$

Comparing this relation with (22) we conclude that only for $k > 1$ the gauge parameters are massless fields while for $k = 1$ they are massive fields. Thus we have constructed equations of motion (10) which respect gauge transformations (23), where the gauge parameter fields Λ_k satisfy the constraints (13), (15) and equations of motion (17). The relevant E_0 and E_0^Λ are given by (22) and (24).

All things above have been done in $SO(d-1,2)$ covariant form. Because sometimes a formulation in terms of intrinsic coordinates is preferable let us transform our result to such the coordinates. Let x^μ, $\mu = 0, 1, \ldots, d-1$ be the intrinsic coordinates in AdS spacetime and let $y^A(x)$ be imbedding map, where $y^A(x)$ satisfy (2). The relationship between $SO(d-1,2)$ tensor field $A^{C_1\cdots}$ and the usual tensor field $A^{\mu\cdots}$ is given by

$$A^{\mu_1\cdots}(x) = y^{\mu_1}_{C_1} \ldots A^{C_1\cdots}(y) \,, \qquad y^\mu_C \equiv g^{\mu\nu}\partial_\nu y_C \,,$$

where the intrinsic geometry metric tensor is given by $g_{\mu\nu} = \partial_\mu y^A \partial_\nu y^A$ while its inverse is $g^{\mu\nu} = \nabla^A x^\mu \nabla^A x^\nu$. The $x^\mu = x^\mu(y)$ is a certain representation of intrinsic coordinates. There are useful relations

$$\theta^{AB} = g^{\mu\nu}\partial_\mu y^A \partial_\nu y^B \,, \qquad \partial_\mu y^A \nabla^A x^\nu = \delta^\nu_\mu \,,$$

$$\nabla^A x^\mu = g^{\mu\nu}\partial_\nu y^A \,, \quad \nabla^2 x^\mu = -\Gamma^\mu_{\rho\sigma} g^{\rho\sigma} \,, \quad D_\mu y^A_\nu = g_{\mu\nu} y^A \,.$$

With these relation at hand and with the help of relations

$$y^{\mu_1}_{C_1}\cdots y^{\mu_s}_{C_s}\nabla^2 A^{C_1\ldots C_s} = D^2 A^{\mu_1\ldots\mu_s} + sA^{\mu_1\ldots\mu_s}\,, \qquad D^2 A^{C_1\ldots} = \nabla^2 A^{C_1\ldots}$$

where $D^2 \equiv D_\mu D^\mu$, $D_\mu = \partial_\mu + \Gamma^{\cdot}_{\mu\cdot}$, we can immediately transform equation of motion (10) to the desired form

$$(D^2 - (h_k - k - 1)(h_k - k - 2 + d) + \sum_{l=1}^{\nu} h_l)A^{\mu_1\cdots} = 0\,. \qquad (25)$$

It turns out that in order to write gauge transformation it is convenient to transform spacetime tensors into the tangent space tensors $A^{a_1\cdots} \equiv e^{a_1}_{\mu_1}\ldots A^{\mu_1\cdots}$ where the e^a_μ is a einbein of AdS geometry, introduce new creation and annihilation operators a^a_l and $\overline{a}^a_l$, $l = 1,\ldots\nu$, $a = 0,1,\ldots,d-1$, and construct Fock space vector

$$|a\rangle = a^{a_1}\ldots A^{a_1\cdots}|0\rangle\,, \qquad [\overline{a}^a_i, a^b_j] = \delta_{ij}\eta^{ab}, \qquad \eta^{ab} = (-,+,\ldots,+)\,.$$

Now the equation, constraints and gauge transformation take the form

$$(D^2_L - (h_k - k - 1)(h_k - k - 2 + d) + \sum_{l=1}^{\nu} h_l)|a\rangle = 0\,, \qquad (26)$$

$$\delta_{(k)}|a\rangle = \sum_{j=0}^{k-1}(-)^j \sum_{l_1,\ldots,l_{j+1}=1}^{k} \delta_{kl_{j+1}} \prod_{i=1}^{j} \frac{\varepsilon^{l_i l_{i+1}}(a_{l_{i+1}}\overline{a}_{l_i})}{k+1-l_i} a^b_{l_1} e^\mu_b D_{\mu L}|\lambda_k\rangle\,,$$

$$\overline{a}^b_i e^\mu_b D_{\mu L}|a\rangle = 0\,, \quad (a^b_i\overline{a}^b_i - h_i)|a\rangle = 0\,, \quad \overline{a}^b_i\overline{a}^b_i|a\rangle = 0\,, \quad \varepsilon^{ij}a^b_i\overline{a}^b_j|a\rangle = 0\,, \quad (27)$$

$$D_{\mu L} \equiv \partial_\mu + \frac{1}{2}\omega^{ab}_\mu M^{ab}\,, \qquad M^{ab} \equiv \sum_{l=1}^{\nu}(a^a_l\overline{a}^b_l - a^b_l\overline{a}^a_l)\,.$$

The ω^{ab}_μ is a Lorentz connection of AdS spacetime. The equation and constraints for the gauge parameter field λ_k are obtainable from (26) and (27) by making there the substitutions $|a\rangle \to |\lambda_k\rangle$, $h_i \to h_{i(k)}$ and $k \to k-2$. In order to demonstrate how our results are working let us consider some particular cases.

Totally antisymmetric fields. In this case $\mathbf{h} = (1,\ldots,1,\ldots,0)$ where unit occurs s-times in this sequence. Therefore we have $k = s$ and $h_k = \varepsilon^{ks+1}$. For this case $E_0 = d - 1 - s$ and we get the equations

$$(D^2 + s(d-s))A^{\mu_1\ldots\mu_s} = 0\,,$$

where the relevant constraint is $D_\mu A^{\mu\mu_2\ldots\mu_s} = 0$. By making use of this gauge the equations above can be easily derived from well known equations

$$D_\mu F^{\mu\mu_1\ldots\mu_s} = 0\,, \qquad F_{\mu_1\ldots\mu_n} \equiv n\partial_{[\mu_1}A_{\mu_2\ldots\mu_n]}\,.$$

Totally symmetric fields. In this case $\mathbf{h} = (s, 0 \ldots, 0)$. Therefore we have $k = 1$ and $h_k = s\varepsilon^{k2}$, the E_0 is given by $E_0 = s + d - 3$ and we get the equations

$$(D^2 - s^2 + (6-d)s + 2d - 6)A^{\mu_1 \ldots \mu_s} = 0\,, \tag{28}$$

where the relevant constraints are

$$A_\mu{}^{\mu\mu_3 \ldots \mu_s} = 0\,, \qquad D_\mu A^{\mu\mu_2 \ldots \mu_s} = 0\,. \tag{29}$$

For $d = 4$ these equations can be obtained from those discovered in (Fronsdal (1978)) by making use of the gauge (29). The *graviton* is a particular case when $s = 2$. From (28) we get the equation $(D^2 + 2)h_{\mu\nu} = 0$ which should be supplemented by constraints like (29). This equation coincides with that obtained from Einstein equation for excitation of metric tensor

$$R_{\mu\nu} = -(d-1)G_{\mu\nu}\,, \quad G_{\mu\nu} = g_{\mu\nu} + h_{\mu\nu}\,, \quad R_{\mu\nu\rho\sigma}(g) = -(g_{\mu\rho}g_{\nu\sigma} - g_{\mu\sigma}g_{\nu\rho})$$

where $g_{\mu\nu}$ is a metric tensor of AdS geometry. Note that the value $s = 2$ is the only when the dependence on d in (28) is cancelled. Thus we have demonstrated that our results cover all previously known particular cases and solve problem for arbitrary spin $\mathbf{h}$ massless bosonic fields in d - dimensional AdS spacetime.

In conclusion let us discuss masslessness in d - dimensional AdS spacetime by using the requirement of conformal invariance. By conformal invariant representations we will understand those irreducible representations of anti-de Sitter group that can be realized as irreducible representations of the conformal group $SO(d,2)$. It turns out (for details see Metsaev (1995))) that this requirement leads to representations whose h_i satisy the constaints $h_1 = \ldots = h_\nu \equiv h$, while $E_0 = h + \nu$. These E_0 and h_i are in accordance with (22), i.e. conformal invariance respects the gauge invariance, but because of constraints above-mentioned the conformal representations constitute only a subset of all massless states for $d > 4$, i.e. conformal group for $d > 4$ cannot be used for defining all massless representations. In this respects the situation in AdS spacetime (Metsaev (1995)) is similar to that in Minkowski spacetime (Siegel (1989)).

Acknowledgements. This work was supported in part by the RFBR, Grant 96-02-17314a and by RFBR Grant for Leading Scientific Schools N 96-15-96463.

References

Vasiliev, M. A. (1990): Phys. Lett. B **243**, 378
Vasiliev, M. A. (1991): Phys. Lett. B**257**, 111
Fronsdal, C. (1978): Phys. Rev. D **12**, 3624
Fang, J., Fronsdal, C. (1978): Phys. Rev. D**18**, 3630
Lopatin, V. E., Vasiliev, M. A. (1988): Mod. Phys. Lett. A**3**, 257

Vasiliev, M. A. (1988): Nucl. Phys. **B301**, 26
Evans, N. T. (1967): J. Math. Phys. **8**, 170
H. Nicolai, (1984): Representations of supersymmetry in anti-de Sitter space, in: Supersymmetry and supergravity, '84, Ed. B. de Wit, P. Fayet and P. van Nieuwenhuizen (World Scientific, Singapore, 1984)
Mack, G. (1977): Comm. Math. Phys. **55**, 1
Metsaev, R. R. (1994): Class. Quant. Grav. **11**, L141
Metsaev, R. R. (1995): Phys. Lett. **B354**, 78
Metsaev, R. R. (1997): Class. Quant. Grav **14**, L115
Metsaev, R. R. (1995): Mod. Phys. Lett. **A10**, 1719
Siegel, W. (1989): Int. J. Mod. Phys. **A4**, 2015

Direct Mode Summation for the Casimir Energy of Spherical Shell and Compact Ball

V.V. Nesterenko and I.G. Pirozhenko

Joint Institute for Nuclear Research, Dubna, 141980, Russia

Abstract. A simple method for calculating the Casimir energy for sphere and compact ball is developed which is based on a direct mode summation by means of contour integration in a complex plane of eigenfrequencies.

1 Introduction

The Casimir effect can be generally defined as an influence of the boundness of the configuration space on the physical characteristics of the quantum field system.

When considering the Casimir effect different methods are used: stress-tensor method, the Green's function formalism, multiple scattering expansion, zeta regularization technique, heat-kernel series (Milton et al. (1978)). In all the approaches to calculation of the Casimir effect a vague point is the procedure of unique separation and subsequent removal of the divergences. The lack of universal mathematically rigorous prescription for this purpose leads in some problems to different results when different methods are applied.

With allowance for all this, the most simple, from mathematical point of view, method of direct mode summation(Nesterenko and Pirozhenko (1997)) has an obvious advantage because it right away allows one to reveal the difficulties generated by divergences. The main goal of this paper is to show the simplicity and efficiency of this method when calculating the Casimir energy for perfectly conducting and infinitely thin spherical shell and for a solid ball placed in an infinite medium. This approach is completely based on using the classical frequencies of quantum field system concerned, and the main tool employed is the Cauchy theorem from complex analysis.

2 Perfectly Conducting Spherical Shell

The starting point of our approach is the following definition of the Casimir energy

$$E = \frac{1}{2}\sum_{s}(\omega_s - \overline{\omega}_s)\,. \tag{1}$$

Here ω_s are the eigenfrequencies of the system under consideration, and $\overline{\omega}_s$ are those of the same system, when the parameters determining its boundaries take on some limiting values. There are two modes of oscillations of the electromagnetic field inside and outside the perfectly conducting sphere with radius a : transverse-electric modes and transverse magnetic ones (TE-modes and TM-modes, respectively). The eigenfrequencies of the TE-modes are defined by the equations (Stratton (1941))

$$j_l(\omega a) = 0, \quad h_l^{(1)}(\omega a) = 0, \tag{2}$$

and the eigenfrequencies for the TM-modes are given by

$$\frac{d}{dr}\left[r j_l(\omega r)\right]\Big|_{r=a} = 0, \quad \frac{d}{dr}\left[r h_l^{(1)}(\omega r)\right]\Big|_{r=a} = 0. \tag{3}$$

In formulae (2) and (3) $j_l(z)$ and $h_l^{(1)}(z)$ are the spherical Bessel functions (Abramowitz and Stegun (1964))

$$j_l(z) = \sqrt{\frac{\pi}{2z}} J_{l+1/2}(z), \quad h_l^{(1)}(z) = \sqrt{\frac{\pi}{2z}} H_{l+1/2}^{(1)}(z), \tag{4}$$

and $l = 1, 2, \ldots$. Only positive roots of these equations $\omega_{nl} > 0$, $n = 1, 2, \ldots$ should be considered. The first (second) equations (2) and (3) specify the frequencies of the electromagnetic oscillations inside (outside) the sphere (Stratton (1941)).

In the case of spherical boundary the sum $\sum_s$ in (1) can be written as

$$\frac{1}{2}\sum_s \omega_s = \frac{1}{2}\sum_{l=1}^{\infty}\sum_{m=-l}^{l}\sum_{n=1}^{\infty}\omega_{nl} = \sum_{l=1}^{\infty}(l+1/2)S_l, \tag{5}$$

where $S_l = \sum_{n=1}^{\infty}\omega_{nl}$, and each frequency equation (2)–(3) generates its own partial sum S_l^{α}, $\alpha = 1, \ldots, 4$, so that $S_l = \sum_{\alpha=1}^{4} S_l^{(\alpha)}$.

For the partial sums $S_l^{(\alpha)}$ we use integral representation that follows from the Cauchy theorem (Abramowitz and Stegun (1964))

$$S_l^{(\alpha)} = \frac{1}{2\pi i}\oint_C dz\, z \frac{d}{dz}\ln f^{(\alpha)}(z, a). \tag{6}$$

Here $f^{(\alpha)}(z, a)$ are the functions defining the frequency equations (2), (3) in the form $f^{(\alpha)}(\omega, a) = 0$, $\alpha = 1, 2, 3, 4$. The contour C encloses counterclockwise positive roots of these equations and consists of the segment $[-i\Lambda, \ i\Lambda]$ of the imaginary axis and a semicircle of radius Λ with $\Lambda \to \infty$ in the right half-plane. When Λ is fixed, the contour integral (6) gives the regularized value of corresponding frequency sum. For negative values of the argument ω the functions $f^{(\alpha)}(\omega, a)$ have to be defined by a condition $f^{(\alpha)}(-\omega, a) = f^{(\alpha)}(\omega, a)$, $\omega > 0$.

In accordance with the definition (1) it is necessary to perform the subtraction in order to obtain a finite (observable) value of the Casimir energy. As usual, we shall subtract the contribution of the Minkowski space that corresponds to the limit $a = \infty$ in Eq. (6). Letting $\overline{S}_l^{(\alpha)}$ represents the value of the partial sum $S_l^{(\alpha)}$ which is to be subtracted from (6) we get

$$S_l^{(\alpha)} - \overline{S}_l^{(\alpha)} = \frac{1}{\pi} \int_0^\infty dy \ln \left[\frac{f^{(\alpha)}(iy, a)}{f^{(\alpha)}(iy, a \to \infty)} \right]. \tag{7}$$

Integration along the semicircle of radius Λ does not give contribution into the difference (7) when $\Lambda \to \infty$. Now we proceed to substituting into Eq. (7) the concrete expressions for the functions $f^{(\alpha)}$ defined by frequency equations (2), (3). ¿From Eq. (2) we obtain

$$\frac{f^{(1)}(iy, a)}{f^{(1)}(iy, a \to \infty)} = \frac{J_\nu(iya)}{\lim\limits_{a \to \infty} J_\nu(iya)} = \frac{I_\nu(ay)}{\lim\limits_{a \to \infty} I_\nu(ay)} = \sqrt{2\pi ay}\, e^{-ay} I_\nu(ay), \tag{8}$$

$$\frac{f^{(2)}(iy, a)}{f^{(2)}(iy, a \to \infty)} = \frac{H_\nu^{(1)}(iay)}{\lim\limits_{a \to \infty} H_\nu^{(1)}(iay)} = \frac{K_\nu(ay)}{\lim\limits_{a \to \infty} K_\nu(ay)} = \sqrt{\frac{2ay}{\pi}} e^{ay} K_\nu(ay). \tag{9}$$

$I_\nu(z)$ is the modified Bessel function $J_\nu(iz) = i^\nu I_\nu(z)$ and $H_\nu^{(1)} = J_\nu(z) + iN_\nu(z)$ is the Hankel function of the first kind. We have used here the asymptotics of the functions $I_\nu(z)$ and $K_\nu(z)$ for fixed value of ν and large z (Abramowitz and Stegun (1964)).

In the same way we deduce from the frequency equation (3)

$$\frac{f^{(3)}(iy, a)}{f^{(3)}(iy, a \to \infty)} = \frac{J_\nu/2 + iyaJ_\nu'}{\lim\limits_{a \to \infty} [J_\nu/2 + iyaJ_\nu']} = \sqrt{\frac{2\pi}{ay}} e^{-ay} [I_\nu/2 + ay\, I_\nu'], \tag{10}$$

$$\frac{f^{(4)}(iy, a)}{f^{(4)}(iy, a \to \infty)} = \frac{K_\nu/2 + ayK_\nu'}{\lim\limits_{a \to \infty} [K_\nu/2 + ayK_\nu']} = -\sqrt{\frac{2}{\pi ya}} e^{ay} [K_\nu/2 + ayK_\nu'] \tag{11}$$

The prime over the Bessel functions $I_\nu(ay)$ and $K_\nu(ay)$ means the differentiation with respect to their arguments.

Finally summing up the contributions of the TE- and TM-modes to the Casimir energy we obtain from (1), (5), (7)–(11)

$$E = \frac{1}{\pi a} \sum_{l=1}^\infty \left(l + \frac{1}{2} \right) \int_0^\infty dy \ln \left[1 - (\sigma_l'(y))^2 \right] \equiv \frac{1}{a} \sum_{l=1}^\infty Q_l, \tag{12}$$

where the notation $\sigma_l(y) = yI_\nu(y)K_\nu(y), \nu = l + 1/2$, is introduced. The integral in (12) converges. This follows from the asymptotics of $\sigma_l'(y)$ for large y and fixed $\nu = l + 1/2$ (Abramowitz and Stegun (1964)). To carry out the summation with respect to l in (12) one needs the behavior of the integral Q_l at large l. Applying the uniform with respect to z asymptotics for the modified Bessel functions at large ν (Milton et al. (1978), Abramowitz and Stegun (1964))

$$I_\nu(\nu z)K_\nu(\nu z) \simeq \frac{1}{2\nu}\frac{1}{(1+z^2)^{1/2}}, \tag{13}$$

we obtain from (12)

$$Q_l \simeq \frac{\nu^2}{\pi}\int_0^\infty dz \ln\left[1-\frac{1}{4\nu^2(1-z^2)^3}\right] \simeq -\frac{1}{4\pi}\int_0^\infty \frac{dz}{(1+z^2)^3} = -\frac{3}{64}, \quad l\to\infty. \tag{14}$$

Thus, the sum (12) at large l diverges as $\sum_{l=1}^{\infty}(l+1/2)^0$. To determine the finite value for this sum we rewrite (12) in the following way

$$E = \frac{1}{a}\sum_{l=1}^{\infty}\left[Q_l+\frac{3}{64}-\frac{3}{64}\right] = \frac{1}{a}\sum_{l=1}^{\infty}\overline{Q}_l - \frac{3}{64a}\sum_{l=1}^{\infty}\left(l+\frac{1}{2}\right)^0, \tag{15}$$

where $\overline{Q}_l = Q_l + 3/64$.

The sum $\sum_{l=1}^{\infty}\overline{Q}_l$ converges because $\overline{Q}_l = -9/(16384\nu^2) + \mathcal{O}(\nu^{-4})$ at large l. The last divergent sum in (15) can be defined by using the Hurwitz zeta function $\zeta(z,q)$ (Abramowitz and Stegun (1964))

$$-\frac{3}{64a}\sum_{l=1}^{\infty}\left(l+\frac{1}{2}\right)^0 = -\frac{3}{64a}\left(\zeta(0,1/2)-1\right) = \frac{3}{64a} \tag{16}$$

since $\zeta(0,1/2)=0$.

Finally we obtain

$$E = \frac{1}{a}\sum_{l=1}^{\infty}\overline{Q}_l + \frac{3}{64a}, \tag{17}$$

The sum $\sum_l \overline{Q}_l$ can be estimated with allowance for the asymptotics of $\overline{Q}_l$ at large l

$$\sum_{l=1}^{\infty}\overline{Q}_l \simeq -\frac{9}{16384}\sum_{l=1}^{\infty}\frac{1}{(l+1/2)^2} = -\frac{9}{2^{14}}[\zeta(2,1/2)-4] = -0.000514\ldots. \tag{18}$$

Thus, the main contribution to (17) is given by the second term and to a good approximation one can put for the Casimir energy $E \simeq 3/(64a) = 0.046875/a$. Taking into account (18) we get with greater accuracy $E \simeq 0.046361/a$. It is worth comparing our calculations with those using other methods (Milton et al. (1978)).

3 Solid Ball in an Infinite Medium When $\varepsilon\mu = 1$

Let us consider the Casimir effect for a ball made of the material with dielectric constant ε_1 and magnetic constant μ_1. The ball is assumed to be surrounded by an infinite medium with dielectric and magnetic constants ε_1 and μ_2, respectively.

The central role in our consideration is again played by the equations defining the eigenfrequencies of the electromagnetic oscillations (Stratton (1941)). The TE-modes are determined by the equation

$$\frac{[\omega_1 r j_l(\omega_1 r)]'}{\mu_1 j_l(\omega_1 r)} = \frac{\left[\omega_2 r h_l^{(1)}(\omega_2 r)\right]'}{\mu_2 h_l^{(1)}(\omega_2 r)}, \quad r = a, \tag{19}$$

where $\omega_i = \sqrt{\varepsilon_i \mu_i}\omega$, $i = 1, 2$. The prime means the differentiation with respect to the arguments of the Bessel functions $\omega_i r$, $i = 1, 2$. For the TM-oscillations we have analogous frequency equations

$$\frac{[\omega_1 r j_l(\omega_1 r)]'}{\varepsilon_1 j_l(\omega_1 r)} = \frac{\left[\omega_2 r h_l^{(1)}(\omega_2 r)\right]'}{\varepsilon_2 h_l^{(1)}(\omega_2 r)}, \quad r = a. \tag{20}$$

The parameter l in Eqs. (19) and (20) takes the values 1,2,.... Under the exchange $\varepsilon_i \leftrightarrow \mu_i$ Eq. (19)turns into Eq. (20) and vice versa.

If the characteristics of the mediums ε_i, μ_i, $i = 1, 2$ satisfy the condition $\varepsilon_i \mu_i = 1$, $i = 1, 2$, the frequency equations are considerably simplified. With allowance for this condition Eqs. (19) and (20) can be written as follows

$$[\omega a j_l(\omega a)]' h_l^{(1)}(\omega a) - \mu^{\pm 1} j_l(\omega a)[\omega a h^{(1)}(\omega a)]' = 0, \tag{21}$$

where $\mu = \mu_1/\mu_2$. Following the calculations in the previous section, we arrive at the formula for the Casimir energy of a solid ball

$$E_{ball} = \frac{1}{\pi a}\sum_{l=1}^{\infty}(l+1/2)\int_0^{\infty} dx \ln \frac{\xi_\nu^+(x)\xi_\nu^-(x)}{\xi_\nu^-(\infty)\xi_\nu^+(\infty)}, \quad \nu = l + 1/2, \tag{22}$$

where

$$\xi_\mu^{\pm}(x) = \left(x I_\nu'(x) + \frac{1}{2} I_\nu(x)\right) K_\nu(x) - \mu^{\pm 1} I_\nu(x)\left(x K_\nu'(x) + \frac{1}{2} K_\nu(x)\right). \tag{23}$$

Using the asymptotics of the modified Bessel functions $I_\nu(x)$ and $K_\nu(x)$ for large x and fixed ν we obtain $\xi^{\pm}(\infty) = (1 + \mu^{\pm 1})/2$. In view of this Eq. (22) assumes the form

$$E_{ball} = \frac{1}{\pi a}\sum_{l=1}^{\infty}\left(l + \frac{1}{2}\right) Q_l, \text{ where } Q_l = \frac{1}{\pi}\int_0^{\infty} dx \ln\left[\frac{4\mu}{(1+\mu)^2}\xi_\nu^+(x)\xi_\nu^-(x)\right]. \tag{24}$$

The formula for Q_l can be rewritten in a compact way

$$Q_l = \frac{1}{\pi}\int_0^\infty dx \, \ln\left[1 - (\eta\sigma_l'(x))^2\right], \; \eta = (1-\mu)/(1+\mu) \tag{25}$$

with the same function $\sigma_l(x)$ as in Eq. (12). Thus the only difference from the case of perfectly conducting sphere is the multiplier η in front of $\sigma_l(x)$. We find the sum over l in (24) again by making use of the Hurwitz zeta function technique. It gives

$$E_{ball} = \frac{1}{a}\sum_{l=1}^{\infty}\overline{Q}_l + \frac{3}{64a}\eta^2, \text{ where } \overline{Q}_l = Q_l + 3\eta^2/64\ldots. \tag{26}$$

Applying the uniform over x asymptotics of the Bessel functions $I_\nu(x)$ and $K_\nu(x)$ at large ν (Abramowitz and Stegun (1964)) we obtain

$$\overline{Q}_l \simeq \frac{9}{16384}\frac{\eta^2}{\nu^2}(6-7\eta^2) + \mathcal{O}(\nu^{-4}). \tag{27}$$

The sum in (26) converges. It can be estimated with the use of the asymptotics (27)

$$\sum_{l=1}^{\infty}\overline{Q}_l \simeq \frac{9}{16384}\eta^2(6-7\eta^2)\sum_{l=1}^{\infty}\frac{1}{(l+1/2)^2} = \frac{9}{2^{14}}\eta^2(6-7\eta^2)\left(\frac{\pi^2}{2}-4\right). \tag{28}$$

Thus the basic contribution into Eq. (26) is due to the second term. Therefore with a fairly good accuracy (a few percents) one can put $E_{ball} \simeq 3\eta^2/(64a)$. As it was shown in the previous section, with the same accuracy we have $E_{shell} = 3/(64a)$, therefore $E_{ball} \simeq \eta^2 E_{shell}$. Taking into account (28) we obtain a more precise formula for the Casimir energy of a solid ball $E_{ball} \simeq 3\eta^2(1.066 - 0.077\eta^2)/(64a)$ (cf. with (Brevik and Kolbenstvedt (1982))).

4 Conclusion

The calculation of the Casimir effect for nonflat boundaries (specifically for sphere) by a direct summation of eigenfrequencies has been used only in pioneer paper by Boyer (Boyer (1968)). The fact that done by us is actually a development and maximum simplification of the Boyer method and bringing it to such a form when numerical calculations are practically not required, and, what is more important, cut-off functions are not used.

This work was accomplished with financial support of Russian Foundation of Fundamental Research (grant 97-01-00745).

References

Milton, K. A., DeRaad Jr., L. L., and Schwinger, J. (1978): Ann. Phys. (N.Y.), vol. 115, 338–403; Balian, R., and Duplantier, B. (1978): Ann. Phys. (N.Y.), vol. 112, 165–208; Leseduarte, S., and Romeo, A. (1996): Ann. Phys. (N.Y.), vol. 250, 448–484; Bordag, M., Elizalde, E., and Kirsten, K. (1996): J. Math. Phys. (N.Y.), vol. 37, 895–916

Nesterenko, V. V., and Pirozhenko, I. G. (1997): J. Math. Phys., vol. 38, 6265–6280

Stratton, J. A. (1941): *Electromagnetic Theory* (New York: McGraw-Hill)

Abramowitz, M., and Stegun, I. A. (1964): *Handbook of Mathematical Functions* (National Bureau of Standards, Washington, D. C..)

Boyer, T. H. (1968): Phys. Rev., vol. 174, 1764–1776

Brevik, I., and Kolbenstvedt, H. (1982): Ann. Phys. (N. Y.), vol. 143, 179–190

On Different BRST Constructions for a Given Lie Algebra

A. Pashnev[1] and M. Tsulaia[2]

[1] JINR–Bogoliubov Laboratory of Theoretical Physics, 141980 Dubna, Moscow Region, Russia
E-mail:pashnev@thsun1.jinr.ru

[2] JINR–Bogoliubov Laboratory of Theoretical Physics, 141980 Dubna, Moscow Region, Russia
E-mail: tsulaia@thsun1.jinr.ru

Abstract. The method of the BRST quantization is considered for the system of constraints, which form a Lie algebra. When some of the Cartan generators do not imply any conditions on the physical states, the system contains the first and the second class constraints. After the introduction auxiliary bosonic degrees of freedom for these cases, the corresponding BRST charges with the nontrivial structure of nonlinear terms in ghosts are constructed.

1 Introduction

The BRST quantization procedure for a system of the first class constraints is straightforward. By the definition, the first class constraints form a closed algebra with respect to the commutators (the Poisson brackets). For simplicity we consider only linear algebras – Lie algebras of constraints.

More general systems include the second class constraints as well, whose commutators contain terms which are nonzero on mass shell (on the subspace where all constraints vanish). In the simplest cases these terms are a numbers or central charges, but sometimes, they are operators which act nontrivially on the space of the physical states. Moreover, the commutators between these operators and the constraints can be nontrivial. In some cases the total system of the constraints and the operators mentioned above form a Lie algebra.

So, in such cases we have a system of operators which form a Lie algebra, but the physical meaning of different operators is different. Some of them play the role of constraints and annihilate the physical states, others are nonzero and simply transform the physical states into other ones. It means, that in the BRST approach for the description of the corresponding physical system we can not use the standard BRST charge for the given Lie algebra. Instead, we have to construct the nilpotent BRST charge in a manner, that some of the operators play the role of the first class constraints, others are second class constraints and the others do not imply any conditions on the physical space of the system.

In this paper we demonstrate the possibility of a different BRST constructions for the system of generators, which form a given Lie algebra and have different physical meaning. In the second part we describe some algebraic approaches, leading to the description of massless (or massive) irreducible representations of the Poincare group in any dimensions. As the simplest example we show, that the same algebra of operators leads to either massless, or massive spectra in the cases when physical meaning of some operators is different. In the third part we discuss the general method of the BRST quantization, when some of the Cartan generators are excluded from the total system of constraints. In the fourth part we describe the construction of auxiliary representations of the algebra by means of the Gel'fand–Tsetlin method. In Sect.5 we give the simple example.

2   The Description of Constraints

In order to kill ghosts the field theoretical lagrangians, describing irreducible Poincare representations must possess some gauge invariance. Along with the basic fields such lagrangians in general include additional ones. The role of these fields is to single out the irreducible representation of the Poincare group. Some of them are auxiliary, others can be gauged away. After a gauge fixing and solving the equations of motion for auxiliary fields one is left with the only essential field, describing the irreducible representation of the Poincare group. This field corresponds to the Young tableaux with k rows

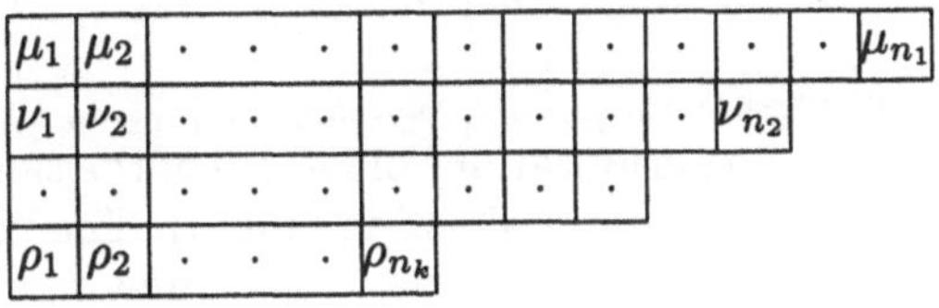

$$
\begin{array}{|c|c|c|c|c|c|c|c|c|c|c|c|}
\hline
\mu_1 & \mu_2 & \cdot & & \cdot & \cdot & \cdot & \cdot & \cdot & \cdot & \cdot & \mu_{n_1} \\
\hline
\nu_1 & \nu_2 & \cdot & & \cdot & \cdot & \cdot & \cdot & \cdot & \nu_{n_2} \\
\cline{1-9}
\cdot & \cdot & \cdot & & \cdot & \cdot & \cdot & \cdot \\
\cline{1-8}
\rho_1 & \rho_2 & \cdot & & \cdot & \rho_{n_k} \\
\cline{1-5}
\end{array}
\tag{2.1}
$$

and is described by $\Phi^{(k)}_{\mu_1\mu_2\cdots\mu_{n_1},\nu_1\nu_2\cdots\nu_{n_2},\cdots,\rho_1\rho_2\cdots\rho_{n_k}}(x)$ which is the $n_1 + n_2 + \cdots + n_k$ rank tensor field symmetrical with respect to the permutations of each type of indices. In addition, this field is subject to the following system of equations, namely the mass shell and transversality conditions for each type of indices. In the massless case we have

$$
p^2_\mu \Phi^{(k)}_{\mu_1\mu_2\cdots\mu_{n_1},\nu_1\nu_2\cdots\nu_{n_2},\cdots,\rho_1\rho_2\cdots\rho_{n_k}}(x) = 0,
\tag{2.2}
$$

$$
p_\mu \Phi^{(k)}_{\mu\mu_2\cdots\mu_{n_1},\nu_1\nu_2\cdots\nu_{n_2},\cdots,\rho_1\rho_2\cdots\rho_{n_k}}(x) = 0,
\tag{2.3}
$$

$$
\cdots\cdots\cdots\cdots\cdots\cdots\cdots\cdots\cdots\cdots
$$

$$
p_\rho \Phi^{(k)}_{\mu_1\mu_2\cdots\mu_{n_1},\nu_1\nu_2\cdots\nu_{n_2},\cdots,\rho\rho_2\cdots\rho_{n_k}}(x) = 0.
\tag{2.4}
$$

Further, all traces of the basic field must vanish:

$$\Phi^{(k)}_{\mu\mu\mu_3\cdots\mu_{n_1},\nu_1\nu_2\cdots\nu_{n_2},\cdots,\rho_1\rho_2\cdots\rho_{n_k}}(x) = 0, \tag{2.5}$$

$$\Phi^{(k)}_{\mu\mu_2\mu_3\cdots\mu_{n_1},\mu\nu_2\cdots\nu_{n_2},\cdots,\rho_1\rho_2\cdots\rho_{n_k}}(x) = 0, \tag{2.6}$$

....

$$\Phi^{(k)}_{\mu_1\mu_2\cdots\mu_{n_1},\nu_1\nu_2\cdots\nu_{n_2},\cdots,\rho\rho\rho_3\cdots\rho_{n_k}}(x) = 0. \tag{2.7}$$

The correspondence with a given Young tableaux implies, that after symmetrization of all indices of one type with one index of another type, the basic field vanishes, for example

$$\Phi^{(k)}_{\{\mu_1\mu_2\cdots\mu_{n_1},\nu_1\}\nu_2\cdots\nu_{n_2},\cdots,\rho_1\rho_2\cdots\rho_{n_k}}(x) = 0. \tag{2.8}$$

To describe all irreducible representations of the Poincare group simultaneously it is convenient to introduce an auxiliary Fock space generated by the creation and annihilation operators a^{i+}_μ, a^j_μ with Lorentz index $\mu = 0, 1, 2, ..., D-1$ and additional internal index $i = 1, 2, ..., k$. These operators satisfy the following commutation relations

$$\left[a^i_\mu, a^{j+}_\nu\right] = -g_{\mu\nu}\delta^{ij}, \; g_{\mu\nu} = diag(1, -1, -1, ..., -1), \tag{2.9}$$

where δ^{ij} is usual Cronecker symbol.

The general state of the Fock space depends on the space-time coordinates x_μ

$$|\Phi\rangle = \sum \Phi^{(k)}_{\mu_1\mu_2\cdots\mu_{n_1},\nu_1\nu_2\cdots\nu_{n_2},\cdots,\rho_1\rho_2\cdots\rho_{n_k}}(x) \times \tag{2.10}$$
$$a^{1+}_{\mu_1}a^{1+}_{\mu_2}\cdots a^{1+}_{\mu_{n_1}}a^{2+}_{\nu_1}a^{2+}_{\nu_2}\cdots a^{2+}_{\nu_{n_2}}\cdots a^{k+}_{\rho_1}a^{k+}_{\rho_2}\cdots a^{k+}_{\rho_{n_k}}|0\rangle$$

and the components $\Phi^{(k)}_{\mu_1\mu_2\cdots\mu_{n_1},\nu_1\nu_2\cdots\nu_{n_2},\cdots,\rho_1\rho_2\cdots\rho_{n_k}}(x)$ are automatically symmetrical under the permutations of indices of the same type (Ouvry and Stern (1986), Labastida (1987), Labastida (1989)). The norm of states in this Fock space is not positively definite due to the minus sign in the commutation relation (2.1) for the time components of the creation and annihilation operators. The transversality conditions (2.3)-(2.4) for the components are equivalent to the following constraints on the physical vectors of the Fock space

$$L^i|\Phi\rangle = 0, \tag{2.11}$$

where

$$L^i = a^i_\mu p_\mu. \tag{2.12}$$

These operators along with their conjugates

$$L^{i+} = a^{i+}_\mu p_\mu. \tag{2.13}$$

and mass shell operator p^2_μ form the following algebra with only nonvanishing commutator

$$[L^i, L^{j+}] = -p^2_\mu\delta^{ij}, \tag{2.14}$$

This simple algebra was considered in (Ouvry and Stern (1986)) in the framework of the BRST approach. The constraints are of the first class and nilpotent BRST charge can be constructed without problems. As a result the description of mixed symmetry fields was obtained. However, all these fields describe the reducible representations of the Poincare group due to the absence of additional conditions (2.5)-(2.7) and (2.8) in the initial system of the constraints.

On the other hand, the same algebra of operators arises in the case of massive particles. The only difference is that the right hand side of the relation (2.14) is now nonvanishing operator. Instead, this operator can have different eigenvalues $p_\mu^2 = m_n^2$ for the different physical states. This situation was analyzed in (Pashnev and Tsulaia (1997)), where corresponding BRST charge, obviously different from the one for the massless case, was constructed using the method of dimensional reduction. As an artifact of this method the construction automatically includes some additional auxiliary variables. So, (2.14) produces the nontrivial example for the different BRST constructions, corresponding to different physical meaning of the generators (p_μ^2 in our case). In what follows we describe the method of alternative constructions of BRST charges which is valid not only for a simple algebras like (2.14).

The tracelessness conditions (2.5)-(2.7) correspond in the Fock space to the constraints

$$L^{ij}|\Phi\rangle = 0, \tag{2.15}$$

with

$$L^{ij} = a^i_\mu a^j_\mu, \qquad L^{ij+} = a^{j+}_\mu a^{i+}_\mu . \tag{2.16}$$

while the symmetry properties (2.8) follow from the constraints

$$T^{ij}|\Phi\rangle = 0, \qquad i < j, \tag{2.17}$$

having the explicit form

$$T^{ij} = a^{i+}_\mu a^j_\mu, \quad T^{ij+} = a^{j+}_\mu a^i_\mu = T_{ji}. \tag{2.18}$$

The operators L^{ij}, L^{ij+} (i, j are arbitrary) and T^{ij}, ($i \neq j$), along with the additional operators

$$H^i = -T^{ii} + \frac{D}{2} = -a^{i+}_\mu a^i_\mu + \frac{D}{2}, \tag{2.19}$$

form the Lie algebra $SO(k+1,k)$. The rank of this algebra is k and corresponding Cartan subalgebra contains all operators H^i. One can choose the operators L^{11} and $T^{i,i+1}$ as k simple roots. The positive and negative roots are, correspondingly, L^{ij}, T^{rs}, $(1 \leq r < s \leq k)$ and L^{ij+}, T^{rs}, $(1 \leq s < r \leq k)$. It means, that the conditions (2.5)-(2.7) and (2.8) are equivalent to annihilation of physical states in the total Fock space by the positive roots of the Lie algebra $SO(k+1,k)$.

As it can be easily seen, the Cartan generators (2.19) are strictly positive in the Fock space and therefore the standard BRST charge has to be modified for the given realization of the $SO(k+1,k)$ algebra.

The BRST approach to the construction of the lagrangians, from which all the equations (2.2)-(2.8) follow, is very powerful. It automatically leads to appearance of all auxiliary fields in the lagrangian. In the massless case the BRST charge for the system of only first class constraints, corresponding to the equations (2.2) - (2.4) was constructed in (Ouvry and Stern (1986)). The methods of such construction were discussed in (Faddeev and Shatashvili (1986), Batalin and Fradkin (1987), Egoryan and Manvelyan (1993)). With the help of additional variables one can modify the second class constraints in such a way that they become commuting, i.e. the first class. At the same time the number of physical degrees of freedom for both systems does not change if the number of additional variables coincides with the number of second class constraints.

On the other hand, the BRST charge for the second class constraints in some cases can be constructed using the method of dimensional reduction. In (Pashnev and Tsulaia (1997)) the system of massive higher spins satisfying equations

$$(p_\mu^2 - m^2)\Phi_{\mu_1\mu_2\cdots\mu_{n_1}}(x) = 0 \qquad (2.20)$$

and

$$p_\mu \Phi_{\mu\mu_2\cdots\mu_{n_1}}(x) = 0 \qquad (2.21)$$

was described in the framework of the BRST approach. From the point of view of D dimensions, where constraint (2.21) is of second class, the $D+1$-st components of the creation and annihilation operators appear in the consideration as additional operators. The corresponding BRST charge is nilpotent and has a very special structure. In particular, the modified constraints have the algebra, which is not closed. Nevertheless, the nontrivial structure of trilinear terms in ghosts in the BRST charge compensates this defect and makes the BRST charge to be nilpotent. Another example of BRST charge for the system, including second class constraints was obtained in (Pashnev and Tsulaia (1998)). It reproduces some properties of the BRST charge of (Pashnev and Tsulaia (1997)): the algebra of modified constraints is not closed and trilinear terms in ghosts are nontrivial as well. Moreover, the BRST charge contains terms up to the seventh degree in ghosts. In the next Section we will describe the simple method, which allows to one to construct the nilpotent BRST charges for a given Lie algebra, when some generators are treated as the second class constraints. As a particular case, this method reproduces the results obtained in (Pashnev and Tsulaia (1997)) and (Pashnev and Tsulaia (1998)).

3 The General Method

In this section, we describe the method of the BRST construction, which leads to the desirable division of the generators of a given Lie algebra into the first and second class constraints. Let H^i, $(i = 1, ..., k)$ and E^α be the Cartan generators and root vectors of the algebra with the following commutation relations

$$[H^i, E^\alpha] = \alpha(i) E^\alpha, \tag{3.1}$$

$$[E^\alpha, E^{-\alpha}] = \alpha^i H^i, \tag{3.2}$$

$$[E^\alpha, E^\beta] = N^{\alpha\beta} E^{\alpha+\beta}. \tag{3.3}$$

Roots $\alpha(i)$ and parameters α^i, $N^{\alpha\beta}$ are structure constants of the algebra in the Cartan - Weyl basis. Our goal is to construct nilpotent BRST charge, which after quantization leads to the following conditions: all positive root vectors E^α $(\alpha > 0)$ of the algebra annihilate the physical states. Contrary, the operators H^i which form the Cartan subalgebra may or may not be constraints, depending on the physical nature of these operators.

The simplest case, when all Cartan generators annihilate the physical states, is well known. We introduce the set of anticommuting variables η_i, η_α, $\eta_{-\alpha} = \eta_\alpha^+$, having ghost number one and corresponding momenta $\mathcal{P}_i, \mathcal{P}_{-\alpha} = \mathcal{P}_\alpha^+, \mathcal{P}_\alpha$, with the commutation relations:

$$\{\eta_i, \mathcal{P}_k\} = \delta_{ik}, \ \{\eta_\alpha, \mathcal{P}_{-\beta}\} = \{\eta_{-\alpha}, \mathcal{P}_\beta\} = \delta_{\alpha\beta} \tag{3.4}$$

we define the "ghost vacuum" as

$$\eta_\alpha|0\rangle = \mathcal{P}_\alpha|0\rangle = \mathcal{P}_i|0\rangle = 0 \tag{3.5}$$

for positive roots α. The BRST charge for the Cartan - Weyl decomposition of the algebra has a standard form

$$Q = \sum_i \eta_i H^i + \sum_{\alpha>0} (\eta_\alpha E^{-\alpha} + \eta_{-\alpha} E^\alpha) - \frac{1}{2} \sum_{\alpha\beta} N^{\alpha\beta} \eta_{-\alpha}\eta_{-\beta}\mathcal{P}_{\alpha+\beta} +$$
$$\sum_{\alpha>0,i} \{\alpha(i)(\eta_i\eta_\alpha \mathcal{P}_{-\alpha} - \eta_i \eta_{-\alpha}\mathcal{P}_\alpha) + \alpha^i \eta_\alpha \eta_{-\alpha} \mathcal{P}_i\} \tag{3.6}$$

The physical states are then the cohomology classes of the BRST operator.

The quantization in this case is equivalent to the quantization *à* la Gupta - Bleuler, because physical states satisfy equations $H^i|Phys\rangle = 0$ and $E^\alpha|Phys\rangle = 0$ only for positive values of α.

The situation changes when some of the Cartan operators H^i, say H^{i_l}, $l = 1, 2, ... N$ are nonvanishing from the physical reasons. In this case the following method can be used.

First of all we construct some auxiliary representation for the generators H^i, E^α of the algebra in terms of additional creation and annihilation operators. The only condition for this representation is that it depends on some parameters h^n. The total number of these parameters is equal to the number of the Cartan generators, which are nonzero in the physical sector. In what follows, we consider the realizations of the algebra with a linear dependence of the Cartan generators on these parameters: $\hat{H}^m(h) = \widetilde{H}^m + c^m_n h^n$, where c^m_n are some constants. The h^n dependence of other generators can be arbitrary. In the next section we describe the general method of construction of such representations. Here we simply assume that they exist.

The next step is to consider the realization of the algebra as a sum of "old" and "new" generators

$$\mathcal{H}^i = H^i + \hat{H}^i(h), \quad \mathcal{E}^\alpha = E^\alpha + \hat{E}^\alpha(h).$$

The BRST charge for the total system has the same form as (3.6), with modified generators:

$$\mathcal{Q} = \sum_i \eta_i \mathcal{H}^i + \sum_{\alpha>0} \left(\eta_\alpha \mathcal{E}^{-\alpha} + \eta_{-\alpha}\mathcal{E}^\alpha\right) - \frac{1}{2}\sum_{\alpha\beta} N^{\alpha\beta}\eta_{-\alpha}\eta_{-\beta}\mathcal{P}_{\alpha+\beta} +$$
$$\sum_{\alpha>0,i} \{\alpha(i)\left(\eta_i\eta_\alpha\mathcal{P}_{-\alpha} - \eta_i\eta_{-\alpha}\mathcal{P}_\alpha\right) + \alpha^i\eta_\alpha\eta_{-\alpha}\mathcal{P}_i\} \tag{3.7}$$

The ghost variables η_{i_l}, correspond to the set of nonvanishing generators H^{i_l} and therefore one needs to remove the η_{i_l} dependence

$$Q_{i_l} = \eta_{i_l}\{H^{i_l} + \widetilde{H}^{i_l} + c^{i_l}_n h^n + \sum_{\beta>0}\beta(i_l)\left(\eta_\beta\mathcal{P}_{-\beta} - \eta_{-\beta}\mathcal{P}_\beta\right)\}. \tag{3.8}$$

from the BRST charge. For this purpose consider an auxiliary N - dimensional space with coordinates x_{i_l} and conjugated momenta p^{i_l}, where $c^{i_l}_n h^n = p^{i_l}$:

$$[x_{i_l}, p^{i_n}] = i\delta^{i_n}_{i_l}. \tag{3.9}$$

After the similarity transformation, which corresponds to the dimensional reduction (Pashnev and Tsulaia (1997))

$$\widetilde{\mathcal{Q}} = e^{i\pi^{i_l}x_{i_l}}\mathcal{Q}e^{-i\pi^{i_l}x_{i_l}}, \tag{3.10}$$

where

$$\pi^{i_l} = H^{i_l} + \widetilde{H}^{i_l} + \sum_{\beta>0}\alpha(i_l)\left(\eta_\beta\mathcal{P}_{-\beta} - \eta_{-\beta}\mathcal{P}_\beta\right) \tag{3.11}$$

the transformed BRST charge $\widetilde{\mathcal{Q}}$ does not depend on the ghost variables η_{i_l}. All parameters p^{i_l} in the BRST charge are replaced by the corresponding operators $-\pi^{i_l}$. The transformation (3.10) does not change the nilpotency property of the BRST charge. It means that the $\mathcal{P}_{i_l}$ independent part $\widetilde{\mathcal{Q}}_0$ of the total charge $\widetilde{\mathcal{Q}}$ is nilpotent as well. Moreover, as a consequence of the nilpotency of $\widetilde{\mathcal{Q}}$ all coefficients at the corresponding antighost operators $\mathcal{P}_{i_l}$ commute with $\widetilde{\mathcal{Q}}_0$. One can show that the quantization with the help of the BRST operator $\widetilde{\mathcal{Q}}_0$ will lead to the desirable reduced system of constraints on the physical states.

4 Construction of Auxiliary Representations of the Algebra

Consider the highest weight representation of the algebra under consideration with the highest weight vector $|\Phi\rangle$ annihilated by the positive roots

$$E^{\alpha}|\Phi\rangle = 0 \tag{4.1}$$

and being the proper vector of the Cartan generators

$$H^{i}|\Phi\rangle = h^{i}|\Phi\rangle. \tag{4.2}$$

As it was shown by Gelfand and Tsetlin (Gel'fand and Tsetlin (1950)), each of the vectors of the irreducible representation with a given highest weight can be associated with the so called Gelfand-Tsetlin scheme. Corresponding scheme for $U(k)$ algebra has the following form:

$$\left|\begin{array}{ccccccccc} m_{1,k} & & m_{2,k} & & & & m_{k-1,k} & & m_{k,k} \\ & m_{1,k-1} & & & \cdot & & & m_{k-1,k-1} & \\ \cdot & \cdot & \cdot & \cdot & \cdot & \cdot & \cdot & \cdot & \cdot \\ & & & m_{12} & & m_{22} & & & \\ & & & & m_{11} & & & & \end{array}\right|$$

The first row of the scheme is defined by the highest weight components:

$$m_{i,k} = h^{i}. \tag{4.3}$$

This row is fixed for a given irreducible representation of $U(k)$. All other rows contain arbitrary numbers under the following conditions:

$$m_{ij} \geq m_{i,j-1} \geq m_{i+1,j},\; j = 2, ..., k;\; i = 1, ..., k-1. \tag{4.4}$$

Any choice of this numbers, consistent with (4.4), corresponds to a fixed vector in the irreducible representation. All these vectors are orthonormal. The total number of $m_{i,j}$ coincides with the number of positive roots of the algebra $U(k)$. This is true for any semisimple algebra. It means, that all such vectors can be represented in an auxiliary Fock space, generated by the oscillators $b_{i,j}, b^{+}_{i,j}$, which are in one to one correspondence with numbers $m_{i,j}$. So the vector, which corresponds to the Gelfand - Tsetlin scheme given above has the following form

$$|\{m_{i,j}\}\rangle = \prod_{i,j} \frac{1}{((m_{i,j} - \kappa_{i,j})!)^{1/2}} (b^{+}_{i,j})^{(m_{i,j}-\kappa_{i,j})}|0\rangle, \tag{4.5}$$

where $\kappa_{i,j}$ are some fixed parameters, connected with the weights h^{i}. Having at hands all matrix elements of generators (Gel'fand and Tsetlin (1950)) one can easily reconstruct the expressions for the generators in terms of oscillators $b_{i,j}, b^+_{i,j}$. They will depend on k parameters h^i and give the needed auxiliary representation of the algebra. Below we construct such representation for the simplest example of $SO(2,1)$ to illustrate the method of the BRST construction described in the previous section.

5 The Simple Example

In this section we consider $SO(2,1)$ algebra of constraints. The Fock space introduced in Sect.2 is spanned by only one oscillator. This system describes higher spin irreducible representations of the Poincare algebra corresponding to the Young tableaux with only one row. The total system of generators includes

$$L_1 = a_\mu p_\mu, \quad L_{-1} = a^+_\mu p_\mu, \quad L_0 = p^2_\mu, \quad [L_1, L_{-1}] = -L_0, \tag{5.1}$$

$$L_2 = \frac{1}{2} a_\mu a_\mu, \quad L_{-2} = \frac{1}{2} a^+_\mu a^+_\mu, \quad [L_2, L_{-2}] = -a^+_\mu a_\mu - \frac{D}{2} \equiv G_0, \tag{5.2}$$

where D is the dimensionality of the space – time. The last three operators L_2, L_{-2} and G_0 form an $SO(2,1)$ algebra. The first three operators in (5.1) transform as a representation of this algebra and they can be included in the BRST charge rather trivially. So, the main problem is to construct the BRST charge for $SO(2,1)$ algebra under the condition, that the operator G_0 is not a constraint in the physical subspace, since G_0 is positively definite in the whole Fock space.

Using the results of the previous Section one can easily construct the auxiliary representation of the algebra $SO(2,1)$ in terms of one additional timelike oscillator $[b, b^+] = -1$:

$$\hat{L}_2 = \sqrt{h + b^+ b}\, b,$$
$$\hat{L}_{-2} = b^+ \sqrt{h + b^+ b}, \tag{5.3}$$
$$\hat{G}_0 = \widetilde{G}_0 - h, \quad \widetilde{G}_0 = -2b^+ b. \tag{5.4}$$

The BRST charge for modified generators

$$\mathcal{L}_2 = L_2 + \hat{L}_2, \quad \mathcal{L}_{-2} = L_{-2} + \hat{L}_{-2}, \quad \mathcal{G}_0 = G_0 + \hat{G}_0 \tag{5.5}$$

takes the following form

$$\mathcal{Q} = \eta_0 \mathcal{G}_0 + \eta_2 \mathcal{L}_{-2} + \eta_{-2} \mathcal{L}_2 - 2\eta_0 \eta_2 \mathcal{P}_{-2} + 2\eta_0 \eta_{-2} \mathcal{P}_2 + \eta_2 \eta_{-2} \mathcal{P}_0 \tag{5.6}$$

As the result of similarity transformation described in Sect.3, the coefficient at ghost variable η_0 vanishes and the following replacement of parameter h takes place:

$$h \to G_0 - 2b^+ b + 2\mathcal{P}_{-2}\eta_2 + 2\eta_{-2}\mathcal{P}_2 \tag{5.7}$$

The resulting nilpotent BRST charge after removing of the antighost variable $\mathcal{P}_0$ looks as follows

$$\mathcal{Q}_0 = \eta_2\{L_{-2} + b^+ \sqrt{G_0 - b^+b + 2\mathcal{P}_{-2}\eta_2 + 2\eta_{-2}\mathcal{P}_2}\} + \eta_{-2}\{L_2 + \sqrt{G_0 - b^+b + 2\mathcal{P}_{-2}\eta_2 + 2\eta_{-2}\mathcal{P}_2}\, b\}. \tag{5.8}$$

the inclusion of the constraints L_0, L_1 and L_{-1} into the BRST charge is trivial. After the corresponding gauge fixing and solving the equations of motion for some of the auxiliary fields, one obtains the lagrangian, given in (Fronsdal (1978)).

6 Conclusions

In this paper we have demonstrated the possibility for the various constructions of the nilpotent BRST charges for a given algebra of constraints. The identification of the generators of the algebra with the constraints on the physical states is model–dependent and therefore after the quantization these BRST charges lead to the different spectrum of the physical states.

Acknowledgments. This investigation has been supported in part by the Russian Foundation of Fundamental Research, grants 96-02-17634 and 96-02-18126, joint grant RFFR-DFG 96-02-00180G, and INTAS, grants 93-127-ext, 96-0308, 96-0538, 94-2317 and grant of the Dutch NWO organization.

References

Ouvry, S. and Stern, J. (1986): *Phys.Lett.* **B177**, 335.
Labastida, J.M.F. (1987): *Phys.Rev.Lett.* **58**, 531.
Labastida, J.M.F. (1989): *Nucl.Phys.* **B322**, 185.
Pashnev, A. and Tsulaia, M. (1997): *Mod.Phys.Lett.* **A12**, 861.
Faddeev, L.D. and Shatashvili, S.L. (1986): *Phys.Lett.* **B167**, 225.
Batalin, I.A. and Fradkin, E.S. (1987): *Nucl.Phys.* **B279**, 514.
Egoryan, E.T. and Manvelyan, R.P. (1993): *Theor. Math.Phys.* **94**, 241.
Pashnev, A. and Tsulaia, M. (1998): *Mod.Phys.Lett.* **A13**, 1853.
Gel'fand, I.M. and Tsetlin, M.L. (1950): *Dokl.Akad.Nauk SSSR* **71**, 825. *ibid* **71**, 1017.
Fronsdal, C. (1978): *Phys.Rev.* **D18**, 3624.

Higher-Spin Gauge Theories – Integrability Versus Locality

Sergey Prokushkin and Mikhail Vasiliev

I.E.Tamm Department of Theoretical Physics, Lebedev Physical Institute, Leninsky Prospect 53, 117924 Moscow, Russia

Abstract. We discuss properties of non-linear equations of motion which describe higher-spin gauge interactions for massive spin-0 and spin-1/2 matter fields in 2+1 dimensional anti-de Sitter space. An integrating flow is found which reduces the full non-linear system to the free field equations via a non-local Bäcklund-Nicolai–type mapping.

1 Introduction and Preliminaries

Viktor Isaakovich Ogievetsky has made a fundamental contribution to modern quantum field theory. Since seventies his main scientific interest was focused on supersymmetric models. Harmonic superspace approach to models with extended supersymmetry developed by Ogievetsky with his group (Galperin, Ivanov, Kalitzin, Ogievetsky, and Sokatchev, 1984) is one of the most beautiful achievements in this field. Another prominent result obtained by Ogievetsky and Polubarinov, 1966 is a dual formulation of massless scalar via two-form, "notoph". The deepness of the scientific intuition of Viktor Isaakovich is now fully manifested by numerous applications of his results at the modern stage of searching a unified theory of fundamental interactions. During last years the main attention of Viktor Isaakovich was focused on the study of models with self-dual gauge fields of higher spins ($s > 2$) in $d = 4$ (Devchand and Ogievetsky, 1996).

The models with higher-spin (HS) gauge fields are responsible for infinite-dimensional gauge symmetries and may serve as an alternative way towards a fundamental theory, M-theory. Although a route to M-theory passes through models in higher dimensions, particularly $d = 11$ and $d = 12$, it is instructive to study a $d = 3$ model which has much simpler dynamics because $d3$ HS gauge fields do not propagate (Blencowe, 1989). In many respects these models are analogous to those with self-dual $d = 4$ HS gauge fields (M. V., 1992) which are expected to be related to the models of Devchand and Ogievetsky.

In this talk, we focus on some recent results in the study of the HS interactions of matter fields in 2+1 dimensional space-time. The main new result consists of the constructive definition of a flow inducing a non-local mapping of the non-linear problem to the free one. Due to the analogy of the $d = 3$ HS models with the self-dual $d = 4$ HS models we speculate that the existence of the integrating flow may be an indication of integrability of the model.

It is convenient to describe HS gauge fields within approach similar to the so-called "geometric approach" to gravity (Kibble, 1961; MacDowell and Mansouri, 1977) with vielbein $h_\mu{}^a$ and Lorentz connection $\omega_\mu{}^{ab}$ identified with the connection 1-forms of an appropriate space-time symmetry algebra g. For example, one can use gauge fields $A_\mu^{BC} = -A_\mu^{CB}$ of the anti-de Sitter (AdS) algebra $g = o(d-1,2)$ to describe the geometry of the d-dimensional AdS space-time (indices $B, C = 0, ..., d$ are raised and lowered by the flat metrics $\eta^{BC} = diag(+ - \cdots - +)$), setting $\omega_\mu{}^{ab} = A_\mu^{ab}$ and $h_\mu{}^a = \lambda^{-1} A_\mu^{a\cdot}$ with the conventions $a, b = 0, ..., d-1$, $B = (b, \cdot)$. Here $\lambda \neq 0$ is some constant. The respective $o(d-1,2)$ gauge curvatures have the form

$$R_{\mu\nu}{}^{ab} = \partial_\mu \omega_\nu{}^{ab} + \omega_\mu{}^a{}_c \,\omega_\nu{}^{cb} - \lambda^2 h_\mu{}^a \, h_\nu{}^b - (\mu \leftrightarrow \nu)\,, \tag{1}$$

$$R_{\mu\nu}{}^a = \partial_\mu h_\nu{}^a + \omega_\mu{}^a{}_c h_\nu{}^c - (\mu \leftrightarrow \nu)\,. \tag{2}$$

Lorentz connection $\omega_\mu{}^{ab}$ is expressed via vielbein $h_\mu{}^a$ with the aid of the constraint $R_{\mu\nu}{}^a = 0$ ($h_\mu{}^a$ is assumed to be non-degenerate). Substituting $\omega_\mu{}^{ab} = \omega_\mu{}^{ab}(h)$ into (1), one can see that the equation $R_{\mu\nu}{}^{ab} = 0$ is equivalent to $\mathcal{R}_{\mu\nu}{}^{ab} = \lambda^2 (h_\mu{}^a \, h_\nu{}^b - h_\nu{}^a \, h_\mu{}^b)$, where $\mathcal{R}_{\mu\nu}{}^{ab} = \partial_\mu \omega_\nu{}^{ab} + \omega_\mu{}^a{}_c \, \omega_\nu{}^{cb} - (\mu \leftrightarrow \nu)$ is the Riemann tensor, and therefore describes AdS space-time with radius λ^{-1}. This is how AdS space appears as a vacuum solution of the HS equations considered below. A role of the algebra $o(d-1,2)$ is twofold: its connection 1-forms are identified with the dynamical fields of the theory and it serves as the symmetry algebra of the most symmetric vacuum solution, space-time symmetry.

In the case of $d = 2+1$, the AdS algebra is $o(2,2)$, and the gravitational action is the Chern-Simons action, $S^W = \int_{M_3} str(w \wedge dw + \frac{2}{3} w \wedge w \wedge w)$, where w is the $o(2,2)$ connection 2-form (Witten, 1989).

The $d = 2+1$ HS superalgebras used below can be described as follows (M. V., *JETP Lett.* 1989, *Int. J. Mod. Phys.* 1991). Consider an associative algebra $Aq(2;\nu)$ with a general element of the form

$$f(\hat{y}, k) = \sum_{\substack{n=0 \\ A=0,1}}^{\infty} \frac{1}{n!} f^{A\,\alpha_1 \ldots \alpha_n}(k)^A \hat{y}_{\alpha_1} \cdots \hat{y}_{\alpha_n}\,, \tag{3}$$

under condition that the coefficients $f^{A\,\alpha_1 \ldots \alpha_n}$ are symmetric with respect to the indices $\alpha_j = 1, 2$, while the generating elements $\hat{y}_\alpha$, k satisfy the relations

$$[\hat{y}_\alpha, \hat{y}_\beta] = 2i\epsilon_{\alpha\beta}(1 + \nu k)\,, \quad k\hat{y}_\alpha = -\hat{y}_\alpha k\,, \quad k^2 = 1\,, \tag{4}$$

where ν is an arbitrary number (α, β, $\gamma = 1, 2$ are $d = 2+1$ spinor indices lowered and raised with the aid of the symplectic form $\epsilon_{\alpha\beta} = -\epsilon_{\beta\alpha}$, $\epsilon_{12} = \epsilon^{12} = 1$, $A^\alpha = \epsilon^{\alpha\beta} A_\beta$, $A_\alpha = A^\beta \epsilon_{\beta\alpha}$.)

Then, the $d3$ HS superalgebra $hs(2;\nu)$ is the Lie superalgebra canonically related to the associative algebra $Aq(2;\nu)$ (with the $\mathbf{Z}_2$ grading counting a number of spinor indices, i.e. $\pi(\hat{y}_\gamma) = 1$, $\pi(k) = 0$). To describe a doubling of the elementary algebras in $g = hs(2;\nu) \oplus hs(2;\nu)$ analogous to $o(2,2) \sim sp(2) \oplus sp(2)$ we introduce an additional central involutive generating element ψ,

$$[\psi, \hat{y}_\alpha] = 0\,, \quad [\psi, k] = 0\,, \quad \psi^2 = 1\,. \tag{5}$$

The two simple subalgebras of g are singled out by the projection operators $P_\pm = \frac{1}{2}(1 \pm \psi)$. Field strengths for g are

$$R(\hat{y}, \psi, k|x) = d\omega(\hat{y}, \psi, k|x) - \omega(\hat{y}, \psi, k|x) \wedge \omega(\hat{y}, \psi, k|x)\,, \tag{6}$$

where $d = dx^\nu \frac{\partial}{\partial x^\nu}$ and the gauge fields $\omega(\hat{y}, \psi, k|x)$ of g are of the form similar to (3). x^μ, $\mu = 0, 1, 2$ are the commuting space-time coordinates.

The $d3$ AdS space-time symmetry algebra $o(2,2) \sim sp(2) \oplus sp(2)$ is the subalgebra of g spanned by the bilinears

$$L_{\alpha\beta} = \frac{1}{4i}\{\hat{y}_\alpha, \hat{y}_\beta\}\,, \qquad P_{\alpha\beta} = \frac{1}{4i}\{\hat{y}_\alpha, \hat{y}_\beta\}\psi\,. \tag{7}$$

The pure gauge HS action has the Chern-Simons form with supertrace defined in M. V., *JETP Lett.* 1989, *Int. J. Mod. Phys.* 1991. It reduces to the Witten gravity action (Witten, 1989) in the spin 2 sector and to the Blencowe's HS action (Blencowe, 1989) in the case of $\nu = 0$.

An important question we address below at the level of equations of motion is how to introduce interactions of HS gauge fields with propagating matter fields. We will follow the so-called "unfolded formulation" approach (M. V., 1994), rewriting dynamical equations in a form of certain zero-curvature conditions and covariant constancy conditions

$$d\omega = \omega \wedge \omega\,, \qquad dB^A = \omega^i (t_i)^A{}_B B^B\,, \tag{8}$$

supplemented with some gauge invariant constraints

$$\chi(B) = 0 \tag{9}$$

which do not contain space-time derivatives. Here $\omega(x) = dx^\nu \omega^i_\nu(x) T_i$ is a gauge field of some Lie superalgebra l ($T_i \in l$), and $B^A(x)$ is a set of 0-forms which take values in a representation space of some representation $(t_i)^B{}_A$ of l.

An interesting property of this form of equations is that their dynamical content is hidden in the constraints (9). Indeed, locally one can integrate out (8) as $\omega = dg(x)g^{-1}(x)$, $B(x) = t_{g(x)}(B_0)$, where $g(x)$ is an arbitrary invertible element, while B_0 is an arbitrary x - independent representation element and $t_{g(x)}$ is the exponential of the representation t of l. Since the constraints $\chi(B)$ are gauge invariant one is left with the only condition $\chi(B_0) = 0$. Let $g(x_0) = I$ for some point of space-time x_0. Then $B_0 = B(x_0)$. To understand how restrictions on values of some 0-forms at a fixed point of space-time can lead to a non-trivial dynamics one should take into account that in the interesting examples (see M. V., 1996), the set of 0-forms B is reach enough to describe all space-time derivatives of dynamical fields, while the constraints (9) just impose all restrictions on the space-time derivatives required by the dynamical equations under consideration. Given solution of (9) one knows all derivatives of the dynamical fields compatible with the field equations and can therefore reconstruct these fields by analyticity in some neighborhood of x_0. The specificity of the HS dynamics which makes such an approach adequate is that HS symmetries mix all orders of derivatives which therefore should be contained in a representation space of HS symmetries.

2 Nonlinear System

Now let us turn to massive matter fields interacting via HS gauge potentials in $d = 2{+}1$. The full nonlinear system of equations, which is a particular realization of the equations (8) and (9), is formulated in terms of the generating functions $W(z, y; \psi_{1,2}, k, \rho|x)$, $B(z, y; \psi_{1,2}, k, \rho|x)$, and $S_\alpha(z, y; \psi_{1,2}, k, \rho|x)$ which depend on the space-time coordinates x^ν ($\nu = 0, 1, 2$), auxiliary commuting spinors z_α, y_α ($\alpha = 1, 2$), $[y_\alpha, y_\beta] = [z_\alpha, z_\beta] = [z_\alpha, y_\beta] = 0$, a pair of Clifford elements $\{\psi_i, \psi_j\} = 2\delta_{ij}$ ($i = 1, 2$) that commute to all other generating elements, and another pair of Clifford-type elements k and ρ which have the properties

$$k^2 = 1\,,\ \rho^2 = 1\,,\ k\rho + \rho k = 0\,,\ k y_\alpha = -y_\alpha k\,,\ k z_\alpha = -z_\alpha k\,,$$
$$\rho y_\alpha = y_\alpha \rho\,,\ \rho z_\alpha = z_\alpha \rho\,. \qquad (10)$$

The space-time 1-form $W = dx^\nu W_\nu(z, y; \psi_{1,2}, k, \rho|x)$,

$$W_\mu(z, y; \psi_{1,2}, k, \rho|x) = \sum_{A,B,C,D=0}^{1} \sum_{n=0}^{\infty} \frac{1}{m!n!} W^{ABCD}_{\mu,\,\alpha_1\ldots\alpha_m\beta_1\ldots\beta_n}(x)$$
$$\times\, k^A \rho^B \psi_1^C \psi_2^D z^{\alpha_1} \ldots z^{\alpha_m} y^{\beta_1} \ldots y^{\beta_n}\,. \qquad (11)$$

is the generating function for HS gauge fields. $B = B(z, y; \psi_{1,2}, k, \rho|x)$ is the generating function for the matter fields. The components of its expansion analogous to (11) are identified with the $d3$ matter fields and all their on-mass-shell non-trivial derivatives. $S_\alpha(z, y; \psi_{1,2}, k, \rho|x)$ describes auxiliary and pure gauge degrees of freedom. The generating functions are treated as elements of an associative algebra with the product law

$$(f * g)(z,y;\psi,k,\rho)=$$
$$\frac{1}{(2\pi)^2}\int d^2u d^2v \exp(iu_\alpha v^\alpha) f(z+u, y+u; \psi, k, \rho) g(z-v, y+v; \psi, k, \rho)\,, \qquad (12)$$

where the integration variables u and v satisfy the commutation relations similar to those of y and z in (10). This product law yields a particular realization of the Weyl algebra, $[y_\alpha, y_\beta]_* = -[z_\alpha, z_\beta]_* = 2i\epsilon_{\alpha\beta}$, $[y_\alpha, z_\beta]_* = 0$ ($[a, b]_* \equiv a * b - b * a$).

The full system of equations has the form

$$dW = W * \wedge W\,, \quad dB = W * B - B * W\,, \quad dS_\alpha = W * S_\alpha - S_\alpha * W\,, \tag{13}$$

$$S_\alpha * S_\beta - S_\beta * S_\alpha = -2i\epsilon_{\alpha\beta}(1 + B * K)\,, \quad S_\alpha * B = B * S_\alpha\,. \tag{14}$$

Here $K = ke^{i(zy)}$ and $(zy) = z_\alpha y^\alpha$. With the aid of the involutive automorphism $\rho \to -\rho$, $S_\alpha \to -S_\alpha$ one can truncate the system (13), (14) to the one with the fields W and B independent of ρ and S_α linear in ρ,

$$W(z, y; \psi_{1,2}, k, \rho|x) = W(z, y; \psi_{1,2}, k|x),$$

$$B(z, y; \psi_{1,2}, k, \rho|x) = B(z, y; \psi_{1,2}, k|x), \tag{15}$$

$$S_\alpha(z, y; \psi_{1,2}, k, \rho|x) = \rho s_\alpha(z, y; \psi_{1,2}, k|x)\,. \tag{16}$$

From now on we consider this reduced system. Eqs. (13), (14) are invariant under the infinitesimal HS gauge transformations

$$\delta W = d\varepsilon - W * \varepsilon + \varepsilon * W\,, \quad \delta B = \varepsilon * B - B * \varepsilon\,, \quad \delta S_\alpha = \varepsilon * S_\alpha - S_\alpha * \varepsilon\,, \tag{17}$$

where $\varepsilon = \varepsilon(z, y; \psi_{1,2}, k|x)$ is an arbitrary gauge parameter.

To elucidate the dynamical content of the system (13), (14), one first of all has to find an appropriate vacuum solution. There exists a class of vacuum solutions (S. P. and M. V., 1998). The simplest one has a form

$$B_0 = \nu = const\,, \tag{18}$$

$$S_{0\alpha} = \rho\left(z_\alpha + \nu(z_\alpha + y_\alpha)\int_0^1 dt t e^{it(zy)} k\right), \tag{19}$$

$$W_0(z, y; \psi_{1,2}, k|x) = W_0(\widetilde{y}; \psi_{1,2}, k|x)\,, \tag{20}$$

where

$$\widetilde{y}_\alpha = y_\alpha + \nu(z_\alpha + y_\alpha)\int_0^1 dt(t-1)e^{it(zy)}k \tag{21}$$

are the elements with the defining property $[\widetilde{y}_\alpha\,, S_{0\beta}]_* = 0$. An arbitrary "function" $W_0(\widetilde{y}; \psi_{1,2}, k|x)$ on the r.h.s. of (20) contains star-products of $\widetilde{y}_\alpha$. It is interesting to note that $\widetilde{y}_\alpha$ possess the deformed oscillator algebra commutation relations (4)

$$[\widetilde{y}_\alpha, \widetilde{y}_\beta]_* = 2i\epsilon_{\alpha\beta}(1 + \nu k)\,, \quad \widetilde{y}_\alpha k = -k\widetilde{y}_\alpha\,. \tag{22}$$

Eqs. (18)-(20) solve all the equations (13), (14) except for

$$dW_0 = W_0 \wedge W_0\,, \tag{23}$$

which requires a further specification of W_0. An appropriate ansatz is

$$W_0 = \omega_0 + \lambda h_0 \psi_1\,, \quad \omega_0 = \frac{1}{8i}\omega_0^{\alpha\beta}(x)\{\widetilde{y}_\alpha, \widetilde{y}_\beta\}_*\,, \quad h_0 = \frac{1}{8i}h_0^{\alpha\beta}(x)\{\widetilde{y}_\alpha, \widetilde{y}_\beta\}_*\,, \tag{24}$$

where $\omega_0^{\alpha\beta}(x)$ and $h_0^{\alpha\beta}(x)$ are identified with Lorentz connection and dreibein of the background space and are required to solve (23). Here the properties of the deformed oscillators (22) play a crucial role, guaranteeing that the anticommutators $\{y_\alpha, y_\beta\}_* = y_\alpha * y_\beta + y_\beta * y_\alpha$ satisfy the $sp(2)$ commutation relations for all ν (M. V., *JETP Lett.* 1989, *Int. J. Mod. Phys.* 1991). As a result, the gauge fields (24) take values in the $d3$ AdS algebra $o(2,2) \sim sp(2) \oplus sp(2)$ and (23) describes AdS background.

Once a vacuum solution is known, one can study the system (13), (14) perturbatively expanding the fields as

$$B = B_0 + B_1 + \dots , \qquad S_\alpha = S_{0\alpha} + S_{1\alpha} + \dots , \qquad W = W_0 + W_1 + \dots . \tag{25}$$

Substitution of these expansions into (13), (14) gives in the lowest order

$$D_0 W_1 = 0 , \tag{26}$$

$$D_0 C = 0 , \tag{27}$$

$$D_0 S_{1\alpha} = [W_1 , S_{0\alpha}]_* , \tag{28}$$

$$[S_{0\alpha} , S_{1\beta}]_* - [S_{0\beta} , S_{1\alpha}]_* = -2i\epsilon_{\alpha\beta} C * K , \tag{29}$$

$$[S_{0\alpha} , C]_* = 0 , \tag{30}$$

where we denote $C = B_1$ and D_0 is the background covariant derivative which acts on a r–form P as $D_0 P = dP - W_0 \wedge P + (-)^r P \wedge W_0$.

To analyze the system (26)-(30) one proceeds as follows. From (30), one concludes that C has a form similar to (20), i.e. $C = C(\widetilde{y}; \psi_{1,2}, k|x)$. Expanding C as $C = C^{aux}(\widetilde{y}; \psi_1, k|x) + C^{dyn}(\widetilde{y}; \psi_1, k|x)\psi_2$, one identifies (Barabanschikov et al, 1997; M. V., 1997) C^{aux} with some topological fields which carry no degrees of freedom, and C^{dyn} with the generating function for the spin 0 and spin 1/2 matter fields. Namely, in accordance with the normal spin-statistics, $\widetilde{y}$-even (odd) part of C^{dyn} identifies with the generating function for spin 0 (1/2) matter fields, along with all their on-mass-shell non-trivial derivatives (Barabanschikov et al, 1997). The equation (27) amounts to free field equations. Resolving the constraints (29), one reconstructs the auxiliary field $S_{1\alpha}$ as a linear functional of C, $S_{1\alpha} = S_{1\alpha}(C)$, up to a gauge ambiguity. Then, (28) allows one to express a part of degrees of freedom in W_1 via C, while the rest modes, which belong to the kernel of the mapping $[S_{0\alpha}, \dots]_*$, remain free. These free modes are again arbitrary functions of $\widetilde{y}_\alpha$, i.e.

$$W_1 = \omega(\widetilde{y}; \psi_{1,2}, k|x) + \Delta W_1(C) , \tag{31}$$

where $\omega(\widetilde{y}; \psi_{1,2}, k|x)$ corresponds to the HS gauge fields, and the dynamical equations for them are imposed by eq. (26) after (28) is solved. Eq. (26) describes the C-dependent first order corrections to the HS strengths for ω, which are argued below to vanish. One proceeds similarly in the highest orders.

An important fact about HS dynamical systems is that they allow non-Abelian internal (Yang-Mills) symmetries, as was first discovered in the $d4$ case (M. V., *Ann. Phys.*, 1989; Konstein and M. V., 1990) and analyzed in detail for the $d3$ case (S. P. and M. V., 1998). The key observation is that the system (13)-(14) remains consistent if components of all fields take their values in an arbitrary associative algebra M with a unit element I_M, i.e. the fields W, B, and S_α take values in $A^{ext} = A \otimes M$, where A is the associative algebra with the general element (11). The gravitational sector is associated with $A \sim A \otimes I_M$ and commutes with $M \sim I_A \otimes M$, where I_A is the unit element of A. Therefore, M describes internal symmetries in the model. For the case of semisimple finite-dimensional inner symmetries, M is identified with some matrix algebra, i.e. $W \to W_i{}^j$, $B \to B_i{}^j$, and $S_\alpha \to S_{\alpha,i}{}^j$.

3 Integrating Flow

A remarkable property of eqs. (13), (14) is that they admit a flow which allows one to express constructively solutions of the full system in terms of free fields. Since our perturbation expansion is just an expansion in powers of the physical fields which are identified with the deviation C of B from its vacuum value ν, let us introduce a formal perturbation expansion parameter η as follows

$$B(\eta) = \nu + \eta \mathcal{B}(\eta) . \tag{32}$$

Simultaneously, the rest of the fields acquire a dependence on η, $W = W(\eta)$ and $S_\alpha = S_\alpha(\eta)$. The system (13), (14) takes a form

$$dW = W * W , \quad d\mathcal{B} = W * \mathcal{B} - \mathcal{B} * W , \quad dS_\alpha = W * S_\alpha - S_\alpha * W , \tag{33}$$

$$S_\alpha * S^\alpha = -2i(1 + \nu K + \eta \mathcal{B} * K) , \quad S_\alpha * \mathcal{B} = \mathcal{B} * S_\alpha . \tag{34}$$

Now, one observes that for the limiting case $\eta = 0$ the system (33), (34) reduces to the free one. Indeed, setting

$$\nu = B_0 , \quad \mathcal{B}(0) = B_1 \equiv C , \quad W(0) = W_0 \equiv \omega , \quad S_\alpha(0) = S_{0\alpha} ,$$

we see that at $\eta = 0$ the system (33), (34) acquires the form of the vacuum system which is solved in terms of $S_\alpha = S_{0\alpha}$ with $W = \omega(\widetilde{y}; \psi_{1,2}, k|x)$ and $\mathcal{B} = C(\widetilde{y}; \psi_{1,2}, k|x)$ satisfying the free field equations $d\omega = \omega \wedge \omega$, $dC = \omega * C - C * \omega$. This situation is similar to the one with contractions of Lie algebras. For all values of $\eta \neq 0$, the systems of equations (33), (34) are pairwise equivalent since the field redefinition (32) is non-degenerate. On the other hand, although the field redefinition (32) degenerates at $\eta = 0$, eqs. (33), (34) still make sense for $\eta = 0$. This limiting system describes the free field dynamics.

Remarkably, the two inequivalent systems are still related to each other. To show this let us define a flow with respect to η as follows,

$$\frac{\partial X}{\partial \eta} = (1-\mu)\,\mathcal{B} * \frac{\partial X}{\partial \nu} + \mu\,\frac{\partial X}{\partial \nu} * \mathcal{B}\,, \tag{35}$$

where $X = W, B$ or S_α and μ is an arbitrary parameter. This flow can be easily shown to be compatible with (33), (34). Therefore, solving the system (35) with the initial data $\mathcal{B}(\eta=0) = C\,,\ W(\eta=0) = \omega\,,\ S_\alpha(\eta=0) = S_{0\alpha}$, we can express solutions of the full nonlinear system at $\eta = 1$ via solutions of the free system at $\eta = 0$. The existence of the integrating flow (35) takes its origin in the fact that from the point of view of the system (13), (14), B behaves like a constant: it commutes to S_α and satisfies covariant constancy condition. Knowledge of the vacuum solution with $B = \nu$ can be used to reconstruct the full dependence of B. Indeed, the meaning of (35) is that a derivative with respect to $\eta\mathcal{B}$ is the same as that with respect to ν. (All fields acquire a non-trivial dependence on ν via the vacuum solution as it follows e.g. from eqs. (31), (21).) The flows (35) at different μ are equivalent modulo gauge transformations.

Such an approach allows one to derive order by order the relevant field redefinitions since the r.h.s.-s of (35) contain one extra power of $\mathcal{B}$. In particular, one can easily derive in the first order a field redefinition necessary to show that the HS gauge field strengths do not admit nontrivial sources linear in fields. In the first order this field redefinition can be shown to be local. This result is expected since in the lowest order the non-trivial r.h.s.-s of the equations for HS gauge fields are the HS currents bilinear in the matter fields.

Remarkably, the method works in all higher orders, thus reducing the full non-linear problem to the free one. The point, however, is that beyond the first order one has to be careful in making statements on the locality of the mapping induced by the flow (35). Actually, although it does not contain explicitly space-time derivatives, it contains them implicitly via highest components $C_{\alpha_1\ldots\alpha_n}$ of the generating function $C(\widetilde{y})$ which are identified with the highest derivatives of the matter fields due to the equations (27) (Barabanschikov et al, 1997). For example, at $\mu = 0$ in the second order in C one gets $\frac{\partial}{\partial\eta}B(z,y) = C(\widetilde{y}) * \frac{\partial}{\partial\nu}C(\widetilde{y})$. Because of the properties of the $*$-product, for each fixed rank multispinorial component of the l.h.s. of this formula, its r.h.s. is an infinite series involving bilinear combinations of the components $C_{\alpha_1\ldots\alpha_n}$ with all n. Therefore, the r.h.s.-s of (35) effectively contain all orders of the space-time derivatives, i.e. the resulting transformation are non-local. This means that one cannot treat the system (13), (14) as locally equivalent to the free system.

To illustrate this issue it is instructive to consider an example of some matter field C interacting with the gravitational field fluctuating near the AdS vacuum solution. Schematically, the mechanism is as follows. Linearized Einstein equations have a form (with appropriate gauge fixings)

$$(L^C - \Lambda^2)h_{\mu\nu} = T_{\mu\nu}(C)\,, \tag{36}$$

where $h_{\mu\nu}$ is the fluctuational part of the metric tensor, L^C is the linear operator corresponding to the l.h.s. of the free field equations of the matter fields $L^C C = 0$, while $\Lambda = \alpha\lambda$ with some $\alpha \neq 0$. It is important that when the cosmological constant is non-vanishing, the term with Λ^2 turns out to be non-vanishing too. This property allows one to solve formally (36) by a field redefinition

$$h_{\mu\nu} \to h'_{\mu\nu} = h_{\mu\nu} - (L^C - \Lambda^2)^{-1} T_{\mu\nu}(C) = h_{\mu\nu} + \Lambda^{-2} \sum_{n=0}^{\infty} (\Lambda^{-2} L^C)^n T_{\mu\nu}(C) \,. \tag{37}$$

Clearly, a non-vanishing dimensionful constant, the cosmological constant, plays an important role in this analysis. This field redefinition admits (S. P. and M. V., 1998) a natural realization in terms of the generating function C in agreement with the general analysis above.

4 Conclusions

The main conclusion is that dynamical systems based on infinite sets of HS gauge fields admit deep structures which allow their constructive perturbative solvability. The fact that some system can be integrated order by order with the help of a non-local field redefinition is not unusual, of course. What is special about the HS systems is that such a field redefinition is described in a systematic way by some flow with respect to an additional evolution parameter. As a result, one can reconstruct solutions of the non-linear HS equations in terms of the solutions of the free system by integrating ordinary differential equations. We believe that this fact can be interpreted as some sort of integrability of the $d = 3$ HS equations, although a rigorous proof of this statement remains to be given. Moreover, the very concept of integrability may need to be modified in application to HS models. Indeed, as mentioned at the end of sect. 2, the part of the equations that contain space-time derivatives has a form of zero-curvature conditions (typical feature of integrable systems) and can be integrated explicitly. As a result, one is to solve only constraint part of the system (in some parallelism with Hamiltonian reduction). This problem is not of an evolution type however, because the equations (14) are some integral equations in the twistor space in which the star product acts. The main result reported in this talk is how the problem of solving these constraints can be reduced to ordinary differential equations which describe evolution with respect to an auxiliary parameter η provided that a particular solution with $B = \nu = const$ is found.

We argue that the resulting solution is essentially non-local in the space-time sense. A non-local character of the transformation manifests itself in the appearance of infinite series in the inverse cosmological constant. Taking into account that HS gauge interactions are known (Fradkin and M. V., 1987) to require non-analyticity in the cosmological constant, we conjecture that

HS gauge theories are indeed non-local in a certain sense. This conjecture agrees with the light-cone analysis by Metsaev, 1991 and fits the ideology of M-theory.

This research was supported in part by INTAS, Grant No.96-0538 and by the RFBR Grant No.96-01-01144. S. P. acknowledges a partial support from the Landau Scholarship Foundation, Forschungszentrum Jülich.

References

A. Galperin, E. Ivanov, V. Ogievetsky, and E. Sokatchev, *JETP Lett.* **40** (1984) 155; A. Galperin, E. Ivanov, S. Kalitzin, V. Ogievetsky, and E. Sokatchev, *Class. Quant. Grav.* **1** (1984) 469.

V. I. Ogievetsky and I. V. Polubarinov, *Yad. Fiz.* **4** (1966) 216.

Ch. Devchand and V. Ogievetsky, *Nucl. Phys.* **B481** (1996) 188.

M. A. Vasiliev, *Phys. Lett.* **B285** (1992) 225

M. A. Vasiliev, *Nucl. Phys. B (Proc. Supplement)* **56B** (1997) 241-252.

M. P. Blencowe, *Class. Quant. Grav.* **6** (1989) 443.

T. W. B. Kibble, *J. Math. Phys.* **2** (1961) 212.

S. W. MacDowell and F. Mansouri, *Phys. Rev. Lett.* **38** (1977) 739.

K. Stelle and P. West, *Phys. Rev.* **D21** (1980) 1466.

E. Witten, *Nucl. Phys.* **B311** (1989) 46.

M.A. Vasiliev, *Class. Quant. Grav.* **11** (1994) 649.

M. A. Vasiliev, *Int. J. Mod. Phys.* **D5** (1996) 763.

S. F. Prokushkin and M. A. Vasiliev, `hep-th/9806236`.

M. A. Vasiliev, *JETP Lett.* **50** (1989), 374; *Int. J. Mod. Phys.* **A6** (1991) 1115.

A. V. Barabanschikov, S. F. Prokushkin, and M. A. Vasiliev, *Rus. Theor. Math. Phys.* **110** (1997) 295, `hep-th/9609034`.

M. A. Vasiliev, *Class. Quantum Grav.* **8** (1991) 1387.

M. A. Vasiliev, *Ann. Phys.* (N.Y.) **190** (1989) 59.

S. E. Konstein and M. A. Vasiliev, *Nucl. Phys.* **B331** (1990) 475.

E. Bergshoeff, B. de Wit, and M. A. Vasiliev, *Nucl. Phys.* **B366** (1991) 315

E. S. Fradkin and M. A. Vasiliev, *Phys. Lett.* **B189** (1987) 89.

R. R. Metsaev, *Mod. Phys. Lett.* **A4** (1991) 359.

Global Quantum Anomaly

Andrei Slavnov

Steklov Mathematical Institute, Gubkina 8, 117966 Moscow, Russia

Abstract. New representation for the phase of a chiral determinant is constructed. This representation is used to modify the action of $SU(2)$ chiral gauge model removing the global anomaly.

1 Introduction

In the present talk I discuss the models affected by quantum global anomalies. The well known example is the $SU(2)$ gauge model with an odd number of Weyl fermion doublets (Witten (1982)). The global anomaly is related to an ambiguity in the definition of the phase of a chiral determinant. It was shown in the paper (Witten (1982)) that one cannot define the phase of the determinant of a single chiral $SU(2)$ fermion in a gauge invariant way.

In the paper (Alvarez-Gaume, Della Pietra, Della Pietra (1986)) an exact formula for the phase of a chiral determinant was derived and it's geometrical meaning was discussed. Below I will get a different representation for the phase of a chiral determinant which allows to write it as a path integral of exponent of a local action. Having this explicit representation one can modify the action of chiral $SU(2)$ fermions in such a way that the global anomaly disappears. My construction in some sence is reminiscent to the Wess-Zumino construction for local quantum anomalies (Wess and Zumino (1971)).

2 Lagrangian Representation for the Phase of a Chiral Determinant and Global $SU(2)$ Anomaly

Let us start with the massless Euclidean Dirac operator

$$\hat{D} = \gamma_\mu(\partial_\mu + iA_\mu(x)) \tag{1}$$

which can be written in the form

$$\hat{D} = \begin{pmatrix} 0 & C \\ -C^+ & 0 \end{pmatrix} \tag{2}$$

where

$$C = e_\mu(\partial_\mu + iA_\mu(x)) \tag{3}$$

with $e_i = -i\sigma_i, \sigma_i(i = 1, 2, 3)$ being the Pauli matrices, $e_0 = -I$. The field $A_\mu(x)$ for each x belongs to the Lie algebra of a gauge group. The matrices C, C^+ represent fermions of opposite chiralities.

It follows from the eq.(2) that

$$\det D = \det C \det C^{+} \tag{4}$$

Hence the determinant of the Dirac operator is equal to the square of the modulus of the determinant of the Weyl operator. In case of the $SU(2)$ group due to pseudoreality of representation

$$\det C = \det C^{+} \tag{5}$$

Using this fact Witten proposed to define the regularized determinant of a single Weyl fermion as a square root of regularized Dirac determinant. Dirac determinant can be regularized in a gauge invariant way by means of a standard Pauli-Villars regularization. However there is a sign ambiguity

$$\det C_R = \pm(\det D_R)^{1/2} \tag{6}$$

For a given gauge field $A_\mu(x)$ one can choose a sign arbitrary. Defined in such a way the Weyl determinant is obviously invariant with respect to infinitesimal gauge transformations and therefore in the framework of perturbation theory one has a consistent gauge invariant definition of the determinant of the chiral $SU(2)$ operator.

However if one allows topologically nontrivial gauge transformations, then one can choose a transformation U which changes the sign of the square root of the determinant

$$[\det D(A_\mu)]^{1/2} = -[\det D(A_\mu^U)]^{1/2} \tag{7}$$

That means the phase of a chiral determinant is not invariant with respect to "large" gauge transformations and leads to inconsistensy of the theory.

The definition of the chiral $SU(2)$ determinant as a square root of the regularized Dirac determinant is perfectly consistent in the framework of perturbation theory, however it does not allow a representation of the Weyl determinant as a path integral of a local action as taking a square root of a determinant is a nonlocal operation. To get such a representation for regularized chiral determinant we use the idea proposed in our paper (Frolov, Slavnov (1993)).

We introduce the infinite set of Pauli-Villars fields with masses Mr, $r = 1, 2 \ldots$. Now the regularized Weyl determinant may be written as follows

$$\det C_R = \int \exp\{\int L_R dx\} d\overline{\psi}_+ d\psi_+ d\overline{\psi}_r d\psi_r L_R =$$
$$= \overline{\psi}_+ C \psi_+ + \sum_{r=1}^{\infty} \overline{\psi}_r (\hat{D} + Mr)\psi_r \tag{8}$$

Here ψ_r are Pauli-Villars fields having Grassmanian parity $(-1)^{r+1}$.

Integrating over ψ we get

$$\det C_R = \det C \prod_{r=1}^{\infty} \det(\hat{D}+Mr)^{(-1)^r} \tag{9}$$

Using the representation like eq.(2) for the Dirac operator one can rewrite it as follows

$$\begin{aligned}\det C_R = \det C \prod_{r=1}^{\infty} \det(|C|^2+M^2r^2)^{(-1)^r} = \\ = \prod_{r=0}^{\infty} \prod_i C_i^{-1}(|C_i|^2+M^2r^2)^{(-1)^r}\end{aligned} \tag{10}$$

Here C_i are diagonal elements of the matrix $C_{ij} =< u_i C v_j >$, where u_i and v_j form orthonormal bases in the spaces of left and right handed fermions respectively.

The produçt over r can be calculated explicitely using the representation

$$\prod_{r=0}^{\infty}(|C_i|^2+M^2r^2)^{(-1)^r} = \exp\{\sum_{r=0}^{\infty}\ln(|C_i|^2+M^2r^2)(-1)^r\} \tag{11}$$

Differentiating the exponent with respect to $|C^i|^2$ one has

$$\frac{\partial}{\partial|C_i|^2}\sum_{r=0}^{\infty}(-1)^r\ln(|C_i|^2+M^2r^2) = 1/2[\sum_{r=-\infty}^{\infty}(-1)^r(|C_i|^2+M^2r^2)^{-1}+|C_i|^{-2}]$$

$$= 1/2[\pi(M|C_i|\sinh(\pi|C_i|M^{-1}))^{-1}+|C_i|^{-2}] \tag{12}$$

Integrating over $|C_i|^2$ one gets

$$\ln(\det C_R) = \ln(\tan(\frac{\pi|C_i|}{2M}))+\ln(|C_i|)+\ln(B) \tag{13}$$

where B is a field independent constant which in the following is assumed to be included into normalization factor. Therefore up to normalization factor we have the following representation for the regularized determinant

$$\det C_R = \prod_i \frac{|C_i|}{C_i}\tan(\frac{\pi|C_i|}{2M}) \tag{14}$$

In the framework of perturbation theory for a given A_μ one can fix the signs of C_i at will, in particular take all C_i positive. Then it follows from the eq.(14) that

$$\det C_R = \prod_i \tan(\frac{\pi|C_i|}{2M}) \tag{15}$$

One sees that all $|C_i| >> M$ are cutted and therefore eqs.(8, 15)indeed provide the gauge invariant regularization of the chiral determinant.

However beyond perturbation theory the expression (14) is not well defined. Although large eigenvalues are cutted, the phase factor is not regularized. It is the source of the global anomaly in our approach. As was mentioned in the beginning, topologically nontrivial gauge transformations may change the sign of C_i and one cannot fix it in arbitrary way.

It is known that in the case of local anomalies one can restore gauge invariance by adding to the action a local term depending on new fields (Wess and Zumino (1971)). Below I show that a similar construction is possible for the global anomaly as well.

Let us modify the Lagrangian (7) by adding to it a gauge invariant term describing the interaction of new fields χ_r

$$L_R \to L'_R = L_R + \Delta L$$

$$\Delta L = \overline{\chi}_+ C \chi_+ + \sum_{r=1}^{\infty} \overline{\chi}_r (\hat{D} + m^2 M^{-1} r) \chi_r \tag{16}$$

where χ_r are again the fields with alternating Grassmanian parity $(-1)^{r+1}$. Here m is some fixed parameter with the dimension of mass. When $M \to \infty$ the masses of the χ fields vanish.

The integral over χ can be calculated as above giving the result

$$\Delta = \prod_i \frac{|C_i|}{C_i} \tan\left(\frac{\pi |C_i|}{2m^2 M^{-1}}\right) \tag{17}$$

Assuming that $C_i \neq 0$ we see that when $M \to \infty$

$$\Delta \to \prod_i \frac{|C_i|}{C_i} \tag{18}$$

It shows that the integral

$$\lim_{M\to\infty} \int \exp\{\int \Delta L dx\} d\overline{\chi}_+ d\chi_+ d\overline{\chi}_r d\chi_r \tag{19}$$

gives the representation for the phase of a chiral determinant as a path integral of the local action. Note that this representation is valid for any gauge group, not necessary $SU(2)$.

One sees that the Δ exactly compensates the indefinite phase factor in the eq.(14) and the integral

$$\det (C_R)' = \int \exp\{\int L'_R dx\} d\overline{\psi}_+ d\psi_+ d\overline{\chi}_+ d\chi_+ d\overline{\psi}_r d\psi_r d\overline{\chi}_r d\chi_r \tag{20}$$

provides a well defined expression which is gauge invariant not only with respect to infinitesimal gauge transformations but with respect to topologically nontrivial transformations as well.

In the case of the $SU(2)$ gauge group the new fields χ do not influence the results of perturbative calculations as in this case one can fix arbitrary the signs of C_i and $\lim_{M\to\infty} \Delta = 1$.

Analogous construction is valid for other models with global anomalies, provided the fermions belong to a pseudoreal representation.

3 Discussion

We constructed above a representation for the phase of a chiral determinant in terms of a path integral of the exponent of the local action. Using this expression we were able to modify the action of $SU(2)$ chiral fermions in such a way that the global anomaly disappears. This construction has some similarity to the Wess-Zumino construction for local anomalies. There are however important differences. The Wess-Zumino term restores quantum gauge invariance, but the classical action including this term is not gauge invariant, where as our modified action (16) is gauge invariant. Another difference is that contrary to the Wess-Zumino case, variation of our additional term under topologically nontrivial gauge transformation is discrete.

It would be interesting to analyze a possible physical meaning of the modified action. Although the new fields are not seen in perturbation theory, they certainly influence nonperturbative configurations. Even if one starts with the configuration which does not include the χ fields, they will be produced in pairs in a final state. One can speculate that the true vaccuum of the modified model includes infinite number of pairs of massless fermions which may change drastically a physical content of the theory.

Acknowledgements This research was supported in part by Russian Basic Research Foundation under grant 96-01-00551 and Presidential program for support of leading scientific schools.

References

Witten, E. (1982), Phys. Lett. **117B**, 324-330.
Alvarez-Gaume, L., Della Pietra, S., Della-Pietro, V. (1986), Phys. Lett. **166B**, 176-182.
Wess, J., Zumino, B. (1971), Phys. Lett. **37B**, 95-98.
Frolov, S., Slavnov, A. (1993), Phys. Lett. **309B**, 344-350.

Non-Commutative Space-Time Algebra and the Spectrum of Its Operators

Julius Wess

Sektion Physik der Ludwig-Maximilians-Universität,
Theresienstraße 17, D-80333 München and
Max-Planck-Institut für Physik (Werner-Heisenberg-Institut),
Föhringer Ring 6, D-80805 München

Let me discuss an example of a noncommutative space-time structure to see what we can learn from it.

Usually we consider space-time as a geometrical object to begin with and then we try to formulate a physical system with differential equations and quantization rules. For short, we start from e.g. $\mathbf{R}^1$ as a geometrical object, learn to differentiate and quantize with the rule $p = -i\frac{\partial}{\partial x}$. This leads to the Heisenberg algebra:

$$[p, x] = -i, \tag{1}$$

with the conjugation properties

$$\overline{x} = x, \qquad \overline{p} = p. \tag{2}$$

We study the Hilbertspace representations of this algebra and we have a well-established scheme for the physical interpretation of the mathematical formalism.

For less trivial structures than $\mathbf{R}^1$ it is not as easy to formulate consistent quantization rules, deformation quantization for example addresses this problem.

For studying noncommutative space-time structures it seems to be more natural to start from an algebra, study its Hilbertspace representations, define the configuration space as the spectrum of position variables and, following the rules of quantum mechanics, interpret the eigenvalues of the position operators as the possible results of a position measurement. Thus we would start from (1) and (2), where (2) would be a mere algebraic conjugation consistent with (1). We would study Hilbertspace representations, e.g. $p = -i\frac{\partial}{\partial x}$, demand that the algebraic conjugation corresponds to the conjugation of linear operators in Hilbertspace and that the algebraic property of being selfadjoint really becomes selfadjoint and not only hermitean for the linear operators in Hilbertspace.

In three dimensions we start from the algebra

$$[x^i, x^j] = 0, \qquad [p^i, p^j] = 0, \tag{3}$$

$$[p^i, x^j] = -i\delta^{ij} \tag{4}$$

and the conjugation properties:

$$\overline{p^i} = p^i, \qquad \overline{x^i} = x^i. \tag{5}$$

These relations are $SO(3)$ covariant. The generators of $SO(3)$ form the algebra

$$[L^i, L^j] = i\varepsilon^{ijk}L^k \tag{6}$$

with the conjugation properties

$$\overline{L^i} = L^i. \tag{7}$$

The variables x^i and p^i form representations of the $SO(3)$ algebra (vectors):

$$\begin{aligned} [L^i, x^r] &= i\varepsilon^{irk}x^k \\ [L^i, p^r] &= i\varepsilon^{irk}p^k \end{aligned} \tag{8}$$

The Heisenberg relation (4) is covariant under $SO(3)$ as well, as we have written the Kronecker symbol at the right hand side of the Heisenberg relation (4).

$\mathbf{R}^3$ turns out to be the configuration space if we represent p^i in the usual way:

$$p^i = -i\frac{\partial}{\partial x^i} \tag{9}$$

We find that the L^is can be represented by differential operators as well

$$L^i = \varepsilon^{ijk}x^j p^k \tag{10}$$

To treat an example of a noncommutative space we start from a q-deformed algebra to replace eqns (6) and (7). We demand that position and momentum variables are modules of this algebra and replace (8) by the corresponding relations. Next we generalize (3) by relations that are consistent with the module property and that are of the homogeneous form.

$$\begin{aligned} &P_{ij}x^i x^j = 0, \qquad \widetilde{P}_{ij}p^i p^j = 0 \\ &P_{ij} \in \mathbf{C}, \qquad \widetilde{P}_{ij} \in \mathbf{C} \end{aligned} \tag{11}$$

These algebraic relations should all be consistent with a conjugation operation that should allow us to represent position operators and momentum operators by selfadfoint operators in a Hilbertspace.

Finally we try to generalize the Heisenberg algebra (4) as well - consistent with the module structure and the conjugation properties. But we demand one more property. We want that a basis of homogeneous polynomials in x and p has the same dimension as in the case of commuting variables x and p. We call this the Poincare-Birkhoff-Witt (PBW) property. Though it is formulated very mathematically it has very important physical implications. A measurement of the observables x and p should not be more restricted than in the commutative case.

It turns out that PBW and the conjugation property restrict the possible algebras very much. Indeed we were not able to find any algebra in terms of x^i, p^i and L^i (q-deformed) that would satisfy all our demands. We had to enlarge the algebra by new elements. If we try to stay with elements that lend themselves to a physical interpretation (e.g. not three coordinates and six momenta) we are able to do it with one additional element Λ that could be interpreted as an operator of a scale transformation. In terms of these algebraic elements there are exactly two algebras.

Before writing down these algebras let me point out an important porperty of quantum groups. The representations have exactly the same pattern as the corresponding non-deformed ($q = 1$) groups. For instance, the product of two vector representations of $SU(2)$ decompose into a singlet, a triplet and a quintuplet.

$$3 \otimes 3 = 1 \oplus 3 \oplus 5 \tag{12}$$

Exactly the same is true for $su_q(2)$, and the Clebsch-Gordan coefficients are known. We shall assume $q \in \mathbf{R}$, $q > 1$.

The Clebsch-Gordan coefficients that combine two vectors to a singlet define a metric

$$\begin{aligned} &X \circ Y = g_{AB} X^A Y^B \\ &g_{33} = 1, \quad g_{+-} = -q, \quad g_{-+} = -\frac{1}{q} \end{aligned} \tag{13}$$

The notation X^3, X^+, X^- is adapted to the quantum group notation where it is natural to introduce $X^{\pm} = X^1 \pm iX^2$.

The Clebsch-Gordan coefficients that combine two vectors to a triplet define a generalized ε-tensor:

$$\begin{aligned} &Z^A = X^C Y^B \varepsilon_{BC}{}^A = X \times Y \\ &\varepsilon_{+-}{}^3 = q, \quad \varepsilon_{-+}{}^3 = -q, \quad \varepsilon_{33}{}^3 = 1 - q^2, \\ &\varepsilon_{+3}{}^+ = 1, \quad \varepsilon_{3+}{}^+ = -q^2, \\ &\varepsilon_{-3}{}^- = -q^2, \quad \varepsilon_{3-}{}^- = 1. \end{aligned} \tag{14}$$

More generally, we can compute from the Clebsch-Gordan coefficients the projectors on the invariant subspaces spanned by the product of two vectors:

$$P_1 + P_3 + P_5 = 1 \tag{15}$$

These are 9 by 9 matrices. There is a combination of these matrices that satisfies the Yang-Baxter equation:

$$\hat{R} = P_5 - \frac{1}{q^4} P_3 + \frac{1}{q^6} P_1 \tag{16}$$

With $\hat{R}$, $\hat{R}^{-1}$ will satisfy the Yang-Baxter equation as well. These are the only two solutions of the Yang-Baxter equation for 9×9 $\hat{R}$ matrices that can be decomposed into the three projectors. The choice of $\hat{R}$ or $\hat{R}^{-1}$ in the definition of the algebra will distinguish the two algebras we have mentioned above. One of them is:

$$\begin{aligned} X^C X^B \varepsilon_{BC}{}^A = 0: \; X^3 X^+ &= q^2 X^+ X^3 \\ X^3 X^- &= \frac{1}{q^2} X^- X^3 \\ X^- X^+ &= X^+ X^- + (q - \frac{1}{q}) X^3 X^3 \end{aligned} \tag{17}$$

This demonstrates the noncommutative space structure.

The $su_q(2)$ algebra:

$$\begin{aligned} L \times L &= -\frac{1}{q^2} W L \\ WL &= LW \end{aligned} \tag{18}$$

The element W is related to the Casimir operator of $su_q(2)$:

$$W^2 = 1 + q^4 (q^2 - 1)^2 L \circ L \tag{19}$$

The module structure of X and P is:

$$\begin{aligned} L^A X^B &= -\frac{1}{q^4} \varepsilon^{ABC} X_C W - \frac{1}{q^2} \varepsilon_{KC}{}^A \varepsilon^{KBD} X^C L_D \\ W X^A &= (q^2 - 1 + q^{-2}) X^A W + (q^2 - 1)^2 \varepsilon^{ABC} X_C L_B \end{aligned} \tag{20}$$

The elements X can be replaced by P.

Heisenberg relation:

$$\begin{aligned} & P^A X^B - \hat{R}^{-1AB}{}_{CD} X^C P^D \\ & \quad = -\frac{i}{2} \Lambda^{-\frac{1}{2}} \left\{ (1 + q^{-6}) g^{AB} W - (1 - q^{-4}) \varepsilon^{ABC} L_C \right\} \end{aligned} \tag{21}$$

Here the new element $\Lambda^{\frac{1}{2}}$ had to be introduced:

$$\begin{aligned} \Lambda^{\frac{1}{2}} X^A &= q^2 X^A \Lambda^{\frac{1}{2}} \\ \Lambda^{\frac{1}{2}} P^A &= q^{-2} P^A \Lambda^{\frac{1}{2}} \\ \Lambda^{\frac{1}{2}} L^A &= L^A \Lambda^{\frac{1}{2}} \\ \Lambda^{\frac{1}{2}} W &= W \Lambda^{\frac{1}{2}} \end{aligned} \tag{22}$$

The conjugation properties are:

$$\begin{aligned} & \overline{X^A} = g_{AB} X^B, \quad \overline{P^A} = g_{AB} P^B \\ & \overline{L^A} = g_{AB} L^B, \quad \overline{W} = W, \quad \overline{\Lambda^{\frac{1}{2}}} = q^{-6} \Lambda^{-\frac{1}{2}}. \end{aligned} \tag{23}$$

It should be noted that the Heisenberg algebra (21) cannot be separated into an algebra of the X and P elements and an algebra that defines L. This is only possible for $q = 1$. In the deformed case ($q \neq 1$) it is not possible to express L as an ordered polynomial in X and P as it was the case in eqn (10).

To incorporate time into the noncommutative structure we start with a q-deformed Lorentz algebra. Again, there is the same Clebsch-Gordan structure

$$4 \times 4 = 1 + 9 + 6 \tag{24}$$

where 6 decomposes into a selfdual and an antiselfdual antisymmetric tensor. The singlet defines a metric η. The projectors can again be combined to a solution of the Yang-Baxter equation. This time, however, there are two independent combinations $\hat{R}_I$ and $\hat{R}_{II}$ with their inverse.

$$\begin{aligned} \hat{R}_I &= P_S + P_T - q^2 P_+ - q^{-2} P_- \\ \hat{R}_{II} &= q^{-2} P_S + q^2 P_T - P_+ - P_- \end{aligned} \tag{25}$$

Here the notation is P_T (trace) for the singlet, P_S (symmetric, traceless) for the nonet, P_+ (antisymmetric, selfdual) for the one triplet and P_- (antisymmetric, antiselfdual) for the other triplet.

We only list a few of the algebraic relations. A complete description of the algebra is given in ref Lorek, Weich and Wess (1997).

$$\begin{aligned} X^0 X^A &= X^A X^0 \\ X^C X^D \varepsilon_{DC}{}^A &= (1 - q^2) X^0 X^A. \end{aligned} \tag{26}$$

The "time" X^0 commutes with the space variables X^A ($A = 3, +, -$). The space variables, however, produce the time variable.

The momenta are subject to the identical relations (26).

The q-Lorentz group can be split into the Pauli decomposition:

$$\begin{aligned} R^C R^D \varepsilon_{DC}{}^A &= (1 + q^2)^{-1} U R^A \\ S^C S^D \varepsilon_{DC}{}^A &= -(1 + q^2)^{-1} U S^A \\ R^A S^B &= q^2 \hat{R}^{AB}{}_{CD} S^C R^D \end{aligned} \tag{27}$$

The $\hat{R}$ matrix here is the 9×9 matrix defined by eqn (16). It is related to the Casimir:

$$U^2 = 1 + \frac{1}{2}(q^4 - 1)^2 (R \circ R + S \circ S) \tag{28}$$

The important relation is the Heisenberg relation:

$$P^a X^b - q^{-2} \hat{R}_{II}^{-1ab}{}_{cd} X^c P^d = -\frac{i}{2} \Lambda^{-\frac{1}{2}} \left\{ (1 + q^4) \eta^{ab} U + q^2 (1 - q^4) V^{ab} \right\} \tag{29}$$

Again a scaling operator $\Lambda^{-\frac{1}{2}}$ had to be introduced. The indices a, b take the values $(0, 3, +, -)$. V^{ab} is the tensor notation for the generators of the q-Lorentz algebra.

Finally the conjugation properties

$$\begin{aligned}
&\overline{X^0} = X^0, \quad \overline{X^A} = g_{AB}X^B \\
&\overline{P^0} = P^0, \quad \overline{P^A} = g_{AB}P^B \\
&\overline{R^A} = -g_{AB}S^B \\
&\overline{U} = U \quad \overline{\Lambda^{\frac{1}{2}}} = q^4\Lambda^{-\frac{1}{2}}
\end{aligned} \tag{30}$$

We are now ready to discuss Hilbertspace representations of this algebra. A complete set of commuting observables is:

$$X^0, \quad r^2 = g_{AB}X^AX^B, \quad L\circ L, \quad L_3 \tag{31}$$

L is the three-dimensional angular momentum

$$L^A = \frac{q^2+1}{q^2}(US^A - UR^A + (q^4-1)\varepsilon_{CB}{}^A R^B S^C) \tag{32}$$

that commutes with the time variable X^0. The eigenvalues of these observables (31) completely characterize a state in the Hilbertspace as they would do in the commuting case.

The representations of the algebra have been constructed in ref Cerchiai and Wess (1998).

Here I only would like to plot the eigenvalues of the observable X^0 (time) and r (three-dimensional radius) for a particular value of q ($q = 1.1$). A scale has to be chosen, it characterizes the representations. We chose $t_0 = 1$.

The representations have an interesting pattern. In fig 1 forward light cone, backward light cone and space-like region are clearly visible. Each of these furnishes a representation of the algebra. If, however, X^a and P^a have to be represented by selfadjoint linear operators then we have to glue together all three representations. The light cone consists of limit points that again have a limit point at the origin. The hyperbola corresponding to the invariant length $\eta_{ab}X^aX^b$ are clearly visible. For $q \to 1$ the points come closer and closer and have the continuum as a limit.

We learn that a q-deformation of the Minkowski space latticizes the space-time manifold.

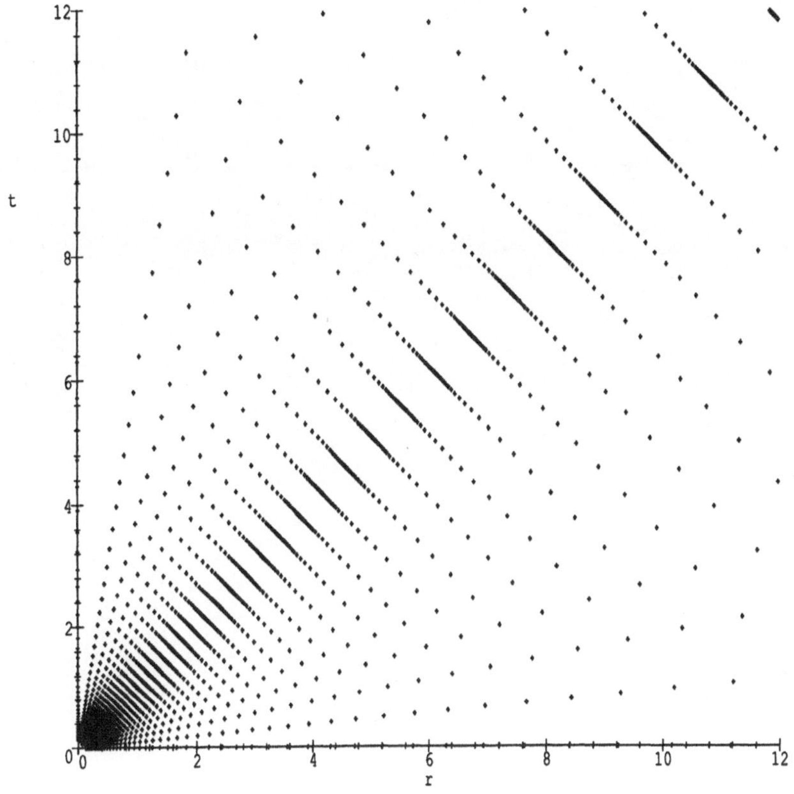

Fig. 1. $q = 1.1,\ t_0 = 1$

References

Lorek, A., Weich, W., Wess J. (1997): *Non-commutative Euclidean and Minkowski structures* Z. Phys. C **76**, 375–386

Cerchiai, B.L., Wess, J. (1998): *q-Deformed Minkowski Space based on a q-Lorentz Algebra* Preprint math.QA/9801104, accepted for publication Eur. Phys. J. C

Part V

Selected Works and List of Main Publications of V.I. Ogievetsky

SOVIET PHYSICS JETP VOLUME 21, NUMBER 6 DECEMBER, 1965

SPINORS IN GRAVITATION THEORY

V. I. OGIEVETSKIĬ and I. V. POLUBARINOV

Joint Institute for Nuclear Research

Submitted to JETP editor December 3, 1964

J. Exptl. Theoret. Phys. (U.S.S.R.) **48**, 1625-1636 (June, 1965)

Spinors in gravitation theory are treated as four-component objects which transform according to a nonlinear representation of the group of general covariant transformations. Interactions of a spinor field with gravitational, electromagnetic and other fields are constructed in accordance with the spinor transformation law thus derived. The interactions are expanded in a series in powers of the gravitational field, and this is convenient for the application of perturbation theory.

1. INTRODUCTION

1. Even now it is already of interest to discuss quantum gravitational effects not only in the weak-field approximation, but also in higher orders in terms of the gravitational coupling constant, and also the problem of renormalizations and the removal of divergences in gravitational interactions. According to Gupta[1] the problem of the quantization of the gravitational field within the framework of perturbation theory is essentially solved by expanding the nonlinear Einstein equations in an infinite series in the gravitational constant. Such an approach as applied to the gravitational field and to gravitational interactions of boson fields was adopted in reference[1].

2. But the gravitational interactions of fermions have not until now been discussed within the framework of this approach. The point is that the gravitational interactions of fermions are essentially more complicated[2-22]. Fock and Ivanenko[4,5] were the first to construct a theory of fermions in a gravitational field utilizing the tetrad formalism. In the majority of subsequent papers the same method is also used. However, within the framework of the tetrad formalism it is not clear how Gupta's program should be carried out, and how the gravitational interaction of fermions should be represented in explicit form in terms of the gravitational field as an expansion in terms of the gravitational coupling constant.

3. This has motivated us to seek another method of describing spinors in a gravitational field. We have used the group-theoretic approach and we have introduced spinors as objects transforming in accordance with a representation of that group according to which the fundamental tensor $g^{\mu\nu}$ is transformed. In this sense spinors turn out to be objects of the same type as tensors (scalar, vector etc.). At the same time there also exists an essential difference: although the law of transformation of spinors obtained below is linear and homogeneous in the spinor field, in contrast to the tensor case it depends on the gravitational field (the metric), and does so in a complicated nonlinear manner. In other words, spinors transform according to a nonlinear representation of the group mentioned previously.

On the other hand such an approach corresponds to Schrödinger's[10] idea of doing without orthogonal basis vectors, but he utilized γ-fields which, like tetrads, are related to the gravitational field only implicitly. On the other hand the "square root of the metric tensor" $r^{\mu\nu}$ appearing in our case can be regarded as a modification of a tetrad. In contrast to a tetrad, $r^{\mu\nu}$ is explicitly expressed in terms of the gravitational field, both its indices can be treated on an equal footing and refer to the same general basis (in the case of tetrads one index refers to a general basis, and the other one to a locally orthogonal one).

4. In the approach proposed here the gravitational interaction of fermions is expressed explicitly in terms of the gravitational field and, in accordance with Gupta's program, can be represented in the form of an infinite series in terms of the gravitational coupling constant (in the same way as the "self-interaction" of a gravitational field and the gravitational interactions between bosons in reference[1]). The interaction obtained in this manner in principle enables one to calculate gravitational effects involving fermions to any arbitrary order in the gravitational coupling constant. We emphasize that even for such simple

effects as the gravitational self-energy of the electron or the Compton-effect of a graviton on a fermion the weak field approximation is insufficient, and it is necessary to take into account interaction terms of the second order in the gravitational coupling constant.

2. THE GROUP PROPERTY OF GENERALLY COVARIANT TRANSFORMATIONS

1. In Riemannian geometry the law of transformation of the fundamental tensor $g^{\mu\nu}(x)$ can be represented in infinitesimal form by means of a local variation (cf., reference [23], p. 323)

$$\delta^* g^{\mu\nu} = g'^{\mu\nu}(x) - g^{\mu\nu}(x) = a(\partial_\rho\lambda^\mu g^{\rho\nu} + \partial_\rho\lambda^\nu g^{\mu\rho} - \lambda^\rho\partial_\rho g^{\mu\nu}), \quad (1)$$

where $\lambda^\mu(x)$ are four arbitrary infinitesimal functions. For convenience these functions are brought to dimensionless form by factoring out a constant a which has the dimension of length (in units of $\hbar = c = 1$). In future it will play the role of the gravitational coupling constant and will turn out to be related to the gravitational constant k in Newton's law by the expression $a^2 = 32\pi k$.

Local variations[24] can be regarded as transformations of the functions only, without a change in coordinates (of the type of gauge transformations in electrodynamics) and this, in particular, is useful in interpreting Einstein's theory in the language of a flat space. At the same time the use of local variations is equivalent to the use of substantive variations

$$\delta g^{\mu\nu} = g'^{\mu\nu}(x') - g^{\mu\nu}(x),$$

the definition of which involves the transformation of coordinates $x'^\mu = x^\mu + a\lambda^\mu(x)$.

Following Gupta[1] we shall describe the gravitational field by the quantity $h^{\mu\nu}(g^{\mu\nu} = \delta_{\mu\nu} + ah^{\mu\nu})$.[1)] From (1) it follows that

$$\delta^* h^{\mu\nu} = \partial_\mu\lambda^\nu + \partial_\nu\lambda^\mu + a(h^{\rho\nu}\partial_\rho\lambda^\mu + h^{\mu\rho}\partial_\rho\lambda^\nu - \lambda^\rho\partial_\rho h^{\mu\nu}). \quad (2)$$

We see that (2) differs from the tensor law by the additional term $\partial_\mu\lambda^\nu + \partial_\nu\lambda^\mu$ which is analogous to the gradient in gauge transformations of the electromagnetic field or of the Young-Mills field.

2. In future it will be convenient for us to use the matrix notation

$$g = \| g^{\mu\nu} \|; \quad g^{-1} = \| g_{\mu\nu} \|; \quad h = \| h^{\mu\nu} \|;$$
$$\Lambda = \| \Lambda_{\alpha\beta} \| \equiv \| \partial_\alpha\lambda^\beta \|, \quad (3)$$

so that[2)]

$$g = 1 + ah, \quad g^{-1} = \sum_{n=0}^{\infty}(-ah)^n;$$
$$(h^n)^{\mu\nu} = h^{\mu\sigma_1}h^{\sigma_1\sigma_2}\ldots h^{\sigma_{n-1}\nu}. \quad (3')$$

In terms of this notation we can write (2) in the following form:

$$\delta^* h = \Lambda + \tilde{\Lambda} + a(\tilde{\Lambda}h + h\Lambda - \lambda^\rho\partial_\rho h), \quad (4)$$

$$\delta^* g = a(\tilde{\Lambda}g + g\Lambda - \lambda^\rho\partial_\rho g),$$
$$\delta^* g^{-1} = -a(\Lambda g^{-1} + g^{-1}\tilde{\Lambda} + \lambda^\rho\partial_\rho g^{-1}). \quad (4')$$

3. It is well known that transformations of the fundamental tensor form a group. In the language of infinitesimal transformations (1) this has the following meaning. We denote the result of transformation (1) with $\lambda^\mu = \lambda_i^\mu$ by $\delta^*_{\lambda_i} g^{\mu\nu}$ and we consider the bracket operation well-known in the theory of continuous Lie groups

$$\delta_{\lambda_2}{}^*\delta_{\lambda_1}{}^* g^{\mu\nu} - \delta_{\lambda_1}{}^*\delta_{\lambda_2}{}^* g^{\mu\nu}.$$

Corresponding to the fact that the transformations form a group the result of the bracket operation must be capable of being represented in the form (1) with a new "bracket" λ^μ_{br}:

$$(\delta^*_{\lambda_2}\delta^*_{\lambda_1} - \delta^*_{\lambda_1}\delta^*_{\lambda_2})g^{\mu\nu} = \delta^*_{\lambda_{br}} g^{\mu\nu}. \quad (5)$$

It is not difficult to verify that this is indeed the case, and that

$$\lambda_{br}{}^\mu = -a(\lambda_1{}^\rho\partial_\rho\lambda_2{}^\mu - \lambda_2{}^\rho\partial_\rho\lambda_1{}^\mu). \quad (6)$$

The "not entirely" tensor quantity $h^{\mu\nu}$, and also all the tensors, transform in accordance with the representations of the same group. For all these objects relation (5) is satisfied, i.e.,

$$(\delta^*_{\lambda_2}\delta^*_{\lambda_1} - \delta^*_{\lambda_1}\delta^*_{\lambda_2})T^{\mu_1\mu_2\cdots}_{\nu_1\nu_2\cdots} = \delta^*_{\lambda_{br}} T^{\mu_1\mu_2\cdots}_{\nu_1\nu_2\cdots}, \quad (7)$$

where λ_{br} is always expressed by formula (6). This can be easily verified in any special case. And conversely, by solving the relation of the structure (7) to λ^μ_{br} (6) one can infer the law of transformation of a tensor with a given number of indices if in addition we require that the law should not depend on other tensors, that it should

[1)]In this article the imaginary time coordinate x^4 = it is used, and the Kronecker symbol $\delta_{\mu\nu}$ serves as the metric tensor of the special theory of relativity.

[2)]Here the upper indices turn out to be contracted with other upper indices, and this is related to our use of nontensor quantities. (Such contractions are unacceptable when one is constructing tensors from other tensors). In future we shall often use quantities with indices $\mu, \nu, \ldots$, which are not tensors.

be linear and homogeneous in terms of the given tensor and, of course, that under Lorentz transformations it should reduce to the usual Lorentz law.

3. THE TRANSFORMATION LAW FOR A SPINOR

In this section we shall turn to just such an inference of the transformation law of an object which differs from a tensor, viz., a spinor. We shall take a spinor to mean a four-component object which transforms in accordance with a representation of the group of transformations of $g^{\mu\nu}$ under discussion in such a way that this law in the special case of Lorentz rotations and translations should, when

$$a\lambda^\mu = 2\omega_\nu{}^\mu x^\nu + c^\mu,$$
$$\omega_\nu{}^\mu = -\omega_\mu{}^\nu = \text{const.} \quad c^\mu = \text{const}, \tag{8}$$

go over into the Lorentz law for a Dirac spinor:

$$\delta^* \psi = -a\lambda^\rho \partial_\rho \psi + {}^1/_4 ia\langle \Lambda \cdot \sigma \rangle \psi \text{ with } a\lambda^\mu = 2\omega_\nu{}^\mu x^\nu + c^\mu, \tag{9}$$

where $\langle \Lambda \cdot \sigma \rangle = \Lambda_{\alpha\beta}\sigma_{\beta\alpha}$, i.e., we use angular brackets to denote (here and later) the operation of contraction with respect to the vector indices not involving the spinor indices. Further we have

$$\sigma_{\beta\alpha} = -i(\gamma_\beta \gamma_\alpha - \delta_{\beta\alpha}),$$

where γ_α are the usual Dirac matrices:

$$\gamma_\alpha \gamma_\beta + \gamma_\beta \gamma_\alpha = 2\delta_{\alpha\beta}. \tag{10}$$

In seeking the transformation law we impose the following restrictions:

a) we assume in analogy with tensors that the desired law contains the spinor in a linear and homogeneous manner;

b) for simplicity we assume that the desired law differs from (9) only by a modification of the term containing the matrices $\sigma_{\beta\alpha}$, and that no new terms containing other matrices occur in it;

c) we assume that the desired law can contain in addition to the spinor field only the gravitational field $h^{\mu\nu}$, but must not contain any other fields, nor derivatives of $h^{\mu\nu}$. In other words, we assume that the spinor representation is nonlinear in $h^{\mu\nu}$.

Assumption c) is weaker than the corresponding assumption for tensors: "the desired law must not involve any other fields in addition to the given field." This relaxation is necessary, for if it is not made then it is not possible to construct spinors,[3] and this agrees with Cartan's[13] assertion. Indeed, if the spinor law does not contain any other fields, then, under assumptions a) and b) and with the necessary condition of reduction to the usual transformation law for a Dirac spinor with λ^μ given by (8), the desired law must necessarily have the form (9). But direct calculations show that for functions λ^μ having a general form different from (8) the transformations (9) do not form a group. Indeed, the use of the bracket operation utilizing the formula[4]

$$[\langle A \cdot \sigma \rangle, \langle B \cdot \sigma \rangle] = 2i\langle (A - \tilde{A}) \cdot (B - \tilde{B}) \cdot \sigma \rangle \tag{11}$$

yields

$$(\delta^*_{\lambda_2}\delta^*_{\lambda_1} - \delta^*_{\lambda_1}\delta^*_{\lambda_2})\psi = -a\lambda^\rho_{\text{br}}\partial_\rho\psi + {}^1/_4 ia\langle \Lambda_{\text{br}} \cdot \sigma \rangle \psi$$
$$+ {}^1/_8 ia^2\langle (\Lambda_1 + \tilde{\Lambda}_1) \cdot (\Lambda_2 + \tilde{\Lambda}_2) \cdot \sigma \rangle \psi. \tag{12}$$

The whole expression in the second line of (12) is superfluous; only for the special choice of λ^μ in the form (8) when Λ_1 and Λ_2 are antisymmetric matrices does this expression vanish. Thus, the desired law cannot coincide with (9) and must contain additional terms which would compensate for the extra term arising in (12) and which would vanish if Λ were antisymmetric.

In accordance with c) we introduce the field $h^{\mu\nu}$ into the law. By a direct analysis it can be shown that the addition to (9) of the term

$$-{}^1/_4 ia^2\langle (\Lambda + \tilde{\Lambda}) \cdot h \cdot \sigma \rangle \psi$$

compensates [because of the additive part of the variation of h(2)] for the extra term in (12), but in turn gives rise (as a result of the multiplicative part of the variation of h) to unnecessary additional terms in the bracket operation, which now contain a^3. In order to cancel the newly arisen additional terms it is necessary to add to (9) (at first with undetermined multipliers) terms which contain h quadratically, then cubically, etc. As a result of this frontal attack procedure involving repeated application of the bracket operation in such a way that the group property would be satisfied up to ever higher powers of the constant a, we found the first few terms of the desired transformation law:

$$\delta^*\psi = -a\lambda^\rho\partial_\rho\psi + {}^1/_4 ia\langle \Lambda \cdot \sigma \rangle \psi - {}^1/_{16} ia^2\langle (\Lambda + \tilde{\Lambda}) \cdot h \cdot \sigma \rangle \psi$$
$$+ {}^1/_{32} ia^3\langle (\Lambda + \tilde{\Lambda}) \cdot h^2 \cdot \sigma \rangle \psi - {}^1/_{256} ia^4\langle h \cdot (\Lambda + \tilde{\Lambda}) \cdot h^2 \cdot \sigma \rangle \psi$$
$$- {}^5/_{256} ia^4\langle (\Lambda + \tilde{\Lambda}) \cdot h^3 \cdot \sigma \rangle \psi + \ldots \tag{13}$$

It is now natural to seek the exact infinitesimal transformation law for a spinor in the form

[3]This assertion will hold even if we drop assumption b) and modify the desired law by including in it all 16 Dirac matrices.

[4]Formula (11) is obtained by multiplying $A_{\nu\mu}$ and $B_{\rho\lambda}$ by the commutation relation for the matrices $\sigma_{\mu\nu}$:

$$[\sigma_{\mu\nu}, \sigma_{\lambda\rho}] = 2i(\delta_{\mu\lambda}\sigma_{\nu\rho} + \delta_{\nu\rho}\sigma_{\mu\lambda} - \delta_{\mu\rho}\sigma_{\nu\lambda} - \delta_{\nu\lambda}\sigma_{\mu\rho}).$$

$$\delta^*\psi = -a\lambda^\rho\partial_\rho\psi + {}^1\!/_4 ia\langle[\Lambda + \Delta(\Lambda)]\cdot\sigma\rangle\psi, \qquad (14)$$

where $\Delta(\Lambda)$ is a matrix of the form

$$\Delta(\Lambda) = \sum_{m,\,n=0}^{\infty} a^{m+n} c_{mn} h^m (\Lambda + \tilde{\Lambda}) h^n. \qquad (15)$$

Since the matrix Δ is contracted with the antisymmetric matrix σ ($\langle\Delta\cdot\sigma\rangle = \Delta_{\mu\nu}\sigma_{\nu\mu}$ occurs in (14)), it is useful from the outset to take the matrix Δ as being antisymmetric: $\Delta = -\tilde{\Delta}$, so that

$$c_{mn} = -c_{nm}. \qquad (16)$$

Applying the bracket operation and at the same time utilizing (11) we obtain

$$(\delta_{\lambda_2}{}^*\delta_{\lambda_1}{}^* - \delta_{\lambda_1}{}^*\delta_{\lambda_2}{}^*)\psi = -a\lambda_{br}{}^\rho\partial_\rho\psi - {}^1\!/_4 ia^2(\lambda_1{}^\rho\partial_\rho\langle\Lambda_2\cdot\sigma\rangle$$
$$-\lambda_2{}^\rho\partial_\rho\langle\Lambda_1\cdot\sigma\rangle)\psi - {}^1\!/_8 ia^2\langle[\Lambda_1 - \tilde{\Lambda}_1 + 2\Delta(\Lambda_1)]\,[\Lambda_2$$
$$-\tilde{\Lambda}_2 + 2\Delta(\Lambda_2)]\cdot\sigma\rangle\psi + {}^1\!/_4 ia\langle[\delta_{\lambda_2}{}^*\Delta(\Lambda_1)$$
$$-\delta_{\lambda_1}{}^*\Delta(\Lambda_2) - a\lambda_1{}^\rho\partial_\rho\Delta(\Lambda_2) + a\lambda_2{}^\rho\partial_\rho\Delta(\Lambda_1)]\cdot\sigma\rangle\psi, \qquad (17)$$

where $\delta^*\Delta$ denotes the variation of Δ due to the variation of the h appearing in it which are varied in accordance with (2). Since we want the transformations (14) to form a representation of the group of transformations of $g^{\mu\nu}$, we must equate the result of the bracket operation to the variation (14):

$$(\delta_{\lambda_2}{}^*\delta_{\lambda_1}{}^* - \delta_{\lambda_1}{}^*\delta_{\lambda_2}{}^*)\psi = \delta^*_{\lambda_{br}}\psi, \qquad (18)$$

where λ_{br} is again given by formula (6). If in the left hand side we substitute expression (17) and write out the right hand side in detail, then in such form (18) will serve as a source for obtaining recurrence relations between the coefficients c_{mn}. (The principal steps in the calculation can be found in Appendix I to our paper[25].) From the recurrence relations it follows that c_{mn} can be defined by means of the generating function

$$G(x, y) = \sum_{m,\,n=0}^{\infty} c_{mn} x^m y^n,$$

$$G(x, y) = \frac{1}{2}\frac{(1+x)^{1/2} - (1+y)^{1/2}}{(1+x)^{1/2} + (1+y)^{1/2}}. \qquad (19)$$

We reproduce the first few coefficients:

$$c_{01} = -\frac{1}{8};\quad c_{02} = \frac{1}{16};\quad \ldots;\quad c_{0n} = (-1)^n\frac{(2n-1)!!}{(2n+2)!!},$$

$$c_{12} = -\frac{1}{128};\quad c_{13} = \frac{1}{128};\quad \ldots;$$

$$c_{1n} = (-1)^{n+1}\frac{(2n-1)!!}{(2n+4)!!}(n-1), \qquad (20)$$

. . . .

The appearance in (19) of square roots indicates that an important role must be played by the matrix $r = (1 + ah)^{1/2}$ which is uniquely defined by its expansion in series[5]:

$$= 1 + \frac{1}{2}ah - \frac{1}{8}a^2h^2 + \ldots + \frac{1}{2}\left(\frac{1}{2} - 1\right)\ldots$$
$$\times\left(\frac{1}{2} - n + 1\right)\frac{1}{n!}a^n h^n + \ldots$$
$$= (1 + ah)^{1/2} = \sqrt{g}. \qquad (21)$$

Representing the denominator of the generating function (19) as an integral of an exponential one can define the matrix Δ in the form

$$\Delta = {}^1\!/_2\int_0^\infty d\alpha\, e^{-\alpha r}[r, \Lambda + \tilde{\Lambda}]\,e^{-\alpha r}, \qquad (22)$$

where the square brackets denote a commutator.

Thus, it has been shown that the law of transformation of a spinor according to the representation of the group of transformations of the tensor $g^{\mu\nu}$ can be written in the form

$$\delta^*\psi = -a\lambda^\rho\partial_\rho\psi + {}^1\!/_4 ia\,\langle\Lambda\cdot\sigma\rangle\,\psi + {}^1\!/_8 ia\int_0^\infty d\alpha$$
$$\times\langle e^{-\alpha r}[r, \Lambda + \tilde{\Lambda}]\,e^{-\alpha r}\cdot\sigma\rangle. \qquad (23)$$

We emphasize that in (23) the field $h^{\mu\nu}$ appears in an essentially nonlinear manner, so that the spinor transforms according to a nonlinear representation of the group of generally covariant transformations.

We note that if we give up assumption b), then it is possible to introduce into (14) terms involving other matrices [over and above the terms which are always needed and which are present in (14)]. Thus, in seeking the law of transformation one could introduce additional terms which are multiples of the matrices 1 and γ_5. This lack of uniqueness can be interpreted as a transition to other objects. Thus, we introduce the quantity ψ_{vw} which transforms like $(\text{Det } g)^{(v+w\gamma_5)/2}\psi$. We shall call this quantity ψ_{vw} a spinor of weight $v + w\gamma_5$. Its variation may be written in the form

$$\delta^*\psi_{vw} = a(v + w\gamma_5)\langle\Lambda\rangle\psi_{vw} - a\lambda^\rho\partial_\rho\psi_{vw}$$
$$+ {}^1\!/_4 ia\langle[\Lambda + \Delta]\cdot\sigma\rangle\psi_{vw}. \qquad (24)$$

The first term on the right hand side is an additional term arising from the weight. It is not difficult to verify that the group property is again

---

5) I.e., in terms of indices:

$$r^{\mu\nu} = \delta_{\mu\nu} + {}^1\!/_2 ah^{\mu\nu} - {}^1\!/_8 a^2 h^{\mu\rho}h^{\rho\nu} + \ldots.$$

obeyed. Weighted spinors bear the same relationship to spinors as the well known weighted tensors do in relation to tensors[26].

In conclusion of this section we note that the law of transformation of the quantity r defined by formula (21) can be written in the form

$$\delta^* r = -a\lambda^\rho \partial_\rho r + a\tilde{\Lambda} r + {}^1/_2 ar(\Lambda - \tilde{\Lambda} + 2\Delta). \quad (25)$$

With the aid of relation

$$r\Delta + \Delta r = {}^1/_2 r(\Lambda + \tilde{\Lambda}) - {}^1/_2(\Lambda + \tilde{\Lambda})r \quad (26)$$

we can write the right hand side of (25) in different forms and, in particular, we can make it explicitly symmetric. For the derivation of (25) and (26) cf. Appendix 2 in reference [25].

It contrast to the usual tetrads in which one index refers to the usual basis system, while the second one to the locally orthogonal one, the indices of $r^{\mu\nu}$ have equal standing and refer to one general basis system. At the same time, with the aid of $r^{\mu\nu}$ one can also introduce a specific locally orthogonal basis with the differentials $dy^\mu = r^{\mu\nu} dx_\nu$, in which $dy^\mu dy^\mu = g^{\mu\nu} dx_\mu dx_\nu$. This remark enables us to give another derivation of the spinor law in accordance with the following outline:

1) $r^{\mu\nu}$ is introduced by means of the equation $r^2 = g = 1 + ah$ and is represented in the form (21);

2) $\delta^* r$ in (25) is evaluated. As a result of this the matrix Δ (15) or (22) is produced;

3) taking into account the fact that $dx'_\mu = dx_\mu - a\partial_\mu \lambda^\rho dx_\rho$ we find that under general transformations dy^μ undergo an induced local orthogonal transformation

$$dy'^\mu = r'^{\mu\nu}(x')dx'_\nu = dy^\mu - {}^1/_2 a(\Lambda - \tilde{\Lambda} + 2\Delta)^{\mu\nu} dy^\nu;$$

4) replacing in the usual law for the Dirac spinor (9) $a\Lambda$ by the "parameters" of the last transformation $\frac{1}{2}a(\Lambda - \tilde{\Lambda} + 2\Delta)$ we again obtain the spinor law (14).[6)]

4. COVARIANT DERIVATIVE OF A SPINOR

Fock and Ivanenko[4] and Fock[5] have defined a spinor as a geometric object on the basis of its behavior under parallel translation and directly from this have obtained the covariant derivative of a spinor.

Our definition of the covariant derivative of a spinor will be the direct consequence of the law obtained above for the transformation of the spinor (23) under general transformations. The covariant derivative of a tensor with respect to $g^{\mu\nu}$ can be defined to be such a modification of the ordinary derivative the application of which to a tensor again leads to a tensor of rank higher by unity[26,27]. Such an approach is equivalent to the approach utilizing parallel translation, but in contrast to the latter it is more explicitly related to the group property of tensors. Similarly, we shall define the covariant derivative of a spinor $\nabla_\mu \psi$ as an object which transforms in accordance with the direct product of a vector (with respect to the index μ) and a spinor (with respect to the spinor index of ψ) representation, i.e.,

$$\delta^* \nabla_\mu \psi = -a\lambda^\rho \partial_\rho (\nabla_\mu \psi) - a\partial_\mu \lambda^\rho \nabla_\rho \psi + {}^1/_4 ia \langle (\Lambda + \Delta(\Lambda)) \cdot \sigma \rangle \nabla_\mu \psi. \quad (27)$$

We represent the symbol for the covariant derivative in the form

$$\nabla_\mu = \partial_\mu - \Gamma_\mu. \quad (28)$$

We now substitute (28) into (27) and utilize the law of transformation of a spinor (14). We then obtain the following transformation law:

$$\delta^* \Gamma_\mu = -a\lambda^\rho \partial_\rho \Gamma_\mu - a\partial_\mu \lambda^\rho \Gamma_\rho + {}^1/_4 ia[\langle (\Lambda + \Delta) \cdot \sigma \rangle, \Gamma_\mu] + {}^1/_4 ia \langle \partial_\mu (\Lambda + \Delta) \cdot \sigma \rangle. \quad (29)$$

On the assumption that Γ_μ is completely expressed in terms of $h^{\mu\nu}$ (similarly to the affine relation for tensors in Riemannian geometry) (29) represents an inhomogeneous equation for the determination of Γ_μ.

We expand Γ_μ in terms of the complete system of Dirac matrices:

$$\Gamma_\mu = a_\mu I + a_\mu{}^\alpha \gamma_\alpha + a_\mu{}^{\alpha\beta} \sigma_{\beta\alpha} + a_\mu{}^{\alpha 5} i\gamma_\alpha \gamma_5 + a_\mu{}^5 \gamma_5. \quad (30)$$

If we substitute the expansion (30) into (29) we can obtain a particular solution of the inhomogeneous equation (29) in the form (cf., Appendix 3 to reference [25])

$$\Gamma_\mu = -{}^1/_4 ir^{\alpha\beta}[\partial_\beta g_{\mu\gamma} + (r^{-1}\partial_\mu r^{-1})_{\beta\gamma}] r^{\gamma\delta} \sigma_{\delta\alpha}$$
$$= -{}^1/_4 ir^{\alpha\beta}[-[\beta, \gamma\mu] + (r^{-1}\partial_\mu r^{-1})_{\beta\gamma}] r^{\gamma\delta} \sigma_{\delta\alpha}. \quad (31)$$

The last expression is written in terms of the three index Christoffel symbol of the first kind[26,27] in order to demonstrate the relationship of the expression obtained above for Γ_μ with the expression which was obtained in tetrad formalism [cf., reference [19], formula (8)]. We emphasize that Γ_μ (31) does not contain any new quantities and represents a series in powers of the gravi-

6) If by means of such an approach one generalizes the tensors of the special theory of relativity, one obtains not the usual tensors, but quantities which also transform according to nonlinear representations, laws of the type (42) – (44), cf., below. For example, instead of the vector law one would obtain (42).

tational field (cf., expansion (21) for r):

$$\Gamma_\mu = -{}^1/_4 i\{-a\partial_\alpha h^{\mu\beta} + a^2[h^{\mu\sigma}\partial_\alpha h^{\sigma\beta} + {}^1/_2\partial_\alpha h^{\mu\sigma}h^{\sigma\beta} - {}^1/_2 h^{\alpha\sigma}\partial_\sigma h^{\mu\beta} - {}^1/_4\partial_\mu h^{\alpha\sigma}h^{\sigma\beta}] + O(a^3)\}\sigma_{\beta\alpha}. \quad (32)$$

The general solution of (29) consists of (31) and of the general solution of the homogeneous equation corresponding to (29):

$$\Gamma^\mu_{\text{gen}} = \Gamma_\mu + a_\mu I + b_{\mu\alpha}r^{\alpha\beta}\gamma_\beta + c_{\mu\alpha\beta}r^{\alpha\gamma}r^{\beta\delta}\sigma_{\delta\gamma} + d_{\mu\alpha}r^{\alpha\beta}i\gamma_\beta\gamma_5 + a_\mu{}^5\gamma_5, \quad (33)$$

where a_μ and a^5_μ, $b_{\mu\alpha}$, $d_{\mu\alpha}$ and $c_{\mu\alpha\beta}$ are covariant vectors and tensors of the second and the third rank respectively (cf., Appendix 3 to reference[25]). In contrast to Γ_μ (the simplest possible solution of the inhomogeneous equation), all the remaining terms in (33) can either be considered equal to zero, or different from zero and constructed from some suitable fields. The introduction of the covariant derivative is useful for the construction of invariant interactions. The inclusion into the covariant derivative of the solution of the inhomogeneous equation, for example (31), is necessary for this. As regards the remaining terms appearing in (33), they are not necessary, and the corresponding interactions can always be written in invariant form separately.

Let us make an analogy. The electromagnetic interactions are always also included through the "covariant derivative" $\partial_\mu - ieA_\mu$. However, the choice of this covariant derivative is not unique to the same extent. Nothing prevents us from taking in place of $\partial_\mu - ieA_\mu$ the "covariant derivative" $\partial_\mu - ieA_\mu + f\gamma_\nu F_{\nu\mu}$, etc. The lack of uniqueness in (33) is of exactly the same type. For example, the tensor $c_{\mu\alpha\beta}$ can be realized in the form

$$c_{\mu\alpha\beta} = g_{\mu\alpha}\partial_\beta R - g_{\mu\beta}\partial_\alpha R,$$

where R is the scalar curvature. This would lead to an additional "non-minimal" interaction with the gravitational field through its higher derivatives.

We shall assume that the covariant derivative of a spinor is $\nabla_\mu = \partial_\mu - \Gamma_\mu$, where the affine relation of Γ_μ is given by expression (31). The interactions to which this leads we shall call "minimal". Such a choice of a covariant derivative is also convenient because in this case

$$\nabla_\mu(\bar\psi\psi) = \partial_\mu(\bar\psi\psi), \qquad \nabla_\mu(\bar\psi\gamma_5\psi) = \partial_\mu(\bar\psi\gamma_5\psi), \quad (34)$$

$$\nabla_\mu(r^{\nu\lambda}\bar\psi\gamma_\lambda\psi) = (\delta_\rho{}^\nu\partial_\mu + \Gamma_{\mu\rho}^\nu)(r^{\rho\lambda}\bar\psi\gamma_\lambda\psi), \quad (35)$$

$$\nabla_\mu(r^{\nu\lambda}\bar\psi\gamma_\lambda\gamma_5\psi) = (\delta_\rho{}^\nu\partial_\mu + \Gamma_{\mu\rho}^\nu)(r^{\rho\lambda}\bar\psi\gamma_\lambda\gamma_5\psi). \quad (36)$$

$$\nabla_\mu(r^{\nu\lambda}r^{\sigma\tau}\bar\psi\sigma_{\lambda\tau}\psi) = (\delta_\alpha{}^\nu\delta_\beta{}^\sigma\partial_\mu + \delta_\alpha{}^\nu\Gamma_{\mu\beta}^\sigma + \delta_\beta{}^\sigma\Gamma_{\mu\alpha}^\nu)(r^{\alpha\lambda}r^{\beta\tau}\bar\psi\sigma_{\lambda\tau}\psi). \quad (37)$$

The covariant derivatives in (34)–(37) are calculated taking into account the distributive property, for example,

$$\nabla_\mu(r^{\nu\lambda}\bar\psi\gamma_\lambda\psi) = (\nabla_\mu r^{\nu\lambda})\bar\psi\gamma_\lambda\psi + r^{\nu\lambda}(\nabla_\mu\bar\psi)\gamma_\lambda\psi + r^{\nu\lambda}\bar\psi\gamma_\lambda\nabla_\mu\psi.$$

In this calculation the covariant derivative of $r^{\nu\lambda}$ as a function of $g^{\alpha\beta}$ is equal to zero:

$$\nabla_\mu r^{\nu\lambda} = 0, \quad (38)$$

and this can be regarded as a consequence of the easily verified important identity

$$\partial_\mu r^{\sigma\rho} + \Gamma_{\mu\lambda}{}^\sigma r^{\lambda\rho} - ir^{\sigma\alpha}\,\mathrm{Sp}(\sigma_{\alpha\rho}\Gamma_\mu) = 0. \quad (39)$$

The form of the covariant derivatives (34)–(37) agrees with the fact that the combinations written out in this form transform like scalars, vectors and a tensor (cf., the next section). At the corresponding point of the formalism of orthogonal basis vectors Fock[5] has fixed the form of the covariant derivative with the aid of conditions (34) and (35). A requirement of this nature leaves in (33) an arbitrariness only in the choice of a_μ, while $b = c = d = a^5 = 0$. However, we note that the terms eliminated from the covariant derivative can be introduced into the invariant Lagrangian as new independent interactions.

In conclusion we briefly discuss weighted spinors ψ_{vw} (24). For these quantities additional terms will appear in (29)

$$a(v + w\gamma_5)\partial_\mu\langle\Lambda\rangle - aw\langle\Lambda\rangle[\Gamma_\mu, \gamma_5].$$

As a result for a weighted spinor one should take for the simplest affine relation

$$\Gamma_\mu{}^{(v,w)} = \Gamma_\mu - (v + w\gamma_5)\Gamma_{\mu\alpha}{}^\alpha \quad (40)$$

incomplete analogy with what occurs for tensors (cf., reference[26], p. 55). The application of the covariant derivative to the bilinear combinations of ψ_{vw} will yield results different from (34)–(37).

5. PROPERTIES OF BILINEAR COMBINATIONS

From (14) it follows that

$$\delta^*(\bar\psi\psi) = -a\lambda^\rho\partial_\rho(\bar\psi\psi), \quad (41)$$

$$\delta^*(\bar\psi\gamma_\mu\psi) = -a\lambda^\rho\partial_\rho(\bar\psi\gamma_\mu\psi) - {}^1/_2 a(\Lambda - \tilde\Lambda + 2\Delta)_{\mu\beta}\bar\psi\gamma_\beta\psi, \quad (42)$$

$$\delta^*(\bar\psi\sigma_{\mu\nu}\psi) = -a\lambda^\rho\partial_\rho(\bar\psi\sigma_{\mu\nu}\psi) - {}^1/_2 a(\Lambda - \tilde\Lambda + 2\Delta)_{\mu\beta}\bar\psi\sigma_{\beta\nu}\psi - {}^1/_2 a(\Lambda - \tilde\Lambda + 2\Delta)_{\nu\beta}\bar\psi\sigma_{\mu\beta}\psi, \quad (43)$$

$$\delta^*(\bar\psi\gamma_\mu\gamma_5\psi) = -a\lambda^\rho\partial_\rho(\bar\psi\gamma_\mu\gamma_5\psi) - {}^1/_2 a(\Lambda - \tilde\Lambda + 2\Delta)_{\mu\beta}\bar\psi\gamma_\beta\gamma_5\psi, \quad (44)$$

$$\delta^*(\bar\psi\gamma_5\psi) = -a\lambda^\rho\partial_\rho(\bar\psi\gamma_5\psi), \quad (45)$$

$$\delta^*(\bar\psi\gamma_\mu\nabla_\nu\psi) = -a\lambda^\rho\partial_\rho(\bar\psi\gamma_\mu\nabla_\nu\psi) - a\partial_\nu\lambda^\rho\bar\psi\gamma_\mu\nabla_\rho\psi - {}^1/_2 a(\Lambda - \tilde\Lambda + 2\Delta)_{\mu\beta}\bar\psi\gamma_\beta\nabla_\nu\psi. \quad (46)$$

Consequently, the quantities $(\overline{\psi}\psi)$ and $\overline{\psi}\gamma_5\psi$ transform like scalars, while the remaining quantities do not transform according to tensor laws. Combinations obtained by multiplying by a nontensor quantity $r^{\mu\nu}$ will transform like tensors; the quantities $r^{\mu\nu}\overline{\psi}\gamma_\nu\psi$ and $r^{\mu\nu}\overline{\psi}\gamma_\nu\gamma_5\psi$ are, in terms of the usual terminology, contravariant vectors, $r^{\mu\alpha}r^{\nu\beta}\psi\sigma_{\alpha\beta}\psi$ is a contravariant antisymmetric tensor of the second rank, while $r^{\mu\alpha}\overline{\psi}\gamma_\alpha\nabla_\nu\psi$ is a mixed tensor of the second rank; the contraction of the latter tensor is a scalar which appears in the Lagrangian for the spinor field.

6. INTERACTIONS OF A SPINOR FIELD

The Lagrangian density must be not a scalar which transforms in accordance with $\delta^*\varphi = -a\lambda^\rho\partial_\rho\varphi$, but a relative scalar of weight 1 which under general transformations changes infinitesimally by a divergence:

$$\delta^*\mathcal{L} = -a\partial_\rho(\lambda^\rho\mathcal{L}).$$

Then the total Lagrangian will be an invariant. Knowing the transformation properties of bilinear combinations of spinors one can easily construct a Lagrangian density for the interaction of a spinor field simultaneously with the gravitational field a and an electromagnetic field A

$$\mathcal{L} = -\{{}^1/_2 r^{\mu\nu}[\overline{\psi}\gamma_\mu(\partial_\nu - \Gamma_\nu - ieA_\nu)\psi - \overline{\psi}(\overleftarrow{\partial}_\nu + \Gamma_\nu + ieA_\nu)\gamma_\mu\psi] + m\overline{\psi}\psi\}(\mathrm{Det}\, g)^{-1/2}. \quad (47)$$

By means of integration by parts this Lagrangian can be reduced to a simpler form, which, however, is not selfconjugate:

$$\mathcal{L} = -\{r^{\mu\nu}\overline{\psi}\gamma_\mu(\partial_\nu - \Gamma_\nu - ieA_\nu)\psi + m\overline{\psi}\psi\}(\mathrm{Det}\, g)^{-1/2}, \quad (48)$$

and from this the Dirac equation in a gravitational field immediately follows:

$$r^{\mu\nu}\gamma_\mu(\partial_\nu - \Gamma_\nu - ieA_\nu)\psi + m\psi = 0. \quad (49)$$

The Lagrangian density (47) (in contrast to the Lagrangian density in the formalism of an orthogonal basis system) is explicitly expressed in terms of the gravitational and other fields and represents an infinite series in powers of the gravitational field $h^{\mu\nu}$. We obtain the initial terms of the series. In order to do this we utilize the expansions of $r^{\mu\nu}$ (21), Γ_μ (32) and [7)]

$$\mathrm{Det}\, g = 1 + ah_1 + {}^1/_2 a^2(h_1^2 - h_2) + {}^1/_6 a^3(h_1^3 - 3h_1h_2 + 2h_3) + {}^1/_{24}a^4(h_1^4 - 6h_1^2h_2 + 8h_1h_3 + 3h_2^2 - 6h_4), \quad (50)$$

$$(\mathrm{Det}\, g)^{-1/2} = 1 - {}^1/_2 ah_1 + {}^1/_8 a^2(h_1^2 + 2h_2) + \ldots;$$
$$h_1 = h^{\alpha\alpha},\ h_2 = h^{\alpha\beta}h^{\beta\alpha},\ h_3 = h^{\alpha\beta}h^{\beta\gamma}h^{\gamma\alpha},\ h_4 = h^{\alpha\beta}h^{\beta\gamma}h^{\gamma\delta}h^{\delta\alpha}. \quad (51)$$

Then

$$\mathcal{L} = -{}^1/_2[\overline{\psi}\gamma_\mu(\partial_\nu - ieA_\nu)\psi - (\partial_\nu + ieA_\nu)\overline{\psi}\gamma_\mu\psi]\{\delta_{\mu\nu} + {}^1/_2 a(h^{\mu\nu} - \delta_{\mu\nu}h_1) + {}^1/_8 a^2[\delta_{\mu\nu}(h_1^2 + 2h_2) - h^{\mu\nu}h_1 - h^{\mu\rho}h^{\rho\nu}]\} - m\overline{\psi}\psi[1 - {}^1/_2 ah_1 + {}^1/_8 a^2(h_1^2 + 2h_2)] + {}^1/_{16}a^2\varepsilon_{\mu\nu\lambda\rho}\overline{\psi}\gamma_5\gamma_\mu\psi h^{\nu\sigma}\partial_\lambda h^{\sigma\rho} + O(a^3). \quad (52)$$

Continuing with the expansion of (47) one can also easily obtain further terms of the series. In the linear approximation in $ah^{\mu\nu}$ the interaction of fermions with gravitons has been considered already[1,28]. The terms of the second order in a written out in (52) enable us to calculate the gravitational self-energy, the Compton-effect of gravitons, etc. One can also easily write down the interactions with other fields. For example, pseudoscalar coupling with a pseudoscalar field φ can be written down in the following manner:

$$\mathcal{L} = if\overline{\psi}\gamma_5\psi\varphi(\mathrm{Det}\, g)^{-1/2}. \quad (53)$$

The four-fermion interactions also have a very simple appearance; the presence of a gravitational field leads only to the appearance of $(\mathrm{Det}\ g)^{-1/2}$:

$$\mathcal{L} = \{f_S(\overline{\psi}_1\psi_2)(\overline{\psi}_3\psi_4) + f_V(\overline{\psi}_1\gamma_\mu\psi_2)(\overline{\psi}_3\gamma_\mu\psi_4) + f_T(\overline{\psi}_1\sigma_{\mu\nu}\psi_2)(\overline{\psi}_3\sigma_{\mu\nu}\psi_4) + \ldots\}(\mathrm{Det}\, g)^{-1/2}. \quad (54)$$

We emphasize that the γ-matrices in (54) are general numerical Dirac matrices:

$$\gamma_\mu\gamma_\nu + \gamma_\nu\gamma_\mu = 2\delta_{\mu\nu}.$$

In conclusion we shall state the result of squaring the Dirac equation in the electromagnetic and the gravitational fields (49):

$$[(\mathrm{Det}\, g)^{1/2}(\partial_\mu - \Gamma_\mu - ieA_\mu)(\mathrm{Det}\, g)^{-1/2}g^{\mu\nu}(\partial_\nu - \Gamma_\nu - ieA_\nu) + {}^1/_4 R + {}^1/_2 e r^{\mu\lambda}r^{\nu\rho}F_{\mu\nu}\sigma_{\lambda\rho} - m^2]\psi = 0,$$

where R is the scalar curvature. This equation was first derived by Fock (in the orthogonal basis formalism), while a convenient system for the intermediate calculations needed for its derivation may be found in Schrödinger's paper[10]. In our formalism the calculations are carried out in an analogous manner.

The authors are grateful to M. A. Markov for discussion.

[7)] Formula (50) follows from the relation
$$\mathrm{Det}\,\|g^{\mu\nu}\| = {}^1/_{24}\{(g^{\alpha\alpha})^4 - 6(g^{\alpha\alpha})^2g^{\beta\gamma}g^{\gamma\beta} + 3(g^{\alpha\beta}g^{\beta\alpha})^2 + 8g^{\alpha\alpha}g^{\beta\gamma}g^{\gamma\delta}g^{\delta\beta} - 6g^{\alpha\beta}g^{\beta\gamma}g^{\gamma\delta}g^{\delta\alpha}\}.$$

[1] S. Gupta, Proc. Phys. Soc. (London) A65, 161, 608 (1952); Rev. Mod. Phys. 29, 334 (1957).

[2] H. Tetrode, Z. Physik 50, 336 (1928).

[3] E. Wigner, Z. Physik 53, 592 (1929).
[4] V. A. Fock and D. D. Ivanenko, Compt. rend. 188, 1470 (1929); Physik. Z. 30, 648 (1929).
[5] V. A. Fock, Compt. Rend. 189, 25 (1929); Z. Physik 57, 261 (1929); ZhRFKhO(Journal of the Russian Physico-chemical Society), Physics Section, 62, 133 (1930).
[6] H. Weyl, Proc. Natl. Acad. Sci. U. S. 15, 323 (1929); Z. Physik 56, 330 (1929).
[7] R. Zaycoff, Ann. Physik 7, 650 (1930).
[8] P. Podolsky, Phys. Rev. 37, 1398 (1931).
[9] J. A. Schouten, J. Math. & Phys. 10, 239 (1931).
[10] E. Schrödinger, Berl. Ber., 1932, p. 105.
[11] V. Bargmann, Berl. Ber., 1932, p. 346.
[12] L. Infeld and B. L. van der Waerden, Berl. Ber., 1933, p. 380.
[13] E. Cartan, Lecons sur la Theorie des Spineurs, Hermann, Paris, 1938 (Russ. Transl., IIL, 1947).
[14] F. Belinfante, Physica 7, 305 (1940).
[15] W. Bade and H. Jehle, Revs. Modern Phys. 25, 714 (1953).
[16] M. Riesz, Lund Univ. Math. Sem. 12, (1954).
[17] Yu. B. Rumer, Issledovaniya po 5-optike (Investigations in 5-optics), Gostekhizdat, 1956.
[18] P. G. Bergmann, Phys. Rev. 107, 624 (1957).
[19] D. Brill and J. A. Wheeler, Rev. Modern Phys. 29, 465 (1957).
[20] P. A. M. Dirac, Max-Planck Festschrift, Berlin, 1958, p. 339.
[21] J. G. Fletcher, Nuovo Cimento 8, 451 (1958).
[22] A. Peres, Nuovo Cimento 28, 865 (1963); J. Math. Phys. 5, 720 (1964); Nuovo Cimento Suppl. 24, 389 (1962).
[23] L. D. Landau and E. M. Lifshitz, Teoriya polya (Field Theory), Fizmatgiz, 1960.
[24] W. Pauli, Relativitätstheorie, Teubner Berlin, 1921 (Russ. Transl. IIL, 1947).
[25] V. Ogievetskiǐ and I. Polubarinov, JINR Preprint R-1890, 1964.
[26] O. Veblen, Invariants of Quadratic Differential Forms, Cambridge, 1933 (Russ. Transl. IIL, 1948).
[27] L. P. Eisenhart, Riemannian Geometry, Princeton, 1926 (Russ. Transl. IIL, 1948).
[28] I. Yu. Kobzarev and L. B. Okun', JETP 43, 1904 (1962), Soviet Phys. JETP 16, 1343 (1963).
[29] C. Moller, Mat.-Fys. Skr. Dan. Vid. Selskab. 1, 10 (1961).

Translated by G. Volkoff
235

The Notoph and its Possible Interactions

V. I. Ogievetskiĭ and I. V. Polubarinov
Joint Institute for Nuclear Research

Submitted to JNP editor November 5, 1965
J. Nucl. Phys. (U.S.S.R.) 4, 216–223 (July, 1966)

For particles with zero mass, unlike particles with mass, the value of the helicity (the projection of the total angular momentum along the momentum) is a relativistic invariant. We discuss a massless particle with zero helicity; its properties are complementary to those of the photon (helicity ±1), and therefore it is called the "notoph." In interactions the notoph, like the photon, transfers the spin 1. The notoph is described by an antisymmetric tensor potential, and the field strength is a four-vector (instead of the vector potential and the electromagnetic field strength). Possible interactions of the notoph are discussed.

1. INTRODUCTION

For particles with nonzero mass m the projection of the spin in the direction of the momentum (the helicity) is not a relativistic invariant. A particle with spin s has $2s \pm 1$ states of polarization, which get transformed in terms of each other in Lorentz transformations. In the limit $m \to 0$ (or $v \to c$) the helicity becomes a relativistic invariant (see Appendix), and the concept of spin loses its meaning.[1] The system of $2s + 1$ states is no longer irreducible; it decomposes and describes a set of different particles with zero mass and helicities $\pm s, \pm(s-1), \ldots, \pm 1, 0$ (for integer spin and if parity is conserved; the situation is analogous for half-integer spins). The frequent statement "the particle with zero mass and spin s has only two spin states with spin projections $\pm s$" is only jargon. Particles with zero mass are characterized by the value of the helicity alone, and the concept of spin does not apply to them.

In the field theory of a particle with integer (half-integer) spin s is as a rule described by a completely symmetric tensor $A_{\mu_1 \ldots \mu_s}$ (a spin-tensor $\psi_{\mu_1 \ldots \mu_{s-1/2}}$) which satisfies an equation which fixes the spin s (cf., e.g., [2]). When we go over to zero mass such a formalism describes only the particle with helicity $\pm s$. In fact, when we go over to $m = 0$ the equation for a particle with mass m and spin s must decompose into $s + 1$ (if parity is conserved) equations, which describe different massless particles with helicities $\pm s, \pm(s-1), \ldots, \pm 1$, 0 (and analogously in the case of half-integer spins).

The question arises: what happens to the other, non-maximum, values of the helicity, which disappear in the standard formalism? How are we to describe the particles with zero mass and helicity less than s, into which massive particles with spin s also go over in the limit $m \to 0$? It turns out that for this we must use not only the completely symmetric tensors (or spin-tensors), as is the usual practice, but also tensors of various ranks symmetrized according to other Young diagrams, capable of describing spin s for nonzero mass. In the present paper we demonstrate this with the example of spin 1. Here in the limit $m \to 0$ we get besides the photon (helicity ±1) also a particle with zero helicity. The properties of this particle are complementary to those of the photon, and we shall give it the name "notoph." The notoph is described by an antisymmetric field tensor $f_{\mu\nu}(x)$. In interactions this field has all three polarization states of spin 1, just as the electromagnetic field does. The point is that a virtual notoph does not have zero mass, and therefore acquires an additional state with helicity ±1 (in complete analogy with the virtual photon, which has an additional state with zero helicity).

In this paper we discuss possible interactions of the notoph, and also give brief consideration to the equivalent vector and tensor formalisms for nonzero mass.

2. ANALYSIS OF THE EQUATIONS OF MOTION FOR THE NOTOPH

1. A well known example of a particle with zero mass is the photon, which has two states of polarization (helicity ±1) and is described by a vector potential A_μ obeying the Maxwell equations

$$\Box A_\mu - \partial_\mu \partial_\nu A_\nu = -j_\mu. \tag{1}$$

If a mass term $-m^2 A_\mu$ were included in this equation it would describe a particle with mass, with spin 1, and with three helicity values: ±1 and 0. In the limit $m \to 0$ the state with zero helicity disappears.

2. Our purpose is to call attention to a theory in which a complementary situation is realized; when the mass is different from zero we have a description of a particle with spin 1, and in the limit $m \to 0$ there remains one state, with zero helicity.

It turns out that instead of the four-vector potential with which one describes the photon, it is convenient to describe the notoph with an antisymmetric field tensor $f_{\mu\nu}(x)$ (tensor potential). We adopt for $f_{\mu\nu}$ an equation analogous to the Maxwell equation:

$$\Box f_{\mu\nu} - \partial_\mu \partial_\lambda f_{\lambda\nu} + \partial_\nu \partial_\lambda f_{\lambda\mu} = -J_{\mu\nu};$$

$$J_{\mu\nu} = -J_{\nu\mu}, \quad \partial_\mu J_{\mu\nu} = 0. \tag{2}$$

Equations (1) and (2) are invariant under the respective gauge transformations of the vector potential and the tensor potential:

$$\delta A_\mu = \partial_\mu \Lambda, \quad \delta f_{\mu\nu} = \partial_\mu \lambda_\nu - \partial_\nu \lambda_\mu, \tag{3}$$

where $\Lambda(x)$ and $\lambda_\mu(x)$ are completely arbitrary functions. These equations can be derived from the requirement of invariance under these transformations.

3. We shall show that the free notoph has only one state of polarization. For added clarity it is interesting to consider the free notoph in parallel with the well known case of the free photon. Using the Lorentz gauge, we can write free-particle equations equivalent to (1) and (2) in the forms

$$\Box A_\mu = 0, \quad \Box f_{\mu\nu} = 0, \tag{4}$$

$$\partial_\mu A_\mu = 0, \quad \partial_\mu f_{\mu\nu} = 0 \tag{5}$$

(in the quantum case the supplementary conditions (5) are imposed on the physical state vectors). Equations (4) and (5) are still invariant under the gauge transformations (3), but with the following restrictions on the forms of the gauge functions:

$$\Box \Lambda = 0, \quad \Box \lambda_\mu - \partial_\mu \partial_\nu \lambda_\nu = 0. \tag{6}$$

Let us go over to momentum space:

$$A_\mu(x) = \int d\mathbf{p}\, A_\mu(\mathbf{p}) e^{ipx} + \text{h. c.},$$

$$f_{\mu\nu}(x) = \int d\mathbf{p}\, f_{\mu\nu}(\mathbf{p}) e^{ipx} + \text{h. c.},$$

$$px = \mathbf{px} - p_0 x_0, \quad p_0 = |\mathbf{p}|. \tag{7}$$

To enumerate the states we expand $A_\mu(\mathbf{p})$ and $f_{\mu\nu}(\mathbf{p})$ in terms of a complete basis p_μ, $e_\mu^{(1)}$, $e_\mu^{(2)}$, n_μ with the properties

$$(e^{(i)}p) = 0, \quad (e^{(i)}e^{(j)}) = \delta_{ij},$$

$$(e^{(i)}n) = 0, \quad p^2 = 0, \quad n^2 = -1. \tag{8}$$

Since this basis contains an isotropic vector it is necessarily nonorthogonal. The expansions are written in the following way:

$$A_\mu(\mathbf{p}) = \sum_{i=1}^{2} \alpha_i e_\mu^{(i)} + \beta p_\mu + \gamma n_\mu,$$

$$f_{\mu\nu}(\mathbf{p}) = \delta(e_\mu^{(1)} e_\nu^{(2)} - e_\nu^{(1)} e_\mu^{(2)})$$

$$+ \sum_{i=1}^{2} \varepsilon_i (e_\mu^{(i)} p_\nu - e_\nu^{(i)} p_\mu)$$

$$+ \sum_{i=1}^{2} \eta_i (e_\mu^{(i)} n_\nu - e_\nu^{(i)} n_\mu) + \xi (p_\mu n_\nu - p_\nu n_\mu). \tag{9}$$

The supplementary conditions (5) exclude all terms containing n_μ (i.e., $\gamma = \eta_i = \xi = 0$), and the gauge invariance makes the components containing p_μ unessential. Accordingly the free notoph actually possesses one polarization state, and the free photon two:

$$A_\mu(\mathbf{p}) = \sum_{i=1}^{2} \alpha_i e_\mu^{(i)}(\mathbf{p}),$$

$$f_{\mu\nu}(\mathbf{p}) = \delta\,[e_\mu^{(1)}(\mathbf{p}) e_\nu^{(2)}(\mathbf{p}) - e_\nu^{(1)}(\mathbf{p}) e_\nu^{(2)}(\mathbf{p})]. \tag{10}$$

It is easily verified that the helicity of the notoph is zero, and that of the photon ± 1; the photon and notoph complement each other, as it were.

4. We note one further curious manifestation of the complementary relation of these particles. The respective potentials in the theories of the photon and the notoph are the vector potential A_μ and the tensor potential $f_{\mu\nu}$. The field strengths, which are independent of the gauge, are (for the notoph the field strength is a four-vector!)

$$F_{\mu\nu} = \partial_\mu A_\nu - \partial_\nu A_\mu, \quad a_\mu = {}^1\!/_2 i \varepsilon_{\mu\nu\lambda\rho} \partial_\nu f_{\lambda\rho}. \tag{11}$$

In terms of the field strengths the equations of motion can be written in the following way (the first two equations are Maxwell's equations):

$$\partial_\mu F_{\mu\nu} = -j_\nu, \quad \partial_\mu \check{F}_{\mu\nu} = 0,$$

$$\partial_\mu a_\nu - \partial_\nu a_\mu = -\frac{i}{2} \varepsilon_{\mu\nu\lambda\rho} J_{\lambda\rho}, \quad \partial_\mu a_\mu = 0. \tag{12}$$

3. EQUIVALENT VECTOR AND TENSOR FORMALISMS FOR NONZERO MASS

Vector and tensor formalisms for the description of a particle with mass and with spin 1, in interaction with an external current, have been considered by Kemmer.[3] In this case the equations

$$\Box A_\mu - \partial_\mu \partial_\nu A_\nu - m^2 A_\mu = -j_\mu,$$

$$\Box f_{\mu\nu} - \partial_\mu \partial_\lambda f_{\lambda\nu} + \partial_\nu \partial_\lambda f_{\lambda\mu} - m^2 f_{\mu\nu} = -J_{\mu\nu} \tag{13}$$

are equivalent,[3] under the necessary condition that the external currents satisfy the relations

$$\partial_\mu j_\mu = 0, \quad J_{\mu\nu} = -\frac{i}{m} \varepsilon_{\mu\nu\lambda\rho} \partial_\lambda j_\rho, \tag{14}$$

In fact, the equations reduce to the same system of first-order equations

$$\partial_\mu F_{\mu\nu} - mA_\nu = -\frac{1}{m} j_\nu, \tag{15a}$$

$$\partial_\mu A_\nu - \partial_\nu A_\mu - mF_{\mu\nu} = 0;$$

$$F_{\mu\nu} = \frac{i}{2}\varepsilon_{\mu\nu\lambda\rho} f_{\lambda\rho}. \tag{15b}$$

We note the dual relation of the vector and tensor formalisms. In the former (15a) is the equation of motion, and (15b) is the definition of $F_{\mu\nu}$. In the second case, on the other hand, (15b) is the equation of motion, and (15a) is the definition of A_ν. In correspondence with this, of the two supplementary conditions

$$\partial_\mu A_\mu = 0, \qquad \partial_\mu f_{\mu\nu} = 0 \tag{16}$$

one is a consequence of the equations of motion, and the other is a trivial consequence of a definition.

Subject to these supplementary conditions the vector and tensor fields describe only one spin 1, and the energy is positive definite (without them the vector field would describe spins 1 and 0, and the tensor field, two spins 1).[4]

If a field $f_{\mu\nu}$ with mass interacts not with an external current but with other fields, the Kemmer procedure leads to essentially nonlinear interactions. Let us consider three cases of interaction of the vector field.

1. The theory of a neutral vector field interacting with spinors through a conserved current (in the limit $m = 0$ this is spinor electrodynamics):

$$\mathcal{L} = -\frac{1}{4}(\partial_\mu A_\nu - \partial_\nu A_\mu)^2 - \frac{m^2}{2} A_\mu A_\mu + ie\bar{\psi}\gamma_\mu\psi A_\mu$$
$$-\bar{\psi}(\gamma\partial + M)\psi. \tag{17}$$

2. The corresponding theory of interaction with a scalar field:

$$\mathcal{L} = -\frac{1}{4}(\partial_\mu A_\nu - \partial_\nu A_\mu)^2 - \frac{m^2}{2} A_\mu A_\mu -$$
$$-(\partial_\mu + ieA_\mu)\varphi^*(\partial_\mu - ieA_\mu)\varphi - M^2\varphi^*\varphi. \tag{18}$$

3. The theory of a massive Yang-Mills field in interaction also with a spinor field (for notations see the last of the papers [4]):

$$\mathcal{L} = -\frac{1}{4}(\partial_\mu b_\nu{}^i - \partial_\nu b_\mu{}^i - a_{ijk}\, b_\mu{}^j\, b_\nu{}^k)^2 - \frac{m^2}{2} b_\mu{}^i\, b_\mu{}^i$$
$$-\bar{\psi}[\gamma_\mu(\partial_\mu + iT_j b_\mu{}^j) + M]\psi. \tag{19}$$

We transform the corresponding equations of motion into equations of motion for a tensor field and equations for other fields interacting with it. Then in the tensor formalism the following are the respective Lagrangian densities for the three cases of interaction of the vector field

$$\mathcal{L} = -\frac{1}{4}\partial_\lambda f_{\mu\nu}\partial_\lambda f_{\mu\nu} + \frac{1}{2}\partial_\lambda f_{\lambda\nu}\partial_\sigma f_{\sigma\nu} - \frac{m^2}{4} f_{\mu\nu} f_{\mu\nu}$$
$$-\frac{1}{2} g\varepsilon_{\mu\nu\lambda\rho}\bar{\psi}\gamma_\nu\psi\partial_\mu f_{\lambda\rho} - \frac{1}{2} g^2\bar{\psi}\gamma_\mu\psi\bar{\psi}\gamma_\mu\psi$$
$$-\bar{\psi}(\gamma\partial + M)\psi, \tag{17'}$$

$$\mathcal{L} = \frac{1}{2}(1 + 2g^2\varphi^*\varphi)^{-1}$$
$$\times\left[\frac{i}{2}\varepsilon_{\nu\mu\lambda\rho}\partial_\nu f_{\lambda\rho} - ig(\varphi^*\partial_\mu\varphi - \partial_\mu\varphi^*\varphi)\right]^2$$
$$-\frac{m^2}{4} f_{\mu\nu} f_{\mu\nu} - \partial_\mu\varphi^*\partial_\mu\varphi - M^2\varphi^*\varphi, \tag{18'}$$

$$\mathcal{L} = \frac{1}{4}(\partial_\sigma G^i_{\sigma\mu} - i\bar{\psi}T'^i\gamma_\mu\psi)(1 + \beta^k G^k)^{-1}_{\mu\nu,\, ij}$$
$$\times(\partial_\tau G^j_{\tau\nu} - i\bar{\psi}T'^j\gamma_\nu\psi)$$
$$+\frac{m}{4} G^i_{\mu\nu} G^i_{\mu\nu} - \bar{\psi}(\gamma\partial + M)\psi, \tag{19'}$$

where in the first and second expressions $g = e/m$, and in the last

$$G^k = \|G_{\mu\nu}{}^k\|, \quad \beta^k = \frac{1}{m}\|a_{ijk}\|, \quad T'^i = \frac{1}{m} T^i.$$

For brevity the last Lagrangian has been written in terms of a field $G_{\mu\nu}{}^i$ analogous to $F_{\mu\nu}$, and not in terms of $f^{\mu\nu}$.[1)]

As we see, in the tensor formalism essentially nonlinear terms get added to the Yukawa couplings. Furthermore in the cases (18') and (19') the interactions are infinite series in the coupling constants, a situation similar to what occurs in the theory of gravitation. An important point is the appearance of an interaction of the current-current type. In particular, in the case (17') a four-fermion vector interaction has appeared.

4. EXAMPLES OF INTERACTIONS OF THE NOTOPH

1. In accordance with the form of the equation of

[1)] The field $G_{\mu\nu}{}^i$ satisfies the nonlinear supplementary condition

$$\partial_\mu \check{G}_{\mu\nu}{}^i - \beta_{ijk}(1 + \beta^n G^n)^{-1}_{\mu\lambda,\, jl}(\partial_\sigma G_{\sigma\lambda}{}^l - i\bar{\psi}T'^l\gamma_\nu\psi)\check{G}_{\mu\nu}{}^k = 0,$$

where $G_{\mu\nu} = (i/2)\epsilon_{\mu\nu\lambda\rho}G_{\lambda\rho}$. With a suitable change of the field variables, however, one can also go over to the Lorentz condition.

motion, the propagator for the notoph potential is written in the form

$$\overline{f_{\mu\nu}(x)\, f_{\lambda\rho}}(y) = -\frac{i}{(2\pi)^4}(\delta_{\mu\lambda}\delta_{\nu\rho} - \delta_{\mu\rho}\delta_{\nu\lambda})\int \frac{d^4p\, e^{ip(x-y)}}{p^2 - i\varepsilon}, \tag{20}$$

and that for the notoph field strength $a_\mu(x)$ [cf. (11)] in the form

$$\overline{a_\mu(x)\, a_\nu}(y) = \frac{i}{(2\pi)^4}\int d^4p \left(\delta_{\mu\nu} - \frac{p_\mu p_\nu}{p^2 - i\varepsilon}\right) e^{ip(x-y)}. \tag{21}$$

2. In looking for possible interactions it is interesting first to examine the theories (17′)–(19′) for $m = 0$. It is found that all of these theories are theories of the free notoph. For brevity we take up only the case (17′). Here the propagator (21) corresponds to virtual notoph lines. The term $p_\mu p_\nu/p^2$ drops out owing to conservation of current. The line for a free (emitted or absorbed) notoph in a Feynman diagram corresponds to a notoph intensity vector $a_\mu(q)$ with $q^2 = 0$. The structure of the matrix element is then the same as in electrodynamics, with the addition of a four-fermion interaction $\bar\psi\gamma_\mu\psi\bar\psi\gamma_\mu\psi$. Therefore for the matrix element for emission (or absorption) of a notoph, $a_\mu(q)M_\mu$, we have the relation

$$q_\mu M_\mu = 0 \tag{22}$$

(as is also true in electrodynamics). It follows from this that in this version of the theory notophs are not emitted or absorbed. In face, for the free notoph ($q^2 = 0$), owing to (10) and (11) and the fact that the basis contains the isotropic vector q_μ, we find

$$a_\mu(q) = -\varepsilon_{\mu\nu\lambda\rho} q_\nu e_\lambda^1 e_\rho^2 = q_\mu, \tag{23}$$

i.e., the intensity vector of a free notoph is the gradient of a scalar. Comparing (22) and (23) we see that the matrix elements corresponding to processes with emission or absorption of a notoph are equal to zero.

Furthermore, since with current conservation the propagators for the notoph intensities have no poles, the virtual notoph lines collapse, and the interaction through a notoph reduces to a four-fermion interaction. At the same time the Lagrangian (17′) contains the four-fermion interaction separately, and in such a way that the total interaction vanishes for $m = 0$. For example, the sum of the contributions from the diagrams of Fig. 1 or of Fig. 2 is zero. It thus turns out that for $m = 0$ the theories with the Lagrangian densities (17′)–(19′) are theories of the free notoph. This is natural, since in the case of vector theories with conserved currents, Eq. (17′)–(19′), for $m = 0$ the photon takes on itself the entire interaction, which confirms the correctness of the transition to $m = 0$ in these theories.

3. In the discussion of the possible interactions of the notoph and the analysis of effects it is useful to know the general structure of the vertex (Fig. 3). Since the current $J_{\mu\nu}$ must be conserved and be antisymmetric, the general structure of the spinor vertex can be written, for positive parity of the notoph, in the form

$$J_{\mu\nu} = A\{(m_1 - m_2)\bar u(q)\sigma_{\mu\nu}u(p) - l_\mu \bar u(q)\gamma_\nu u(p) + l_\nu \bar u(q)\gamma_\mu u(p)\} + B\varepsilon_{\mu\nu\lambda\rho}k_\lambda \bar u(q)\gamma_\rho\gamma_5 u(p), \tag{24}$$

and for negative parity, in the form

$$J_{\mu\nu} = C\Big\{\bar u(q)\sigma_{\mu\nu}\gamma_5 u(p) + \frac{1}{m_1 + m_2}[l_\mu \bar u(q)\gamma_\nu\gamma_5 u(p) - l_\nu \bar u(q)\gamma_\mu\gamma_5 u(p)]\Big\} + D\varepsilon_{\mu\nu\lambda\rho}k_\lambda \bar u(q)\gamma_\rho u(p), \tag{25}$$

where

$$l_\mu = (p + q)_\mu, \quad k_\mu = (q - p)_\mu,$$

$$p^2 = -m_1^2, \quad q^2 = -m_2^2.$$

There are only two independent form-factors for each parity. For constant form-factors the matrix elements give possible forms of the current $J_{\mu\nu}$ in the Yukawa interaction in the Lagrangian.

4. It can be concluded from this that it is impossible to construct a theory of an interacting notoph with dimensionless coupling constants. As can be seen from (24) and (25), one candidate could be the current obtained by setting $A = B = D = 0$, $C =$ const. Further terms could be added to it so that it would be conserved in the interaction. Such a theory, however, again turns out to be a free one, since in it the inter-

Fig. 1.

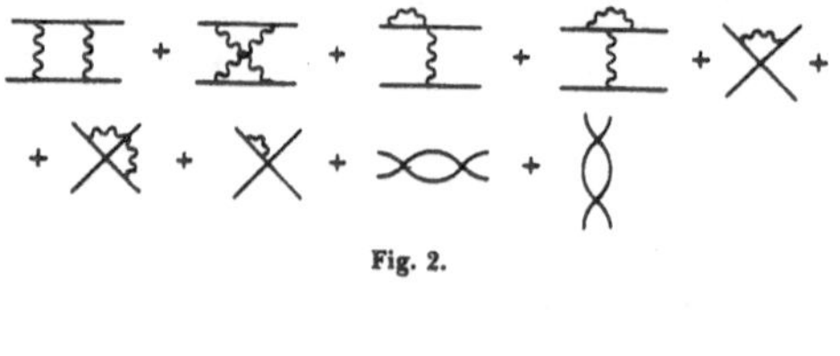

Fig. 2.

Fig. 3.

action can be eliminated by means of a transformation of the field variables, $\psi' = \exp(i\alpha\sigma_{\mu\nu}\gamma_5 f_{\mu\nu})\psi$ or a generalization of this. Another current $J_{\mu\nu}$ conserved owing to the equations of motion, obtained with $B = C = D = 0$, $A = \text{const}$, leads to an "interaction" which can also be eliminated, by means of a change of the field variables of the type $\psi' = \exp(i\beta\sigma_{\mu\nu}f_{\mu\nu})\psi$. There now remain to be considered only interactions with automatically conserved currents, corresponding to $A = C = D = 0$ in the case of positive parity and to $A = B = C = 0$ for negative parity, i.e., interactions of the type $\epsilon_{\mu\nu\lambda\rho}j_\nu\partial_\mu f_{\lambda\rho}$.

5. The first nontrivial theory of interaction of the notoph is obtained if we omit the current-current interaction in (17′)—that is, if we consider a theory with Yukawa coupling of the form

$$\mathcal{L}_{\text{int}} = -{}^1\!/_2 g\varepsilon_{\mu\nu\lambda\rho}\bar{\psi}\gamma_\nu\psi\partial_\mu f_{\lambda\rho}. \qquad (26)$$

It is obvious that here the current $\bar{\psi}\gamma_\mu\psi$ is conserved. Owing to this, in this version of the theory there are again no notophs emitted. The elements of the S matrix for no-notoph processes are not zero, however. Because of current conservation the virtual notoph lines again collapse and lead to a four-fermion coupling. It is curious that the theory of a vector four-fermion interaction can be represented as a theory with a Yukawa type of interaction with an intermediate boson of zero mass (the notoph!). Furthermore in this version of the theory there is no need for the actual existence of such a particle, since it is not emitted (a massless analog of an intermediate vector boson with infinitely large mass). If we proceed in an analogous way with the Lagrangian (19′), we get a representation of a theory with direct interaction of charged currents $j_\mu^+ j_\mu$ in the form of a theory with a charged intermediate boson with zero mass (a charged notoph), which again does not get emitted.

6. Let us now consider the most important case, in which the notoph exists as a real particle. This class of theories includes all those with the interaction

$$\mathcal{L}_{\text{int}} = \varepsilon_{\mu\nu\lambda\rho}j_\nu\partial_\mu f_{\lambda\rho}, \qquad (27)$$

where the current, unlike that in (26), is not conserved: $\partial_\nu j_\nu \neq 0$. For example, j_ν could be the pseudovector current $ig\bar{\psi}\gamma_\mu\gamma_5\psi$, vector currents $g\bar{\psi}_1\gamma_\mu\psi_2 - g^+\bar{\psi}_2\gamma_\mu\psi_1$, where ψ_1 and ψ_2 are spinors with different masses m_1 and m_2, and so on. In these theories the notoph is emitted and absorbed and leads to nontrivial interactions. Depending on the version of the theory chosen, the spatial and charge parities of the notoph can be arbitrary. We note that this sort of theory is equivalent to the theory of a zero-mass one-component field φ with an interaction of the form $j_\mu\partial_\mu\varphi$, accompanied by a corresponding four-fermion interaction $1/2 j_\nu j_\nu$. A specific feature of this combination of interactions is that the pair produced in the process of Fig. 1 will have total angular momentum (spin) 1.

If we do not concern ourselves with the dynamics of the interaction, in the framework of the formal theory of reactions the notoph is simply a particle with zero mass and zero helicity. The peculiarity of the notoph (as of the photon) is precisely in the dynamics. The equations of motion for the interacting notoph are so constructed that a virtual notoph (like a virtual photon) transfers the spin 1. Whereas for the virtual photon there is in addition to the states with helicities ± 1 (belonging to the free photon) also a state with zero helicity (Coulomb quanta), a virtual notoph has besides the state with zero helicity (belonging to the free notoph) also states with helicities ± 1. By the definition of the spin of an interacting field[4] the notoph field $f_{\mu\nu}$, like the electromagnetic field A_μ, has spin 1. In particular, if a notoph is converted into a pair, the total angular momentum of the pair will be equal to 1, as in the case of a photon. Along with this, a difference from the case of the photon is that 0–0 transitions with emission of a notoph are allowed. There is a temptation to associate K_2^0–K_1^0 transitions, and weak interactions in general, with the notoph. Obviously we cannot do this, since then the notoph would have been discovered experimentally long ago, and besides this a weak interaction through the notoph could not lead to decay of particles with spin 0, for example $\pi^+ \to \mu^+ + \nu$. Nevertheless, in discussions of new particles and interactions the notoph should be kept in mind, together with its generalizations to cases of nonzero helicities.

The author takes pleasure in expressing his hearty gratitude to M. A. Markov and B. N. Valuev for valuable comments and discussions.

APPENDIX

The helicity operator is

$$\frac{(\mathbf{sp})}{|\mathbf{p}|} = (\mathbf{sn}),$$

where $\mathbf{s}$ is the spin matrices and $\mathbf{p}$ is the momentum of the particle ($\mathbf{n} = \mathbf{p}/|\mathbf{p}|$). The quantities $\Gamma_0 = (\mathbf{s}\cdot\mathbf{p})$ and $\Gamma = m\mathbf{s} + \mathbf{p}(\mathbf{s}\cdot\mathbf{p})/(p_0 + m)$ form a four-vector.[5] Knowing this, we are in a position to transform the helicity to a Lorentz reference system moving with the velocity $\beta = v/c$:

$$(\mathbf{sn}') = \left[\gamma\left(1 - \frac{\beta\mathbf{p}}{p_0 + m}\right)(\mathbf{sp}) - m\gamma(\mathbf{s}\beta)\right] \Big/ \Big| \mathbf{p} + \beta\gamma \times \left[(\beta\mathbf{p})\frac{\gamma}{\gamma+1} - p_0\right]\Big|$$
$$= (\mathbf{sn}) + \frac{m}{|\mathbf{p}|[1 - (\beta\mathbf{n})]}[(\beta\mathbf{n})(\mathbf{sn}) - (\mathbf{s}\beta)] + o\left(\frac{m}{|\mathbf{p}|[1 - (\beta\mathbf{n})]}\right),$$

where $\gamma = (1 - \beta^2)^{1/2}$. It can be seen from this that at high energies, for which $m/|\mathbf{p}|[1 - (\boldsymbol{\beta} \cdot \mathbf{n})] \ll 1$, the helicity is approximately invariant even for particles with nonzero mass. In the limit $m = 0$ the helicity is rigorously invariant and is one of the invariants of the inhomogeneous Lorentz gourp.[1] The passage to the limit has been discussed in detail by Ritus.[1]

[1]E. P. Wigner, *Ann. of Math.* **40,** 149 (1939); *Revs. Modern Phys.* **29,** 255 (1957). Yu. M. Shirokov, *JETP* **33,** 1208 (1957), *Soviet Phys. JETP* **6,** 929 (1958). Chou Kuang-chao, *JETP* **36,** 909 (1959), *Soviet Phys. JETP* **9,** 642 (1959). L. G. Zastavenko and Chou Kuang-chao, *JETP* **38,** 134 (1960), *Soviet Phys. JETP* **11,** 97 (1960). V. I. Ritus, *JETP* **40,** 352 (1961), *Soviet Phys. JETP* **13,** 352 (1961).

[2]H. Umezawa, *Quantum Field Theory*, North-Holland and Interscience, New York, 1956.

[3]N. Kemmer, *Helv. Phys. Acta* **33,** 829 (1960).

[4]V. I. Ogievetskiĭ and I. V. Polubarinov, *Nuovo cimento* **23,** 173 (1962); *JETP* **45,** 237 (1963), *Soviet Phys. JETP* **18,** 166 (1964); *Ann. Phys.* (N.Y.) **25,** 358 (1963); *Nuclear Phys.* **76,** 677 (1966).

[5]Chou Kuang-chao and M. I. Shirokov, *JETP* **34,** 1230 (1958), *Soviet Phys. JETP* **7,** 851 (1958).

Translated by W. H. Furry

LETTERE AL NUOVO CIMENTO VOL. 8, N. 17 22 Dicembre 1973

Infinite-Dimensional Algebra of General Covariance Group as the Closure of Finite-Dimensional Algebras of Conformal and Linear Groups.

V. I. OGIEVETSKY

Joint Institute for Nuclear Research - Dubna

(ricevuto il 25 Settembre 1973)

The Einstein theory of general relativity rests on the group of general co-ordinate transformations (or general covariance group)

$$x_i = f_i(x_1, x_2, x_3, x_4)\,, \tag{1}$$

where $f_i(x)$ are arbitrary functions of the co-ordinates $x_1, x_2, x_3, x_4 = ict$. The general covariance group is an infinite-parameter group. The aim of the present paper is to call attention to the simple and remarkable fact that the action of the general covariance group can be reduced to alternating actions of its two finite-parameter subgroups: the special linear group $SL_{4,R}$ and the conformal group C_{15}. Both the linear and conformal groups act on the same manifold, that of co-ordinates, but these groups do not commute with each other. Below we prove that the algebra of the general covariance group (1) turns out to be the closure of algebras of the linear and conformal groups, *i.e.* that any generator of (1) is representable as some linear combination of repeated commutators of generators of $SL_{4,R}$ and C_{15}. In this way, the transformation properties (invariance, in particular) of any quantity under the action of the infinite-dimensional general covariance group are determined by its transformation properties under that of the essentially simpler finite-dimensional groups $SL_{4,R}$ and C_{15}. Some perspectives of this new approach to the general covariance group will be sketched in the concluding remarks.

Now we proceed to prove the main statement. To that end we expand the functions $f_i(x)$ (1) into infinite series in powers of co-ordinates. Coefficients of the series serve as parameters of the general covaiiance group, and its generators can be written as follows:

$${}^{n}L_k^{n_1,n_2,n_3,n_4} = -ix_1^{n_1}x_2^{n_2}x_3^{n_3}x_4^{n_4}\,\partial_k \qquad (n \equiv n_1+n_2+n_3+n_4,\ \partial_k \equiv \partial/\partial x_k)\,. \tag{2}$$

The group $SL_{4,R}$ is formed by all linear transformations $x_i' = a_{ik}x_k$ with the determinant equal to unity, and its generators are

$$SR_{ik} = -i[x_i\partial_k - \tfrac{1}{4}\delta_{ik}(x\partial)]\,. \tag{3}$$

A set of generators of the conformal group C_{15}, as is known, consists of those of four-dimensional rotations (also entering into (3) as a subalgebra), translations P_i, dilatations D and special conformal transformations K_i:

$$P_i = -i\partial_i\,, \tag{4a}$$

$$D = -i(x\partial)\,, \tag{4b}$$

$$K_i = -i(x^2\partial_i - 2x_i(x\partial))\,. \tag{4c}$$

Consider the closure of the linear- and conformal-group algebras, *i.e.* such minimal algebra which would contain all commutators of the generators (3) and (4) and their linear combinations. We will prove that this algebra coincides with that of the general covariance group (2), *i.e.* the following theorem is valid:

Any generator of the general covariance group $^nL_k^{n_1,n_2,n_3,n_4}$ *is representable as some linear combination of the commutators of generators of the special linear and conformal groups.*

The proof is carried out by mathematical induction. The translation generators (4a) give all generators $^nL_k^{\cdots}$ with $n=0$. The generators of $SL_{4,R}$ together with those of dilatations (4b) constitute the algebra $L_{4,R}$ with generators

$$R_{ik} = -ix_i\partial_k \tag{5}$$

and give all $^nL_k^{\cdots}$ with $n=1$. Generators of the special conformal transformations K_i (4c) are quadratic in x. Let us show that all generators $^nL_k^{\cdots}$ with $n=2$ are contained in the closing algebra. Consider the commutator of the dilatation along the m-th axis R_{mm} with K_p, $m\neq p$,

$$[R_{mm}, K_p] = -2x_m^2\partial_p\,. \tag{6}$$

So, we have the generator $-ix_m^2\partial_p$. Further,

$$[R_{pm}, -ix_m^2\partial_p] = -2x_m x_p\partial_p + x_m^2\partial_m\,. \tag{6'}$$

Comparing eqs. (6) and (6′) with K_m (4c) itself we conclude that the closing algebra includes the generators $-ix_m x_p\partial_p$ $(m\neq p)$ and $-ix_m^2\partial_m$, as well. Finally, the generator $-ix_m x_n\partial_p$ $(m\neq n\neq p)$ arises from the commutator

$$[R_{mp}, -ix_n x_p\partial_p] = -x_m x_n\partial_p\,. \tag{6''}$$

Hence, all the generators $^nL_k^{\cdots}$ with $n=2$ are exhausted. Commuting the generators quadratic in x with each other, we arrive at those cubic in x:

$$[-ix_m^2\partial_n, -ix_n^2\partial_n] = -2x_m^2x_n\partial_n\,, \tag{7a}$$

$$[-ix_m x_n\partial_n, -ix_n^2\partial_n] = -x_m x_n^2\partial_n\,, \tag{7b}$$

i.e. we have found the generators $-ix_m^2x_n\partial_n$, $-ix_m x_n^2\partial_n$ $(m\neq n)$. The commutator of the latter generator with R_{nm} (5) is the following:

$$[R_{nm}, -ix_m x_n^2\partial_n] = -x_n^3\partial_n + x_n^2x_m\partial_m\,, \tag{8a}$$

and with (7a) taken into account we obtain the generator of further importance

$$-ix_n^3\,\partial_n\,. \tag{8b}$$

Other generators of third power in x will not be required for our proof by induction. We have shown above that the theorem holds for $n=0, 1, 2$. Suppose that the one is valid for some n, *i.e.* all generators ${}^nL_k^{n_1,n_2,n_3,n_4}$ can be represented as linear combinations of repeated commutators of the generators of linear and conformal groups. Prove that then the theorem will be valid for $n+1$, as well.

All the differentiation indices k of ${}^nL_k^{n_1,n_2,n_3,n_4}$ are on the same status, so it suffices to consider $k=1$. Then, if $n_1\neq 0$, $n_1\neq 3$,

$$ {}^{n+1}L_1^{n_1,n_2,n_3,n_4}=\frac{i}{n_1-3}\,[-ix_1^2\,\partial_1,\,{}^nL_1^{n_1-1,n_2,n_3,n_4}]\,. \tag{9}$$

Consider now the cases $n_1=0$ and $n_1=3$. In these cases, if at least one of n_2, n_3, n_4 is larger than zero, for instance $n_2>0$, then we have

$$ {}^{n+1}L_1^{n_1,n_2,n_3,n_4}=\frac{i}{n_1-1}\,[-ix_1x_2\,\partial_1,\,{}^nL_1^{n_1,n_2-1,n_3,n_4}]\,. \tag{10}$$

Finally, for $n_1=3$, $n_2=n_3=n_4=0$ the corresponding generator is given by (8). So, the theorem is proven.

One can see the validity of this theorem for spaces of any dimension, the generators of the general covariance group being representable as linear combinations of repeated commutators of generators of the corresponding special linear and conformal groups.

Elsewhere we will show that the Einstein equations of general relativity follow from the invariance under the conformal and linear groups, and this is quite natural within the framework of the approach developed. In deep analogy with the fact that pions are connected with nonlinear realizations of the dynamical chiral $SU_2\times SU_2$ symmetry (see, *e.g.*, ref. (1)), gravity field proves to be connected with common nonlinear realizations of the dynamical conformal and affine symmetries (*).

Note also, that the presented approach raises hopes that unitary representations for the infinite-dimensional algebra of the general covariance group can be constructed.

* * *

The author thanks sincerely F. A. BEREZIN, A. B. BORISOV, D. V. VOLKOV and V. TYBOR for useful discussions.

(1) S. WEINBERG: *Proceedings of the 1970 Brandeis Summer Institute in Theoretical Physics*, edited by S. DESER (Cambridge, 1970), p. 287.

(*) The nonlinear realizations of space-time symmetries are discussed in ref. (2-4).

(2) D. V. VOLKOV: *Particles and Nuclei*, **4**, 3 (1973).

(3) C. J. ISHAM, A. SALAM and J. STRATHDEE: *Ann. of Phys.*, **62**, 98 (1971).

(4) V. I. OGIEVETSKY: *Proceedings of the X Winter School on Theoretical Physics* (Karpach, 1973).

THEORY OF DYNAMICAL AFFINE AND CONFORMAL SYMMETRIES AS THE THEORY OF THE GRAVITATIONAL FIELD

A. B. Borisov and V. I. Ogievetskii

Invariance under the infinite-parameter generally covariant group is equivalent to simultaneous invariance under the affine and the conformal group. A nonlinear realization of the affine group (with linearization on the Poincaré group) leads to a symmetric tensor field as Goldstone field. The requirement that the theory correspond simultaneously to a realization of the conformal group as well leads uniquely to the theory of a tensor field whose equations are Einstein's. This shows that the theory of the gravitational field is the theory of spontaneous breaking of affine and conformal symmetries in the same way as chiral dynamics is the theory of spontaneous breaking of chiral symmetry. This analogy brings out new aspects of the role of gravitation in the theory of elementary particles.

1. Introduction

It is well known that the gravitational field is a gauge field that guarantees invariance of Einstein's theory of gravitation under the group of general coordinate transformations. This reveals the deep analogy between the gravitational field and the vector Yang–Mills fields – gauge fields for internal symmetries.

There is further deep analogy – the analogy between gravitons in the theory of gravitation and π-mesons in the $SU(2) \times SU(2)$ chiral dynamics based on nonlinear realizations of the chiral symmetry. Chiral invariance is achieved by corresponding interactions with a (massless) pion field, while general covariance is guaranteed by corresponding interactions with the massless gravitational field. These interactions are introduced by replacing ordinary derivatives by covariant derivatives that depend nonlinearly on the pion field (in chiral symmetry) and on the gravitational field (in the theory of gravitation). Chiral symmetry is spontaneously broken and the (massless) pions are Goldstone bosons.'

In the present paper we shall show that the theory of the gravitational field is the theory of joint nonlinear realizations of the affine and conformal symmetries. Gravitons are Goldstone bosons for these symmetries. We deduce Einstein's equations from the requirement of invariance under the affine and conformal groups realized nonlinearly with linearization on the Poincaré group in the same way as the equations of chiral dynamics are derived in the theory of nonlinear realizations of chiral symmetry [1, 2].

In the group of general coordinate transformations

$$\delta x_\mu = f_\mu(x) \qquad (1)$$

the functions $f_\mu(x)$ depend arbitrarily on $x^1, \ldots, x^4 = ict$. For what follows, it is sufficient to restrict ourselves to functions $f_\mu(x)$ that can be expanded in series in powers of the coordinates x_μ. Then (1) defines an infinite-parameter continuous group of transformations whose parameters can be regarded as coefficients of expansions in series in powers of the coordinates. The algebra of this group contains infinitely many generators

$$L_\mu^{n_1 n_2 n_3 n_4} = -i x_1^{n_1} x_2^{n_2} x_3^{n_3} x_4^{n_4} \partial_\mu \quad (\partial_\mu = \partial/\partial x_\mu). \qquad (2)$$

The theory of nonlinear realizations has been developed for finite-parameter groups [2]. How can we transfer it to an infinite-parameter group (1)? The key to the resolution of this question is a remarkable theorem established by one of the authors of the present paper: the infinite-dimensional algebra (2) is the

Joint Institute for Nuclear Research, Dubna. Translated from Teoreticheskaya i Matematicheskaya Fizika, Vol. 21, No. 3, pp. 329-342, December, 1974. Original article submitted December 29, 1973.

closure of the finite-dimensional algebras of SL(4,R) and the conformal group (it is convenient for us to augment the special linear group SL(4,R) to the affine group of all linear transformations $x'_\mu = a_{\mu\nu}x_\nu + c_\mu$). The generator of strictly conformal transformations in the coordinate space $K_\mu = -i(x^2\partial_\mu - 2x_\mu x\partial)$ is quadratic in the coordinates. The result of commuting it with the SL(4,R)-generator $-ix_\mu\partial_\nu$ is again quadratic in x. Commuting the resulting operators with one another, we arrive at operators of third degree in x, etc. In [3] we showed that any generator $L_\mu^{n_1 n_2 n_3 n_4}$ (2) of the generally covariant group can be represented as a linear combination of repeated commutators of the generators of the special linear group and the conformal group. It follows from this that any theory that is invariant simultaneously under the special linear group and the conformal group will also be invariant under the group of general coordinate transformations. Thus, we arrive naturally at a new approach to the theory of gravitation based on the properties of its invariance under the finite-parameter conformal and special linear groups, whose structure is essentially simpler than that of the infinite-parameter group of general coordinate transformations.

In nature there are no conservation laws corresponding to strictly linear and strictly conformal transformations. Therefore, SL(4,R) and the conformal symmetry must be dynamical, spontaneously broken. Accordingly, we shall consider their nonlinear realizations, so that only their good algebraic subgroup – the Lorentz group (and also translations) – will be represented by linear and homogeneous transformations of fields.

Nonlinear realizations of finite-parameter groups of symmetries, including space–time ones, have been studied on a number of occasions [4-7] and in various papers conformal symmetry has been considered; the linear group has been discussed by Isham, Salam, and Strathdee [5].

The present paper is planned as follows. In the second section we describe nonlinear realizations of the affine group. For this, we require as Goldstone field a symmetric tensor field $h_{\mu\nu}(x)$. In the definition of the covariant derivatives nonminimal terms are possible because the corresponding theory is not fixed sufficiently stringently. In the third section we give the necessary results on the nonlinear realizations of the conformal group. The general theory prescribes two Goldstone fields, a vector field φ_μ and a scalar field φ. However, the specific feature of the conformal group is such that the vector field φ_μ can be represented as the gradient of the scalar field φ, so that there remains only the scalar field φ. In the investigation of nonlinear realizations we shall, following Volkov [6], make essential use of Cartan forms. Then, in the fourth and main section we show that the requirement of joint nonlinear realizations of the affine and conformal groups fixes the form of the nonminimal terms in the covariant derivatives, and we formulate rules for constructing an invariant variational principle. In the section that follows we identify our theory with Einstein's theory of the gravitational field. This analogy between the theory of gravitation and theories of nonlinear realizations of groups of internal symmetries (chiral and other) leads to the formulation of a number of problems whose solution would lead to a deeper understanding of the role of gravitation in the theory of elementary particles, for example, in the formation of the mass spectra of particles. Some of these problems are briefly discussed in the conclusions.

2. Nonlinear Realizations of the Affine Group

We now turn to the description of nonlinear realizations of the affine group A(4) – the group of all linear transformations in four-dimensional space: $x'_\mu = a_{\mu\nu}x_\nu + c_\mu$. The affine group is a semidirect product of L(4,R) and the translation group, $A(4) = P_4 \odot L(4,R)$, and contains the Poincaré group as a subgroup. Its algebra consists of the generators $L_{\mu\nu}$ of the Lorentz group, the generators of strictly linear transformations, including dilations, $R_{\mu\nu}$, and the generators P_μ of translations:

$$\frac{1}{i}[L_{\mu\nu}, L_{\rho\sigma}] = \delta_{\mu\rho}L_{\nu\sigma} - \delta_{\mu\sigma}L_{\nu\rho} - (\mu \leftrightarrow \nu),$$

$$\frac{1}{i}[L_{\mu\nu}, R_{\rho\sigma}] = \delta_{\mu\rho}R_{\nu\sigma} + \delta_{\mu\sigma}R_{\nu\rho} - (\mu \leftrightarrow \nu), \tag{3}$$

$$\frac{1}{i}[R_{\mu\nu}, R_{\rho\sigma}] = \delta_{\mu\rho}L_{\sigma\nu} + \delta_{\mu\sigma}L_{\rho\nu} + (\mu \leftrightarrow \nu),$$

$$\frac{1}{i}[L_{\mu\nu}, P_\rho] = \delta_{\mu\rho}P_\nu - (\mu \leftrightarrow \nu),$$

$$\frac{1}{i}[R_{\mu\nu}, P_\rho] = \delta_{\mu\rho}P_\nu + (\mu \leftrightarrow \nu). \tag{4}$$

In the vector representation, the generators $L_{\mu\nu}$ and $R_{\mu\nu}$ can be specified in the matrix form

$$(L_{\mu\nu})_{\alpha\beta}=-i(\delta_{\mu\alpha}\delta_{\nu\beta}-\delta_{\mu\beta}\delta_{\nu\alpha}),\quad (R_{\mu\nu})_{\alpha\beta}=-i(\delta_{\mu\alpha}\delta_{\nu\beta}+\delta_{\mu\beta}\delta_{\nu\alpha}). \tag{5}$$

In the affine group, only the Poincaré subgroup is associated with true conservation laws – conservation of the total momentum and the total angular momentum – and therefore the affine symmetry can be only dynamical. Therefore, we shall consider nonlinear realizations of A(4) that become linear only on its subgroup – the Poincaré group. We consider realizations of A(4) in the factor space A(4)/L, where L is the Lorentz group. In accordance with the general theory [2-7], we introduce a symmetric tensor field $h_{\mu\nu}(x)$ and define the action of an element of the group in accordance with

$$g: g\exp(ix_\mu P_\mu)\exp({}^1/_2 ih_{\mu\nu}(x)R_{\mu\nu})=\exp(ix_\mu' P_\mu)\exp\left\{\frac{i}{2}h_{\mu\nu}'(x')R_{\mu\nu}\right\}\exp\left\{\frac{i}{2}u_{\mu\nu}(h,g)L_{\mu\nu}\right\}, \tag{6}$$

where x_μ' are the transformed coordinates; $h_{\mu\nu}'(x')$ is the transformed field $h_{\mu\nu}(x)$; and $u_{\mu\nu}(h(x), g)$ depend on the group element g and the field $h_{\mu\nu}$. The action of the group A(4) on an arbitrary field $\Psi(x)$ is defined by

$$g: \Psi'(x')=\exp\left\{\frac{i}{2}u_{\mu\nu}(h(x),g)L_{\mu\nu}^{\Psi}\right\}\Psi(x), \tag{7}$$

where $L_{\mu\nu}^{\Psi}$ are matrices that represent the generators of the Lorentz group L for the considered field $\Psi(x)$. For example, for a scalar $L_{\mu\nu}^{\Psi}=0$, for a spinor $L_{\mu\nu}^{\Psi}=1/2\sigma_{\mu\nu}$, for a vector $(L_{\mu\nu}^{\Psi})_{\alpha\beta}=-i(\delta_{\mu\alpha}\delta_{\nu\beta}-\delta_{\nu\alpha}\delta_{\mu\beta})$, etc., and the corresponding infinitesimal transformations are given by $(\delta\Psi(x)=\Psi'(x')-\Psi(x))$

$$\delta\varphi(x)=0, \tag{8a}$$

$$\delta\Psi(x)=-\frac{i}{4}u_{\mu\nu}(h(x),g)\sigma_{\mu\nu}\Psi(x), \tag{8b}$$

$$\delta a_\mu(x)=u_{\mu\nu}(h(x),g)a_\nu(x). \tag{8c}$$

The group property of the transformations (6), (7), and (8) can be verified directly. It is not difficult to see that the Poincaré group can be represented by standard linear transformations. For example, for shifts we have $g=e^{ic_\mu P_\mu}$, $u_{\mu\nu}=0$, $x_\mu'=x_\mu+c_\mu$, $h_{\mu\nu}'(x')=h_{\mu\nu}(x)$, $\Psi'(x')=\Psi(x)$ and for Lorentz transformations $g=\exp\{(i/2)\beta_{\mu\nu}L_{\mu\nu}\}$, $u_{\mu\nu}=\beta_{\mu\nu}$ and does not depend on $h_{\mu\nu}(x)$, i.e., in accordance with (6) all fields undergo ordinary Lorentz transformations. For strictly linear transformations g contains the factor $\exp\{(i/2)\alpha_{\mu\nu}R_{\mu\nu}\}$ and $u_{\mu\nu}$ depends essentially on $h_{\mu\nu}$, so that all the fields except $h_{\mu\nu}$ transform in accordance with the Lorentz group with parameters $u_{\mu\nu}(h(x), g)$ that depend nonlinearly on $h_{\mu\nu}(x)$. In the lowest order in $\alpha_{\mu\nu}$, the infinitesimal transformations of the fields $h_{\mu\nu}(x)$ and $u_{\mu\nu}(h, g)$ have the form

$$\delta h_{\mu\nu}=h_{\mu\nu}'(x')-h_{\mu\nu}(x)=\sum_{mn}b_{mn}(h^m(x)\alpha h^n(x))_{\mu\nu}, \tag{9}$$

$$u_{\mu\nu}(h(x),g)=\sum_{m,n}c_{mn}(h^m(x)\alpha h^n(x))_{\mu\nu}, \tag{10}$$

where

$$(h^m\alpha h^n)_{\mu\nu}=h_{\mu\sigma_1}\ldots h_{\sigma_{m-1}\sigma_m}\alpha_{\sigma_m\rho_1}h_{\rho_1\rho_2}\ldots h_{\rho_n\nu}$$

and the coefficients b_{mn} and c_{mn} are given by the generating functions

$$\sigma_1(x,y)=\sum_{m,n}b_{mn}x^m y^n=(x-y)\,\mathrm{cth}\,(x-y), \tag{11}$$

$$\sigma_2(x,y)=\sum_{m,n}c_{mn}x^m y^n=-\mathrm{th}\left(\frac{x-y}{2}\right). \tag{12}$$

In the vector representation (5) we introduce the important quantity

$$r_{\mu\nu}(x)=\left(\exp\left\{\frac{i}{2}h_{\alpha\beta}R_{\alpha\beta}\right\}\right)_{\mu\nu}=(e^h)_{\mu\nu}=\delta_{\mu\nu}+h_{\mu\nu}+\frac{h_{\mu\sigma}h_{\sigma\nu}}{2}+\ldots \tag{13}$$

and its inverse

$$r_{\mu\nu}^{-1}(x)=(e^{-h})_{\mu\nu}. \tag{14}$$

The infinitesimal transformations of these Lorentz tensors corresponding to (9) have in accordance with (6) the form

$$\delta r_{\mu\nu}^{\pm 1}(x)=\pm\alpha_{\mu\sigma}r_{\sigma\nu}^{\pm 1}-r_{\mu\sigma}^{\pm 1}u_{\sigma\nu}(h,\alpha). \tag{15}$$

The symmetry of $\delta r_{\mu\nu}^{\pm}$ is guaranteed by the identities

$$\{u(h,g),\, r^{\pm 1}(x)\}_{\mu\nu}=\pm[\alpha,\, r^{\pm 1}]_{\mu\nu}. \tag{16}$$

One can construct functions of the field $h_{\mu\nu}$ which are such that they transform linearly. Such quantities are the squares of the tensors $r_{\mu\nu}$ and $r_{\mu\nu}^{-1}$,

$$g_{\mu\nu}=r_{\mu\sigma}(x)r_{\sigma\nu}(x)=(e^{2h})_{\mu\nu},\quad g^{\mu\nu}=r_{\mu\sigma}^{-1}r_{\sigma\nu}^{-1}=(e^{-2h})_{\mu\nu}, \tag{17}$$

$$\delta g_{\mu\nu}=\alpha_{\mu\sigma}g_{\sigma\nu}(x)+g_{\mu\sigma}\alpha_{\sigma\nu},\quad \delta g^{\mu\nu}=-\alpha_{\mu\sigma}g^{\sigma\nu}-g^{\mu\sigma}\alpha_{\sigma\nu}. \tag{18}$$

In the theory of gravitation, these quantities correspond to the contravariant and covariant metric tensors. This similarity is very deep. By a change of variables of the field one can introduce linearly transforming contra- and covariant quantities for all fields of integral spin. For example, for a vector field $a_\mu(x)$

$$A_\mu(x)=r_{\mu\nu}(x)a_\nu(x),\quad A^\mu(x)=r_{\mu\nu}^{-1}a_\nu(x),\quad A^\mu=g^{\mu\nu}A_\nu \tag{19}$$

are co- and contravariant vectors, respectively. From (8c) with allowance for (15) it follows that

$$\delta A^\mu(x)=-\alpha_{\mu\nu}A^\nu(x),\quad \delta A_\nu(x)=\alpha_{\mu\nu}A_\nu(x). \tag{20}$$

Similarly, for any number of vector indices they can be transformed into co- or contravariant quantities by multiplication by $r_{\mu\nu}$ or $r_{\mu\nu}^{-1}$, respectively.

For fields of half-integral spin, the transition to linearly transforming quantities is impossible, because of the absence of finite-dimensional spinor representations of the affine group. We may mention that the nonlinear spinor law (8b) in gravitation theory was obtained in 1965 [8].

Because of the nonlinear dependence of the transformation laws on the Goldstone field $h_{\mu\nu}(x)$, it is necessary to define covariant derivatives of fields that under affine transformations transform in accordance with representations of the Lorentz group with $h_{\mu\nu}$-dependent parameters like the fields themselves. Ordinary derivatives are not suitable. Covariant derivatives are most readily defined by means of the Cartan forms [9, 6, 7]. We introduce the notation $C=\exp\{ix_\mu P_\mu\}\exp\{(i/2)h_{\alpha\beta}R_{\alpha\beta}\}$ and consider the transformation properties of the expression $C^{-1}dC$, where the differential d acts on x_μ and $h_{\alpha\beta}(x)$. It follows from (6) that

$$C'^{-1}dC'=\exp\left\{\frac{iu_{\mu\nu}L_{\mu\nu}}{2}\right\}C^{-1}gd\left(g^{-1}C\exp\left\{\frac{i}{2}u_{\mu\nu}L_{\mu\nu}\right\}\right)$$

$$=\exp\left\{\frac{i}{2}u_{\mu\nu}L_{\mu\nu}\right\}(C^{-1}dC)\exp\left\{-\frac{i}{2}u_{\mu\nu}L_{\mu\nu}\right\}+\exp\left\{\frac{i}{2}u_{\mu\nu}L_{\mu\nu}\right\}d\exp\left\{-\frac{i}{2}u_{\mu\nu}L_{\mu\nu}\right\}, \tag{21}$$

since the element g does not depend on x. Expanding $C^{-1}dC$ with respect to the generators of the affine group,

$$C^{-1}dC=i\omega_\mu^P(d)P_\mu+\frac{i}{2}\omega_{\mu\nu}^R(d)R_{\mu\nu}+\frac{i}{2}\omega_{\mu\nu}^L(d)L_{\mu\nu}, \tag{22}$$

we obtain the Cartan forms

$$\omega_\mu^P(d)=r_{\mu\nu}dx_\nu, \tag{23a}$$

$$\omega_{\mu\nu}^R(d)={}^1/_2\{r^{-1}(x),\, dr(x)\}_{\mu\nu}, \tag{23b}$$

$$\omega_{\mu\nu}^L(d)={}^1/_2[r^{-1}(x),\, dr(x)]_{\mu\nu}, \tag{23c}$$

where we have used matrix notation, for example, $[r^{-1},dr]_{\mu\nu}\equiv r_{\mu\sigma}^{-1}dr_{\sigma\nu}-dr_{\mu\sigma}r_{\sigma\nu}^{-1}$. The generators $R_{\mu\nu}$, and P_μ, $L_{\mu\nu}$ form a representation of the Lorentz group. It therefore follows from (21) that under the action of the affine group the Cartan forms undergo a Lorentz transformation with the parameters $u_{\mu\nu}\cdot(h(x),g)$, and the form $\omega_{\mu\nu}^L(d)$ also acquires an additive correction:

$$L_{\mu\nu}\omega_{\mu\nu}{}^{L}(d)' = \exp\left\{\frac{i}{2}u_{\alpha\beta}L_{\alpha\beta}\right\}\omega_{\mu\nu}{}^{L}(d)L_{\mu\nu}\exp\left\{-\frac{i}{2}u_{\alpha\beta}L_{\alpha\beta}\right\} - 2i\exp\left\{\frac{i}{2}u_{\alpha\beta}L_{\alpha\beta}\right\}d\exp\left\{-\frac{i}{2}u_{\alpha\beta}L_{\alpha\beta}\right\}. \tag{24}$$

This addition enables one to obtain a covariant differential for an arbitrary field $\Psi(x)$. Namely, it follows from (24) and (9) that in contrast to the ordinary differential d the covariant differential

$$D\Psi(x) = \left(d + \frac{i}{2}\omega_{\mu\nu}{}^{L}L_{\mu\nu}{}^{\Psi}\right)\Psi(x) \tag{25}$$

transforms in accordance with the same representation as the field $\Psi(x)$, $(D\Psi(x))' = \exp\{(i/2)u_{\mu\nu}L^{\Psi}_{\mu\nu}\}(D\Psi)$. The form $\omega^{R}_{\mu\nu}(d)$ in (23b) can be identified with the covariant differential of the field $h_{\mu\nu}$. To construct covariant derivatives one must use the Cartan form $\omega_{\mu}(d)$ (23a), since it (and not dx_{μ}) has the appropriate transformation properties. The covariant derivative of the distinguished field $h_{\mu\nu}(x)$ is defined as ($\partial_{\mu} \equiv \partial/\partial x_{\mu}$)

$$\nabla_{\lambda}h_{\mu\nu} = \frac{\omega_{\mu\nu}{}^{R}(d)}{\omega_{\lambda}{}^{P}(d)} = {}^{1}\!/_{2}\,r^{-1}_{\lambda\tau}(x)\{r^{-1}(x), \partial_{\tau}r(x)\}_{\mu\nu}. \tag{26}$$

The minimal covariant derivative of an arbitrary field can be written in the form

$$\nabla_{\lambda}\Psi(x) = \frac{D\Psi(x)}{\omega_{\lambda}{}^{P}(d)} = r^{-1}_{\lambda\tau}\partial_{\tau}\Psi(x) + \frac{i}{2}v^{\min}_{\mu\nu,\lambda}L_{\mu\nu}{}^{\Psi}\Psi, \tag{27a}$$

where

$$v^{\min}_{\mu\nu,\lambda}(x) = \frac{r^{-1}_{\lambda\tau}(x)}{2}[r^{-1}(x), \partial_{\tau}r(x)]_{\mu\nu}. \tag{27b}$$

The transformation properties of the covariant derivative $\nabla_{\lambda}\Psi$ are not changed if $v^{\min}_{\mu\nu,\lambda}$ is replaced by

$$V_{\mu\nu,\lambda} = v^{\min}_{\mu\nu,\lambda} + c_1(\nabla_{\mu}h_{\nu\lambda} - \nabla_{\nu}h_{\mu\lambda}) + c_2(\delta_{\mu\lambda}\nabla_{\nu}h_{\sigma\sigma} - \delta_{\nu\lambda}\nabla_{\mu}h_{\sigma\sigma}) + c_3(\delta_{\mu\lambda}\nabla_{\tau}h_{\nu\tau} - \delta_{\nu\lambda}\nabla_{\tau}h_{\mu\tau}). \tag{28}$$

The added terms contain the first derivative of the field $h_{\mu\nu}$, as also $v^{\min}_{\mu\nu,\lambda}$. Therefore, the general form of the covariant derivative:

$$\nabla_{\lambda}\Psi(x) = r^{-1}_{\lambda\tau}\partial_{\tau}\Psi(x) + \frac{i}{2}V_{\mu\nu,\lambda}L_{\mu\nu}{}^{\Psi}\Psi(x) \tag{29}$$

is not fixed sufficiently stringently and contains the nonminimal constants c_1, c_2, c_3. Below we shall show that the values of these constants are determined by the requirement of conformal invariance.

The invariant element of volume is given by the outer product

$$dV = -i\omega_1{}^{P}(d)\wedge\omega_2{}^{P}(d)\wedge\omega_3{}^{P}(d)\wedge\omega_4{}^{P}(d) = \det\|r_{\mu\nu}(x)\|d^4x = e^{\operatorname{sp}\|h_{\mu\nu}\|}d^4x. \tag{30}$$

The action $\int \mathcal{L}\det\|r_{\mu\nu}\|d^4x$ is invariant under the affine group if the Lagrangian density $\mathcal{L}(\Psi(x), \nabla_{\lambda}\Psi(x), \nabla_{\lambda}h_{\mu\nu}(x))$ is an ordinary scalar with respect to the Lorentz group. Indeed, all the fields $\Psi(x)$, their covariant derivatives, and the covariant derivative of the field $h_{\mu\nu}(x)$ transform in accordance with representations of the Lorentz group (albeit with parameters $u_{\mu\nu}(h, g)$ that depend on $h_{\mu\nu}$). A further restriction on the theory is achieved by imposing the requirement of conformal symmetry.

3. Dynamical Conformal Symmetry

Realizations of the conformal group in the factor space with respect to the Lorentz group were considered in [4-6]. We give the necessary information. The algebra of the conformal group includes the generators $L_{\mu\nu}$ and P_{ν} of the Poincaré group (the commutators between them were given above, see (3) and (4)) and the generators of scale and strictly conformal transformations D and K_{μ}:

$$\begin{gathered}[L_{\mu\nu}, D] = 0, \quad [L_{\mu\nu}, K_{\rho}] = i\delta_{\mu\rho}K_{\nu} - (\mu \leftrightarrow \nu), \quad [K_{\mu}, K_{\nu}] = 0,\\ [P_{\mu}, D] = -iP_{\mu}, \quad [K_{\mu}, D] = iK_{\mu}, \quad [P_{\mu}, K_{\nu}] = 2i(\delta_{\mu\nu}D - L_{\mu\nu}).\end{gathered} \tag{31}$$

We represent an element of the conformal group $\bar{g}$ in the form

$$\bar{g} = e^{ia_{\mu}P_{\mu}}e^{i\beta_{\mu}K_{\mu}}e^{i\beta D}e^{i\beta_{\mu\nu}L_{\mu\nu}/2}, \tag{32}$$

where a_{μ}, β_{μ}, β, and $\beta_{\mu\nu}$ are parameters of the transformation. We introduce the Goldstone fields $\varphi_{\mu}(x)$ and $\sigma(x)$ and denote

$$\bar{C}(x) = e^{ix_{\mu}P_{\mu}}e^{i\varphi_{\mu}(x)K_{\mu}}e^{i\sigma(x)D}. \tag{33}$$

The action of the conformal group is determined (in complete analogy with (6)) from

$$\bar{g}:\ \bar{g}\bar{C}(x)=\bar{C}'(x')e^{iu_{\mu\nu}(x,\varphi_\mu\sigma)L_{\mu\nu}/2}. \tag{34}$$

The infinitesimal strictly conformal and scale transformations of x, $\varphi_\mu(x)$, and $\sigma(x)$ have the form

$$\begin{aligned}
&\delta x_\mu = x^2\beta_\mu - 2(\beta x)x_\mu - \lambda x_\mu,\\
&\delta\sigma(x) = 2(\beta x)+\lambda,\\
&\delta\varphi_\mu(x) = [1+2(x\varphi(x))]\beta_\mu + 2(x\beta)\varphi_\mu(x) - 2(\beta\varphi(x))x_\mu + \lambda\varphi_\mu.
\end{aligned} \tag{35}$$

Note that $\varphi_\mu(x)$ transforms in accordance with the same law as $1/2\partial_\mu\sigma(x)$. All the remaining fields transform in accordance with their representations of the Lorentz group (see formula (7) but with the parameters $\bar{u}_{\mu\nu}$). An infinitesimal transformation (with $\bar{g}$ in the general form (32)) leads to

$$\bar{u}_{\mu\nu} = \beta_{\mu\nu} + 2(\beta_\mu x_\nu - \beta_\nu x_\mu), \tag{36}$$

for example, for the vector field

$$\delta a_\mu(x) = a_\mu'(x') - a_\mu(x) = \bar{u}_{\mu\nu}a_\nu(x) = \beta_{\mu\nu}a_\nu(x) + 2\beta_\mu(xa(x)) - 2x_\mu(\beta a(x)). \tag{37}$$

We expand $\bar{C}^{-1}(x)d\bar{C}(x)$ with respect to the generators P_μ, $L_{\mu\nu}$, D, and K_μ and find the Cartan forms:

$$\bar{\omega}_\mu{}^P(d) = e^{\sigma(x)}dx^\mu, \tag{38a}$$

$$\bar{\omega}_\mu{}^K(d) = e^{-\sigma(x)}(d\varphi_\mu(x) + \varphi^2(x)dx_\mu - 2(\varphi(x)dx)\varphi_\mu(x)), \tag{38b}$$

$$\bar{\omega}^D(d) = d\sigma(x) - 2dx_\mu\varphi_\mu(x), \tag{38c}$$

$$\bar{\omega}_{\mu\nu}{}^L(d) = 2(dx_\mu\varphi_\nu(x) - dx_\nu\varphi_\mu(x)). \tag{38d}$$

The covariant derivative of the field $\sigma(x)$ is defined in accordance with

$$\bar{\nabla}_\mu\sigma(x) = \frac{\bar{\omega}^D(d)}{\bar{\omega}_\mu{}^P(d)} = e^{-\sigma(x)}(\partial_\mu\sigma(x) - 2\varphi_\mu(x)) \tag{39}$$

Let us now make an important observation. It follows from (39) that the Goldstone field $\varphi_\mu(x)$ is unimportant and one can do without it; for let us make the covariant derivative $\bar{\Delta}_\mu\sigma(x)$ in (39) vanish – this is a covariant operation since $\bar{\Delta}_\mu\sigma(x)$ transforms as a Lorentz vector (with parameters $\bar{u}_{\mu\nu}$). Then the field $\varphi_\mu(x)$ is transformed into the gradient of the field $\sigma(x)$*

$$\bar{\Delta}_\mu\sigma(x) = 0 \rightarrow \varphi_\mu(x) = {}^1/_2\partial_\mu\sigma(x). \tag{40}$$

Then $\bar{\omega}_\mu^k(d)/\bar{\omega}_\nu^P(d)$ gives a Lorentz tensor composed of $\partial_\mu\partial_\nu\sigma(x)$, $\partial_\mu\sigma(x)$, $\sigma(x)$.

The covariant derivative of the arbitrary field $\Psi(x)$ that transforms in accordance with a representation of the Lorentz group with generators $L_{\mu\nu}^\Psi$ is constructed on the basis of the Cartan form (38d):

$$\nabla_\lambda\Psi(x) = \frac{d\Psi(x) + \frac{i}{2}\bar{\omega}_{\mu\nu}{}^L L_{\mu\nu}\Psi(x)}{\omega_\lambda{}^P(d)} = e^{-\sigma(x)}(\partial_\lambda\Psi(x) + 2i\varphi_\nu L_{\lambda\nu}\Psi). \tag{41}$$

or, after the substitution $\varphi_\nu = 1/2\partial_\nu\sigma$,

$$\bar{\nabla}_\lambda\Psi(x) = e^{-\sigma(x)}(\partial_\lambda\Psi(x) + i\partial_\nu\sigma L_{\lambda\nu}{}^\Psi\Psi(x)). \tag{42}$$

For example, for the tensor field $h_{\alpha\beta}(x)$

$$\bar{\nabla}_\lambda h_{\alpha\beta}(x) = e^{-\sigma(x)}\{\partial_\lambda h_{\alpha\beta} + \partial_\nu\sigma(x)(\delta_{\alpha\lambda}h_{\nu\beta}(x) + \delta_{\beta\lambda}h_{\alpha\nu}(x) - \delta_{\alpha\nu}h_{\lambda\beta}(x) - \delta_{\beta\nu}h_{\alpha\lambda}(x))\}. \tag{43}$$

The scalar element of volume, $d\bar{V}'(x') = d\bar{V}(x)$, is written as

$$d\bar{V}(x) = -i\bar{\omega}_1{}^P\wedge\bar{\omega}_2{}^P\wedge\bar{\omega}_3{}^P\wedge\bar{\omega}_4{}^P = e^{4\sigma(x)}d^4x. \tag{44}$$

4. Joint Realizations of Affine and Conformal Symmetries

We now require invariance under the conformal and affine groups simultaneously. Then, as we discussed in the introduction, it follows from the theorem proved in [3] that there must arise invariance of

*If we do not make $\bar{\nabla}_\mu\sigma(x)$ vanish, then $\varphi_\mu = (1/2)\partial_\mu\sigma(x) + e^{\sigma(x)}v_\mu(x)$, where $v_\mu(x)$ transforms in accordance with (37) and accordingly is not a Goldstone field.

the theory under the group of general coordinate transformations, and we arrive at Einstein's theory of gravitation. Let us see how this happens.

It is obvious that the trace of the affine generator $R_{\mu\nu}$ is related to the generator D of scale transformations by the equation* $R_{\mu\mu} = 2D$. Therefore, we must identify the trace of the affine Goldstone field $h_{\mu\nu}$ with the conformal Goldstone field:

$$\sigma(x) = {}^1\!/\!_4 h_{\mu\mu}(x) \tag{45}$$

and set

$$h_{\mu\nu}(x) = \bar{h}_{\mu\nu}(x) + \delta_{\mu\nu}\sigma(x). \tag{46}$$

The affine element of volume (30) then coincides with the conformal one (44). We have seen that in the dynamical affine symmetry the covariant derivative (28) contains the free parameters c_1, c_2, c_3. We now show that for a definite value of these parameters the expression (28) will serve as the covariant derivative for not only the affine but also simultaneously for the conformal symmetry.

The conformal covariant derivative $\bar{\nabla}_\lambda \bar{h}_{\mu\nu}$ is calculated in accordance with (43). We express the affine covariant derivative $\nabla_\lambda h_{\mu\nu}$ in terms of $\nabla_\lambda \bar{h}_{\mu\nu}(x)$, $\sigma(x)$, and $\partial_\chi \sigma(x)$. Using (43), we rewrite the expression (26) for $\nabla_\lambda h_{\mu\nu}$ in the form

$$\begin{aligned}\nabla_\lambda h_{\mu\nu} = {}&{}^1\!/\!_2 (e^{-\bar{h}})_{\lambda\gamma} e^{-\sigma} \{e^{-\bar{h}}, \partial_\gamma e^{\bar{h}}\}_{\mu\nu} + e^{-\sigma}(e^{-\bar{h}})_{\lambda\tau}\partial_\tau\sigma\delta_{\mu\nu} = {}^1\!/\!_2 (e^{-\bar{h}})_{\lambda\gamma}\{e^{-\bar{h}}, \bar{\nabla}_\gamma e^{\bar{h}}\}_{\mu\nu} \\ &+ \frac{\partial_\tau\sigma}{2} e^{-\sigma}[(e^{-\bar{h}})_{\mu\tau}\delta_{\lambda\nu} - (e^{-2\bar{h}})_{\tau\lambda}(e^{\bar{h}})_{\tau\nu} + (e^{-\bar{h}})_{\lambda\tau}\delta_{\mu\nu} + (\mu \leftrightarrow \nu)].\end{aligned} \tag{47}$$

We express similarly $v^{\min}_{\mu\nu,\lambda}$, which occurs in (28),

$$\begin{aligned}v^{\min}_{\mu\nu,\lambda} = {}&{}^1\!/\!_2 r_{\lambda\gamma}{}^{-1}[r^{-1}, \partial_\gamma r]_{\mu\nu} = {}^1\!/\!_2 e^{-\sigma}(e^{-\bar{h}})_{\lambda\gamma}[e^{-\bar{h}}, \partial_\gamma e^{\bar{h}}]_{\mu\nu} \\ &= {}^1\!/\!_2 (e^{-\bar{h}})_{\lambda\gamma}[e^{-\bar{h}}\bar{\nabla}_\gamma e^{-\bar{h}}]_{\mu\nu} + {}^1\!/\!_2 \partial_\tau\sigma e^{-\sigma}[(e^{-\bar{h}})_{\mu\tau}\delta_{\lambda\nu} \\ &+ 2(e^{-\bar{h}})_{\mu\lambda}\delta_{\tau\nu} + (e^{-2\bar{h}})_{\tau\lambda}(e^{\bar{h}})_{\mu\tau} - (\mu \leftrightarrow \nu)]\end{aligned} \tag{48}$$

and we represent $r^{-1}_{\lambda\tau}\partial_\tau \Psi$ in (27) in the form

$$r_{\lambda\tau}{}^{-1}\partial_\tau \Psi = (e^{-\bar{h}})_{\lambda\tau}\bar{\nabla}_\tau \Psi - i\partial_\nu\sigma e^{-\sigma}(e^{-\bar{h}})_{\lambda\mu} L_{\mu\nu}{}^{\Psi}\Psi. \tag{49}$$

We substitute (47), (48), and (49) into (29) and require in accordance with what we said above that $\nabla_\lambda \Psi$ depend on $\sigma(x)$ and $\partial_\nu \sigma(x)$ solely through the conformally covariant operations $\bar{\nabla}$. This requirement can be implemented and gives four equations for the parameters c_1, c_2, c_3, from which it follows that $c_1 = -1$, $c_2 = c_3 = 0$. Thus, simultaneously for the affine and the conformal symmetry the covariant derivative of any field $\Psi(x)$ is

$$\nabla_\lambda \Psi(x) = r^{-1}_{\lambda\tau}\partial_\tau \Psi(x) + \frac{i}{2} V_{\mu\nu,\lambda} L_{\mu\nu}{}^{\Psi}\Psi(x), \tag{50}$$

where the connection $V_{\mu\nu,\lambda}$ is uniquely defined as

$$V_{\mu\nu,\lambda} = {}^1\!/\!_2 \{r_{\lambda\gamma}{}^{-1}[r^{-1}, \partial_\gamma r]_{\mu\nu} - r_{\mu\gamma}{}^{-1}\{r^{-1}, \partial_\gamma r\}_{\lambda\nu} + r_{\nu\gamma}{}^{-1}\{r^{-1}, \partial_\gamma r\}_{\lambda\mu}\}. \tag{51}$$

At the same time one can see that no combination of $\nabla_\lambda h_{\mu\nu}$, $\nabla_\sigma h_{\tau\tau}$, $\nabla_\tau h_{\tau\tau}$ and Kronecker symbols can be expressed solely in terms of the conformal derivatives $\bar{\nabla}$. It follows from this that in a joint realization of affine and conformal symmetries not only the covariant derivative of the field $\sigma(x)$ but the whole Goldstone tensor field $h_{\mu\nu}$ vanishes. However, one can uniquely specify a covariant expression that includes in addition to $h_{\mu\nu}$ and $\partial_\lambda h_{\mu\nu}$ the second derivatives of the tensor field $h_{\mu\nu}$. This is done most readily by considering the commutator of the covariant derivatives of any field $\Psi(x)$. We have

$$(\nabla_\lambda\nabla_\rho - \nabla_\rho\nabla_\lambda)\Psi = \frac{i}{2} R_{\mu\nu,\lambda\rho} L_{\mu\nu}{}^{\Psi}\Psi, \tag{52}$$

where

$$R_{\mu\nu,\lambda\rho} = r_{\lambda\gamma}{}^{-1}\partial_\gamma V_{\mu\nu,\rho} + V_{\mu\nu,\gamma}V_{\rho\gamma,\lambda} + V_{\mu\gamma,\rho}V_{\nu\gamma,\lambda} - (\lambda \leftrightarrow \rho). \tag{53}$$

Under the action of the affine and conformal groups, $R_{\mu\nu,\lambda\rho}$ transforms as a tensor under Lorentz transformations with the parameter $u_{\mu\nu}$ in (10) and $\bar{u}_{\mu\nu}$ in (36), respectively. Its contraction

$$R = R_{\mu\nu,\mu\nu} = 2r_{\mu\gamma}{}^{-1}\partial_\gamma V_{\mu\nu,\nu} + V_{\mu\nu,\gamma}V_{\nu\gamma,\mu} - V_{\mu\gamma,\mu}V_{\nu\gamma,\nu} \tag{54}$$

*For example, in space–time $R_{\mu\nu} = -i(x_\mu\partial_\nu + x_\nu\partial_\mu)$, $D = -ix_\lambda\partial_\lambda$.

is a scalar with respect to the affine and conformal groups. It is clear that any expression $L(\Psi, \nabla_\mu\Psi, R_{\mu\nu,\lambda\rho})$ composed of different fields Ψ, their covariant derivatives $\nabla_\mu\Psi$ and $R_{\mu\nu,\lambda\rho}$ such that it is a scalar with respect to the Lorentz group is necessarily a scalar as well with respect to the affine and conformal groups simultaneously. An invariant action can be obtained by integrating such an expression over the scalar volume dV (30). The minimal interaction with the field $h_{\mu\nu}$ is described by the action integral

$$\int\left[\mathcal{L}(\Psi, \nabla_\mu\Psi)+\frac{1}{4f^2}R\right]\det r\, d^4x, \tag{55}$$

where $\mathcal{L}(\Psi, \nabla_\mu\Psi)$ is obtained from the free Lagrangian for the fields Ψ by replacing the ordinary derivatives $\partial_\mu\Psi(x)$ by the covariant derivatives $\nabla_\mu\Psi$ (50), and the term $R/4f^2$ describes the self-interaction of the field $h_{\mu\nu}$. To guarantee the correct dimensions one must introduce a universal coupling constant f with the dimensions of a length (in the units $\hbar = c = 1$) and everywhere make the substitution $h_{\mu\nu} \to fh_{\mu\nu}$ (like the introduction of the constant F_π in chiral dynamics [1]). We emphasize that the Goldstone field $h_{\mu\nu}$ itself can enter the Lagrangian only through the covariant derivatives of different fields $\nabla_\mu\Psi$, through $R_{\mu\nu,\lambda\rho}$, and through det r in the scalar volume.

5. Identification with the Theory of the Gravitational Field

We now show that the theory constructed in this way is identical with Einstein's theory of gravitation (see, for example, [10]). We have seen above (see (19)) that for fields with integral spin $a_{\mu\nu\ldots}$ one can introduce linearly transforming co- and contravariant quantities by multiplying by $r_{\mu\nu}$ or $r^{-1}_{\mu\nu}$ with respect to each index, respectively. Thus, $A_\mu = r_{\mu\bar{\nu}}a_{\bar{\nu}}$ is a covariant vector; $A_{\mu\nu} = r_{\mu\bar{\mu}}r_{\nu\bar{\nu}}a_{\bar{\mu}\bar{\nu}}$ is a covariant tensor $A_\mu^\nu = r_{\mu\bar{\mu}}r^{-1}_{\nu\bar{\nu}}a_{\bar{\mu}\bar{\nu}}$ is a mixed tensor, etc. A similar operation can also be performed with the covariant derivatives and one can define a linearly transforming covariant derivative of a covariant vector A_μ as

$$D_\lambda A_\mu = r_{\lambda\bar{\lambda}}r_{\mu\bar{\mu}}\nabla_{\bar{\lambda}}a_{\bar{\mu}}, \tag{56a}$$

of a contravariant vector as

$$D_\lambda A^\mu = r_{\lambda\bar{\lambda}}r^{-1}_{\mu\bar{\mu}}\nabla_{\bar{\lambda}}a_{\bar{\mu}}, \tag{56b}$$

of a covariant tensor of second rank as

$$D_\lambda A_{\mu\nu} = r_{\lambda\bar{\lambda}}r_{\mu\bar{\mu}}r_{\nu\bar{\nu}}\nabla_{\bar{\lambda}}a_{\bar{\mu}\bar{\nu}}$$

and so forth. After some simple calculations one can see that these definitions are identical to the standard definitions in the theory of gravitation; for example,

$$D_\lambda A_\mu = r_{\lambda\bar{\lambda}}r_{\mu\bar{\mu}}\nabla_{\bar{\lambda}}a_{\bar{\mu}} = r_{\lambda\bar{\lambda}}r_{\mu\bar{\mu}}[r_{\bar{\lambda}\tau}{}^{-1}\partial_\tau(r_{\bar{\mu}\sigma}A_\sigma + V_{\bar{\mu}\bar{\rho},\bar{\lambda}}r_{\bar{\rho}\sigma}{}^{-1}A_\sigma] = \partial_\lambda A_\mu - \Gamma_{\lambda\mu}{}^\sigma A_\sigma, \tag{57}$$

where $\Gamma^\sigma_{\mu\lambda}$ is the Christoffel symbol,

$$\Gamma_{\mu\lambda}{}^\sigma = -(r\partial_\lambda r^{-1})_{\mu\sigma} - r_{\mu\bar{\mu}}r_{\lambda\bar{\lambda}}V_{\bar{\mu}\bar{\sigma},\bar{\lambda}}\, r_{\bar{\sigma}\sigma}{}^{-1} = {}^1\!/_2 g^{\sigma\tau}(\partial_\mu g_{\tau\lambda} + \partial_\lambda g_{\mu\tau} - \partial_\tau g_{\mu\lambda}). \tag{58}$$

Note that the vanishing of the covariant derivative D_λ of the metric tensor is due to the vanishing of the covariant derivative ∇_μ of the constant tensor $\delta_{\mu\nu}$ (or $\eta_{\mu\nu}$ if $\mu = 1, 2, 3, 0$):

$$D_\lambda g_{\mu\nu} = r_{\lambda\bar{\lambda}}r_{\mu\bar{\mu}}r_{\nu\bar{\nu}}\nabla_{\bar{\lambda}}(g_{\sigma\tau}r^{-1}_{\sigma\bar{\mu}}r^{-1}_{\tau\bar{\nu}}) = r_{\lambda\bar{\lambda}}r_{\mu\bar{\mu}}r_{\nu\bar{\nu}}\nabla_{\bar{\lambda}}\delta_{\bar{\mu}\bar{\nu}} = 0. \tag{59}$$

The definition of the covariant derivative of the spinor $D_\lambda\Psi = r_{\lambda\bar{\lambda}}\nabla_{\bar{\lambda}}\Psi$ is exactly the same as in the tetrad formalism with relativistic symmetric gauge of tetrads [8]. Further, the covariant curvature tensor $R_{\mu\nu,\lambda\rho}$ can be expressed in terms of $R_{\mu\nu,\lambda\rho}$:

$$R_{\mu\nu\lambda\rho} = r_{\mu\bar{\mu}}r_{\nu\bar{\nu}}r_{\lambda\bar{\lambda}}r_{\rho\bar{\rho}}R_{\bar{\mu}\bar{\nu}\bar{\lambda}\bar{\rho}}. \tag{60}$$

For the Ricci tensor we have

$$R_{\mu\lambda} = r_{\mu\bar{\mu}}r_{\lambda\bar{\lambda}}R_{\bar{\mu}\nu\bar{\lambda}\nu}, \tag{61}$$

and the curvature R coincides with the complete contraction:

$$R = R_{\mu\nu,\mu\nu}. \tag{62}$$

The above rule for forming scalars is identical to the corresponding rule in Riemannian geometry; for example,

$$a_\mu a_\mu = A^\mu A_\mu, \quad \nabla_\mu\varphi\nabla_\mu\varphi = g^{\mu\nu}\partial_\mu\varphi\partial_\nu\varphi,$$
$$\nabla_\mu a_\nu\nabla_\mu a_\nu = g^{\sigma\tau}D_\sigma A^\mu D_\tau A_\mu.$$

The expressions for the scalar volume are also equal, since $\det r = \sqrt{\det g}$. Finally, assuming that the constant f of coupling with the field $h_{\mu\nu}$ is expressed in terms of the Newtonian gravitational constant k ($k = 6.67 \cdot 10^{-8}\ \mathrm{cm}^3 \cdot \mathrm{g}^{-1} \cdot \mathrm{sec}^{-2}$) by the equation

$$\frac{1}{4\pi} f^2 = K, \tag{63}$$

we completely identify our theory with Einstein's. The invariant element of length is constructed on the basis of the differential forms $w_{\lambda}^{\rho}(d)$ in (23a):

$$ds^2 = w_{\lambda}^{\rho}(d)\, w_{\lambda}^{\rho}(d) = r_{\mu\lambda} dx^{\mu} r_{\nu\lambda} dx^{\nu} = g_{\mu\nu} dx^{\mu} dx^{\nu}. \tag{64}$$

We may point out that we have arrived at the theory of gravitation in the tetrad formalism (see, for example, [11]) but with the relativistic symmetric gauge of tetrads* used first by Polubarinov and one of the authors of the present paper in 1965 [8].

6. Conclusions

Thus, we have shown that the theory of joint nonlinear realizations of the affine and conformal symmetries is Einstein's theory of the gravitational field. These symmetries are spontaneously broken to the Poincaré group and are dynamical. Gravitons are the corresponding Goldstone particles (as is well known they are simultaneously gauge fields for the group of general coordinate transformations).

This attractive analogy between the theory of the gravitational field and the essentially simpler theories of nonlinear realizations of internal symmetries (chiral, unitary, etc.) would appear promising. The analogy suggests new ways of searching for connections between the theory of gravitation and the theory of elementary particles. For example, in theories of nonlinear realizations of internal symmetries there arises asymptotic algebraic symmetry if one requires that the tree diagrams behave reasonably at high energies [13]; it must be possible to classify the particles with respect to linear representations of the original symmetry ($SU(2) \times SU(2)$ in chiral symmetry, SU(3) in unitary symmetry, etc.) and the mass operator must have simple transformation properties, which in all cases have been found to be reasonable [13]. Making a similar requirement in the theory of the gravitational field, one can expect algebrization of the affine and conformal symmetries. Because of the absence of finite-dimensional spinor representations of the group SL(4,R), infinite-dimensional representations of SL(4,R) must arise. This group contains SL(3,R) as a three-dimensional subgroup. Note that the use of infinite-dimensional representations of SL(3,R) was advocated by Gell-Mann et al. [14] to describe orbital excitations of hadrons. Recently, Biedenharn [15] has demonstrated that primitive infinite-dimensional representations of SL(3,R) reproduce the Regge sequences of hadrons. The condition of a reasonable interaction between elementary particles and gravitons gives a justification for using SL(3,R) and even the larger group SL(4,R). It would appear natural that gravitation, which determines interaction of masses, could also determine certain qualitative aspects of the manifold of particle masses. Note that there then arise nonminimal gravitational couplings that describe the transitions $s_1 \to s_2$ + graviton, where s_1 and s_2 are different particles† (for example, $g\varphi_{\mu\nu}\varphi R^{\mu\nu}$, where φ and $\varphi_{\mu\nu}$ are scalar and tensor fields and $R_{\mu\nu}$ is the Ricci tensor) with fixed coupling constants. Evidently, it is of interest to study the algebra of affine and conformal currents (cf. the chiral algebra of currents). Note that tree graphs in the theory of gravitation depend on the momenta like graphs in nonlinear realizations of chiral symmetry (the coupling constant has the same dimensions) and not at all as in gauge vector theories. One could imagine that Einstein's theory of gravitation is an effective Lagrangian theory (cf. effective Lagrangians in chiral dynamics) valid for the description of classical effects and in the longwave limit. The problems that arise deserve, despite all their difficulty, the closest attention.

---

*The tetrad $L_{\mu a}$ ($L_{\mu a} L_{\nu a} = g_{\mu\nu}$) is defined to within an orthogonal matrix. In a polar decomposition we have $L_{\mu\tau} = r_{\mu\nu}(e^{\Omega})_{\nu\tau}$, where $r_{\mu\nu}$ is a definite symmetric matrix and $\Omega_{\nu\tau}$ is an arbitrary antisymmetric matrix. Under the action of the Weyl SL(2,C) gauge group, $r_{\mu\nu}$ is not affected while $\Omega_{\nu\tau}$ is changed arbitrarily. By a relativistic symmetric gauge of a tetrad we understand a gauge with respect to SL(2,C) for which $\Omega_{\nu\tau} = 0$. Then the tetrad (in our case $r_{\mu\nu}$) is symmetric and 10-component, like the metric tensor itself.

†We assume that such couplings are important for the analysis of single-loop divergences carried out recently in the instructive paper [16] and could correct the situation with respect to the renormalization of the gravitational interaction of a scalar field in the single-loop approximation.

We are very grateful to F. A. Berezin, B. N. Valuev, D. V. Volkov, A. N. Zaslavskii, M. A. Markov, I. V. Polubarinov, and V. Tybor for helpful discussions.

LITERATURE CITED

1. S. Weinberg, 1970 Brandeis Lectures, Cambridge (1970).
2. S. Coleman, J. Wess, and B. Zumino, Phys. Rev., 177, 2239 (1969); D. V. Volkov, Preprint ITF 69-75 [in Russian], Kiev (1969); C. Isham, Nuovo Cim., 59A, 356 (1969).
3. V. I. Ogievetsky, Lett. Nuovo Cim., 8, 988 (1973).
4. A. Salam and J. Strathdee, Phys. Rev., 184, 1750 (1969).
5. C. Isham, A. Salam, and J. Strathdee, Ann. Phys. (New York), 62, 3 (1971).
6. D V. Volkov, ÉChAYa, 4, 3 (1973); D. V. Volkov and V. P. Akulov, Zh. Éksp. Teor. Fiz. Pis'ma Red., 16, 621 (1972).
7. V. Ogievetsky, 10th Winter School of Theor. Phys., Karpach, Poland (1973).
8. V. I. Ogievetskii and I. V. Polubarinov, Zh. Éksp. Teor. Fiz., 48, 1625 (1965); V. Ogievetsky and I. Polubarinov, Ann. Phys. (New York), 35, 167 (1965).
9. É. Cartan, Geometry of Lie Groups and Symmetric Spaces [Russian translation], IL, (1949).
10. L. D. Landau and E. Lifshitz, The Classical Theory of Fields, Addison-Wesley, Cambridge. Mass. (1951).
11. A. Trautman, Reports of Math. Phys., No. 1, 29 (1970).
12. C. Isham and A. Salam, Lett. Nuovo Cim., 5, 969 (1972).
13. S. Weinberg, Phys. Rev., 177, 2064 (1969); V. Ogievetsky, Phys. Lett., 33B, 227 (1970); Problems of Modern Physics. Collection of Papers Dedicated to the Memory of I. E. Tamm [in Russian], Nauka (1972), p. 224.
14. M. Gell-Mann, Phys. Rev. Lett., 14, 77 (1965); Y. Dothan, M. Gell-Mann, and Y. Neeman, Phys. Lett., 17, 148 (1965).
15. L. Biedenharn, R. Cusson, M. Han, and O. Weaver, Phys. Lett., 42B, 257 (1972).
16. G't. Hooft and M. Veltman, Preprint TH-1723, CERN (1973).

STRUCTURE OF THE SUPERGRAVITY GROUP

V. OGIEVETSKY and E. SOKATCHEV
Joint Institute for Nuclear Research, Dubna, USSR

Received 15 August 1978

The supergravity group is found to be the direct product of general covariance groups in complex conjugated left and right handed superspaces. The ordinary space–time coordinate and the axial gravitational superfield.are the real and imaginary part of the complex coordinate, respectively.

Recently several groups of authors [1,2] managed to add to the gravitational supermultiplet (2, 3/2) a minimal set of auxiliary fields and thus to close the supergravity algebra. Their set of fields corresponds to the field content of the axial superfield $H^\mu(x,\theta,\bar\theta)$ which we proposed in 1976 [3] (see also ref. [4]) as the most adequate minimal gravitational superfield. However, the closing transformation algebra has been found by a cumbersome and vague technique. The lack of geometrical meaning and the used component notations could cause difficulties when generalizing these results to extended supergravity and investigating the higher-order counter terms. On the other hand, the superspace approaches (ref. [5] and references therein) are far from being clear. They involve an excessively wide general covariance group in the real superspace $\{(x^\mu,\theta^\alpha,\bar\theta^{\dot\alpha})\}$, and, respectively, too large a number of superfluous fields.

In the present letter we show that the true supergravity group is much simpler. It is represented by a unification of two complex conjugated general covariance groups in two smaller "chiral" superspaces, just in the left-handed one $\{(x^{\mu L},\theta^\alpha)\}$ and in the right-handed one $\{(x^{\mu R},\bar\theta^{\dot\alpha})\}$. The main "metric" object in our approach is the axial superfield $H^\mu(x,\theta,\bar\theta)$ and it is introduced as the imaginary part of the complex space–time coordinate in our complex superspace while its real part is identified with the true space–time coordinate.

As a starting point we have used the earlier results of our supercurrent approach [3,4]. There we have investigated at the linearized level the interaction of the axial superfield $H^\mu(x,\theta,\bar\theta)$ with matter chiral superfields $\Phi_L(x,\theta,\bar\theta)$ and $\Phi_R(x,\theta,\bar\theta)$ through the supercurrent [‡1] and have found two closed transformation groups, one for Φ_L and another for Φ_R (eqs. (21a) and (21b) in ref. [4]). To understand their geometrical meaning we represented them as transformations of the superspace coordinates. It is well known [6] that the left (right) handed chiral superfield is treated most naturally in the so called "left" ("right") handed basis of superspace $\{(x^\mu,\theta^\alpha,\bar\theta^{\dot\alpha})\}$. The left and right bases are connected by a purely imaginary shift $2i\theta\sigma^\mu\bar\theta$ of the space–time coordinate x^μ, thereby making it complex in essence. Therefore it is tempting to introduce two complex conjugated "chiral" superspaces $\{(x^{\mu L},\theta^\alpha)\}$ and $\{(x^{\mu R},\bar\theta^{\dot\alpha})\}$ instead of the real one $\{(x^\mu,\theta^\alpha,\bar\theta^{\dot\alpha})\}$, $x^{\mu L}$ and $x^{\mu R}$ being complex conjugated space–time coordinates in the "left"-handed and "right"-handed superspaces, respectively.

In such a new framework the transformations (21a) and (21b) in ref. [4] prove simply to form the general covariance group of infinitesimal transformations in the left handed and right handed superspaces, respectively,

$$\delta x^{\mu L} = a^\mu(x^L,\theta), \qquad \delta x^{\mu R} = \bar a^\mu(x^R,\bar\theta),$$
$$\delta\theta^\alpha = \epsilon^\alpha(x^L,\theta), \qquad \delta\bar\theta^{\dot\alpha} = \bar\epsilon^{\dot\alpha}(x^R,\bar\theta), \qquad (1)$$

‡1 The concept of supercurrent has been introduced by Ferrara and Zumino [7]. The existence of this object in the general case has recently been proved in our paper [8].

Volume 79B, number 3 PHYSICS LETTERS 20 November 1978

where a^μ, ϵ^α and their complex conjugates $\bar{a}^\mu$, $\bar{\epsilon}^{\dot\alpha}$ are arbitrary vector and spinor superfunction parameters, respectively. The group meaning of these transformations is obvious and the group law is

$$\begin{aligned} a^\mu_{\text{bracket}} &= a^\nu(2)(\partial/\partial x^{\nu L})a^\mu(1) \\ &+ \epsilon^\beta(2)(\partial/\partial\theta^\beta)a^\mu(1) - (1 \leftrightarrow 2), \\ \epsilon_\alpha^{\text{bracket}} &= a^\nu(2)(\partial/\partial x^{\nu L})\epsilon_\alpha(1) \\ &+ \epsilon^\beta(2)(\partial/\partial\theta^\beta)\epsilon_\alpha(1) - (1 \leftrightarrow 2), \end{aligned} \tag{2}$$

and its complex conjugate.

In fact in ref. [4] we have derived the transformations (21a) and (21b) which are rewritten now in the form (1) with the restriction eq. (11b) of ref. [4], which is equivalent to

$$-(\partial/\partial\theta_\alpha)\epsilon_\alpha + (\partial/\partial x^{\mu L})a^\mu = 0, \tag{3}$$

and its complex conjugate. These constraints are compatible with the group law (2) and have a simple geometrical meaning. Restrictions (3) mean that the berezinian of transformations (1) becomes unity or in other words, that the "volume" both in the left and right handed superspaces is conserved. It can be shown that the whole group (1) corresponds to conformal supergravity while its volume-preserving subgroup (eqs. (1), (3)) describes Einstein supergravity.

After establishing this simple group structure underlying supergravity we have to answer two more questions. First, we need the common real superspace $\{(x^\mu, \theta^\alpha, \bar\theta^{\dot\alpha})\}$ to deal with real superfields. Secondly, we have to introduce a "metric" (gauge) object, which will be, as we believe, the axial superfield $H^\mu(x, \theta, \bar\theta)$. These two problems are solved simultaneously. Let us make a change of variables:

$$x^\mu = \tfrac{1}{2}(x^{\mu L} + x^{\mu R}), \qquad H^\mu = (1/2i)(x^{\mu L} - x^{\mu R}). \tag{4}$$

Now the real variable x^μ can be identified with the common physical space–time coordinate.

Further, instead of regarding the imaginary part H^μ as an independent coordinate we express it as a function of the remaining variables [‡2]

$$H^\mu = H^\mu(x^\nu, \theta^\alpha, \bar\theta^{\dot\alpha}). \tag{5}$$

‡2 Note that Volkov and Akulov have tried to identify the spinor coordinate with a Goldstone neutrino field in their pioneering paper [9].

So we introduced a real superspace $\{(x^\mu, \theta^\alpha, \bar\theta^{\dot\alpha})\}$ and an axial gauge superfield H^μ in it with transformation laws following directly from eqs. (1) and (4):

$$\begin{aligned} H^{\mu\prime}(x', \theta', \bar\theta') &= H^\mu(x, \theta, \bar\theta) \\ &- \tfrac{1}{2} i a^\mu(x^\nu + iH^\nu(x, \theta, \bar\theta), \theta^\beta) \\ &+ \tfrac{1}{2} i \bar{a}^\mu(x^\nu - iH^\nu(x, \theta, \bar\theta), \bar\theta^{\dot\beta}), \end{aligned} \tag{6a}$$

$$\begin{aligned} x^{\mu\prime} &= x^\mu + \tfrac{1}{2} a^\mu(x^\nu + iH^\nu(x, \theta, \bar\theta), \theta^\beta) \\ &+ \tfrac{1}{2} \bar{a}^\mu(x^\nu - iH^\nu(x, \theta, \bar\theta), \bar\theta^{\dot\beta}), \end{aligned} \tag{6b}$$

$$\begin{aligned} \theta^{\alpha\prime} &= \theta^\alpha + \epsilon^\alpha(x^\nu + iH^\nu(x, \theta, \bar\theta), \theta^\beta), \\ \bar\theta^{\dot\alpha\prime} &= \bar\theta^{\dot\alpha} + \bar\epsilon^{\dot\alpha}(x^\nu - iH^\nu(x, \theta, \bar\theta), \bar\theta^{\dot\beta}). \end{aligned} \tag{6c}$$

It is not hard to convince oneself that the new transformations (6) obey the same group law (2) as the old ones (1).

Our gravitational superfield $H^\mu(x, \theta, \bar\theta)$ is introduced in a rather unconventional manner. It is an object of dual nature, playing the role of a coordinate in the complex superspace $\{(x^{\mu L}, x^{\mu R}, \theta^\alpha, \bar\theta^{\dot\alpha})\}$ and of a "metric" object of the real subspace $\{(x^\mu, \theta^\alpha, \bar\theta^{\dot\alpha})\}$. In order to understand better its second nature, it is worthwhile to write down the transformation law (6) in the more habitual terms of component fields. We shall do it in this letter for the conformal supergravity case only. The first step is to fix properly the gauge to avoid the complicated nonlinearity of eq. (6). This is done by means of a trick proposed by Wess and Zumino for the supersymmetric Yang–Mills case [10]. The few first components of H^μ are gauged out and its decomposition is reduced to

$$\begin{aligned} H^\mu &= \theta\sigma_a\bar\theta e^{\mu a} + \kappa\theta\theta.\bar\theta\bar\psi^\mu + \kappa\bar\theta\bar\theta.\theta\psi^\mu \\ &+ \theta\theta.\bar\theta\bar\theta(\kappa A^\mu + \tfrac{1}{4}\epsilon^{\mu\nu\lambda\rho}e_{a\lambda}\partial_\nu e^a{}_\rho). \end{aligned} \tag{7}$$

Here the fields are: $e^{\mu a}(x)$ (and its inverse $e_{a\mu}(x)$) is the vierbein field; $\psi^{\mu\alpha}(x)$, $\bar\psi^{\mu\dot\alpha}(x)$ is the "gravitino" field; $A^\mu(x)$ is the gauge field for the local chiral invariance. After the gauge has been fixed there remains some class of transformations (6) which preserve the gauge. Omitting all the details, we formulate the final results. The fields in eq. (7) undergo the following transformations: (1) general covariance transformations (with μ being a contravariant index); (2) local Lorentz ones; (3) local scale ones; (4) local chiral ones; (5) local supersymmetry (Q- and S-supersymmetries with param-

Volume 79B, number 3 PHYSICS LETTERS 20 November 1978

eters $\epsilon_\alpha(x)$ and $\eta_\alpha(x)$ in four-component notation):

$$\begin{aligned}
&\delta e^{\mu a} = \kappa \bar{\epsilon}\gamma^a \psi^\mu ,\\
&\delta \psi^\mu{}_\alpha = \kappa^{-1}[-(\gamma^\mu \eta)_\alpha - 2\mathrm{i}\hat{\nabla}^\mu \epsilon_\alpha + \tfrac{1}{2}\mathrm{i}(\gamma^\mu\gamma^\nu \hat{\nabla}_\nu \epsilon)_\alpha],\\
&\delta A^\mu = \mathrm{i}\bar{\eta}\gamma_5\psi^\mu - \tfrac{1}{4}\epsilon^{\mu\nu\lambda\rho}\nabla_\nu(\bar{\epsilon}\gamma_\lambda\psi_\rho)\\
&\quad - \nabla_\nu\bar{\epsilon}\gamma_5\gamma^\mu\psi^\nu + \tfrac{1}{2}\bar{\epsilon}\gamma_5\gamma^\nu\hat{\nabla}_\nu\psi^\mu + \tfrac{1}{2}\bar{\epsilon}\gamma^\nu\psi^\mu A_\nu .
\end{aligned} \tag{8}$$

Here $\hat{\nabla}_\mu = \nabla_\mu - \kappa A_\mu \gamma_5$, and ∇_μ is the standard general covariant derivative. These transformations form a closing algebra because they are derived from transformations (6) having an obvious group character. The simplest way to find the bracket parameters is to use the general formula (2) properly adapted to the gauge-fixing procedure. We shall not discuss this business here. Of course, one can check, if one prefers, the closure of the algebra by direct tedious calculations.

The transformations we have obtained are similar to those of conformal supergravity [2]. The difference in local supersymmetry transformations is probably due to the fact that we combine into a supermultiplet (2, 3/2) the contravariant vierbein, while in ref. [2] there appears a covariant vierbein. Besides, on the component field level there remains a great freedom of redefining the field variables and also the parameter function $\eta_\alpha(x)$.

In conclusion we wish to point out that a number of questions concerning the formalism remains open. One has to know how to define superfields with external indices, supercovariant derivatives and invariants of the group, etc. However, now we can say with certainty that this extremely simple and clear geometrical picture of the supergravity group will provide an adequate basis for the supergravity theory. A generalization to extended supergravity is straightforward: one has simply to supply the Grassmann variables θ^α, $\bar{\theta}^{\dot{\alpha}}$ with internal symmetry indices a and to consider the left handed and right handed extended superspaces $\{(x^{\mu L}, \theta^{\alpha a})\}$, $\{(x^{\mu R}, \bar{\theta}^{\dot{\alpha} a})\}$. In any case, our approach to supergravity as the theory of an axial superfield generated by the supercurrent [4] turns out to be true.

More details on this subject will be given elsewhere.

It is a pleasure for us to thank E.A. Ivanov for many useful discussions.

References

[1] S. Ferrara and P. van Nieuwenhuizen, Phys. Lett. 74B (1978) 333;
K.S. Stelle and P.C. West, Phys. Lett. 74B (1978) 330;
E.S. Fradkin and M.A. Vasiliev, Prepr. IAS-778-3PP (1978).

[2] S. Ferrara, M.T. Grisaru and P. van Nieuwenhuizen, CERN preprint TH2467 (1978).

[3] V. Ogievetsky and E. Sokatchev, Proc. 4th Intern. Conf. on Non-local and non-linear field theory (Alushta, 1976) Dubna publ. D2-9788, p. 183.

[4] V. Ogievetsky and E. Sokatchev, Nucl. Phys. B124 (1977) 309.

[5] V. Akulov, D. Volkov and V. Soroka, Zh. Eksp. Teor. Fiz. Pis'ma 22 (1975) 396; Theor. Math. Phys. 31 (1977) 12;
J. Wess and B. Zumino, Phys. Lett. 66B (1977) 361;
J. Wess, Supersymmetry–supergravity, Lectures given at the VIII G.I.F.T. Seminar (Salamanca, June 1977), Karlsruhe preprint;
Y. Ne'eman and T. Regge, preprint ORO 3992 328 (1977);
L. Brink, M. Gell-Mann, P. Ramond and J. Schwarz, Phys. Lett. 74B (1978) 336.

[6] S. Ferrara, J. Wess and B. Zumino, Phys. Lett. 51B (1974) 239;
V. Ogievetsky and L. Mezincescu, Usp. Fiz. Nauk. 117 (1975) 637; Engl. transl. Sov. Phys. Usp. 18 (1976) 960.

[7] S. Ferrara and B. Zumino, Nucl. Phys. B87 (1975) 207.

[8] V. Ogievetsky and E. Sokatchev, JINR preprint E2-11528 (1978).

[9] D. Volkov and A. Akulov, Zh. Eksp. Teor. Fiz. Pis'ma 16 (1972) 621.

[10] J. Wess and B. Zumino, Nucl. Phys. B78 (1974) 1.

Grassmann analyticity and extension of supersymmetry

A. Gal'perin,[1] E. Ivanov, and V. Ogievetskiĭ
Joint Institute for Nuclear Research

(Submitted 1 December 1980)

Pis'ma Zh. Eksp. Teor. Fiz. **33**, No. 3, 176–181 (5 February 1981)

The concept of analyticity of Grassmann spinor variables is introduced. It is shown that this concept makes it possible to realize $N=2$ supersymmetry on the ordinary, complex, $N=1$ superpole outside the mass shell.

PACS numbers: 11.30.Pb

1. The Majorana spinor is regarded as a real Grassmann variable in the theory of supersymmetries

$$\Theta_a = \begin{pmatrix} \Theta_a \\ \bar{\Theta}^{\dot{a}} \end{pmatrix}, \qquad \Theta = C\bar{\Theta}^T, \tag{1}$$

where Θ_α and $\bar{\Theta}^{\dot{\alpha}}$ are adjoint Weyl spinors and C is the charge-conjugation matrix. A direct approach in the extended N supersymmetries gives N Grassmann variables. A simple superfield $\Phi(x,\Theta^1,\Theta^2,..\Theta^N)$ contains 2^{4N} field components.

We shall introduce the concept of Grassmann analyticity and demonstrate that the number of Grassmann variables can be reduced with its help by using a simple example of $N=2$ supersymmetry. The Cauchy-Riemann analyticity condition

$$\frac{\partial}{\partial z} f(x,y) = \frac{1}{2}\left(\frac{\partial}{\partial x} + i\frac{\partial}{\partial y}\right) f(x,y) = 0$$

has the following analog for Grassmann variables:

$$\left(\tilde{D}_a^\Theta + i\tilde{D}_a^\eta\right)\Phi(x,\Theta,\eta) = 0, \tag{2}$$

where $\tilde{D}_\alpha^\Theta$ ($\tilde{D}_\alpha^\eta$) is a covariant spinor derivative of the Grassmann variable Θ (η). This condition, which represents the superpole's "independence" of the Θ-$i\eta$ variable, makes it possible to turn to the complex scalar superpole of a single Grassman variable $V(x,\Theta)$. We should point out that the condition (2) was used in the two-dimensional case in the theory of supersymmetric strings.[1] Further, the chirality is explained the same way as analyticity in the ordinary $N=1$ supersymmetry. In fact, the chirality

$$\bar{D}_{\dot{a}}\Psi(x,\Theta) = 0 \tag{3}$$

means that $\Psi(x,\Theta)$ depends only on the left-handed Weyl spinor Θ^α and is independent of the right-handed adjoint spinor $\bar{\Theta}^\alpha$. The solution of Eq. (3) (see, for example, Ref. 2) is

$$\Psi(x,\Theta) = a(x_L) + \Theta^{\alpha}\Psi_{\alpha}(x_L) + \Theta^{\alpha}\Theta_{\alpha} b(x_L), \qquad x_L^a = x^a + \frac{1}{2}\bar{\Theta}\gamma^a\gamma_5\Theta\,. \tag{4}$$

The solution of the analyticity condition (2) given below will be used for reducing the superpole of the two Majorana Grassmann variables to the superpole defined by a single Grassmann variable.

2. We proceed to our problem. We arrived at Grassmann analyticity by starting from the known fact that the complexified $N=1$ supermultiplets in the mass shell are $N=2$-supersymmetry representations with a central charge (see, for example, Ref. 3). This also seems to apply to superfields outside the mass shell. We now consider the complex superfield $V(x,\Theta)$. Its expansion in the terms

$$V(x,\Theta) = \frac{1}{m}M(x) - i\bar{\Theta}\frac{\Psi^2}{m}(x) + \frac{1}{2}\bar{\Theta}\Theta\frac{P^{12}}{m}(x) + \frac{1}{2}\bar{\Theta}\gamma_5\Theta S(x)$$

$$+ \frac{1}{2}\bar{\Theta}i\gamma^a\gamma_5\Theta V_a(x) + \bar{\Theta}\Theta\bar{\Theta}\left[\Psi^1(x) - \frac{1}{2m}\partial\!\!\!/\Psi^2(x)\right] + \frac{1}{4}(\bar{\Theta}\Theta)^2$$

$$\times\left[iP^{11}(x) - \left(\frac{\Box}{2m} + m\right)M(x)\right] \tag{5}$$

includes the complex vector and scalar fields $V_a(x)$, $M(x)$, and $S(x)$, which produce the 0(2) singlets; $P^{ij}(x)$ is a complex, spurless, symmetric tensor 0(2) and $\Psi^i(x)$–0(2) is a doublet of the Dirac spinors. The action for $V(x,\Theta)$

$$S = \int d^4x\, d^4\Theta\,\frac{1}{2}V^*(x,\Theta)\left[\Box + \frac{(\bar{D}D)^2}{16} + m^2\right]V(x,\Theta) = \tag{6a}$$

$$= \int d^4x\left\{-\frac{1}{2}F^*_{ab}(x)F^{ab}(x) + m^2V^*_a(x)V^a(x) + \partial_a M^*(x)\partial^a M(x) - m^2M^*(x)M(x)\right.$$

$$\left. + m^2S^*(x)S(x) + \frac{1}{2}P^{ij*}(x)P^{ij}(x) + i\bar{\Psi}^k(x)\partial\!\!\!/\Psi^k(x) + im\epsilon^{kl}\bar{\Psi}^k(x)\Psi^l(x)\right\} \tag{6b}$$

is invariant with respect to the 0(2)-supersymmetry transformations

$$\delta V_a = \bar{\epsilon}^k i\gamma_a\gamma_5\Psi^k + \frac{i}{m}\epsilon^{kl}\bar{\epsilon}^k\gamma_5\partial_a\Psi^l,$$

$$\delta\Psi^i = iP^{ij}\epsilon^j + \left(-mM + \frac{i}{2}\sigma_{ab}\gamma_5F^{ab}\right)\epsilon^i$$

$$+ \epsilon^{ik}\left(-im\gamma_5 S + m\gamma^a\gamma_5V_a - \partial\!\!\!/M\right)\epsilon^k\,, \tag{7a}$$

$$\delta M = -i\epsilon^{ij}\bar{\epsilon}^i\Psi^j, \quad \delta S = \bar{\epsilon}^i\gamma_5\Psi^i - \epsilon^{kl}\bar{\epsilon}^k\gamma_5\frac{\not\partial}{m}\Psi^l,$$

$$\delta P^{ij} = -\bar{\epsilon}^i(\not\partial\Psi^j + m\Psi^k\epsilon^{jk}) + \frac{1}{2}\delta^{ij}\bar{\epsilon}^l(\not\partial\Psi^l + m\Psi^k\epsilon^{lk}) + (i\leftrightarrow j). \quad (7b)$$

$F_{ab} = \partial_a V_b - \partial_b V_a$, $\epsilon^{12} = -\epsilon^{21} = 1 = \not\partial = \partial^m\gamma_m$ in these formulas and the indices of the internal symmetry were intentionally written on one level in order to emphasize that we are now dealing with 0(2) rather than with $SU(2)$. The parameter of the ordinary supersymmetry is represented by ϵ^1 and the parameter of the second supersymmetry is ϵ^2. These transformations are written in the superfield form as follows:

$$\delta V = -i\bar{\epsilon}^i Q^i V$$

moreover,

$$Q^1_{\bar{\beta}} = i\frac{\partial}{\partial\bar{\Theta}^\beta} - (\not\partial\Theta)_\beta, \quad (8a)$$

$$Q^2_\beta = -2m\Theta_\beta + \frac{1}{4m}\bar{D}DD_\beta; \qquad \left(D_\beta = \frac{\partial}{\partial\bar{\Theta}^\beta} - i(\not\partial\Theta)_\beta\right) \quad (8b)$$

and

$$\{\bar{Q}^i, Q^k\} = 2\delta^{ik}\gamma^a P_a + 2i\epsilon^{ik}Z. \quad (9)$$

It is important that our 0(2) superalgebra has a central charge Z proportional to the mass

$$ZV = mV, \qquad ZV^* = -mV^*. \quad (10)$$

Like the electric charge, it has opposite values for the particles and anitparticles. It is somewhat surprising that there is a term with three spinor derivatives in Eq. (8b). We shall show that these transformations occur naturally.

3. We shall analyze the complex superfield $\Phi(x,\Theta,\eta)$, which satisfies the Cauchy-Riemann condition (2), i.e., it is analytic. The supersymmetry generators for it, which are defined in the form

$$Q^1_a = i\frac{\partial}{\partial\bar{\Theta}^a} - (\not\partial\Theta)_a, \quad Q^2_a = i\frac{\partial}{\partial\bar{\eta}^a} - (\not\partial\eta)_a - 2\Theta_a Z \quad (11)$$

satisfy the commutation relations (9). The influence of Z on Φ and Φ^* is defined as in (10) with the substitution of Φ for V. Thus, the operators of the spinor derivatives in the condition (2) can be written as follows:

$$\tilde{D}^\Theta_a = \frac{\partial}{\partial\bar{\Theta}^a} - i(\not\partial\Theta)_a + 2i\eta_a Z, \quad \tilde{D}^\eta_a = \frac{\partial}{\partial\bar{\eta}^a} - i(\not\partial\eta)_a. \quad (12)$$

The Cauchy-Riemann condition can be resolved [analog (4)]:

$$\Phi(x, \Theta, \eta) = e^{-m\bar{\eta}\eta} \phi(x^m - \bar{\Theta}\gamma^{m\eta}, \Theta + i\eta) \tag{13a}$$

$$= e^{-m\bar{\eta}\eta} e^{-\bar{\Theta}\not{\partial}\eta} e^{i\bar{\eta}\frac{\partial}{\partial\bar{\Theta}}} \phi(x, \Theta) . \tag{13b}$$

It appears that the superfield $V(x,\Theta)$ introduced above can be expressed in terms of $\phi(x,\Theta)$ in the following way:

$$V(x, \Theta) = \phi(x, \Theta) + \frac{1}{4m} \bar{D}D\phi(x, \Theta) . \tag{14}$$

Thus,

$$\delta\Phi(x, \Theta, \eta) = -i\epsilon^i Q^i \Phi(x, \Theta, \eta) \tag{15}$$

correspond exactly to the transformations (8) for $V(x,\Theta)$. It is interesting that the action (6) can be written in the form

$$S = \frac{1}{32} \int d^4x d^4\Theta d^4\eta \, \Phi^*(x, \Theta, \eta) \Phi(x, \Theta, \eta) \tag{16a}$$

$$= \frac{1}{2} \int d^4x d^4\Theta \, \phi^*(x, \Theta) \left[\Box + \frac{(\bar{D}D)^2}{8} + \frac{m}{2}\bar{D}D + m^2\right] \phi(x, \Theta)$$

$$= \frac{1}{2} \int d^4x d^4\Theta \, V^*(x, \Theta) \left[\Box + \frac{(\bar{D}D)^2}{16} + m^2\right] V(x, \Theta) . \tag{16b}$$

The representation (16) demonstrates the existence of a strong analogy with the ordinary chiral Lagrangian.

4. The representation (13a) suggests a presence of "analytic" and "antianalytic" bases

$$x^m - \bar{\Theta}\gamma^m\eta, \quad \Theta + i\eta, \tag{17a}$$

$$x^m + \bar{\Theta}\gamma^m\eta, \quad \Theta - i\eta, \tag{17b}$$

which resemble the right-hand and the left-hand bases in $N = 1$ (Ref. 4) and which are transformed into them as $\eta \rightarrow \gamma_5^\Theta$ [with an appropriate change in the x scale because of differentiation of Eq. (2). In fact, (17a) and (17b) for invariant spaces of the $N = 2$ supersymmetry. It is expected that $N = 2$ supergravitation, which does not contain external constraints and is analogous to that proposed for $N = 1$ in Ref. 4, can be formulated. To do this, 0(2) must first be expanded to $SU(2)$ and the analytic properties of the supermultiplets in Ref. 5 must be analyzed. We can assume that there is an internal coupling with the systems of hypercomplex numbers in the case of higher supersymmetries.[6]

We have verified that the hypercomplex coordinates

$$\tilde{x}^m = x^m + i\bar{\Theta}\gamma^m\eta_k e_k - \frac{i}{2}\bar{\eta}_k\gamma^m\eta_l f_{klp}e_p, \quad \tilde{\Theta} = \Theta + e_k\eta_k$$

$$(e_k e_l = -\delta_{kl} + f_{klp}e_p) \tag{18}$$

form invariant spaces of the $N = 4$ supersymmetry (e_k are the quaternion units) and of the $N = 8$ supersymmetry (e_k are the octave units). These problems will be discussed in our future papers.

[1]Institute of Nuclear Physics, Academy of Sciences, Uzbek SSR.

[1]M. Ademollo, L. Brink *et al.*, Nucl. Phys. **B111**, 77 (1976).
[2]A. Salam and J. Strathdee, Phys. Rev. **11D**, 1521 (1975); V. Ogievetskiĭ and L. Mezinchesku, Usp. Fiz. Nauk **117**, 637 (1975) [Sov. Phys. Uspekhi **18**, 960 (1975)].
[3]P. Fayet, Preprint LPTENS 80/7.
[4]V. Ogievetskiĭ and E. Sokatchev, Phys. Lett. **79B**, 222 (1978).
[5]P. Breitenlohner and M. F. Sohnius, Nucl. Phys. **B165**, 483 (1980); M. F. Sohnius, K. S. Stelle, and P. C. West, Preprint ICTP/79-80/51.
[6]J. Lukierski, Preprint UGVA-DPT 1979/09-216.

Translated by S. J. Amoretty
Edited by Robert T. Beyer

Harmonic superspace: key to $N = 2$ supersymmetry theories

A. Gal'perin, E. Ivanov, V. Ogievetskiĭ, and É. Sokachev
Joint Institute for Nuclear Research; Institute of Nuclear Physics, Tashkent; and Institute of Nuclear Research and Nuclear Energy, Sofia

(Submitted 29 May 1984)
Pis'ma Eksp. Teor. Fiz. **40**, No. 4, 155–158 (25 August 1984)

The concept of a harmonic $N = 2$ superspace with additional coordinates related to the SU(2)/U(1) sphere is introduced. This concept leads to an adequate geometric description of the $N = 2$ theories of matter, the Yang-Mills theory, and supergravity in terms of superfields without coupling. A new effect has been discovered: the unboundedness of the number of gauge and auxiliary degrees of freedom.

1. So far, it has not been found possible to construct expanded supersymmetry theories (Yang-Mills theories, theories of supergravity, or theories of matter) by an explicitly covariant geometric approach in terms of superfields not subject to constraints. Partial success has been achieved in a nongeometric approach to the Abelian Yang-Mills theory[1] and its generalizations (in the form of a recursive procedure) to the non-Abelian case.[2] The need for such a construction is urgent in many regards, primarily in connection with the remarkable property of finiteness of several supersymmetry models of field theory.

2. We propose such a construction in the present letter. It is based on the introduction of harmonic variable $u_i^{\pm}$, the coordinates of the SU(2)/U(1) sphere, in addition to the ordinary even coordinates. The variables $u_i^{\pm}$ are dyads; they are assigned indices of two types, an SU(2) index (i) and a U(1) index ($\pm$), and they form a bridge between these groups. Making use of the normalization condition

$$u^{+i}u_i^- = 1, \tag{1}$$

we can reversibly convert the ordinary spinor coordinates $\theta^i_\alpha, \bar\theta^i_{\dot\alpha}$ with the indices i of the SU(2) group into spinor coordinates $\theta^{\pm}_\alpha, \bar\theta^{\pm}_{\dot\alpha}$ with the indices $\pm$ of the U(1) group:

$$\theta^{\pm}_\alpha = \theta^i_\alpha u_i^{\pm}, \qquad \bar\theta^{\pm}_{\dot\alpha} = \bar\theta^i_{\dot\alpha} u_i^{\pm} \tag{2}$$

$$\theta^i_\alpha = u^{+i}\theta^-_\alpha - u^{-i}\theta^+_\alpha, \qquad \bar\theta^i_{\dot\alpha} = u^{+i}\bar\theta^-_{\dot\alpha} - u^{-i}\bar\theta^+_{\dot\alpha}. \tag{3}$$

A key point here is that the subspace

$$\{\zeta_A = (x^m_A, \theta^+_\alpha, \bar\theta^+_{\dot\alpha}), u^{\pm}\}, \quad x^m_A = x^m - 2i\theta^{(i}\sigma^m\bar\theta^{j)}u_k^+u_j^- \tag{4}$$

(not including the coordinates $\theta^-_\alpha, \bar\theta^-_{\dot\alpha}$) is closed under the transformations of the $N=2$ supersymmetry:

$$\delta x^m_A = -2i(\epsilon^k\sigma^m\bar\theta^+ + \theta^+\sigma^m\bar\epsilon^k)u_k^-, \tag{5}$$

$$\delta\theta^+_\alpha = \epsilon^i_\alpha u_i^+, \quad \delta\bar\theta^+_{\dot\alpha} = \bar\epsilon^i_{\dot\alpha}u_i^+, \quad \delta u^{\pm} = 0.$$

We call (4) the "analytic superspace of $N=2$ supersymmetry," and the superfields $\Phi^{(q)}(\zeta_A,u)$ defined on it are "analytic." A superfield $\Phi^{(q)}(\zeta_A u)$ as a whole is a representation of U(1) with a charge q. Its components have charges ranging from q to $q-4$, depending on the number θ^+, θ^+ in the given term of the expansion. Each component is in turn a function of the new coordinates $u_i^{\pm}$ which permits an infinite expansion of the type

$$F^{(q)}(x_A, u^{\pm}) = \sum_{n=0}^{\infty} f^{(i_1\ldots i_{n+q}j_1\ldots j_n)}(x_A)\, u^+_{i_1}\ldots u^+_{i_{n+q}}u^-_{j_1}\ldots u^-_{j_n}. \tag{6}$$

In the expansion of $\Phi^{(q)}$ the charges of the U(1) group carry only harmonic variables, the dyads $u_i^{\pm}$ (and the spinor coordinates θ^+, θ^+, which contain them). Remarkably, the components fields $f^{(i_1\cdots j_n)}(x_A)$ are representations of the SU(2) group and scalars under U(1)! The reason is that expansion (6) is actually a harmonic expansion on a homogeneous space, the SU(2)/U(1) sphere.[3] Analysis shows that of the two quantum numbers of the $N=2$ supersymmetry the first—the superspin—is identical for all terms of the expansion and vanishes for scalar analytic superfields with the U(1) charge q, while the second—the superisospin I—takes on the values

$$\Phi^{(q)}(\zeta_A, u): \quad I = \left|\frac{q}{2} - 1\right| + n, \quad n = 0, 1, 2, \ldots \tag{7}$$

We also wish to emphasize that one could define a covariant harmonic derivative D^{++} of such a nature that in acting on an analytic superfield it leaves this field analytic and is compatible with normalization (1):

$$D^{++}\Phi^{(q)} = \left(u^{+i}\frac{\partial}{\partial u^{-i}} - 2i\theta^{+}\sigma^{m}\bar{\theta}^{+}\frac{\partial}{\partial x^{m}_{A}} \right)\Phi^{(q)} . \tag{8}$$

We have thus been able to generalize the Grassmann analyticity[4] in such a way that the SU(2) symmetry is conserved! The analytic variables from Ref. 4 can be found from (2) by making the particular choices $u^{+1} = -(1/\sqrt{2})$, $u^{+2} = -(i/\sqrt{2})$. We have already found a coordinate of the type x^{m}_{A} from (4) (see the bases in Ref. 5). It is important to note that we have the operation* which is the product of a complex conjugation and the operation*,

$$* \quad u^{\pm i} \to \pm u^{\mp i} , \tag{9}$$

which maps an analytic subspace into itself. The operation* can be used to determine analytic real superfields. In constructing an action we need an integral over the analytic superspace. An integral over the harmonic coordinates is defined by the following rules:

$$\int du \cdot 1 = 1, \int du\, u^{\left(+i_1 \right.} \ldots u^{+i_p} u^{-j_1} \ldots u^{\left. -j_q\right)} = 0, \quad p + q > 0. \tag{10}$$

The volume element is $d\zeta_A^{(-4)}du = d^4x_A d^2\theta^+ d^2\bar{\theta}^+ du$.

3. We turn now to the construction of $N=2$ supersymmetry theories. We begin with the Fayet-Sohnius matter hypermultiplet.[6,7] This hypermultiplet has a superspin 0 and a superisospin 1/2. According to (7), the simplest superfield which contains it is the analytic superfield q^+. For it, the action is written in the form[2)]

$$S = \int d\zeta_A^{(-4)} du \,[\overset{*}{\bar{q}}{}^{+} D^{++} q^{+} + \frac{\lambda}{2}(\overset{*}{\bar{q}}{}^{+})^2 (q^{+})^2] + \text{H.a.} \tag{11}$$

and implies the equations of motion

$$D^{++} q^{+} = -\overset{*}{\bar{q}}{}^{+} (q^{+})^2 . \tag{12}$$

In studying the superfield q^+, we find a fundamentally new phenomenon. According to (7), q^+ contains an infinite number of auxiliary fields, which form supermultiplets with superspin 0 and infinitely increasing superisospins 1/2,3/2. . . . It follows from equations of motion (12) that all fields with a superisospin exceeding 1/2 are zero in the free case ($\lambda = 0$) and can be expressed in terms of fields with a superisospin 1/2 at $\lambda \neq 0$.

An analogous situation arises for a hypermultiplet of another type.[8] It has a superisospin 1 and a superspin 0 and is described by an analytic superfield ω which contains, according to (7), superspin 1, 2,. . . . The action for ω in the free case is written

$$S_0 = \int d\zeta_A^{(-4)} du\, D^{++}\omega D^{++}\omega , \tag{13}$$

and analysis of the equations of motion $(D^{++})^2\omega = 0$ shows that an infinite set of auxiliary supermultiplets with superisospins $\geqslant 2$ on the mass shell vanishes. It is easy to construct an interaction, e.g., of the type

$$S_{int+0} = \int d\zeta_A^{(-4)} du g^{ab}(\kappa\omega) D^{++}\omega_a D^{++}\omega_b . \tag{14}$$

where $g^{ab}(\kappa\omega)$ is the "metric," and κ is the coupling constant. Howe *et al.*[8] actually use the coupling (in our notation) $(D^{++})^3\omega=0$, which is compatible only with free equations of motion.

4. An $N=2$ Yang-Mills theory has previously been known in component form,[9,6] in the form of a superfield theory with coupling[2)] (Ref. 10), and in a nongeometric description with prepotentials of higher dimensionality.[1,2] The Yang-Mills $N=2$ supermultiplet contains the vector field $A_a(x)$, the scalar fields $M(x)$ and $N(x)$, the triplet $D_{ij}(x)$, and the Majorana isodoublet $\Psi^i_\alpha(x),\bar{\Psi}_{\dot{\alpha}i}(x)$. Its superspin and superisospin are both zero. A supermultiplet of this sort could be described by a superfield V^{++} (superspin 0, superisospins 0, 1, . . .). The extraneous superisospins are "decontaminated" by gauge transformations:

$$(V^{++})'=\frac{1}{ig}e^{i\omega}(D^{++}+igV^{++})e^{-i\omega}, \qquad \begin{aligned} V^{++}&=V_i^{++}T_i,\\ \omega&=\omega_iT_i, \end{aligned} \tag{15}$$

where the T_i are the generators of the internal-symmetry group. Transformation (15) literally copies a transformation from the ordinary Yang-Mills theory ($N=0$): The derivative $\partial/\partial x^m$ is replaced by the derivative D^{++}, the V_m coupling is replaced by a V^{++} coupling, and the gauge function $\omega(x)$ is replaced by the gauge analytic superfield $\omega(\zeta_A,u)$. A remarkable new phenomenon is that we are now dealing with an infinite number of gauge degrees of freedom. We will postpone a discussion of the action to a more detailed paper; at this point, we simply note that it is also a simple matter to incorporate an interaction with material fields: It is necessary to lengthen the harmonic derivative, $D^{++}\to D^{++}+igV^{++}$, in (11), (13), and (14).

5. We now consider the $N=2$ Einstein supergravity. A fundamental gauge group is chosen by requiring conservation of the analytic representations, in the manner in which it is chosen from the requirement of the conservation of chirality in the $N=1$ case.[13] The analytic coordinates must be supplemented in this case by the coordinate of the central charge, x_A^5 (in order to describe the graviphoton). The gauge group is realized as the group of coordinate-independent representations in $\{\zeta_A^M,x_A^5,u^\pm\}$ which leave invariant the subspace $\{\zeta_A^M,u^\pm\}$:

$$\delta\zeta_A^M=\lambda^M(\zeta_A,u),\quad \delta x_A^5=\lambda^5(\zeta_A,u). \tag{16}$$

As before, the harmonic variables $u_i^\pm$ undergo only global SU(2) and U(1) rotations. The basic geometric quantities of the theory are the $++$ components of the analytic frame of reference

$$V^{++M}=V^{++M}(\zeta_A,u),\qquad V^{++5}=V^{++5}(\zeta_A,u) \tag{17}$$

with the transformation laws

$$\delta V^{++M,5}=V^{++M,5\prime}(\zeta_A',u)-V^{++M,5}(\zeta_A,u)=D^{++}\lambda^{M,5}(\zeta_A,u), \tag{18}$$

where

$$D^{++}=u^{+i}\frac{\partial}{\partial u^{-i}}+V^{++M}(\zeta_A,u)\frac{\partial}{\partial\zeta_A^M}+V^{++5}(\zeta_A,u)\frac{\partial}{\partial x_A^5} \tag{19}$$

is the covariant version of derivative (8). It can be shown that in the Wess-Zumino gauge the $V^{++M,5}$ component composition is precisely the same as the composition of the multiplet of the $N=2$ Einstein supergravity (in its original version[14]). A differential geometry can be constructed by analogy with the $N=2$ Yang-Mills case. All the constraints are again solved in terms of prepotentials $V^{++M,5}(\zeta_A, u)$.

6. In summary, all the expanded $N=2$ supersymmetry theories allow an explicitly covariant formulation without the imposition of constraints of any sort. While the $N=1$ supersymmetry requires complexification, a "harmonization" arises in the $N=2$ case. A fundamentally new phenomenon which emerges in this case is the infinite number of gauge and/or auxiliary degrees of freedom. It appears to us that general phenomenon can explain the sources of the known theorems stating that it is impossible to find a finite number of auxiliary fields in $N=4$ theories, which await development.

We sincerely thank S. N. Kalitsin and B. M. Zupnik for useful discussions.

[1)]The self-action may also include other terms. It is easily generalized to the case of several hypermultiplets.
[2)]Roslyĭ[11] uses variables of the type u_i in analyzing the constraints of an $N=2$ Yang-Mills theory in the spirit of Ward's paper.[12]

[1]L. Mezincescu, Preprint R2-12572, Joint Institute for Nuclear Research, Dubna, 1979.
[2]P. S. Howe, K. S. Stelle, and P. K. Townsend, Preprint ICTP/82-83/20.
[3]A. Salam and J. Strathdee, Ann. Phys. **141**, 316 (1982).
[4]A. S. Gal'perin, E. A. Ivanov, and V. I. Ogievetskiĭ, Pis'ma Zh. Eksp. Teor. Fiz. **33**, 176 (1981) [JETP Lett. **33**, 168 (1981)].
[5]A. S. Gal'perin, E. A. Ivanov, and V. I. Ogievetskiĭ, Yad. Fiz. **35**, 790 (1982) [Sov. J. Nucl. Phys. **35**, 458 (1982)].
[6]P. Fayet, Nucl. Phys. **B113**, 135 (1976).
[7]M. F. Sohnius, Nucl. Phys. **B138**, 109 (1978).
[8]P. S. Howe, K. S. Stelle, and P. K. Townsend, Nucl. Phys. **B214**, 519 (1983).
[9]S. Ferrara and B. Zumino, Nucl. Phys. **B79**, 413 (1974).
[10]R. Grimm, M. Sohnius, and J. Wess, Nucl. Phys. **B133**, 275 (1978).
[11]A. Roslyĭ, in: Trudy Mezhdunarodnogo seminara Teoretiko-gruppovye metody v fizike (Zevenigorod, 1982) (Proceedings of the International Seminar on Group-Theory Methods in Physics), Vol. 1, Nuaka, Moscow, p. 263.
[12]R. S. Ward, Phys. Lett. **61A**, 81 (1977).
[13]V. Ogievetsky and E. Sokatchev, Phys. Lett. **79B**, 222 (1978); Yad. Fiz. **28**, 1631 (1978) [Sov. J. Nucl. Phys. **28**, 840 (1978)]; **31**, 264 (1980) [Sov. J. Nucl. Phys. **31**, 140 (1980)]; W. Siegel and S. J. Gates, Nucl. Phys. **B147**, 77 (1979).
[14]E. S. Fradkin and M. A. Vasiliev, Phys. Lett. **85B**, 47 (1979); B. de Wit, J. W. Holten, and A. Van Proeyen, Nucl. Phys. **B167**, 186 (1980).

Translated by Dave Parsons
Edited by S. J. Amoretty

$N = 3$ SUPERSYMMETRIC GAUGE THEORY

A. GALPERIN, E. IVANOV, S. KALITZIN, V. OGIEVETSKY and E. SOKATCHEV
Joint Institute for Nuclear Research, Laboratory of Theoretical Physics, 141 980 Dubna, USSR

Received 18 October 1984

The harmonic $N = 3$ superspace with the even part $M^4 \times [SU(3)/U(1) \times U(1)]$ is used to build up an unconstrained off-shell superfield formulation of $N = 3$ super Yang–Mills theory. It is defined in an analytic subspace of this $N = 3$ superspace and is described by three analytic gauge connections entering into the harmonic derivatives. Jumping over the "$N = 3$ barrier" becomes possible due to the presence of an infinite set of auxiliary fields.

1. One of the long-standing problems in extended supersymmetry (SUSY) is how to formulate off-shell the theories with $N \geqslant 3$. Such a formulation (on the component level) has been found so far only for $N = 3, 4$ conformal supergravity [1]. There were serious doubts about the existence of off-shell versions of $N = 3, 4$ super Yang–Mills (SYM) theories. The "no-go" theorems stating the impossibility to find (finite!) sets of auxiliary fields for these theories have been proved [2]. A partial way out is to allow for central charges [3].

In the present paper we circumvent the "$N = 3$ barrier" within the harmonic superspace (SS) approach. The principle of harmonization of the SS has been recently used to obtain an unconstrained superfield (SF) description of all the $N = 2$ SUSY theories [4]. Here we apply it to construct the first off-shell SF formulation of the $N = 3$ SYM theory (without central charges).

The harmonic $N = 3$ SS relevant to the off-shell $N = 3$ SYM has a ten-dimensional even part $M^4 \times SU(3)/U(1) \times U(1)$. The theory is described by three unconstrained SF's living in an analytic subspace of this SS. These analytic SF's have a clear geometric interpretation as the gauge connections for harmonic derivatives preserving the analyticity. They contain an infinite number of auxiliary fields (along with an infinite tower of gauge degrees of freedom). It is just the point where the standard "no-go" arguments fail. The SF equations of motion and the relevant action are written entirely in the analytic SS in an unexpectedly simple form. On-shell the theory can be transformed to the standard SF description in terms of constrained spinor connections [5,6].

2. Let us harmonize the standard $N = 3$ SS

$$z^M = (x^{\alpha\dot\alpha}, \theta^\alpha_i, \bar\theta^{\dot\alpha i}) \quad (i = 1, 2, 3)\,, \tag{1}$$

by extending its even part M^4 to ten-dimensional space [‡1]

$$M^4 \times [SU(3)/U(1) \times U(1)]\,. \tag{2}$$

Here, SU(3) is the automorphism group of $N = 3$ SUSY and two U(1) generators can be represented by the following 3×3 matrices:

$$H_1 = \begin{pmatrix} 1 & & \\ & -1 & \\ & & 0 \end{pmatrix}, \quad H_2 = \begin{pmatrix} 1 & & \\ & 1 & \\ & & -2 \end{pmatrix}, \tag{3}$$

We describe the homogeneous space $SU(3)/U(1) \times U(1)$ by the harmonic variables $u_i^{(a,b)}$, $\bar u^{(a,b),i}$ which are naturally combined into the 3×3 matrices (a couple of indices (a, b) represents the U(1)-changes)

$$u = \begin{pmatrix} u_1^{(1,1)} & u_2^{(1,1)} & u_3^{(1,1)} \\ u_1^{(-1,1)} & u_2^{(-1,1)} & u_3^{(-1,1)} \\ u_1^{(0,-2)} & u_2^{(0,-2)} & u_3^{(0,-2)} \end{pmatrix} \in SU(3)\,,$$

$$\bar u = (\bar u^{(-1,-1)i}, \bar u^{(1,-1)i}, \bar u^{(0,2)i}) \quad (i = 1,2,3)\,, \tag{4}$$

‡1 Any other harmonization (e.g. with the internal space SU(3)/U(2)) proves to be non-adequate to $N = 3$ SYM.

0370-2693/85/$ 03.30 © Elsevier Science Publishers B.V.
(North-Holland Physics Publishing Division)

Volume 151B, number 3,4 PHYSICS LETTERS 14 February 1985

They transform under SU(3) and two U(1) as follows

$$u' = g\,u\exp(\mathrm{i}a_1 H_1 + \mathrm{i}a_2 H_2), \quad g \in \mathrm{SU(3)}\,. \tag{5}$$

Note that a_1, a_2 are arbitrary functions of $u_i^{(a,b)}$, $\bar{u}^{(a,b)i}$. These "local" U(1)-invariances, together with the standard unitarity and unimodularity conditions

$$u\bar{u} = \bar{u}u = 1\,, \quad \det u = 1\,, \tag{6}$$

leave in $u_i^{(a,b)}$ 6 essential parameters that is just the dimension of SU(3)/U(1) × U(1). So, $u_i^{(a,b)}, \bar{u}^{(a,b)i}$ are basic harmonics of the harmonic expansion on this coset.

There exist 6 (as many as the dimension of SU(3)/U(1) × U(1), covariant harmonic derivatives consistent with (6):

$$\begin{aligned}
&\mathrm{D}^{(1,3)} = -u_i^{(1,1)}\partial/\partial u_i^{(0,-2)} + \bar{u}^{(0,2)i}\partial/\partial\bar{u}^{(-1,-1)i}\,,\\
&\mathrm{D}^{(-1,3)} = u_i^{(-1,1)}\partial/\partial u_i^{(0,-2)} - \bar{u}^{(0,2)i}\partial/\partial\bar{u}^{(1,-1)i}\,,\\
&\mathrm{D}^{(2,0)} = u_i^{(1,1)}\partial/\partial u_i^{(-1,1)} - \bar{u}^{(1,-1)i}\partial/\partial\bar{u}^{(-1,-1)i}\,,
\end{aligned} \tag{7}$$

and their complex conjugates. An important property of (7) is that these form a closed algebra:

$$\begin{aligned}
&[\mathrm{D}^{(1,3)}, \mathrm{D}^{(-1,3)}] = [\mathrm{D}^{(1,3)}, \mathrm{D}^{(2,0)}] = 0\,,\\
&[\mathrm{D}^{(-1,3)}, \mathrm{D}^{(2,0)}] = \mathrm{D}^{(1,3)}\,,
\end{aligned} \tag{8}$$

thus defining an integrable Cauchy-Riemann (CR) structure [7]. This property is related to the kählerian nature of the space SU(3)/U(1) × U(1): at each point of the latter one may give three independent complex tangent directions.

It is remarkable that in general harmonic $N = 3$ SS (z^M, u) there exists the analytic subspace

$$\{\zeta_A^M = (x_A^{\alpha\dot{\alpha}}, \theta^{(1,-1)\alpha}, \theta^{(0,2)\alpha}, \bar{\theta}^{(1,1)\dot{\alpha}}, \bar{\theta}^{(-1,1)\dot{\alpha}}), u\}\,, \tag{9}$$

$$\begin{aligned}
&x_A^{\alpha\dot{\alpha}} = x^{\alpha\dot{\alpha}} + 2\mathrm{i}\,[\theta^{(0,2)\alpha}\bar{\theta}^{(0,-2)\dot{\alpha}} - \theta^{(-1,-1)\alpha}\bar{\theta}^{(1,1)\dot{\alpha}}]\,,\\
&\theta^{(a,b)\alpha} = \bar{u}^{(a,b)i}\theta_i^{\alpha}\,, \quad \bar{\theta}^{(a,b)\dot{\alpha}} = u_i^{(a,b)}\bar{\theta}^{\dot{\alpha}i}\,,
\end{aligned} \tag{10}$$

closed under the $N = 3$ SUSY transformations. In the analytic basis the spinor derivatives $\mathrm{D}_\alpha^{(1,1)}$ and $\bar{\mathrm{D}}_{\dot{\alpha}}^{(0,2)}$ become

$$\begin{aligned}
&(\mathrm{D}_\alpha^{(1,1)})_{AB} = u_i^{(1,1)}(\mathrm{D}_\alpha^i)_{AB} = \partial/\partial\theta^{(-1,-1)\alpha}\,,\\
&(\bar{\mathrm{D}}_{\dot{\alpha}}^{(0,2)})_{AB} = -\partial/\partial\bar{\theta}^{(0,-2)\dot{\alpha}}\,,
\end{aligned} \tag{11}$$

so one may consider analytic SF's

$$\mathrm{D}_\alpha^{(1,1)}\Phi = \bar{\mathrm{D}}_{\dot{\alpha}}^{(0,2)}\Phi = 0\,, \tag{12}$$

which are unconstrained objects on the analytic SS (9). They can be chosen self-conjugated with respect to the combined conjugation ($\overset{*}{-}$) where new involution $*$ affects only the U(1)-charges of u's. This conjugation takes (9) into itself:

$$\begin{aligned}
&u_i^{(1,1)} \overset{(\overset{*}{-})}{\longleftrightarrow} \bar{u}^{(0,2)i}\,, \quad \theta_\alpha^{(1,-1)} \overset{(\overset{*}{-})}{\longleftrightarrow} -\bar{\theta}_{\dot{\alpha}}^{(-1,1)}\,,\\
&u_i^{(0,-2)} \leftrightarrow \bar{u}^{(-1,-1)i}\,, \quad \theta_\alpha^{(0,2)} \leftrightarrow \bar{\theta}_{\dot{\alpha}}^{(1,1)}\,,\\
&u_i^{(-1,1)} \leftrightarrow -\bar{u}^{(1,-1)i}\,, \quad x_A^{\alpha\dot{\alpha}} \overset{(\overset{*}{-})}{\longleftrightarrow} x_A^{\alpha\dot{\alpha}}\,.
\end{aligned} \tag{13}$$

3. We take the parameters of $N = 3$ gauge group to be unconstrained functions over the analytic $N = 3$ SS (9) (cf. $N = 2$ case [4]). In this $N = 3$ case, there are three harmonic derivatives which commute with spinor derivatives (11) and hence preserve the analytic SF's. These are $\mathrm{D}^{(\pm 1,3)}$, $\mathrm{D}^{(2,0)}$(7). Thus we are led to introduce three harmonic connections $V^{(\pm 1,3)}$, $V^{(2,0)}$:

$$\mathrm{D}^{(a,b)} \to \mathcal{D}^{(a,b)} = \mathrm{D}^{(a,b)} + \mathrm{i}V^{(a,b)}\,, \tag{14}$$

$$\mathrm{D}_\alpha^{(1,1)}V^{(a,b)} = \bar{\mathrm{D}}_{\dot{\alpha}}^{(0,2)}V^{(a,b)} = 0 \quad ((a,b) = (\pm 1, 3), (2,0)), \tag{15}$$

with the following gauge transformation law:

$$V^{(a,b)\prime} = -\mathrm{i}e^{\mathrm{i}\lambda}(\mathcal{D}^{(a,b)}e^{-\mathrm{i}\lambda})\,, \quad \lambda = \lambda(\zeta_A^M, u)\,. \tag{16}$$

A natural requirement that the properties of $\mathcal{D}^{(a,b)}$ under the conjugation ($\overset{*}{-}$) coincide with those of $\mathrm{D}^{(a,b)}$ gives tise to the reality conditions

$$\begin{aligned}
&\overset{*}{\overline{V^{(1,3)}}} = -V^{(1,3)}\,, \quad \overset{*}{\overline{V^{(-1,3)}}} = V^{(2,0)}\,,\\
&\overset{*}{\overline{V^{(2,0)}}} = V^{(-1,3)}\,, \quad \overset{*}{\bar{\lambda}} = \lambda\,.
\end{aligned} \tag{17}$$

Harmonic connections contain an infinite set of components. The physical fields occur, e.g. in $V^{(-1,3)}$ as follows:

$$\begin{aligned}
&V^{(-1,3)}(x_A, \theta_\alpha^{(1,-1)}, \theta_\alpha^{(0,2)}, \bar{\theta}_{\dot{\alpha}}^{(1,1)}, \bar{\theta}_{\dot{\alpha}}^{(-1,1)}, u)\\
&= \ldots + \mathrm{i}\theta_\alpha^{(0,2)}\bar{\theta}_{\dot{\alpha}}^{(-1,1)}A^{\alpha\dot{\alpha}}(x_A)\\
&+ \bar{\theta}_{\dot{\alpha}}^{(-1,1)}\bar{\theta}^{(-1,1)\dot{\alpha}}u_i^{(1,1)}\varphi^i(x_A)\\
&- \bar{\theta}_{\dot{\alpha}}^{(1,1)}\bar{\theta}^{(-1,1)\dot{\alpha}}u_i^{(-1,1)}\varphi^i(x_A) + \ldots\\
&+ \bar{\theta}_{\dot{\alpha}}^{(-1,1)}\bar{\theta}^{(1,1)\dot{\alpha}}\theta^{(0,2)\alpha}\bar{u}^{(-1,-1)i}\chi_{\alpha i}(x_A)\\
&+ \bar{\theta}_{\dot{\alpha}}^{(-1,1)}\bar{\theta}^{(-1,1)\dot{\alpha}}\bar{\theta}_{\dot{\beta}}^{(1,1)}\bar{\psi}^{\dot{\beta}}(x_A) + \ldots\,.
\end{aligned} \tag{18}$$

Even after fixing the WZ gauge in $V^{(\pm 1,3)}$, $V^{(2,0)}$, there survives an infinite number of auxiliary degrees of freedom (while in the $N = 2$ SYM this number is finite). It remains now to find the SF equations of motion and an appropriate action.

Volume 151B, number 3,4 PHYSICS LETTERS 14 February 1985

By commuting covariantized harmonic derivatives between themselves, we define three nonvanishing covariant strength tensors F, G, H:

$$[\mathcal{D}^{(1,3)}, \mathcal{D}^{(-1,3)}] = F^{(0,6)},$$
$$[\mathcal{D}^{(1,3)}, \mathcal{D}^{(2,0)}] = G^{(3,3)},$$
$$[\mathcal{D}^{(-1,3)}, \mathcal{D}^{(2,0)}] = \mathcal{D}^{(1,3)} + H^{(1,3)}. \quad (19)$$

A simple observation that among the physical scalars of the theory there are no dimensionless components with the U(1) × U(1) content of F, G, H implies the latter to vanish on-shell:

$$F^{(0,6)} = G^{(3,3)} = H^{(1,3)} = 0, \quad (20)$$

or

$$\mathrm{D}^{(1,3)} V^{(-1,3)} - \mathrm{D}^{(-1,3)} V^{(1,3)} + \mathrm{i}[V^{(1,3)}, V^{(-1,3)}] = 0,$$
$$\mathrm{D}^{(1,3)} V^{(2,0)} - \mathrm{D}^{(2,0)} V^{(1,3)} + \mathrm{i}[V^{(1,3)}, V^{(2,0)}] = 0,$$
$$\mathrm{D}^{(-1,3)} V^{(2,0)} - \mathrm{D}^{(2,0)} V^{(-1,3)} + \mathrm{i}[V^{(-1,3)}, V^{(2,0)}]$$
$$- V^{(1,3)} = 0. \quad (21)$$

One may directly check that eqs. (21) are true equations of motion. They admit an interesting geometric interpretation as the conditions for the CR-structure (7), (8) to be integrable in the local case. In section 4 we demonstrate that eqs. (21) have the same dynamical content as the familiar constraints in the standard real $N = 3$ SS.

These equations can be obtained from an action principle. The action is given as an integral over the analytic SS (9):

$$S = \int \mathrm{d}\zeta_A^{(-2,-6)} \mathrm{d}u \, \mathrm{Tr}\{V^{(2,0)}(\mathrm{D}^{(1,3)} V^{(-1,3)}$$
$$- \mathrm{D}^{(-1,3)} V^{(1,3)})$$
$$- V^{(-1,3)}(\mathrm{D}^{(1,3)} V^{(2,0)} - \mathrm{D}^{(2,0)} V^{(1,3)})$$
$$+ V^{(1,3)}(\mathrm{D}^{(-1,3)} V^{(2,0)} - \mathrm{D}^{(2,0)} V^{(-1,3)})$$
$$- (V^{(1,3)})^2 + 2\mathrm{i} V^{(1,3)} [V^{(-1,3)}, V^{(2,0)}]\}, \quad (22)$$

where the $\mathrm{D}^{(a,b)}$ are written in the analytic basis,

$$\mathrm{d}\zeta_A^{(-2,-6)} \mathrm{d}u = \mathrm{d}^4 x_A \mathrm{d}^2\theta^{(1,-1)} \mathrm{d}^2\theta^{(0,2)} \mathrm{d}^2\bar{\theta}^{(1,1)}$$
$$\times \mathrm{d}^2\bar{\theta}^{(-1,1)} \mathrm{d}u,$$

is the supervolume element of the SS(9) and the rule of integration over harmonics is:

$$\int \mathrm{d}u \, f^{(a,b)}(u) = \delta^{a,0}\delta^{b,0} f^{(a,b)}(0). \quad (23)$$

Surprisingly, the Lagrange density in (22) is not built out of tensors (19) and changes by total harmonic derivatives under the gauge transformations (16). The action is invariant as the rule (23) permits to integrate by parts. Note an intriguing similarity between (22) and the Chern–Simons terms in $N = 0, d = 3$ and $N = 1, d = 5$ YM theories [8].

4. Let us show that eqs. (21) yield the same on-shell dynamics as the well-known constraints on spinor connections in the standard $N = 3$ SS (1) [5,6]. To this end, we note first that eqs. (21) suggest for $V^{(\pm 1,3)}$, $V^{(2,0)}$ a "pure gauge" form on-shell:

$$V^{(a,b)} = -\mathrm{i}\,\mathrm{e}^{\mathrm{i}v}(\mathrm{D}^{(a,b)}\mathrm{e}^{-\mathrm{i}v}), \quad v = \overset{*}{\bar{v}}, \quad (24)$$

where v is some harmonic $N = 3$ SF (not analytic in general!) subjected to the constraints following from the analyticity of $V^{(a,b)}$. It transforms as

$$\mathrm{e}^{\mathrm{i}v'} = \mathrm{e}^{\mathrm{i}\lambda}\mathrm{e}^{\mathrm{i}v}\mathrm{e}^{-\mathrm{i}\tau}, \quad \tau = \overset{*}{\bar{\tau}} = \bar{\tau}, \quad \mathrm{D}^{(a,b)}\tau = 0. \quad (25)$$

With the help of the "bridge" $\mathrm{e}^{\mathrm{i}v}$ we may go to the new equivalent representation of the on-shell $N = 3$ SYM (we call it "τ-representation" whereas the original one is called "λ-representation") in which $\mathcal{D}^{(a,b)}$(14) have no connections:

$$(\mathcal{D}^{(a,b)})_\tau = \mathrm{e}^{-\mathrm{i}v}\mathcal{D}^{(a,b)}\mathrm{e}^{\mathrm{i}v} = \mathrm{D}^{(a,b)}$$
$$((a,b) = (\pm 1, 3), (2,0)),$$

while the spinor derivatives (11) lengthen

$$(\mathcal{D}_\alpha^{(1,1)})_\tau = \mathrm{e}^{-\mathrm{i}v}\mathrm{D}_\alpha^{(1,1)}\mathrm{e}^{\mathrm{i}v} = \mathrm{D}_\alpha^{(1,1)} + \mathrm{i}A_\alpha^{(1,1)},$$
$$(\bar{\mathcal{D}}_{\dot\alpha}^{(0,2)})_\tau = \mathrm{e}^{-\mathrm{i}v}\bar{\mathrm{D}}_{\dot\alpha}^{(0,2)}\mathrm{e}^{\mathrm{i}v} = \bar{\mathrm{D}}_{\dot\alpha}^{(0,2)} + \mathrm{i}\bar{A}_{\dot\alpha}^{(0,2)}, \quad (26)$$
$$A_\alpha^{(1,1)} = -\mathrm{i}\mathrm{e}^{-\mathrm{i}v}(\mathrm{D}_\alpha^{(1,1)}\mathrm{e}^{\mathrm{i}v}),$$
$$\bar{A}_{\dot\alpha}^{(0,2)} = -\mathrm{i}\mathrm{e}^{-\mathrm{i}v}(\bar{\mathrm{D}}_{\dot\alpha}^{(0,2)}\mathrm{e}^{\mathrm{i}v}) = \overline{\overset{*}{A}{}_\alpha^{(1,1)}}. \quad (27)$$

A simple consideration shows that the analyticity of $V^{(a,b)}$ (eqs. (15)) imposes the following condition on $A_\alpha^{(1,1)}, \bar{A}_{\dot\alpha}^{(0,2)}$:

$$\mathrm{D}^{(a,b)} A_\alpha^{(1,1)} = \mathrm{D}^{(a,b)} \bar{A}_{\dot\alpha}^{(0,2)} = 0 \quad ((a,b) = (\pm 1, 3), (2,0)), \quad (28)$$

the general solution of which is

$$A_\alpha^{(1,1)} = u_i^{(1,1)} A_\alpha^i(z), \quad \bar{A}_{\dot\alpha}^{(0,2)} = \bar{u}^{(0,2)i} \bar{A}_{\dot\alpha i}(z), \quad (29)$$

where $A_\alpha^i(z), \bar{A}_{\dot\alpha i}(z) = \overline{\overset{*}{A}{}_\alpha^i}(z) = \overline{A_\alpha^i}(z)$ do not depend on u in the central basis.

The form of the commutation relations between the covariant derivatives cannot depend on the repre-

Volume 151B, number 3,4 PHYSICS LETTERS 14 February 1985

sentation chosen. So we have

$$\{(\mathcal{D}_{\alpha}^{(1,1)})_{\tau}, (\mathcal{D}_{\beta}^{(1,1)})_{\tau}\} = \{(\mathcal{D}_{\dot{\alpha}}^{(0,2)})_{\tau}, (\mathcal{D}_{\dot{\beta}}^{(0,2)})_{\tau}\}$$
$$= \{(\mathcal{D}_{\alpha}^{(1,1)})_{\tau}, (\bar{\mathcal{D}}_{\dot{\beta}}^{(0,2)})_{\tau}\} = 0 \,, \qquad (30)$$

whence, with taking into account of eq. (29):

$$\{\mathcal{D}_{\alpha}^{(i}, \mathcal{D}_{\beta}^{j)}\} = \{\bar{\mathcal{D}}_{\dot{\alpha}(i}, \bar{\mathcal{D}}_{\dot{\beta}j)}\} = 0 \,,$$
$$\{\mathcal{D}_{\alpha}^{i}, \bar{\mathcal{D}}_{\dot{\beta}j}\} - \tfrac{1}{3}\delta_j^i \{\mathcal{D}_{\alpha}^{k}, \bar{\mathcal{D}}_{\dot{\beta}k}\} = 0 \,, \qquad (31)$$

where $\mathcal{D}_{\alpha}^{i} = D_{\alpha}^{i} + iA_{\alpha}^{i}$. These relations are just the constraints imposed on the spinor connection in the standard SS approach and are known to contain the equations of motion in the $N = 3$ case [5,6]. Thus, we have proved the equivalence of our on-shell description of $N = 3$ SYM with the standard one. In fact, one may reverse the arguments: start from the constraints (31), put them in form (30) [‡2] with the obvious solution (27) and then pass to the λ-representation with $\mathcal{D}_{\alpha}^{(1,1)}$, $\bar{\mathcal{D}}_{\dot{\alpha}}^{(0,2)}$ shortened and $\mathcal{D}^{(\pm 1,3)}$, $\mathcal{D}^{(2,0)}$ lengthened. The analyticity of harmonic connections follows from the property (29) which holds by construction.

This consideration clarifies the group-theoretical meaning of eqs. (21). They guarantee the existence of the τ-representation, i.e. the u-independent τ-gauge group and corresponding SF's. In contrast to the $N = 2$ case [4] the τ-representation of $N = 3$ SYM theory can be defined only on-shell. The primary level dynamics of $N = 3$ SYM manifests itself in the $\bar{\lambda}$-representation which exists off- as well as on-shell.

5. Thus, the unconstrained off-shell $N = 3$ SYM theory exists in a manifestly supersymmetric and Lorentz-invariant form and can be formulated completely in the analytic $N = 3$ SS (9) [‡3]. The central point is the presence of an infinite number of auxiliary fields. Due to this phenomenon, jumping over the $N = 3$ barrier has become possible.

We end with several remarks. The first one concerns the relation to the Kaluza–Klein theories [10]. The fact that $N = 3$ SYM has a simple and natural formulation just in ten-dimensional harmonic space $M^4 \times [SU(3)/U(1) \times U(1)]$ may be one more argument in favour of the relevance of higher dimensions. However, in contrast to the standard KK picture, in the harmonic approach extra dimensions are associated with generators of the automorphism group of SUSY, not with extra components of momenta. They are not introduced "by hand" but are required by the underlying geometric properties of the theory.

The next comment concerns the $N = 4$ SYM. Though $N = 3$ and $N = 4$ SYM theories are the same on-shell, it is not the case off-shell. We have checked that the auxiliary fields of a given dimension cannot be assigned to SU(4)-multiplets. It is an open problem how to construct an off-shell $N = 4$ SYM theory.

Finally, we would like to express the belief that the new formulation of $N = 3$ SYM will provide an adequate framework for an easier and more convincing proof of the famous $N = 4$ (or $N = 3$) finiteness.

[‡2] Such a representation of the $N = 3$ SYM constraints has been proposed for the first time by Rosly [7].

[‡3] Manin [9] has anticipated the fundamental role of SS's of this kind in SYM and supergravity theories.

We are sincerely thankful to R. Kallosh, Yu. Manin, O. Ogievetsky, A. Perelomov, A. Rosly, A. Schwarz, K. Stelle, D. Volkov and B. Zupnik for discussions. Two of us (E.S. and S.K.) are grateful to Professor Abdus Salam and the ICTP for the hospitality extended to them at the ICTP, where part of this work has been done.

[1] E. Bergshoeff, M. de Roo, J.W. van Holten, B. de Wit and A. van Proeyen, In: Superspace and supergravity, eds. S. Hawking and M. Roček (Cambridge, U.P., London, 1981).
[2] M. Roček and W. Siegel, Phys. Lett. 105B (1981) 278; V.O. Rivelles and J.G. Taylor, Phys. Lett. 121B (1983) 37;
P. Howe, K. Stelle and P. Townsend, Nucl. Phys. B236 (1984) 125.
[3] J. Hassoun, A. Restuccia and J.G. Taylor, KCL preprint (1983).
[4] A. Galperin, E. Ivanov, S. Kalitzin, V. Ogievetsky and E. Sokatchev, ICTP, Trieste, Preprint IC/84/43 (April 1984).
[5] M. Sohnius, Nucl. Phys. B136 (1978) 461.
[6] E. Witten, Phys. Lett. 77B (1978) 394.
[7] A.A. Rosly, in: Proc. Intern. Seminar on Group theoretical methods in physics (Zvenigorod, 1982), Vol. 1 (Nauka, Moscow, 1983) p. 263.
[8] S. Deser, R. Jackiw and S. Templeton, Ann. Phys. 140 (1982) 372;
M. Günaydin, G. Sierra and P.K. Townsend, Preprint CALT-68-1123 (April 1984).
[9] Yu.I. Manin, in: Differential geometry, Lie groups and mechanics, VI, eds. A.B. Venkov and L.A. Takhtajan (Nauka, Leningrad, 1984) p. 160.
[10] A. Salam and J. Strathdee, Ann. Phys. 141 (1982) 316; M. Duff, in: Trieste school (1982).

Physics Letters B 297 (1992) 93-98
North-Holland

Super-self-duality as analyticity in harmonic superspace

Ch. Devchand

Laboratory of Theoretical Physics, JINR, 141 980 Dubna, Russian Federation

and

V. Ogievetsky [1]

Theory Division, CERN, CH-1211 Geneva 23, Switzerland

Received 23 September 1992

A twistor correspondence for the self-duality equations for supersymmetric Yang-Mills theories is developed. Their solutions are shown to be encoded in analytic harmonic superfields satisfying appropriate generalised Cauchy-Riemann conditions. An action principle yielding these conditions is presented.

1. There has recently been a revival of interest in self-duality equations, arising from numerous confirmations of a remarkable suggestion [1] that all integrable systems are obtainable by dimensional reduction from 4D self-dual theories. The purpose of this letter is to show that for supersymmetric Yang-Mills theories [2] the self-duality equations can be written in a form in which their integrability becomes manifest and their solutions can be constructed in terms of superfields which are in a definite sense holomorphic. In other words, we shall establish the so-called "twistor correspondence" for the supersymmetric self-duality equations analogous to that for the ordinary ($N = 0$) self-duality equations [3], which in the harmonic space language [4-6] involves a splitting of the coordinate $x^{\alpha\dot{\alpha}}$ into $x^{+\alpha} = u^{+}_{\dot{\beta}} x^{\alpha\dot{\beta}}$ and $x^{-\alpha} = u^{-}_{\dot{\beta}} x^{\alpha\dot{\beta}}$, where $u^{\pm}_{\dot{\beta}}$ are harmonics on the two-sphere, which appears in the harmonisation of the rotation group [4-6], and α and $\dot{\beta}$ are two-spinor indices. The gist of ordinary self-duality is the Cauchy-Riemann-like equation

$$\nabla^{+}_{\alpha}\phi = 0, \tag{1}$$

where ∇^{+}_{α} is the covariant derivative in $x^{-\alpha}$. We shall show that this construction can be extended naturally to supersymmetric gauge theories. In the $N = 1$ case we shall use the harmonic superspace with coordinates

$$x^{\pm\alpha} \equiv u^{\pm}_{\dot{\beta}} x^{\alpha\dot{\beta}}\,,\quad \vartheta^{\alpha}\,,\quad \bar{\vartheta}^{\pm} \equiv u^{\pm}_{\dot{\alpha}}\bar{\vartheta}^{\dot{\alpha}}\,,\quad u^{\pm}_{\dot{\alpha}}\,. \tag{2}$$

Now super-self-duality is the condition for the integrability of the equation

$$\bar{\mathcal{D}}^{+}\phi = 0, \tag{3}$$

where $\bar{\mathcal{D}}^{+}$ is the gauge-covariant spinorial derivative with respect to the variable $\bar{\vartheta}^{-}$. $N = 1$ supersymmetric gauge theories respect chirality, so we may, without loss of generality, take the field ϕ in (3) to be a *chiral* superfield $\phi(x^{+\alpha}, x^{-\alpha}, \bar{\vartheta}^{+}, \bar{\vartheta}^{-})$ independent of ϑ^{α}, i.e.,

$$\mathrm{D}_{\alpha}\phi = 0, \tag{4}$$

where D_{α} is the covariant spinorial derivative with respect to the variables ϑ^{α}, which may be made flat by definition; and we shall set, throughout this letter, the corresponding connections to zero,

$$A_{\alpha} = 0. \tag{5}$$

Of importance is the fact that consistency of (3) and (4) implies (1) in virtue of the algebra of spinorial derivatives, so $N = 1$ self-duality implies the

[1] On leave from the Laboratory of Theoretical Physics, JINR, 141 980 Dubna, Russian Federation.

Elsevier Science Publishers B.V.

usual $N = 0$ self-duality. We shall demonstrate that all solutions of the supersymmetric self-duality conditions are encoded in a "holomorphic" chiral superfield which satisfies generalised Cauchy–Riemann conditions; in other words, we shall establish the famous twistor correspondence [3] for the supersymmetric self-duality equations. By holomorphicity we mean that there is a basis in which this superfield is independent of $x^{-\alpha}, \bar\vartheta^{-}$ and ϑ^α. The resulting formulation greatly simplifies the problem of constructing gauge superpotentials $A_{\alpha\dot\beta}, A_{\dot\beta}$ solving the super-self-duality equations. It also helps in the search for an action principle for super-self-duality.

Although our considerations are rigorous only for 4D euclidean space, we shall remain in the complexified picture with Lorentz group $SL(2,\mathbb{C})_L \times SL(2,\mathbb{C})_R$, with α and $\dot\alpha$ labelling fundamental representations of $SL(2,\mathbb{C})_L$ and $SL(2,\mathbb{C})_R$, respectively. In ref. [6] it was shown that consideration of the complexified picture is required by conformal invariance of the self-duality equations. Reality conditions appropriate for the required signature of four dimensional real space may be imposed; e.g. by identifying undotted and dotted spinors as representations of the two different SU(2)'s in the 4D Lorentz group $SU(2)_L \times SU(2)_R$ corresponding to a euclidean signature; or by identifying them as representations of two different $SL(2,\mathbb{R})$'s, with Lorentz group $SL(2,\mathbb{R})_L \times SL(2,\mathbb{R})_R$ corresponding to a (2, 2) signature. In the latter case we expect intriguing peculiarities due to the non-compactness of $SL(2,\mathbb{R})$ and an appropriate harmonic space needs to be considered (see e.g. ref. [7]).

2. In complexified superspace $\mathcal{M}_{4|4}$ of complex dimension (4|4) with coordinates $(x^{\alpha\dot\beta}, \vartheta^\alpha, \bar\vartheta^{\dot\alpha})$, the $N = 1$ super-Yang–Mills theory is conventionally described in terms of two spinorial field strengths $w_\alpha, \bar w_{\dot\alpha}$ defined by

$$[\bar{\mathcal{D}}_{\dot\beta}, \mathcal{D}_{\alpha\dot\alpha}] = \epsilon_{\dot\beta\dot\alpha} w_\alpha, \quad [\mathcal{D}_\beta, \mathcal{D}_{\alpha\dot\alpha}] = \epsilon_{\beta\alpha} \bar w_{\dot\alpha}, \tag{6}$$

where the gauge-covariant derivatives $\mathcal{D}_A \equiv \partial_A + A_A = (\mathcal{D}_{\alpha\dot\beta}, \mathcal{D}_\alpha, \bar{\mathcal{D}}_{\dot\beta})$ satisfy the familiar constraints

$$\{\mathcal{D}_\alpha, \mathcal{D}_\beta\} = 0, \tag{7a}$$

$$\{\bar{\mathcal{D}}_{\dot\alpha}, \bar{\mathcal{D}}_{\dot\beta}\} = 0, \tag{7b}$$

$$\{\mathcal{D}_\alpha, \bar{\mathcal{D}}_{\dot\beta}\} = 2\mathcal{D}_{\alpha\dot\beta} \tag{7c}$$

and the supertranslations

$$\partial_A = (\partial_{\alpha\dot\beta}, \mathrm{D}_\alpha, \bar{\mathrm{D}}_{\dot\beta}) \equiv \left(\frac{\partial}{\partial x^{\alpha\dot\beta}}, \frac{\partial}{\partial\vartheta^\alpha}, \frac{\partial}{\partial\bar\vartheta^{\dot\beta}} + 2\vartheta^\alpha \frac{\partial}{\partial x^{\alpha\dot\beta}}\right)$$

realise the free superalgebra

$$[\mathrm{D}_\beta, \partial_{\alpha\dot\beta}] = [\bar{\mathrm{D}}_{\dot\alpha}, \partial_{\alpha\dot\beta}] = \{\mathrm{D}_\alpha, \mathrm{D}_\beta\} = \{\bar{\mathrm{D}}_{\dot\alpha}, \bar{\mathrm{D}}_{\dot\beta}\} = 0,$$

$$[\mathrm{D}_\alpha, \mathrm{D}_{\dot\beta}] = 2\partial_{\alpha\dot\beta}.$$

The self-duality equations for the superconnection A_A take the form of the following further [in comparison with (6)] constraints:

$$[\mathcal{D}_\beta, \mathcal{D}_{\alpha\dot\alpha}] = 0, \tag{8}$$

which say that $\bar w_{\dot\alpha}$ vanishes. That this is the supersymmetrisation of the usual ($N = 0$) self-duality condition is evident from the dimension (1|1) Jacobi identity which yields the following superfield equations:

$$f_{\dot\alpha\dot\beta} = 0, \tag{9a}$$

$$\mathcal{D}_\alpha w_\beta = 2f_{\alpha\beta} \tag{9b}$$

for the self- and anti-dual vector field strengths $f_{\dot\alpha\dot\beta}, f_{\alpha\beta}$ appearing in the definition

$$[\mathcal{D}_{\alpha\dot\alpha}, \mathcal{D}_{\beta\dot\beta}] \equiv \epsilon_{\alpha\beta} f_{\dot\alpha\dot\beta} + \epsilon_{\dot\alpha\dot\beta} f_{\alpha\beta}.$$

The dimension $(\frac{3}{2}|1)$ and $(\frac{1}{2}|\frac{3}{2})$ Jacobi identities are then satisfied, respectively, if w_α and $f_{\alpha\beta}$ satisfy the (anti-)chirality equations

$$\mathcal{D}_\gamma f_{\alpha\beta} = 0, \tag{9c}$$

$$\bar{\mathcal{D}}_{\dot\beta} w_\alpha = 0. \tag{9d}$$

All other Jacobi identities are then automatic, requiring the introduction of no further superfield strengths. Eqs. (8) are the superfield self-duality equations. They indeed imply the equations of motion $\epsilon^{\gamma\alpha}\mathcal{D}_{\gamma\dot\gamma} f_{\alpha\beta} = \epsilon^{\gamma\alpha}\mathcal{D}_{\gamma\dot\gamma} w_\alpha = 0$. We have therefore shown that the constraints (7), (8) for A_A imply the superfield equations (9) for the superfield-strength

w_α and superfield vector-potential $A_{\alpha\dot\beta}$. These in turn imply the ordinary space supersymmetric self-duality equations for the component fields (which we denote by the same symbols as the superfields of which they are the leading components)

$$f_{\dot\alpha\dot\beta} = 0, \quad \epsilon^{\gamma\alpha}\mathcal{D}_{\gamma\dot\gamma}w_\alpha = \overline{w}_{\dot\alpha} = 0,$$

on eliminating all gauge degrees of freedom depending on the anticommuting superspace coordinates. The converse, that given a set of component fields satisfying these component equations, one can reconstruct superfields satisfying (9), and in turn the superconnection A_A satisfying (7), (8) is also true. The proof closely follows the methods of ref. [8].

3. Our main purpose here, however, is to introduce yet another piece of data corresponding to the above three: a "holomorphic" prepotential in harmonic superspace. We shall show that the constraints (7), (8) imply generalised Cauchy–Riemann (CR) conditions for a prepotential in harmonic superspace and that any superconnection satisfying (7), (8) may be expressed in terms of such holomorphic prepotentials. (We shall use "holomorphic" in this generalised sense; to describe solutions of these generalised CR conditions.) This construction is a realisation of the twistor construction [3] for supersymmetric self-dual systems in the harmonic superspace framework.

In the present complex setting, the harmonics $u^\pm_{\dot\alpha}$ remain, as usual, S^2 harmonics, the two-sphere being a coset of the $SL(2,\mathbb{C})_R$ part of the Lorentz group with its maximal parabolic subgroup. In this setting, u^+ and u^- are *not* complex conjugates of each other; and are defined up to parabolic subgroup transformations. They obey the usual constraints $u^{+\dot\alpha}u^-_{\dot\alpha} = 1$. Details of this construction, the role of conformal invariance, as well as appropriate reality conditions may be found in ref. [6]. For our $N = 1$ harmonic superspace the derivatives $\partial^\pm_\alpha \equiv u^{\pm\dot\alpha}\partial_{\alpha\dot\alpha}$, $\overline{\mathrm{D}}^\pm = u^{\pm\dot\alpha}\overline{\mathrm{D}}_{\dot\alpha}$, D_α, together with the harmonic ones

$$\mathrm{D}^{\pm\pm} = u^{\pm\dot\alpha}\frac{\partial}{\partial u^{\mp\dot\alpha}}, \quad \mathrm{D}^0 = u^{+\dot\alpha}\frac{\partial}{\partial u^{+\dot\alpha}} - u^{-\dot\alpha}\frac{\partial}{\partial u^{-\dot\alpha}},$$

realise the free superalgebra

$$\{\mathrm{D}_\alpha, \overline{\mathrm{D}}^\pm\} = 2\partial^\pm_\alpha, \tag{10a}$$

$$[\mathrm{D}^{++},\mathrm{D}^{--}] = \mathrm{D}^0, \quad [\mathrm{D}^0,\mathrm{D}^{\pm\pm}] = \pm 2\mathrm{D}^{\pm\pm}, \tag{10b}$$

$$[\mathrm{D}^{\pm\pm},\overline{\mathrm{D}}^\mp] = \overline{\mathrm{D}}^\pm, \quad [\mathrm{D}^{\pm\pm},\partial^\mp_\alpha] = \partial^\pm_\alpha, \tag{10c}$$

with all other commutators vanishing. Now, in terms of the gauge-covariant derivatives

$$\nabla^+_\alpha = \partial^+_\alpha + A^+_\alpha = u^{+\dot\alpha}\mathcal{D}_{\alpha\dot\alpha},$$

$$\overline{\mathcal{D}}^+ = \overline{\mathrm{D}}^+ + \overline{A}^+ = u^{+\dot\alpha}\overline{\mathcal{D}}_{\dot\alpha}, \quad \mathcal{D}_\alpha = \mathrm{D}_\alpha$$

(recall that $A_\alpha = 0$), obeying the commutation relations

$$[\mathrm{D}^{++},\overline{\mathcal{D}}^+] = 0, \tag{11a}$$

$$[\mathrm{D}^{\pm\pm},\mathcal{D}_\alpha] = 0, \tag{11b}$$

$$[\mathrm{D}^{++},\nabla^+_\alpha] = 0, \tag{11c}$$

the constraints (7), (8) are equivalent to the Cauchy–Riemann system

$$[\overline{\mathcal{D}}^+,\nabla^+_\alpha] = 0, \tag{12a}$$

$$[\mathcal{D}_\beta,\nabla^+_\alpha] = 0, \tag{12b}$$

$$[\overline{\mathcal{D}}^+,\mathcal{D}_\alpha] = 2\nabla^+_\alpha. \tag{12c}$$

Remarkably, these are precisely the integrability conditions for eqs. (1), (3) and (4). We therefore have the following pure-gauge-like expressions for A^+_α and $\overline{A}^+$:

$$A^+_\alpha = -\partial^+_\alpha\phi\phi^{-1}, \quad \overline{A}^+ = -\overline{\mathrm{D}}^+\phi\phi^{-1}. \tag{13}$$

4. Eqs. (11), (12) are therefore equivalent to the constraints (7), (8). Let us now choose a coordinate basis, which we shall call the analytic frame, in which the derivatives take the forms

$$\widehat{\mathrm{D}_\alpha} = \phi^{-1}[\mathcal{D}_\alpha]\phi = \frac{\partial}{\partial\vartheta^\alpha},$$

$$\widehat{\nabla^+_\alpha} = \phi^{-1}[\nabla^+_\alpha]\phi = \frac{\partial}{\partial x^{-\alpha}},$$

$$\widehat{\overline{\mathrm{D}}^+} = \phi^{-1}[\overline{\mathcal{D}}^+]\phi = \frac{\partial}{\partial\overline{\vartheta}^-} + 2\vartheta^\alpha\frac{\partial}{\partial x^{-\alpha}},$$

$$\widehat{\mathrm{D}^{++}} = \phi^{-1}[\mathrm{D}^{++}]\phi = \mathrm{D}^{++} + V^{++},$$

$$\widehat{\mathrm{D}^{--}} = \phi^{-1}[\mathrm{D}^{--}]\phi = \mathrm{D}^{--} + V^{--}. \tag{14}$$

Volume 297, number 1,2 PHYSICS LETTERS B 24 December 1992

In this basis the covariant derivatives $\widehat{\overline{D}}^+$ and the $\widehat{\nabla}^+_{\dot\alpha}$ become flat, losing their connections ($\widehat{\mathcal{D}}_\alpha$ remains flat), while the harmonic derivatives $D^{\pm\pm}$ clearly acquire the connections

$$V^{++} = \phi^{-1}D^{++}\phi, \tag{15a}$$

$$V^{--} = \phi^{-1}D^{--}\phi. \tag{15b}$$

In order to preserve the operator D^0 as a charge counting operator, we have used (as usual, see e.g. ref. [3]) the conventional gauge in which it does not acquire a connection:

$$[\widehat{D}^{++}, \widehat{D}^{--}] = D^0. \tag{16}$$

In this basis the dynamical content is contained entirely in (11), the rest of the equations being kinematical. Using the identity

$$A(Bff^{-1}) \equiv f(B(f^{-1}Af))f^{-1} + [A,B]ff^{-1},$$

for arbitrary differential operators A and B, eqs. (11) take the form of generalised CR conditions

$$\frac{\partial}{\partial\overline{\vartheta}^-}V^{++} = 0,$$
$$\frac{\partial}{\partial\vartheta^\alpha}V^{\pm\pm} = 0,$$
$$\frac{\partial}{\partial x^{-\alpha}}V^{++} = 0. \tag{17}$$

V^{++} is therefore holomorphic: it depends on $x^{+\alpha}$, ϑ^+ and $u^\pm$, being independent of $x^{-\alpha}, \overline{\vartheta}^-$ and ϑ^α; both V^{++} and V^{--} are chiral. Note that the third equation is a consequence of the other two. We have shown that to any solution of the super-self-duality constraints (8), there corresponds a chiral holomorphic superfield V^{++} taking values in the gauge algebra and having component expansion

$$V^{++}(x^{+\alpha}, \overline{\vartheta}^+, u^\pm_\alpha)$$
$$= v^{++}(x^{+\alpha}, u^\pm_\alpha) + \overline{\vartheta}^+\chi^+(x^{+\alpha}, u^\pm_\alpha). \tag{18}$$

The superfield V^{++} is defined modulo gauge transformations

$$\delta V^{++} = [\widehat{D}^{++}, \lambda],$$

where λ is an arbitrary holomorphic superfield. Note that due to the presence of the fermion mode there is an important difference with the $N = 0$ case: whereas the $N = 0$ connection was encoded in one function on the two-sphere, in the present case we have *two* functions, v^{++} and χ^+, instead.

5. The converse statement, that any chiral analytic superfield prepotential V^{++} encodes a superconnection A_A satisfying (7), (8), also holds. To reconstruct the superconnection A_A there are two options:

(a) We can start with the chiral superfield ϕ. In this case we need to solve (15a) for ϕ; V^{++} being given. Equations on a two-sphere of this kind are not too easy to solve and they appear in many applications of the harmonic-twistor approach, see e.g. ref. [4]. In order to determine the corresponding superconnection solving (7), (8), we need to insert the ϕ thus obtained into the following formulae:

$$A_{\alpha\dot\alpha} = 2\int d^2u\, u^-_{\dot\alpha}\phi\partial^+_\alpha\phi^{-1},$$
$$A_{\dot\alpha} = 2\int d^2u\, u^-_{\dot\alpha}\phi\overline{D}^+\phi^{-1}, \tag{19}$$

which follow immediately from (13); and A_α is of course zero.

(b) Instead of ϕ, we can start with the harmonic connection V^{--}. It follows from (16) that

$$Z \equiv D^{++}V^{--} - D^{--}V^{++} + [V^{++}, V^{--}]$$
$$= 0. \tag{20}$$

For a given holomorphic V^{++} taking values in the gauge algebra, it contains a set of coupled first-order linear equations for the gauge algebra components of V^{--}. These may be solved [9] somewhat more easily than (15a). Now, as further consequences of gauge-covariantising the harmonic derivatives (14), we have, from (10),

$$[\widehat{D}^{--}, \overline{D}^+] = \overline{\mathcal{D}}^-, \quad [\widehat{D}^{--}, \overline{\mathcal{D}}^-] = 0,$$
$$[\widehat{D}^{--}, \partial^+_\alpha] = \nabla^-_\alpha, \quad [\widehat{D}^{--}, \nabla^-_\alpha] = 0, \tag{21}$$

from which we obtain the superconnections

$$\overline{A}^- = -\overline{D}^+V^{--}, \quad A^-_\beta = -\partial^+_\beta V^{--}, \tag{22}$$

in terms of solutions of (20). Now, superconnections satisfying (7), (8) may be recovered by harmonic integration similar to (19). In fact, from V^{--} we may

Volume 297, number 1,2 PHYSICS LETTERS B 24 December 1992

also directly construct the superfield strength w_α satisfying the superfield equations (9); namely, from (6),

$$w_\alpha = -[\overline{D}^+, \nabla_\alpha^-] = \overline{D}^+ \partial_\alpha^+ V^{--}. \tag{23}$$

An alternative to eq. (20) follows from the following commutators contained in the superalgebra (10) (in the analytic frame):

$$[\overline{D}^-, \nabla_\alpha^-] = 0 = [\nabla_\alpha^-, \nabla_\beta^-].$$

These yield the alternative equations for V^{--}

$$L_\alpha^{--} \equiv -\partial_\alpha^+ \overline{D}^- V^{--} + \partial_\alpha^- \overline{D}^+ V^{--} + [\overline{D}^+ V^{--}, \partial_\alpha^+ V^{--}] = 0, \tag{24}$$

$$L^{--} \equiv \partial^{+\alpha}\partial_\alpha^- V^{--} + [\partial^{+\alpha} V^{--}, \partial_\alpha^+ V^{--}] = 0. \tag{25}$$

The latter equation is in fact the one introduced for the $N = 0$ case in ref. [10]. We note the following interesting interrelations amongst the left-hand-sides of eqs. (20), (24) and (25):

$$\overline{D}^+ L^{--} = \partial^{+\alpha} L_\alpha^{--},$$

$$\nabla_\alpha^- \partial^{+\alpha} Z = \hat{D}^{++} L^{--},$$

$$\nabla_\alpha^- \overline{D}^+ Z = \hat{D}^{++} L_\alpha^{--}.$$

6. We now present an action for super-self-duality. Since all we need to have is the generalised CR conditions (17), with V^{++} expressed in terms of the field ϕ, as a variational equation, we plug this condition into an action functional with the help of a Lagrange multiplier-type auxiliary field. The latter does not propagate if it contains only gauge degrees of freedom. For the chiral superfield ϕ, the action functional

$$S = \int d^4x\, d^2\overline{\vartheta}\, du\, \mathrm{tr}(\overline{D}^+ \zeta^{-3} \phi^{-1} D^{++} \phi) \tag{26}$$

yields, on varying the auxiliary field ζ^{-3}, the CR condition $\overline{D}^+ V^{++} = 0$, V^{++} is chiral by definition; and the final condition $(\partial/\partial x^{-\alpha})V^{++} = 0$ is a consequence. Now, on varying ϕ, we obtain

$$-\phi^{-1} D^{++} [\phi \overline{D}^+ \zeta^{-3} \phi^{-1}] = 0. \tag{27a}$$

It follows (cf. ref. [11]) that

$$\overline{D}^+ \zeta^{-3} = 0. \tag{27b}$$

All solutions of this equation have the form

$$\zeta^{-3} = \overline{D}^+ y^{-4}. \tag{28}$$

However, ζ^{-3} enters the action via $\overline{D}^+ \zeta^{-3}$, so it is only defined modulo the addition of $\overline{D}^- y^{-4}$. ζ^{-3} therefore does not represent any additional physical degree of freedom. For $N = 0$ an analogous action was discussed in refs. [11,12].

Alternatively, we may choose V^{--} (instead of ϕ) as the dynamical field, express V^{++} in terms V^{--} with the help of (20), and construct an action having analyticity conditions for the functional $V^{++}[V^{--}]$ as variational equations.

There also exists the possibility (analogous to the $N = 0$ action considered in ref. [10]) of writing an action for eq. (25) trilinear in V^{--}. It explicitly contains a constant harmonic factor of charge +4, say $(u_1^+ u_2^+)^2$, and is consequently not Lorentz invariant.

7. To conclude we generalise our construction to N-extended harmonic superspace with coordinates $x^{\pm\alpha}, \vartheta^{\alpha i}, \overline{\vartheta}_j^{\pm} \equiv u_{\dot\alpha}^{\pm} \overline{\vartheta}_j^{\dot\alpha}, u_{\dot\alpha}^{\pm}$, where $i, j = 1, \ldots, N$. Solutions of the generalised CR conditions

$$\frac{\partial}{\partial \overline{\vartheta}_j^-}(\phi^{-1} D^{++}\phi) = 0, \quad \frac{\partial}{\partial \vartheta^{\alpha i}}(\phi^{-1} D^{\pm\pm}\phi) = 0,$$

$$\frac{\partial}{\partial x^{-\alpha}}(\phi^{-1} D^{++}\phi) = 0 \tag{29}$$

encode N-extended self-dual superconnections. It may be verified that the superconnection components

$$A_{\alpha\dot\alpha} = \int d^2u\, u_{\dot\alpha}^- \phi \partial_\alpha^+ \phi^{-1},$$

$$A_{\dot\alpha}^j = \int d^2u\, u_{\dot\alpha}^- \phi \overline{D}^{+j} \phi^{-1}, \quad A_{\alpha i} = 0,$$

with ϕ satisfying (29), automatically satisfy the self-dual restrictions of the conventional extended superconnection constraints

$$F_{i\alpha j\beta} = 0 = F^{ij}_{\dot\alpha\dot\beta} + F^{ij}_{\dot\beta\dot\alpha}, \quad F^{i}_{\alpha\dot\beta j} = 0 = F^{i}_{\gamma,\alpha\dot\beta}.$$

Volume 297, number 1,2 PHYSICS LETTERS B 24 December 1992

The fact that integrability conditions for these equations yield (29) follows from a reasoning parallel to that for the $N = 1$ case above.

For the full (non-self-dual) $N = 2$ and 3 theories, for which harmonic superspace formulations (harmonising the internal automorphism group) exist [4,6], the self-duality conditions are also equivalent to "double" analyticity conditions arising from Lorentz as well as internal automorphism group harmonisation. We intend to return to these theories elsewhere.

We should like to sincerely thank A. Galperin, A. Leznov, E. Ivanov, D. Khetselius and E. Sokatchev for valuable discussions. One of us (V.O.) appreciates very much the fruitful and cordial atmosphere of the CERN Theory Division.

References

[1] R.S. Ward, Phil. Trans. R. Soc. A 315 (1985) 451; N.J. Hitchin, Proc. Lond. Math. Soc. 55 (1987) 59.

[2] V. Novikov, M. Shifman, A. Vainstein and V. Zakharov, Nucl. Phys. B 229 (1983) 394, 407; A. Semikhatov, Phys. Lett. B 120 (1983) 171; I. Volovich, Phys. Lett. B 123 (1983) 329; S. Ketov, H. Nishino and S.J. Gates Jr., preprint UMDEPP 92-211.

[3] R.S. Ward and R.O. Wells, Twistor geometry and field theory (Cambridge U.P., Cambridge, 1990).

[4] A. Galperin, E. Ivanov, S. Kalitzin, V. Ogievetsky and E. Sokatchev, in: Quantum field theory and statistics (Hilger, Bristol, 1987) Vol. 2, p. 233; O. Ogievetsky, in: Proc. Conf. on Group theoretical methods in physics (Varna, 1987) (Springer, Berlin, 1988) p. 548; thesis, Lebedev Physical Institute (Moscow, 1988) p. 329; S. Kalitzin and E. Sokatchev, Class. Quantum Grav. 4 (1987) L173.

[5] A. Galperin, E. Ivanov, V. Ogievetsky and E. Sokatchev, Ann. Phys. (NY) 185 (1988) 1.

[6] M. Evans, F. Gürsey and V. Ogievetsky, preprint CERN-TH.6533/92, RU-92-11-B.

[7] F. Delduc, A. Galperin and E. Sokatchev, Nucl. Phys. B 368 (1992) 143.

[8] J. Harnad, J. Hurtubise, M. Légaré and S. Shnider, Nucl. Phys. B 256 (1985) 609; J. Harnad and S. Shnider, Commun. Math. Phys. 106 (1986) 183.

[9] B. Zupnik, Theor. Math. Phys. 69 (1986) 175.

[10] A. Leznov, Theor. Math. Phys. 73 (1987) 302; A. Leznov and M. Mukhtarov, J. Math. Phys. 28 (1987) 2574.

[11] S. Kalitzin and E. Sokatchev, Phys. Lett. B 257 (1991) 151.

[12] N. Markus, Y. Oz and S. Yankielovicz, Nucl. Phys. B 379 (1992) 121.

List of Main Publications of V.I. Ogievetsky

1. V.I. Ogievetsky. On correlations in multiple scattering in magnetic field. *Zh. Exper. Teor. Fiz.* **21** (1951) 312-319.

2. V.I. Ogievetsky. Theory of propagation of gamma-rays. Candidate (PhD) Thesis, Lebedev Physical Institute, Moscow, 1954.

3. V.I. Ogievetsky. The theory of propagation of gamma-rays through matter. *Zh. Exper. Teor. Fiz.* **29** (1955) 454-463; *Soviet Phys. JETP* **1** (1956) 312-319.

4. V.I. Ogievetsky. Angular distribution of gamma rays at great depths of penetration. *Zh. Exper. Teor. Fiz.* **29** (1955) 464-472; *Soviet Phys. JETP* **1** (1956) 319-326.

5. V.I. Ogievetsky. A possible interpretation of the perturbation theory series in QFT. *Proc.Academy of Sci.USSR* **109** (1956) 919-922.

6. V.I. Ogievetsky. Hypernuclei.Theoretical review and recommendations. *Proc. of the Conf. on photoemulsions.* vol.1, p.209-215, Dubna, 1957.

7. A.L. Lubimov, V.I. Ogievetsky, M.I. Podgoretsky and L.G. Zastavenko. On a possibility of K-meson investigation. *Nucl. Phys.* **3** (1957) 549-552.

8. V.I. Ogievetsky. On the range of many-body forces between Λ -hyperon and nucleon. *Zh. Exper. Teor. Fiz.* **33** (1957) 546-547; *Soviet Phys. JETP* **6**(1958) 427-428.

9. V.S. Galishev, V.I. Ogievetsky and A.N. Orlov. Theory of multiple scattering of gamma rays. *Uspekhi Fiz.Nauk* **61** (1957) 161-216.

10. Chou Kuang-Chao and V.I. Ogievetsky . Charge symmetry properties and representations of the extended Lorentz group in the theory of elementary particles. *Zh. Exper. Teor. Fiz.* **36** (1959) 264-270; *Soviet Phys. JETP* **9** (1959) 179-183.

11. V.I. Ogievetsky. Interaction between K-mesons and pions. *Zh. Exper. Teor. Fiz.* **36** (1959) 642-643; *Soviet Phys. JETP* **9** (1959) 447-448.

12. Chou Kuang-Chao and V.I. Ogievetsky . Projective representations and charge symmetry. *Nucl.Phys.* **10** (1959) 235-234.

13. V.I. Ogievetsky and I.V. Polubarinov. Wave equations with zero and nonzero rest masses. *Zh. Exper. Teor. Fiz.* **37** (1959) 470-476; *Soviet Phys. JETP* **10** (1960) 335-338.

14. Chou Kuang-Chao and V.I. Ogievetsky . Electromagnetic mass of K-mesons. *Zh. Exper. Teor. Fiz.* **37** (1959) 866-867; *Soviet Phys. JETP* **10** (1960) 616-621.

15. V.G. Grishin and V.I. Ogievetsky. On the minimal number of partial waves in two-body reactions. *Nucl.Phys.* **18** (1960) 516-520.

16. V.G. Grishin and V.I. Ogievetsky . Estimating minimal radius of two-particle interactions at high energies. *Zh. Exper. Teor. Fiz.* **38** (1960) 1008-1009; *Soviet Phys. JETP* **11** (1960) 725-726.

17. V.I. Ogievetsky and I.V. Polubarinov. On gauge transformation of Green's functions. *Zh. Exper. Teor. Fiz.* **40** (1961) 926; *Soviet Phys. JETP* **13** (1961) 647-651.

18. V.I. Ogievetsky and I.V. Polubarinov. On gauge invariant formulation of neutral vector field theory. *Zh. Exper. Teor. Fiz.* **41** (1961) 247; *Soviet Phys.JETP* **14** (1962) 179-184.

19. V.I. Ogievetsky, E.O. Okonov and M.I. Podgoretsky. Note on the properties of K-mesons pairs. *Zh. Exper. Teor. Fiz.* **43** (1962) 720-723; *Soviet Phys.JETP* **16** (1963) 511-513.

20. V.I. Ogievetsky and M.I. Podgoretsky. Certain interference phenomena in $K, \bar{K}$systems. *Zh. Exper. Teor. Fiz.* **43** (1962) 1362-1364; *Soviet Phys.JETP* **16** (1963) 967-972.

21. V.I. Ogievetsky and I.V. Polubarinov. On a sense of gauge invariance. *Nuovo Cim.* **23** (1962) 173-180.

22. V.I. Ogievetsky and I.V. Polubarinov. Quantum electrodynamics in terms of electromagnetic field strenthes. *Zh. Exper. Teor. Fiz.* **43** 1365-1370; *Soviet Phys.JETP* **16** (1963) 969-972.

23. V.I. Ogievetsky and I.V. Polubarinov. Gauge invariance and vector fields. Proc. Int. Conf. on High Energy Phys. Geneva,(1966) 666-670.

24. V.I. Ogievetsky and I.V. Polubarinov. On interacting fields with definite spin. *Zh. Exper. Teor. Fiz.* **45** (1963) 237-245; *Soviet Phys. JETP* **18** (1964) 166-171.

25. V.I. Ogievetsky and I.V. Polubarinov. On the theory of the neutral vector field with spin 1. *Soviet Phys.JETP* **18** (1964) 487-489.

26. V.I. Ogievetsky and I.V. Polubarinov. Interacting spin 1 fields and symmetry properties. *Zh. Exper. Teor. Fiz.* **45** (1963) 966-977; *Soviet Phys.JETP* **18** (1964) 668-675.

27. V.I. Ogievetsky and I.V. Polubarinov. Interacting fields of spin 1 and symmetry properties. *Ann.of Phys.(N.Y.)* **25** (1963) 358-386.

28. V.I. Ogievetsky and I.V. Polubarinov. Minimal interactions between spin 0, 1/2 and 1 fields. *Zh. Exper. Teor. Fiz.* **46** (1964) 1048-1055; *Soviet Phys.JETP* **19** (1964) 712-716.

29. V.I. Ogievetsky and I.V. Polubarinov. On the choice of propagators for vector fields. *Zh. Exper. Teor. Fiz.* **46** (1964) 2102-2107; *Soviet Phys.JETP***19** (1964) 1418-1421.

30. V.I. Ogievetsky and Xien Ting-Chang. Possibility of hypernuclei of large strangeness. *Phys.Lett.* **9** (1964) 354-358.

31. G.I. Kopylov and V.I. Ogievetsky. Forbidden configurations in many-meson decays. *Nucl.Phys.* **50** (1964) 341-267.

32. V.I. Ogievetsky and Xien Ting-Chang. Does a world exist of extra strange nuclei? Prepr. JINR, P-1161, 1964.

33. V.I. Ogievetsky and I.V. Polubarinov. Spin and symmetries of interactions. Proc. XII Int.Conf. on High. Energy Phys. Dubna, 1964, 755-758.

34. V.I. Ogievetsky. Broken symmetries at high energies. *Zh. Exper. Teor. Fiz.* **47** (1964) 966-969.

35. V.I. Ogievetsky. Techniques of the SU(3) group. Proc. Int. School on Theor. Phys., Dubna (1964) v.2, 5-27.

36. V.I. Ogievetsky and I.V. Polubarinov. Interacting field spin and symmetries. Proc.Int. School on Theor. Phys., Dubna (1964) v.2, 160-179.

37. V.I. Ogievetsky and I.V. Polubarinov. Spin 2 fields in flat space and gravitation. Proc. of II Sov. Grav. Conf. (1965), p. 174-176, Tbilisi, 1967.

38. V.I. Ogievetsky and I.V. Polubarinov. Spinors in gravity theory. *Zh. Exper. Teor. Fiz.* **48** (1965) 1625-1636; *Soviet Phys. JETP* **21** (1965) 1093-1100.

39. V.I. Ogievetsky and I.V. Polubarinov. The group-theoretic approach to spinors in Einstein gravity theory. Proc. of II Sov. Grav. Conf. (1965), p. 454-462, Tbilisi, 1967.

40. V.I. Ogievetsky and I.V. Polubarinov. Photon and notoph. Prepr. JINR P-2330 (1965) 10 p.

41. V.I. Ogievetsky and I.V. Polubarinov. Notoph and its possible interactions. *Yad. Fiz.* **4** (1966) 216-224; *Soviet J. Nucl. Phys.* **4** (1967) 156-161.

42. V.I. Ogievetsky and I.V. Polubarinov. Interacting field of spin 2 and the Einstein equations. *Ann. Phys (NY)* **35** (1965) 167-207.

43. V.I. Ogievetsky and I.V. Polubarinov. Theories of interacting fields with spin 1. *Nucl.Phys.* **76** (1966) 677-684.

44. V.I. Ogievetsky and I.V. Polubarinov. Spinors in the Einstein gravity theory in group theoretical framework. Proc. Int. Conf. on Rel. Theory of Gravity London, v.2, (1965).

45. V.I. Ogievetsky and I.V. Polubarinov. How to get the Einstein equations in flat space. Proc. Int. Conf. on Rel. Theory of Gravity London, v.2, (1965).

46. V.I. Ogievetsky and I.V. Polubarinov. The Einstein equations as those for a massless spin 2 field. *Dokl. Akad. Nauk SSSR* **166** 584-587; *Doklady of Acad.Sci. USSR* **11** (1966) 71-74.

47. V.I. Ogievetsky and I.V. Polubarinov. Theory of a neutral massive spin 2 field. *Dokl. Akad. Nauk SSSR* **166** 839-842; *Doclady of Acad. Sci. USSR* **11** (1966) 135-137.

48. V.I. Ogievetsky and I.V. Polubarinov. SU(3) eightfold formalism and 10 and 27-plets. *Yad. Fiz.* **4** (1966) 853-861; *Soviet J. Nucl. Phys.* **4** (1967) 605-614.

49. V.I. Ogievetsky and I.V. Polubarinov. Conserved tensor currents and relativistic structure of SU(6). *Yad. Fiz.* **4** (1966) 1231-1247; *Soviet J. Nucl. Phys.* **4** (1967) 885-895.

50. V.I. Ogievetsky and I.V. Polubarinov. Quantum theory of gravity in Heisenberg and interaction pictures. Preprint JINR P-2692 (1966) 1-29.

51. V.G. Grishin, V.L. Lyuboshiz, V.I. Ogievetsky and M.I. Podgorezky. Possible experimental tests of the isotopic structure of the electromagnetic interactions. *Yad. Fiz.* **4** (1966) 126-129; *Soviet J.Nucl.Phys.* **4** (1967) 90-92.

52. V.I. Ogievetsky and I.V. Polubarinov. Photons longitudinal and scalar. Phys. Encyclopaedia, Moscow, (1966) v.5.

53. V.I. Ogievetsky and I.V. Polubarinov. Gravity quantum theory. Phys. Encyclopaedia, Moscow, (1966) v.5, p.220.

54. V.I. Ogievetsky. Fields of definite spin in interaction and conserved currents. Doctoral thesis (habilitation) JINR, Dubna, 1966, 220 pp.

55. V.I. Ogievetsky and I.V. Polubarinov. Consistent relativization of $SU(6)_w$ for two-particle processes. *Pis'ma ZhETF* **4** (1966) 325-329.

56. V.I. Ogievetsky and I.V. Polubarinov. Extension of $SU(6)_w$ on noncollinear processes. Prepr.JINR E2-2826 (1966).

57. V.I. Ogievetsky and I.V. Polubarinov. The $SU(6)_x$ group as extension of $SU(6)_w$ on noncollinear processes. Proc. of YII Cracow School of Theor.Phys. (1967) pp. 43-72.

58. V.I. Ogievetsky, W.Tybor and A.N. Zaslavsky. Breakdown of higher symmetries and nonet of pseudoscalar mesons. Proc. of Int. School on High Energy Phys. Popradske Pleso(1967), Bratislava,pp. 325-338.

59. V.I. Ogievetsky, W.Tybor and A.N. Zaslavsky . Broken $SU(6)_w$ symmetry and the ninth pseudoscalar meson. *Acta Phys. Polonica* **33** (1968) 209-215.

60. V.I. Ogievetsky, W.Tybor and A.N. Zaslavsky . The ninth pseudoscalar meson in broken $SU(6)_w$. *Pis'ma ZhETF* **6** (1967) 604-607.

61. V.I. Ogievetsky, W.Tybor and A.N. Zaslavsky. The need for new measu- rements of spin and parity of the X(960)-meson. *Yad. Fiz.* **9** (1969) 852-856; *Soviet J. Nucl. Phys.* **9** (1969) 498-514.

62. V.I. Ogievetsky. Non-linear realizations and breakdown of SU(3) symmetry. Proc. of IX Cracow School of Theor.Phys. (1969) pp. 29-51.

63. V.I. Ogievetsky and B.M. Zupnik. Investigation of non-linear realizations of chiral groups by the method of generating functions. *Teor. Mat. Fiz.* **1** (1969) 19-33; *Theor. Math. Phys.* **1** (1969) 14-16.

64. V.I. Ogievetsky and B.M. Zupnik. On the chiral $SU(2) \times SU(2)$ dynamics for A, ρ and π -mesons. *Nucl.Phys.* **B24** (1970) 612-622.

65. V.I. Ogievetsky and B.M. Zupnik. Closed form of chiral $SU(2) \times SU(2)$ dynamics with ρ and A_1-mesons. Abstracts of talks on 15 inter. conference on high-energy physics (Kiev-1970). v. 2, p. 654, Public.JINR, Dubna,1970 .

66. V.I. Ogievetsky and B.M. Zupnik . Algebraic realization of chiral symmetry for meson sistems and the Yang-Mills type theory. JINR Communications 2-5759, Dubna, 1971.

67. V.I. Ogievetsky. Algebraic realization of unitary symmetry. *Phys.Lett.* **B33**(1970) 227-230.

68. V.I. Ogievetsky and B.M. Zupnik. On chiral dynamics of $A - \rho - \pi-$ meson system. *Pis'ma ZhETF* **12** (1970) 194-197.

69. V.I. Ogievetsky. Mass formulas in an algebraic realization of unitary symmetry. *Yad. Fiz.* **13** (1971) 187-197; *Soviet J. Nucl. Phys.* **13** (1971) 105-110.

70. V. Ogievetsky, W.Tybor and A.Zaslavsky. X- meson: 2- or 0- ? *Phys. Lett.* **B35** (1971) 69-71.

71. V.I. Ogievetsky. Algebraic realisation of chiral $SU(3) \times SU(3)$ symmetry. Problems of Modern Physics,I.E.Tamm memorial volume, Moscow, (1972) 22-40.

72. A. Buyak, A. Filippov, V. Ogievetsky and A.Zaslavsky. Possible spin effects in pp scattering at high energies and the X (960) meson. *Yad. Fiz.* **18** (1973) 894-898; *Soviet J. Nucl. Phys.* **18**(1973) 462-464.

73. R. Lednizki, V. Ogievetsky and A. Zaslavsky. Remarks concerning the anisotropies in the Adair distribution for the $K^- p \mapsto X^0 \Lambda$ reaction. *Yad. Fiz.* **20** (1974) 203-209; *Soviet J. Nucl. Phys.* **20** (1975) 106-109.

74. V.I. Ogievetsky. Non-linear realizations of internal and space-time symmetries. Proc. X Winter School of Theor.Phys. in Karpach, vol. 1, Wroclaw, (1974) 117-141.

75. V.I. Ogievetsky. Infinite-dimensional algebra of general covariance group as the closure of finite - dimensional algebras of conformal and linear groups.*Lett. Nuovo Cim.* **8** (1973) 988-990.

76. A.B. Borisov and V.I. Ogievetsky . Theory of dynamical affine and conformal symmetries as the theory of the gravitational field. *Teor. Mat. Fiz.* **21** (1974) 329-342; *Theor. Math. Phys.* **21** (1974) 1179-1187.

77. L. Mezincescu and V. Ogievetsky . Action principle in superspace. Prepr.JINR E-2-8277 (1974) 1-8.

78. L. Mezincescu and V. Ogievetsky . Boson-fermion symmetries and superfields. *Usp. Fiz. Nauk* **117** (1975) 637-683; *Soviet Phys. Usp.* **18** (1976) 960-982.

79. V. Ogievetsky and E. Sokatchev. Primitive representations of the SL(3,R) algebra. *Theor. Math. Phys.* **25** (1975) 451-461.

80. E.A. Ivanov and V.I. Ogievetsky . Inverse Higgs effect in non-linear realizations.*Theor. Math. Phys.* **25** (1975) 1051-1059.

81. E.A. Ivanov and V.I. Ogievetsky . Gauge theories are also theories of spontaneous breakdown.*Pis'ma ZhETF* **23** (1976) 661-664; *JETP Lett.* **23** (1976) 601.

82. V. Ogievetsky and E. Sokatchev. Equations of motion for superfields. Proc.of 4th Int. conf. on non-local and non-linear field theory (Alushta, April 1976), D2-9788, pp. 183-197.

83. E.A. Ivanov and V.I. Ogievetsky. Gauge theories as theories of spontaneous breakdown. *Lett. Math. Phys.* **1** (1976) 309-315.

84. V. Ogievetsky and E.Sokatchev. On gauge spinor superfield. *Pis'ma ZhETF* **23** (1976) 66-69; *JETP Lett.* **23** (1976) 58.

85. V. Ogievetsky and E. Sokatchev. On a vector superfield generated by supercurrent. *Nucl.Phys.* **B124** (1977) 309-321.

86. V. Ogievetsky and E. Sokatchev. Superfield equations of motion. *J. Phys.* **A10** (1977) 2021-2031.

87. V. Ogievetsky and E. Sokatchev. Superfield equations of motion. *Phys. Lett.* **B78** (1978) 589-592.

88. V. Ogievetsky and V. Tzeitlin. Exceptional gauge theories in 3x3 matrix formalism.*J. Phys.* **A11** (1978) 1419-1425.

89. V. Ogievetsky and E. Sokatchev. The supercurrent. *Yad. Fiz.* **28** (1978) 825-836; *Soviet J. Nucl. Phys.* **28** (1978) 423-428.

90. V. Ogievetsky and V.Tzeitlin. On unifield exceptional gauge models with a stable proton. *Yad. Fiz.* **28** (1978) 1616-1630; *Soviet J. Nucl. Phys.* **28** (1978) 832-840.

91. V. Ogievetsky and E. Sokatchev. On a structure of supergravity group. *Chech. J.of Physics* **B29** (1979) 68-72.

92. V. Ogievetsky and E.Sokatchev. Structure of the supergravity group. *Phys.Lett.* **B79** (1978) 222-227.

93. V. Ogievetsky and E.Sokatchev. Axial superfield and the supergravity group. *Yad. Fiz.* **28** (1978) 1631-1639.

94. V. Ogievetsky and E.Sokatchev. Geometry, axial superfield and supergravity. Proc. of 5th Int.Conf on non-loc. field theory. Alushta, (1979) JINR-P2-12462, pp. 148-159.

95. V. Ogievetsky and E.Sokatchev. The gravitational axial superfield and the formalism of differential geometry. *Yad. Fiz.* **31** (1980) 821-840; *Soviet J. Nucl.Phys.* **31** (1980) 424-434.

96. V. Ogievetsky and E.Sokatchev. The simplest Einstein supergravity group. *Yad. Fiz.* **31** (1980) 264-279; *Soviet J. Nucl.Phys.* **31** (1980) 140-148.

97. V. Ogievetsky and E. Sokatchev. The normal gauge in supergravity. *Yad. Fiz.* **32** (1980) 862-869; *Soviet J. Nucl.Phys.* **32** (1980) 443-446.

98. V. Ogievetsky and E. Sokatchev. Torsion and curvature in terms of axial superfield. *Yad. Fiz.* **32** (1980) 870-879; *Soviet J. Nucl.Phys.* **32** (1980) 447.

99. V. Ogievetsky and E. Sokatchev. Equations of motion for the axial gravitational superfield. *Yad. Fiz.* **32** (1980) 1142-1151; *Soviet J. Nucl.Phys.* **32** (1980) 589.

100. V. Ogievetsky and E. Sokatchev. The simplest geometric approch to supergravity. In Theor.Group Methods in Physics. vol. 2, pp. 126-133, Nauka, Moskva, 1980.

101. A.S. Galperin, E.A. Ivanov and V.I. Ogievetsky. Grassmann analyticity and extended supersymmetries. *Pis'ma ZhETF* **33** (1981) 176-181.

102. A.S. Galperin, E.A. Ivanov and V.I. Ogievetsky. Superfield anatomy of the Fayet-Sohnius multiplet. *Yad. Fiz.* **35** (1982) 790-800; *Soviet J. Nucl. Phys.* **35** (1982) 458-565.

103. A.S. Galperin, V.I. Ogievetsky and E.S. Sokatchev. On U(1) supergravity. *Pis'ma ZhETF* **35** (1982) 263-267; *JETP Lett.* **35** (1982) 330-333.

104. A.S. Galperin, V.I. Ogievetsky and E.S. Sokatchev. Pecularities of N=1 supergravity with local U(1). *J.Phys.* **A15** (1982) 3785-3797.

105. A.S. Galperin, V.I. Ogievetsky and E.S. Sokatchev. Aspects of matter couplings in the U(1) supergravity. Proc.XIX Winter School, Karpach, World Scient. (1983) pp. 167-181.

106. A.S. Galperin, V.I. Ogievetsky and E.S. Sokatchev. Versions of the N=1 supergravity. Proc.XYII Int.Sympos. Ahrenshoop on gauge field theories, Berlin, (1983) pp. 201-230.

107. A.S. Galperin, V.I. Ogievetsky and E.S. Sokatchev. On matter couplings in N=1 supergravity. *Nucl.Phys.* **B252** (1985) 435-457.

108. A.S. Galperin, V.I. Ogievetsky and E.S. Sokatchev. Intrinsic geometries of U(1) supergravity. Proc. II Quantum Grav. Sem. Moscow, (1982) 211-222.

109. V.I. Ogievetsky and E.S. Sokatchev. On supersymmetry and supergravity. Proc. XIY Int. School for young scientists, Dubna, 1982, JINR, D2-4-83-179, pp. 143-222.

110. S. Kalitzin and V. Ogievetsky. An N=1 supergravity extension of classification of the Einstein spaces. *Lett. Math. Phys.* **96** (1982) 175-181.

111. V.I. Ogievetsky. Intrinsic geometry of supergravity. in Symmetries in Part. Physics, ed Bars a.o., Plenum Press,New York, (1984) 177-189.

112. A. Galperin, E. Ivanov, S. Kalitzin, V. Ogievetsky and E. Sokatchev. Intrinsic geometry of N=2 supersymmetry and supergravity. Proc.Int.Seminar on High Energy Phys. and QFT,1984, Protvino, 1984, vol II, pp. 219-233.

113. A. Galperin, E. Ivanov, V. Ogievetsky and E. Sokatchev. Harmonic superspace as a key to the N=2 supersymmetric theories. *Pis'ma ZhETF* **40** (1984) 155-159; *JETP Lett.* **40** (1984) 912-916.

114. A. Galperin, E. Ivanov, S. Kalitzin, V. Ogievetsky and E. Sokatchev. Unconstrained N=2 matter, Yang-Mills and supergravity theories. *Class. Quantum Grav.* **1** (1984) 469-498.

115. A. Galperin, E. Ivanov, S. Kalitzin, V. Ogievetsky and E. Sokatchev. Extended supersymmetry and supergravity in in harmonic superspace. Proc.Trieste Spring School on Supersymmetry and Supergravity, World Scientific, (1984) pp. 449-467.

116. A. Galperin, E. Ivanov, S. Kalitzin, V. Ogievetsky and E. Sokatchev. Harmonic superspace and extended supersymmetry. Proc.YII Int.Conf.on High Energy Physics, Leipzig,1984, Zeuthen, (1985) vol. 1, pp. 37-46.

117. A. Galperin, E. Ivanov, S. Kalitzin, V. Ogievetsky and E. Sokatchev. N=2 supersymmetry in harmonic superspace. Problems of QFT, Alushta, April 1984, JINR D2-84-366, pp. 196-226.

118. A. Galperin, E. Ivanov, S. Kalitzin, V. Ogievetsky and E. Sokatchev. Unconstrained off-shell N=3 supersymmetric Yang- Mills theory. *Class. Quantum Grav.* **2** (1985) 155-166.

119. A. Galperin, E. Ivanov, S. Kalitzin, V. Ogievetsky and E. Sokatchev. N=3 supersymmetric gauge theory. *Phys.Lett.* 151B (1985) 215-218.

120. E. Ivanov, S. Kalitzin, Nguen Ai Viet and V. Ogievetsky. Harmonic superspaces of extended supersymmetries: Calculus of harmonic variables. *J. of Phys.A: Math, Gen.* **18** (1985) 3433-3443.

121. V.I. Ogievetsky and E.S. Sokatchev. Supersymmetry and supergravity (Review). Itogi Nauki i Tekhniki, vol.22,VINITI, Moscow, pp. 137-173.

122. A. Galperin, E. Ivanov, V. Ogievetsky and E. Sokatchev. N=2 theories without constraints in harmonic superspace. Problems Nucl.Phys. and Cosm.Rays, 24 (1985) 7-17.

123. A. Galperin, E. Ivanov, V. Ogievetsky and E. Sokatchev. Harmonic supergraphs: Green functions. *Class. Quantum Grav.***2** (1985) 601-616.

124. A. Galperin, E. Ivanov, V. Ogievetsky and E. Sokatchev. Harmonic supergraphs: Feynman rules and examples. *Class. Quantum Grav.* **2** (1985) 617-630.

125. A. Galperin, E. Ivanov, V. Ogievetsky and E. Sokatchev. Conformal invariance in harmonic superspace. Quantum field theory and quantum statistics. ed. I. Batalin et al., v.2. Adam Hilger, Bristol, (1987) pp. 233-248.

126. A. Galperin, E. Ivanov, V. Ogievetsky and E. Sokatchev. HyperKähler metrics and harmonic superspace. *Comm.Math.Phys.* **103** (1986) 515-526.

127. A. Galperin, E. Ivanov, V. Ogievetsky and E. Sokatchev. Harmonic superspace and extended supersymmetries. Theor. Group Methods in Phys., ed. M.Markov v.1, Moscow (1986) 42-52.

128. A. Galperin, E. Ivanov, V. Ogievetsky and P.Townsend. Eguchi-Hanson type metrics from harmonic superspace.*Class. Quantum Grav.* **3** (1986) 625-633.

129. A. Galperin, E. Ivanov and V. Ogievetsky. Duality transformations and most general matter self-couplings in N=2 supersymmetry. *Nucl.Phys.* **B282** (1987) 74-102.

130. A. Galperin, E. Ivanov and V. Ogievetsky. Superspace actions and duality transformations for tensor N=2 multiplets. *Phys.Scripta* **75** (1987) 176-183.

131. A. Galperin, E. Ivanov and V. Ogievetsky. Interactions and duality transformations of N=2 tensor multiplets. *Yad. Fiz.* **45** (1987) 245-257; *Soviet J.Nucl.Phys.* **45** (1987) 157-163.

132. A. Galperin, E. Ivanov and V. Ogievetsky. Superspaces for N=3 supersymmetry. *Yad. Fiz.* **46** (1987) 948-960; *Soviet J. Nucl. Phys.* **46** (1987) 543-549.

133. A. Galperin, E. Ivanov and V. Ogievetsky. Harmonic superspace in action. General couplings of N=2 matter fields. Supersymmetry, Supergravity, Superstrings,ed. B. de Wit, P.Fayet, World Sci. (1986) pp. 511-565.

134. A. Galperin, E. Ivanov, V. Ogievetsky and E. Sokatchev. N=2 supergravity in superspace:its versions and matter couplings.*Class.Quantum Grav.* **4** (1987) 1255-1265.

135. A. Galperin, E. Ivanov, V. Ogievetsky and E. Sokatchev. Harmonic superspace. Results and prospects. Proc.YIII Int.Conf. on QFT, Alushta, 1987, D2-87-798, pp 267-282.

136. A. Galperin, E. Ivanov, V. Ogievetsky and E. Sokatchev. N=2 supergravity in superspace. Proceedings of int. semin. "Quantum gravity" (Moscow-1987), ed. M. Markov et al., p. 539-547, World Sci., Singapore, 1988.

137. A. Galperin, E. Ivanov, V. Ogievetsky and E. Sokatchev. Gauge field geometry from complex and harmonic analyticities.1.Kähler and self-dual Yang-Mills cases. *Ann.Phys.(N.Y)* **185**(1988) 1-21.

138. A. Galperin, E. Ivanov, V. Ogievetsky and E. Sokatchev. Gauge field geometry from complex and harmonic analyticities. 2. HyperKähler case. *Ann.Phys.(N.Y)* **185** (1988) 22-45.

139. V.I. Ogievetsky. Supersymmetry and supergravity. in "Introduction to Superanalysis" by F. Berezin, Reidel, 1988, pp. 401-413.

140. J. Bagger, A. Galperin, E. Ivanov and V. Ogievetsky. Gauging N=2 sigma-models in harmonic superspace. *Nucl.Phys.* **B303** (1988) 522-542.

141. V.I. Ogievetsky. Discrete symmetries in N=2 supersymmetric theories. *Yad. Fiz.* **49** (1989) 569-578; *Soviet J. Nucl.Phys.* **49** (1989) 147-150.

142. A. Galperin and V. Ogievetsky. Playing with hypermultiplets and gauge multiplets. Selected Problems in QFT and Math. Physics, ed. I.Niederle, World Sci. (1990) 355-359.

143. E. Ivanov and V. Ogievetsky. Harmonic superspace: how it works? in: Gauge Theories of Fundamental Interactions. Warsaw, (1990) v.1, 89-104.

144. A. Galperin and V. Ogievetsky. Spontaneous breaking of supersymmetry in nonminimal couplings of N=2 Maxwell multiplets.*Nucl.Phys. B (Proc. Suppl.)* **15** (1990) 51-60.

145. A. Galperin and V. Ogievetsky. N=2, D=4 supersymmetric sigma models and Hamiltonian mechanics. *Class.Quantum Grav.* **8** (1991) 1757-1764.

146. G. Khelashvili and V. Ogievetsky. Non-renormaliziblity of the massive N=2 Yang- Mills theory. *Mod.Phys.Lett.* **A6** (1991) 2143-2154.

147. A. Galperin and V. Ogievetsky. On isometries relating non-compact sigma models and supergravity. *Class.Quantum Grav.* **9** (1992) 1425-1432.

148. V.I. Ogievetsky. Harmonics: Hamiltonian analogies and quaternionic analyticity. In: Proc. of the First Int. Sakharov Conference on Physics, Vol. 1, p. 59-71, New Science Publishers, Commack, NY, 1992.

149. M. Evans, F. Gürsey and V. Ogievetsky. From 2D conformal to 4D self-dual theories: Quaternionic analyticity. *Phys.Rev* **D47** (1993) 3496.

150. Ch. Devchand and V. Ogievetsky. Super self-duality as analyticity in harmonic superspace. *Phys.Lett.* **B297** (1992) 93.

151. Ch. Devchand and V. Ogievetsky. The structure of all extended supersymmetric self-dual gauge theories. *Nucl.Phys.* **B414** (1994) 763; Erratum: *Nucl.Phys.* **B451** (1995) 768.

152. Ch. Devchand and V. Ogievetsky. The super-self-dual matreoshka. In: Quantum Field Theory and String Theory, ed. by L.Baulieu, V.Dotsenko, V.Kazakov, P.Windey, Plenum, 1995, New York, London, pp. 87-98.

153. Ch. Devchand and V. Ogievetsky. Integrability of N=3 super Yang-Mills equations. Prepr.Bonn-HE-93-33. Topics in statistical and theoretical physics. ed. R.L. Dobrushin et al. p.51-58.

154. Ch. Devchand and V. Ogievetsky. The matreoshka of supersymmetric selfdual theories. *Nucl. Phys.* **B414** (1994) 763-782.

155. Ch. Devchand and V. Ogievetsky. Self-dual gravity revisited. *Class. Quan. Grav.* **13** (1996) 2515-2536.

156. Ch. Devchand and V. Ogievetsky. Four-dimensional integrable theories. in: Strings and Symmetries, ed.by G.Atkas a.o., Springer, Berlin, 1995, pp. 169 - 182.

157. Ch. Devchand and V. Ogievetsky. Self-dual supergravities. *Nucl.Phys.* **B444** (1995) 381.

158. Ch. Devchand and V. Ogievetsky. Unravelling the on-shell constraints of self-dual supergravity theories. Talk at the 29th Intern. Ahrenshop Symposium, Buckow, August 1995. *Nucl. Phys. Proc. Suppl.* **49** (1996) 139-143.

159. Ch. Devchand and V. Ogievetsky. Conserved currents for unconventional supersymmetric couplings of self-dual gauge theories. *Phys.Lett.* **B367** (1996) 140-144.

160. Ch. Devchand and V. Ogievetsky. Consistent equations for interacting fields of arbitrary spin from $N > 4$ super-self-duality. *Nucl. Phys.* **B481** (1996) 188-214.

Lecture Notes in Physics

For information about Vols. 1–487 please contact your bookseller or Springer-Verlag

Vol. 488: B. Apagyi, G. Endrédi, P. Lévay (Eds.), Inverse and Algebraic Quantum Scattering Theory. Proceedings, 1996. XV, 385 pages. 1997.

Vol. 489: G. M. Simnett, C. E. Alissandrakis, L. Vlahos (Eds.), Solar and Heliospheric Plasma Physics. Proceedings, 1996. VIII, 278 pages. 1997.

Vol. 490: P. Kutler, J. Flores, J.-J. Chattot (Eds.), Fifteenth International Conference on Numerical Methods in Fluid Dynamics. Proceedings, 1996. XIV, 655 pages. 1997.

Vol. 491: O. Boratav, A. Eden, A. Erzan (Eds.), Turbulence Modeling and Vortex Dynamics. Proceedings, 1996. XII, 245 pages. 1997.

Vol. 492: M. Rubí, C. Pérez-Vicente (Eds.), Complex Behaviour of Glassy Systems. Proceedings, 1996. IX, 467 pages. 1997.

Vol. 493: P. L. Garrido, J. Marro (Eds.), Fourth Granada Lectures in Computational Physics. XIV, 316 pages. 1997.

Vol. 494: J. W. Clark, M. L. Ristig (Eds.), Theory of Spin Lattices and Lattice Gauge Models. Proceedings, 1996. XI, 194 pages. 1997.

Vol. 495: Y. Kosmann-Schwarzbach, B. Grammaticos, K.M. Tamizhmani (Eds.), Integrability of Nonlinear Systems. VII, 380 pages. 1997.

Vol. 496: F. Lenz, H. Grießhammer, D. Stoll (Eds.), Lectures on QCD. VII, 483 pages. 1997.

Vol. 497: J. P. Greve, R. Blomme, H. Hensberge (Eds.), Stellar Atmospheres: Theory and Observations. VIII, 352 pages. 1997

Vol. 498: Z. Horváth, L. Palla (Eds.), Conformal Field Theories and Integrable Models. Proceedings, 1996. X, 251 pages. 1997.

Vol. 499: K. Jungmann, J. Kowalski, I. Reinhard, F. Träger (Eds.), Atomic Physics Methods in Modern Research, IX, 448 pages. 1997.

Vol. 500: D. Joubert (Ed.), Density Functionals: Theory and Applications, XVI, 194 pages. 1998.

Vol. 501: J. Kertész, I. Kondor (Eds.), Advances in Computer Simulation. VIII, 166 pages. 1998.

Vol. 502: H. Aratyn, T. D. Imbo, W.-Y. Keung, U. Sukhatme (Eds.), Supersymmetry and Integrable Models. Proceedings, 1997. XI, 379 pages. 1998.

Vol. 503: J. Parisi, S. C. Müller, W. Zimmermann (Eds.), A Perspective Look at Nonlinear Media. From Physics to Biology and Social Sciences. VIII, 372 pages. 1998.

Vol. 504: A. Bohm, H.-D. Doebner, P. Kielanowski (Eds.), Irreversibility and Causality. Semigroups and Rigged Hilbert Spaces. XIX, 385 pages. 1998.

Vol. 505: D. Benest, C. Froeschlé (Eds.), Impacts on Earth. XVII, 223 pages. 1998.

Vol. 506: D. Breitschwerdt, M. J. Freyberg, J. Trümper (Eds.), The Local Bubble and Beyond. Proceedings, 1997. XXVIII, 603 pages. 1998.

Vol. 507: J. C. Vial, K. Bocchialini, P. Boumier (Eds.), Space Solar Physics. Proceedings, 1997. XIII, 296 pages. 1998.

Vol. 508: H. Meyer-Ortmanns, A. Klümper (Eds.), Field Theoretical Tools for Polymer and Particle Physics. XVI, 258 pages. 1998.

Vol. 509: J. Wess, V. P. Akulov (Eds.), Supersymmetry and Quantum Field Theory. Proceedings, 1997. XV, 405 pages. 1998.

Vol. 510: J. Navarro, A. Polls (Eds.), Microscopic Quantum Many-Body Theories and Their Applications. Proceedings, 1997. XIII, 379 pages. 1998.

Vol. 511: S. Benkadda, G. M. Zaslavsky (Eds.), Chaos, Kinetics and Nonlinear Dynamics in Fluids and Plasmas. Proceedings, 1997. VIII, 438 pages. 1998.

Vol. 512: H. Gausterer, C. Lang (Eds.), Computing Particle Properties. Proceedings, 1997. VII, 335 pages. 1998.

Vol. 513: A. Bernstein, D. Drechsel, T. Walcher (Eds.), Chiral Dynamics: Theory and Experiment. Proceedings, 1997. IX, 394 pages. 1998.

Vol. 514: F. W. Hehl, C. Kiefer, R. J. K. Metzler, Black Holes: Theory and Observation. Proceedings, 1997. XV, 519 pages. 1998.

Vol. 515: C.-H. Bruneau (Ed.), Sixteenth International Conference on Numerical Methods in Fluid Dynamics. Proceedings. XV, 568 pages. 1998.

Vol. 516: J. Cleymans, H. B. Geyer, F. G. Scholtz (Eds.), Hadrons in Dense Matter and Hadrosynthesis. Proceedings, 1998. XII, 253 pages. 1999.

Vol. 517: Ph. Blanchard, A. Jadczyk (Eds.), Quantum Future. Proceedings, 1997. X, 244 pages. 1999.

Vol. 518: P. G. L. Leach, S. E. Bouquet, J.-L. Rouet, E. Fijalkow (Eds.), Dynamical Systems, Plasmas and Gravitation. Proceedings, 1997. XII, 397 pages. 1999.

Vol. 519: R. Kutner, A. Pękalski, K. Sznajd-Weron (Eds.), Anomalous Diffusion. From Basics to Applications. Proceedings, 1998. XVIII, 378 pages. 1999.

Vol. 520: J. A. van Paradijs, J. A. M. Bleeker (Eds.), X-Ray Spectroscopy in Astrophysics. EADN School X. Proceedings, 1997. XV, 530 pages. 1999.

Vol. 521: L. Mathelitsch, W. Plessas (Eds.), Broken Symmetries. Proceedings, 1998. VII, 299 pages. 1999.

Vol. 522: J. W. Clark, T. Lindenau, M. L. Ristig (Eds.), Scientific Applications of Neural Nets. Proceedings, 1998. XIII, 288 pages. 1999.

Vol. 523: B. Wolf, O. Stahl, A. W. Fullerton (Eds.), Variable and Non-spherical Stellar Winds in Luminous Hot Stars. Proceedings, 1998. XX, 424 pages. 1999.

Vol. 524: J. Wess, E. A. Ivanov (Eds.), Supersymmetries and Quantum Symmetries. Proceedings, 1997. XX, 442 pages. 1999.

Vol. 525: A. Ceresole, C. Kounnas, D. Lüst, S. Theisen (Eds.), Quantum Aspects of Gauge Theories, Supersymmetry and Unification. Proceedings, 1998. X, 511 pages. 1999.

Vol. 526: H.-P. Breuer, F. Petruccione (Eds.), Open Systems and Measurement in Relativistic Quantum Theory. Proceedings, 1998. VIII, 240 pages. 1999.

Monographs

For information about Vols. 1–11 please contact your bookseller or Springer-Verlag

Vol. m 12: G. Esposito, Quantum Gravity, Quantum Cosmology and Lorentzian Geometries. Second Corrected and Enlarged Edition. XVIII, 349 pages. 1994.

Vol. m 13: M. Klein, A. Knauf, Classical Planar Scattering by Coulombic Potentials. V, 142 pages. 1992.

Vol. m 14: A. Lerda, Anyons. XI, 138 pages. 1992.

Vol. m 15: N. Peters, B. Rogg (Eds.), Reduced Kinetic Mechanisms for Applications in Combustion Systems. X, 360 pages. 1993.

Vol. m 16: P. Christe, M. Henkel, Introduction to Conformal Invariance and Its Applications to Critical Phenomena. XV, 260 pages. 1993.

Vol. m 17: M. Schoen, Computer Simulation of Condensed Phases in Complex Geometries. X, 136 pages. 1993.

Vol. m 18: H. Carmichael, An Open Systems Approach to Quantum Optics. X, 179 pages. 1993.

Vol. m 19: S. D. Bogan, M. K. Hinders, Interface Effects in Elastic Wave Scattering. XII, 182 pages. 1994.

Vol. m 20: E. Abdalla, M. C. B. Abdalla, D. Dalmazi, A. Zadra, 2D-Gravity in Non-Critical Strings. IX, 319 pages. 1994.

Vol. m 21: G. P. Berman, E. N. Bulgakov, D. D. Holm, Crossover-Time in Quantum Boson and Spin Systems. XI, 268 pages. 1994.

Vol. m 22: M.-O. Hongler, Chaotic and Stochastic Behaviour in Automatic Production Lines. V, 85 pages. 1994.

Vol. m 23: V. S. Viswanath, G. Müller, The Recursion Method. X, 259 pages. 1994.

Vol. m 24: A. Ern, V. Giovangigli, Multicomponent Transport Algorithms. XIV, 427 pages. 1994.

Vol. m 25: A. V. Bogdanov, G. V. Dubrovskiy, M. P. Krutikov, D. V. Kulginov, V. M. Strelchenya, Interaction of Gases with Surfaces. XIV, 132 pages. 1995.

Vol. m 26: M. Dineykhan, G. V. Efimov, G. Ganbold, S. N. Nedelko, Oscillator Representation in Quantum Physics. IX, 279 pages. 1995.

Vol. m 27: J. T. Ottesen, Infinite Dimensional Groups and Algebras in Quantum Physics. IX, 218 pages. 1995.

Vol. m 28: O. Piguet, S. P. Sorella, Algebraic Renormalization. IX, 134 pages. 1995.

Vol. m 29: C. Bendjaballah, Introduction to Photon Communication. VII, 193 pages. 1995.

Vol. m 30: A. J. Greer, W. J. Kossler, Low Magnetic Fields in Anisotropic Superconductors. VII, 161 pages. 1995.

Vol. m 31 (Corr. Second Printing): P. Busch, M. Grabowski, P.J. Lahti, Operational Quantum Physics. XII, 230 pages. 1997.

Vol. m 32: L. de Broglie, Diverses questions de mécanique et de thermodynamique classiques et relativistes. XII, 198 pages. 1995.

Vol. m 33: R. Alkofer, H. Reinhardt, Chiral Quark Dynamics. VIII, 115 pages. 1995.

Vol. m 34: R. Jost, Das Märchen vom Elfenbeinernen Turm. VIII, 286 pages. 1995.

Vol. m 35: E. Elizalde, Ten Physical Applications of Spectral Zeta Functions. XIV, 224 pages. 1995.

Vol. m 36: G. Dunne, Self-Dual Chern-Simons Theories. X, 217 pages. 1995.

Vol. m 37: S. Childress, A.D. Gilbert, Stretch, Twist, Fold: The Fast Dynamo. XI, 406 pages. 1995.

Vol. m 38: J. González, M. A. Martín-Delgado, G. Sierra, A. H. Vozmediano, Quantum Electron Liquids and High-Tc Superconductivity. X, 299 pages. 1995.

Vol. m 39: L. Pittner, Algebraic Foundations of Non-Com-mutative Differential Geometry and Quantum Groups. XII, 469 pages. 1996.

Vol. m 40: H.-J. Borchers, Translation Group and Particle Representations in Quantum Field Theory. VII, 131 pages. 1996.

Vol. m 41: B. K. Chakrabarti, A. Dutta, P. Sen, Quantum Ising Phases and Transitions in Transverse Ising Models. X, 204 pages. 1996.

Vol. m 42: P. Bouwknegt, J. McCarthy, K. Pilch, The W3 Algebra. Modules, Semi-infinite Cohomology and BV Algebras. XI, 204 pages. 1996.

Vol. m 43: M. Schottenloher, A Mathematical Introduction to Conformal Field Theory. VIII, 142 pages. 1997.

Vol. m 44: A. Bach, Indistinguishable Classical Particles. VIII, 157 pages. 1997.

Vol. m 45: M. Ferrari, V. T. Granik, A. Imam, J. C. Nadeau (Eds.), Advances in Doublet Mechanics. XVI, 214 pages. 1997.

Vol. m 46: M. Camenzind, Les noyaux actifs de galaxies. XVIII, 218 pages. 1997.

Vol. m 47: L. M. Zubov, Nonlinear Theory of Dislocations and Disclinations in Elastic Body. VI, 205 pages. 1997.

Vol. m 48: P. Kopietz, Bosonization of Interacting Fermions in Arbitrary Dimensions. XII, 259 pages. 1997.

Vol. m 49: M. Zak, J. B. Zbilut, R. E. Meyers, From Instability to Intelligence. Complexity and Predictability in Nonlinear Dynamics. XIV, 552 pages. 1997.

Vol. m 50: J. Ambjørn, M. Carfora, A. Marzuoli, The Geometry of Dynamical Triangulations. VI, 197 pages. 1997.

Vol. m 51: G. Landi, An Introduction to Noncommutative Spaces and Their Geometries. XI, 200 pages. 1997.

Vol. m 52: M. Hénon, Generating Families in the Restricted Three-Body Problem. XI, 278 pages. 1997.

Vol. m 53: M. Gad-el-Hak, A. Pollard, J.-P. Bonnet (Eds.), Flow Control. Fundamentals and Practices. XII, 527 pages. 1998.

Vol. m 54: Y. Suzuki, K. Varga, Stochastic Variational Approach to Quantum-Mechanical Few-Body Problems. XIV, 324 pages. 1998.

Vol. m 55: F. Busse, S. C. Müller, Evolution of Spontaneous Structures in Dissipative Continuous Systems. X, 559 pages. 1998.

Vol. m 56: R. Haussmann, Self-consistent Quantum Field Theory and Bosonization for Strongly Correlated Electron Systems. VIII, 173 pages. 1999.